T0188946

Fluid Mechanics and Its Applications

Founding Editor

René Moreau

Volume 130

Series Editor

André Thess, German Aerospace Center, Institute of Engineering Thermodynamics, Stuttgart, Germany

The purpose of this series is to focus on subjects in which fluid mechanics plays a fundamental role. As well as the more traditional applications of aeronautics, hydraulics, heat and mass transfer etc., books will be published dealing with topics, which are currently in a state of rapid development, such as turbulence, suspensions and multiphase fluids, super and hypersonic flows and numerical modelling techniques. It is a widely held view that it is the interdisciplinary subjects that will receive intense scientific attention, bringing them to the forefront of technological advancement. Fluids have the ability to transport matter and its properties as well as transmit force, therefore fluid mechanics is a subject that is particulary open to cross fertilisation with other sciences and disciplines of engineering. The subject of fluid mechanics will be highly relevant in such domains as chemical, metallurgical, biological and ecological engineering. This series is particularly open to such new multidisciplinary domains. The median level of presentation is the first year graduate student. Some texts are monographs defining the current state of a field; others are accessible to final year undergraduates; but essentially the emphasis is on readability and clarity.

Springer and Professor Thess welcome book ideas from authors. Potential authors who wish to submit a book proposal should contact Dr. Mayra Castro, Senior Editor, Springer Heidelberg, e-mail: mayra.castro@springer.com

Indexed by SCOPUS, EBSCO Discovery Service, OCLC, ProQuest Summon, Google Scholar and SpringerLink

More information about this series at https://link.springer.com/bookseries/5980

Erik Dick

Fundamentals
of Turbomachines

Second Edition

 Springer

Erik Dick
Department of Electromechanical, Systems
and Metal Engineering
Ghent University
Gent, Belgium

ISSN 0926-5112   ISSN 2215-0056 (electronic)
Fluid Mechanics and Its Applications
ISBN 978-3-030-93580-1   ISBN 978-3-030-93578-8 (eBook)
https://doi.org/10.1007/978-3-030-93578-8

This Springer imprint is published by the registered company Springer Nature Switzerland AG
The registered company address is: Gewerbestrasse 11, 6330 Cham, Switzerland

Preface

The book is a coursebook on the functioning of turbomachines. The approach is the analysis of all kinds of turbomachines with the same theoretical framework. The building up of theory is mixed in the sense that first derivations are general, but that elaboration of the theoretical concepts is done on a particular machine, however taking into account reuse on other machines or generalisation from constant-density to variable-density formulation.

The book starts with two chapters on general concepts and then follows the analysis of radial and axial fans, because these machines are the simplest ones. The next machines studied are steam turbines and pumps. Subsequent chapters are on hydraulic turbines, wind turbines, gas turbines for power and thrust, axial compressors, radial compressors, axial and radial turbines for gases.

The order in which the different types of turbomachines are treated is chosen by the possibility of gradually building up theoretical concepts. For each of the machine types, a balance is sought between fundamental understanding and knowledge of practical aspects. The main goal is understanding the functioning of the machine and the shaping of the components.

The point of view taken by the author is that readers should be able to understand what they see when a turbomachine is opened. They should also be able to make a reasoned choice of a turbomachine for a specific application and to understand its operation. Design is not a primary objective. Design requires a more specialised study, although basic design of the simplest turbomachines such as a centrifugal fan, an axial fan, an axial steam turbine, a centrifugal pump or an axial pump, is possible with the topics covered in the book.

The text of the first edition has been adapted to new evolutions in research and technology, but keeping the objective of a coursebook on fundamentals of turbomachines. The text of the first edition has been thoroughly revised with a focus on improved readability. The discussion has been extended of a limited number of topics that were treated somewhat too briefly in the first edition for reaching the goal of fundamental understanding. The exercises have been extended to cover more of the course material and to focus on a better understanding of fundamental aspects.

The volume of text is limited to what can be absorbed in a course of 9–12 ETC study points by a mechanical engineering student with basic knowledge of fluid mechanics and thermodynamics. An effort of nine ETC study points is possible by leaving out some less essential topics and some exercises.

Gent, Belgium   Erik Dick
October 2021

Contents

About the Author

Erik Dick was born on December 10, 1950, in Torhout, Belgium. He received his M.Sc. in Mechanical Engineering from Ghent University (Belgium) in 1973 and his Ph.D. in Computational Fluid Dynamics from the same university in 1980. From 1973 to 1991, he worked at Ghent University as a researcher, a senior researcher and the head of research in the turbomachinery division of the department of mechanical engineering. He became an associate professor of thermal turbomachines and propulsion at the University of Liège (Belgium), in July 1991. He returned to Ghent University in September 1992 as an associate professor and became a full professor in 1995, with teaching in turbomachines and computational fluid dynamics. He retired in September 2014 and is now a part-time individual senior researcher, combined with retirement. His area of research is computational methods for flow problems in mechanical engineering and models for laminar-to-turbulent transition in turbomachinery boundary layer flows. He is the author or the co-author of about 130 papers in international scientific journals and about 220 papers at international scientific conferences.

Symbols

a	Acceleration (m/s^2)
	Axial interference factor (–)
A	Through-flow section area (m^2)
b	Rotor width in axial direction (m)
	Tangential interference factor (–)
	Bypass ratio (–)
c	Chord (m)
	Velocity of sound (m/s)
c_a	Axial chord (m)
C_D	Drag coefficient (–)
c_f	Friction coefficient (–)
Cf	Centrifugal force by rotor rotation (N/kg)
C_{Fu}	Tangential force coefficient (2.28) (–)
C_L	Lift coefficient (–)
C_M	Pfleiderer moment coefficient (3.30) (–)
Co	Coriolis force by rotor rotation (N/kg)
c_p	Differential specific heat at constant pressure (J/kgK)
C_p	Pressure coefficient (–)
	Integral specific heat at constant pressure (J/kgK)
C_P	Power coefficient (–)
C_T	Thrust coefficient (–)
Cu	Centrifugal force by curvature (N/kg)
d	Diameter (m)
D	Drag per unit of span (N/m)
DF	Diffusion factor (13.13 (–)
D_{loc}	Local diffusion factor (13.13) (–)
D_s	Specific diameter (7.7) (–)
e	Internal energy per unit of mass (J/kg)
E_k	Kinetic energy per unit of mass (J/kg)
E_m	Mechanical energy per unit of mass (J/kg)
E_p	Pressure energy per unit of mass (J/kg)

f	Force per unit of mass (N/kg)
	Friction factor (2.30) or fuel-air ratio (–)
f_R	Curvature factor (3.29) (–)
g	Gravitational force per unit of mass (N/kg)
h	Enthalpy (J/kg)
	Blade height or scroll height (m)
H_m	Manometric head (m)
I	Rothalpy (J/kg)
L	Lift per unit of span (N/m)
$\dot{m}$	Mass flow rate (kg/s)
M	Rotor moment (Nm)
M_d	Disc or wheel friction moment (Nm)
M_{shaft}	Shaft moment (Nm)
M_{st}	Static moment of meridional section (3.22) (m^3)
$NPSH$	Net positive suction head (8.5) (m)
n	Polytropic exponent (–)
$\vec{n}$	Unit normal (–)
p	Pressure (Pa)
P	Power (W)
Pf	Pfleiderer factor (3.29) (–)
q	Heat transferred per unit of mass (J/kg)
	Dynamic pressure (Pa)
q_{irr}	Heat by dissipation inside flow path (J/kg)
q_{irr}^0	Heat by dissipation outside flow path (J/kg)
Q	Volume flow rate (m^3/s)
	Heat transferred per unit of time (J/s $=$ W)
r	Radius (m)
	Pressure ratio (–)
R	Kinematic degree of reaction (–)
	Radius of curvature (m)
	Gas constant (J/kgK)
Re	Reynolds number (–)
R_p	Pressure degree of reaction (3.1) (–)
R_s	Isentropic degree of reaction (6.16) (–)
s	Entropy (J/kgK)
	Spacing of blades (m)
S	Surface area (m^2)
t	Time (s)
	Thickness of blades (m)
T	Temperature (K or °C)
$\mathbf{T}$	Thrust force (12.1) (N)
u	Blade speed (radius $\times$ angular speed) (m/s)
U	Gravitational potential energy (J/kg)
v_0	Inflow velocity (m/s)
v_e	Energy reference velocity (m/s)

w	Flow velocity in relative frame (m/s)
W	Work per unit of time (J/s = W)
x	Coordinate along streamline (m)
	Coordinate in axial direction (m)
y	Coordinate perpendicular to streamline (m)
	Coordinate in circumferential direction (m)
z	Coordinate in vertical direction (m)
	Coordinate in radial direction (m)
Z	Number of blades (–)
α	Angle of absolute velocity w.r.t. meridional plane (°)
β	Angle of relative velocity w.r.t. meridional plane (°)
Γ	Circulation along a contour (m²/s)
δ	Boundary layer thickness (m)
ΔW	Rotor work per unit of mass (J/kg)
ε	Pfleiderer work reduction factor (3.24) (–)
η_i	Internal efficiency (–)
η_m	Mechanical efficiency (–)
η_p	Polytropic efficiency (–)
	Propulsive efficiency (12.7)
η_s	Isentropic efficiency (–)
η_{sre}	Isentropic re-expansion efficiency (11.7) (–)
η_t	Thermal efficiency (12.8) (–)
η_{td}	Thermodynamic efficiency (12.9) (–)
η_{tt}	Total-to-total isentropic efficiency (–)
η_v	Volumetric efficiency (–)
η_∞	Infinitesimal efficiency (–)
θ	Angular coordinate (rad)
	Flow turning angle (rad)
κ	Heat transfer coefficient (11.19) (J/kJK)
λ	Speed ratio (u/v_0) (–)
	Coefficient in Pfleiderer factor Pf (–)
μ	Dynamic viscosity (Pas)
ν	Kinematic viscosity (Pas)
ξ	Pressure loss coefficient (–)
ρ	Density (kg/m3)
σ	Solidity c/s (–)
	Stodola slip factor (3.20) (–)
	Cavitation number (8.1) (–)
σ_a	Axial solidity c_a/s (–)
σ_M	Moment solidity (3.31) (–)
τ	Shear stress (N/m³)
	Obstruction factor (Fig. 3.15) (–)
ϕ	Flow coefficient v_a/u or v_{2r}/u_2 (–)
Φ	Flow factor (7.4) (–)
ψ	Work coefficient $\Delta W/u_2^2$ (–)

Ψ	Head factor (7.5) (–)
ψ_0	Rotor total pressure coefficient (–)
ψ_r	Rotor static pressure coefficient (–)
ω	Rotor of relative velocity (1/s)
	Enthalpy loss coefficient (–)
Ω	Angular speed (rad/s)
Ω_s	Specific speed (7.6) (–)
Ω_{ss}	Suction specific speed (8.14) (–)

Subscripts

0	Inlet of machine or installation
	Total state
1	Just upstream of rotor inlet
1b	Just downstream of rotor inlet
2	Just downstream of rotor outlet
2b	Just upstream of rotor outlet
3	Outlet of machine or installation
∞	Far away from object
	With infinite number of blades
	On infinitesimal flow path
a	In axial direction
c	Compressor
	Critical value
	Choking value
d	Discharge/delivery side
def	Deflection
id	Ideal
irr	Due to irreversibility
m	In meridional direction
	Mechanical or manometric or mean
mean	Mean value
o	Optimum
p	Pressure side
r	In radial direction or in relative frame
	Rotor or reversible
s	Suction side or stator
	Isentropic
ss	Isentropic for stator
sr	Isentropic for rotor
sre	Isentropic re-expansion value

t	Theoretical value
	Turbine
tt	Total-to-total isentropic
u	In circumferential direction

Superscripts

*	Design value
	Choking value
—	Average
→	Vector quantity
b	Blade value

Chapter 1
Working Principles

Abstract In this chapter, we study the working principles of turbomachines with some examples. We derive the basic laws for energy exchange between a shaft and a fluid and the laws describing energy changes at the fluid side. We also analyse the role of the energy exchanging forces and introduce definitions of efficiency.

1.1 Definition of a Turbomachine

A turbomachine is a machine that exchanges energy between the continuous flow of a fluid and a continuously rotating machine component, with the energy exchange based on forces that are generated by the flow. The rotating component is called the *rotor*. It is a bladed drum, a bladed wheel or a collection of bladed wheels. Energy may be transferred from the flow to the rotor or vice versa. In the first case, energy extracted from the flow is used to drive the rotor, driving on its turn a useful external load. The machine may then be called shaft power delivering, or for short, *power delivering*, but typically, it is termed a *turbine*, irrespective of the fluid.

Possible fluids are:

- water: *water turbine* or hydraulic turbine
- steam (vapour): *steam turbine*
- air in atmospheric wind: *wind turbine*
- gas produced by combustion of a fuel in pressurised air: *gas turbine*
- other fluid, e.g. a refrigerant in a cooling cycle: *expansion turbine*.

When energy is supplied to the fluid by the rotor, the machine has to be driven by an external motor. It may be called shaft power receiving, or for short, *power receiving*. Specific names are used, depending on the fluid and the energy component that is mainly increased. Energy exchanged between a rotor and a fluid is mechanical energy in a technical sense (this is explained in Sect. 1.4.4). Ignoring the small effect of gravitational potential energy, mechanical energy can only take two forms inside the machine itself: velocity-associated energy (kinetic energy) and pressure-associated energy (pressure potential energy in case of a constant-density fluid).

Examples of power receiving machines are:

1

- *incompressible fluid* (water, oil), dominantly a pressure increase: *pump*
- *compressible fluid* (air, gas, vapour)

  – high pressure increase: *compressor*
  – low pressure increase (especially with air): *fan* (puts air into motion).

1.2 Examples of Axial Turbomachines

We consider two examples with water: a hydraulic turbine and a pump.

1.2.1 Axial Hydraulic Turbine

Figure 1.1 is a longitudinal section of a small hydraulic turbine. In the flow sense, the machine consists of three components: a stator part, encompassing an *inlet* and a *guide ring*, a rotor (runner) and a stationary diverging pipe. Both the guide ring and the runner are bladed. Figure 1.2 is a cylindrical section through the blades of the guide ring and the rotor, with the velocities in the absolute (v) and the relative frame (w). Subscript 1 indicates the rotor entry, subscript 2 the rotor exit. The absolute

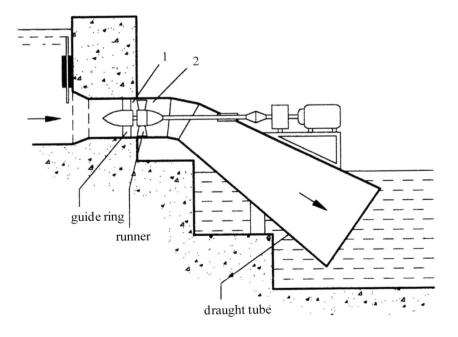

Fig. 1.1 Longitudinal section of a small axial hydraulic turbine

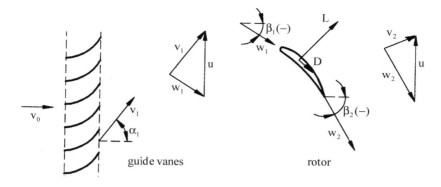

Fig. 1.2 Cylindrical section of guide vane ring and a runner blade

frame is attached to the stationary casing of the machine. The relative frame rotates together with the rotor. The velocity u stands for the relative frame velocity (angular speed Ω multiplied with radius r), called *blade speed*. The absolute flow velocity vector ($\vec{v}$) is the vector sum of the blade speed ($\vec{u}$) and the relative flow velocity vector ($\vec{w}$).

The components have the following functions.

- *Inlet and guide ring*:

  - guiding fluid to the rotor and distributing it over the rotor
  - generating kinetic energy: $v_1^2/2$.

The kinetic energy originates from a pressure drop, with pressure upstream of the rotor built up from gravitational potential energy (see the further Eq. 1.5: height difference between headwater level and rotor entry). Acceleration of the flow is never strong with hydraulic turbines as in Fig. 1.1. Guiding the fluid is the main function. Therefore, the terms *guide ring* and *guide vanes* are used. Stationary objects guiding the flow are typically called *vanes*, while the term *blade* is mostly used for a rotary object, but the term blade is also used for either. In other machines, flow acceleration may be strong, as with steam turbines. The stator is then said to be composed of nozzles and the term *nozzle ring* is used.

- *Rotor* (or *runner*): energy extraction from the flow.

In the relative frame, the kinetic energy increases: $w_2^2/2 > w_1^2/2$. This increase corresponds to a pressure drop (see the further Eq. 1.13). In the absolute frame, the kinetic energy decreases: $v_2^2/2 < v_1^2/2$. Both the decrease of pressure and the decrease of kinetic energy in the absolute frame represent energy transfer from the flow to the rotor. The transfer principle can already be understood. By the profile shape of the blades, resembling an aircraft wing profile, turning of the relative velocity at the rotor entry (w_1) towards a more tangential direction at the rotor exit (w_2), generates a *lift force* (*L*). This lift is approximately perpendicular to the average relative velocity and has the sense indicated in the figure.

The tangential component of the lift is in the sense of the blade speed *u*. This implies that the running blade is driven by the flow, which corresponds to work done by the flow on the rotor. The (much smaller) drag force *D* has a tangential component opposing the motion.

- The outlet stator component has various functions and may get various names, depending on the function emphasised:

  - collecting water leaving the rotor: collector
  - converting a part of the kinetic energy at the rotor exit into pressure potential energy: the pressure at the rotor exit decreases, increasing the pressure difference through the rotor: diffuser
  - utilising the downward height by guiding the water to the downward level (tailwater): the pressure at the rotor exit decreases due to the effect of gravity in the water column: draught tube.

The latter function is often the most important one. For example: $v_2 = 5$ m/s (typical): $v_2^2/2 = 12.5$ J/kg; 2.5 m height difference between rotor exit and tailwater: gravitational potential energy $g \Delta z \approx 25$ J/kg. Extraction of the energy associated to the downward height is easy, while kinetic energy conversion can only partly be realised. So, mostly the term *draught tube* is used.

The turbine discussed here is called *axial*, since streamlines approximately lie on cylinders with an axis coinciding with the centre of the machine shaft (real machines are mostly slightly conic). The velocity vectors mainly have axial and tangential components (radial components are very small). There are also machines with the flow approximately in planes perpendicular to the shaft; in other words, the velocity vectors mainly have radial and tangential components (the axial components are very small). These machines are called *radial*. Their working principle is more complex. These machines are discussed later in this chapter. Also intermediate forms exist, called diagonal machines or *mixed-flow* machines. There are still other types such as tangential (peripheral) and diametral (cross-flow) machines. For the time being, the fundamental discussion is limited to axial machines, as the study of their working principle is the simplest.

1.2.2 Axial Pump

Figure 1.3 (left) shows an axial pump for use as a submerged pump in a pit. The working principle is similar to that of an axial turbine, but with an inverse sense of the energy exchange. Figure 1.3 (right) is a cylindrical section, drawn with horizontal shaft direction.

The following parts can be distinguished.

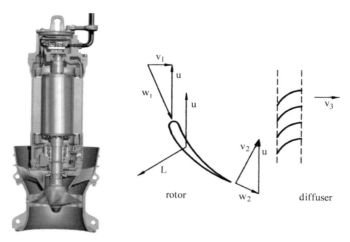

Fig. 1.3 Axial pump (*Courtesy* Flygt, a Xylem company); cylindrical section of a rotor blade and the diffuser ring

- *Inlet*: guiding the fluid to the rotor and accelerating the fluid, as with a turbine. There are no guide vanes in the example. In principle, the flow direction stays axial, but the fluid whirls somewhat due to the rotation of the rotor. In some pumps, pre-whirl is prevented by guide vanes in the axial direction.
- *Rotor* (or *impeller*): transferring energy to the fluid. Turning of the relative velocity at the rotor entry w_1 to a more axial direction at the exit w_2 generates a lift force in the indicated sense. The lift has a tangential component in the sense opposite to the blade speed u, meaning that work has to be supplied to the rotor by a driving motor. Energy is transferred from the rotor to the fluid. The energy transfer is partly noticeable by the kinetic energy increase in the absolute frame $\left(v_2^2/2 > v_1^2/2\right)$ and partly by the kinetic energy decrease in the relative frame $\left(w_2^2/2 < w_1^2/2\right)$ with a corresponding pressure increase.
- *Diffuser*: a stationary ring of vanes downstream of the rotor turns the entering velocity v_2 to the axial direction (tangential component of the velocity becomes zero) while the axial velocity component is reduced by increased through-flow area. Kinetic energy in the flow decreases $\left(v_3^2/2 < v_2^2/2\right)$, causing pressure increase. A component converting kinetic energy into pressure energy is called a *diffuser*. The term literally means that the flow is spread.

The rotating component of a pump is sometimes called the *impeller*. The meaning of the term is to impart motion to the fluid. As such, this is a general term well describing the role of the rotor in a power receiving machine, but it is typically only used with radial pumps and radial compressors. The term *runner*, mentioned in the previous section is, as such, also a general term completely equivalent to the term rotor, but it is typically only used with hydraulic turbines.

1.3 Mean Line Analysis

The fundamental analysis of a turbomachine employs the *average flow concept*. For a machine with periodicity in the circumferential direction (axial, radial, mixed-flow), the flow is calculated by averaging quantities in the circumferential direction (also called tangential direction) for a given axial and radial position. The real flow is always unsteady due to the presence of the running rotor. The average flow is steady for a constant speed of rotation, due to the circumferential periodicity of the rotor and stator blade rows. It is further assumed that the average flow is described by steady flow equations. Strictly, this cannot be correct, as products of flow quantities occur in the flow equations. The average value of a product does not equal the product of the average values. Deviation terms form, which in fluid mechanics are called Reynolds terms.

The Reynolds terms are similar to the terms by averaging a turbulent flow in time. The Reynolds terms are ignored (or replaced by a model) in a fundamental analysis. The approximation is good, as long as circumferential flow variations are not very significant. This is the case with a machine operating not very far away from design conditions. When the flow periodicity is seriously broken, the approximation is less good. In principle, some prudence is called for when applying relations from a circumferentially averaged flow representation. But, generally, the Reynolds terms due to periodic flow variations may be ignored. The following example is an illustration of the attained accuracy.

$$u = u_0 + u_a \sin \omega t \rightarrow \overline{u} = u_0,$$

$$u^2 = u_0^2 + 2u_0 u_a \sin \omega t + u_a^2 \sin^2 \omega t \rightarrow \overline{u^2} = u_0^2 + \frac{1}{2}u_a^2,$$

$$\frac{\sqrt{\overline{u^2}}}{\overline{u}} = \sqrt{1 + \frac{1}{2}\left(\frac{u_a}{u_0}\right)^2} \rightarrow \frac{u_a}{u_0} = 0.2: \ \sqrt{\overline{u^2}} \approx 1.01\overline{u}.$$

Further approximations are introduced in analyses meant for fundamental understanding. With an axial machine, it is assumed that the streamlines of the average flow lie on cylinders, in other words, that there is no radial velocity component. It is further assumed that there is no variation of flow quantities in the radial direction. Due to these simplifications, the flow becomes one-dimensional in the sense that only axial variation of flow quantities is considered. The flow stays multidimensional in the sense that velocity has two components, an axial and a tangential one.

The flow analysis thus achieved is mostly termed *mean line analysis*. It means that relations on a mean streamline in the circumferentially averaged flow are considered as representative for the whole machine.

Similar approximations are introduced with other machine types to come to a one-dimensional flow representation.

For a machine with circumferential periodicity, the streamlines of the average flow lie on surfaces of revolution. With a mean line analysis, the flow is described on a *mean circumferential streamsurface*, assuming that there is no variation of flow quantities in the circumferential direction and in the direction perpendicular to this surface. A mean line analysis is not very accurate and is mainly meant, as already said, for fundamental understanding. Hereafter, we derive the basic laws for one-dimensional flows.

1.4 Basic Laws for Stationary Duct Parts

We first derive some basic laws for one-dimensional flow in an elementary duct part, stationary in an absolute frame. The geometry is less complex than within a turbomachine. Gradually, we will extend to basic laws that can be applied to turbomachines. Figure 1.4 shows an elementary part of a duct. The axis of the duct part is denoted by x. The inlet section A_1 and the outlet section A_2 are perpendicular to this axis. With a one-dimensional representation, the flow is uniform in sections perpendicular to the axis and flow quantities only vary in the x-direction. Density is denoted by ρ and pressure by p.

1.4.1 Conservation of Mass

With a steady flow, the mass entering the duct part during a time interval dt, also leaves it:

$$\rho_2 A_2 v_2 dt = \rho_1 A_1 v_1 dt,$$

or

$$\dot{m} = \rho v A = \text{constant (mass flow rate)}. \tag{1.1}$$

1.4.2 Conservation of Momentum

According to Newton's second law, the change of momentum per time unit of an object with a constant mass equals the sum of the applied forces. We consider the fluid element with constant mass at the time t in the duct part between the sections A_1 and A_2 in Fig. 1.4. At time $t + dt$, the element has moved. The momentum change during the time interval dt is

Fig. 1.4 Elementary duct part for mass and momentum balances

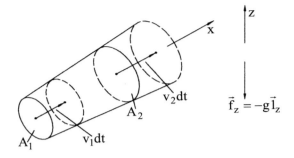

$$\rho_2 v_2 A_2 dt\, \vec{v}_2 - \rho_1 v_1 A_1 dt\, \vec{v}_1,$$

and equals the sum of the forces multiplied by dt.
   So:

$$\rho_2 v_2 A_2 \vec{v}_2 - \rho_1 v_1 A_1 \vec{v}_1 = \sum \vec{F},$$

or with (1.1):

$$\dot{m}(\vec{v}_2 - \vec{v}_1) = \sum \vec{F}. \tag{1.2}$$

The relations (1.1) and (1.2) apply to any part of the duct. The flow even need not be confined by walls and may be part of a streamtube.
   We further consider a duct part (stationary material walls), in which, provisionally, only pressure force and gravity force are internal forces. Gravity per mass unit (N/kg) is denoted by $\vec{f}_z = -g\vec{1}_z$, with the z-axis vertically directed upward and g being the gravity acceleration.
   For an elementary part with length dx, the momentum law projected on the x-direction is

$$\dot{m}(v + dv - v) = pA - (p + dp)(A + dA) + \left(p + \frac{1}{2}dp \right) dA$$
$$- \rho\left(A + \frac{1}{2}dA \right) dx\, g\vec{1}_z.\vec{1}_x.$$

With $dx\,\vec{1}_x.\vec{1}_z = dz$ and $gdz = dU$ (gravitational potential energy) it follows, up to first order, that

$$\rho v A dv = -Adp - \rho A dU. \tag{1.3}$$

After multiplication by the velocity and division by the mass flow rate, follows

$$d\frac{1}{2}v^2 + \frac{1}{\rho}dp + dU = 0. \tag{1.4}$$

Fig. 1.5 Elementary duct
part with active force,
friction and heat exchange

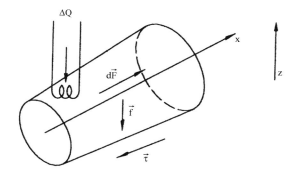

Equation (1.4) is a *work equation*, called *Bernoulli's equation*. With a constant-density fluid (ρ = constant), the equation may be integrated along a streamline to

$$\frac{1}{2}v^2 + \frac{p}{\rho} + U = \text{constant.}$$

We now take friction forces and active forces into account. The friction force on the envelope of the duct part, exerted on the flow per surface unit in the direction of the velocity, but with opposite sense, is denoted by τ (= shear stress): see Fig. 1.5.

With an *active force* is meant a force that exchanges energy between the surroundings and the flow. The active force component in the flow direction, counted as positive in the flow sense, is indicated by dF in Fig. 1.5.

Shear stresses on the inlet and outlet surfaces of the control volume do not contribute to the resulting force in x-direction.

The momentum equation (1.3) has to be completed now with the right-hand side

$$-\tau O dx + dF,$$

with O being the circumference. This gives, after multiplication by velocity and division by mass flow rate:

$$\frac{-\tau O dx\, v}{\rho A v} + \frac{dF\, v}{\rho A v}.$$

The first term may be considered as the work done per time unit on the moving fluid by the friction force, per unit of mass flow rate, or the work per unit of mass. The second term is, similarly, the active force work per unit of mass. We denote the additional terms by

$$-dq_{irr} + dW \ (\text{J/kg}).$$

In the notation we already take into account that the negative work of the friction force on the moving fluid, with friction on a stationary material wall, is equal in magnitude to the heat generated by friction. Irreversibility due to energy dissipation is associated to friction (see Sect. 1.4.5 and Chap. 2, Sect. 2.1.5). The work of the active force is positive when work is done on the fluid $\left(\overrightarrow{dF}.\vec{v} > 0\right)$.

Bernoulli's equation (work equation) becomes

$$d\frac{1}{2}v^2 + \frac{1}{\rho}dp + dU + dq_{irr} = dW. \tag{1.5}$$

1.4.3 Conservation of Energy

The first law of thermodynamics states that, with an object of constant mass, the energy increase equals the sum of work supplied and heat supplied. We denote work supplied per time unit by the active force by W and the heat supplied per time unit by Q (J/s = W). With dW and dq work and heat per unit of mass (J/kg):

$$W = dF\, v = dW\, \dot{m} \quad \text{and} \quad Q = dq\, \dot{m}.$$

The energy law states:

$$\dot{m}(E + dE - E) = pAv - (p + dp)(A + dA)(v + dv)$$
$$- \rho\left(A + \frac{1}{2}dA\right)dx\; g\vec{1}_z.\vec{1}_x v + W + Q.$$

We notice that no work is done by the pressure force and the friction force on the duct wall, as this wall stands still. The symbol E represents mechanical energy in the fundamental sense, i.e. the sum of internal and kinetic energy $\left(E = e + \frac{1}{2}v^2\right)$.

It follows that

$$\dot{m}\, dE = \dot{m}\frac{p}{\rho} - \dot{m}\frac{p + dp}{\rho + d\rho} - \dot{m}\, dU + W + Q.$$

After division by mass flow rate, follows

$$d\left(e + \frac{1}{2}v^2 + \frac{p}{\rho} + U\right) = dW + dq. \tag{1.6}$$

We further use the term *enthalpy* $h = e + p/\rho$, so that

$$d\left(h + \frac{1}{2}v^2 + U\right) = dW + dq. \tag{1.7}$$

1.4.4 Forms of Energy: Mechanical Energy and Head

The *energy* of a mechanical system is, in technical sense, the maximum amount of work and heat that can be produced by it. According to the energy balance (1.7), the energy components of a steady flow are enthalpy, kinetic energy and gravitational potential energy. In fundamental sense, there is only one form of mechanical energy: kinetic energy. With a macroscopic description of a flow, we use macroscopic velocity as the mass-weighted average velocity of the microscopic fluid particles (atoms or molecules), averaged over a small volume. The mass-weighted average kinetic energy of the particles is written as the sum of the macroscopic kinetic energy ($v^2/2$) and the kinetic energy of the motion around the macroscopic average. The latter term is denominated internal energy (e). Potential energy is recoverable work done against a conservative force (in the present case: gravity). A force is *conservative* if the force per mass unit can be noted as the gradient of a scalar that is only space-dependent.

For gravity it is

$$\vec{g} = -g\vec{1}_z = -\nabla(gz).$$

The *displacement work* of this force, for an elementary displacement $dx\vec{1}_x$ is

$$-g\vec{1}_z.dx\vec{1}_x = -gdz = -dU.$$

So, the displacement work done against gravity is the total differential of the term $U = gz$. Work supplied for a displacement between two points is thus independent of the path followed. This implies that no work is required for a displacement with coinciding initial and final points. Consequently, work done against a conservative force is entirely recoverable and may thus be considered as stored energy.

Pressure-related work does not have the same character. The resulting force of the pressure exerted on a fluid particle, with volume V and surface S, is

$$-\int_S p\vec{1}_n dS,$$

with dS representing an elementary surface area and $\vec{1}_n$ the corresponding external normal. According to the gradient integral theorem, the resulting pressure force is

$$-\int_V \nabla p \, dV.$$

The pressure force per volume unit is $-\nabla p$. The pressure force per mass unit is

$$-\frac{1}{\rho}\nabla p.$$

The *displacement work* of the pressure force is

$$-\frac{1}{\rho}\nabla p.dx\,\vec{1}_x = -\frac{1}{\rho}dp,$$

where dp is the pressure change along the infinitesimal path. Differentials are applied in this sense in Eqs. (1.5–1.7). The displacement work of the pressure force constitutes a total differential if density is only a function of pressure: $\rho = \rho(p)$. Strictly, no fluid meets this requirement, as density is always also a function of temperature. With a liquid, density depends only weakly on pressure and temperature. In practice, a constant density is mostly assumed. The fluid is then said to be *incompressible*. Strictly, the term means that density does not depend on pressure. Commonly, *constant density* is meant.

With constant density, $(1/\rho)dp$ constitutes a total differential. The term p/ρ is then, for a steady flow, potential energy, called *pressure potential energy*. For variable density, the term p/ρ cannot be defined as potential energy. However, due to the form of the energy balance (1.6), there is the wish to consider it as energy because it may be the source of work or heat.

Strictly, the term $d(p/\rho)$ constitutes the sign-changed total work of the pressure. Principally, the term $-d(p/\rho)$ should be kept in the right-hand part of Eq. (1.6) and the term $-(1/\rho)dp$ in the right-hand part of Eq. (1.5). We write the terms in the left-hand parts, however, as we wish to formulate statements about the work and the heat exchanged between the flow and its surroundings. The term $-d(p/\rho)$ is termed *pressure work* or *flow work*. The term p/ρ is added to the internal energy and the sum is termed *enthalpy* (see thermodynamics).

Pressure work consists of two parts:

$$-d\left(\frac{p}{\rho}\right) = -\frac{1}{\rho}dp - pd\left(\frac{1}{\rho}\right).$$

The first part is the *displacement work*. The second part is the *volume change work*. More generally, the second part is the *deformation work*, the sum of volume change work and form change work. The form change work of pressure is zero, however (pressure force perpendicular to the surface). The difference between work associated to the gravity force and the pressure force is that the work term of the former is identical in the balances of work (1.5) and energy (1.6). This expresses the complete recoverability of work done against the gravity force.

In technical sense, the *mechanical energy* of a system is the maximum amount of work that can be produced by it. For constant density, mechanical energy of a flow can easily be identified from the work balance (1.5). The maximum work can be produced by a flow in absence of shear forces ($dq_{irr} = 0$) and is the sum of the kinetic energy ($\frac{1}{2}v^2$), the pressure potential energy (p/ρ) and the gravitational potential energy (U), where the potential energies have to be calculated with respect to reference conditions of pressure and position. Mostly, the reference conditions are not relevant, because only changes of potential energies intervene. So, for constant density, the mechanical energy of a flow is defined by

$$E_m = \frac{1}{2}v^2 + p/\rho + U. \tag{1.8}$$

In hydraulics, the term *head* means the mechanical energy (J/Kg) divided by the gravity acceleration (9.81 m/s^2), expressed as a height (m). The term is also used for the rise of this quantity by a pump and the drop of it by a hydraulic turbine. However, with machines exchanging energy, it is more convenient to use changes of energy. It has thus become common in the turbomachinery literature to use the term head for a change of mechanical energy. From now on, we will mostly use the term head in this sense for all types of turbomachines.

For a fluid with variable density, the definition of mechanical energy of a flow remains the maximum amount of work that can be produced by the fluid and its value is still derived from the work balance (1.5). In this Chapter and Chaps. 2, 3 we will only use fluids with constant density and the expression (1.8) thus applies. We will treat the extension to compressible fluids in the later Chaps. 6 and 11.

1.4.5 Energy Dissipation: Head Loss

The term $-dq_{irr}$ in the work equation (1.5), when put at the right-hand side, has been described as work of the friction force on the moving fluid. According to the terminology of the previous section, this is *displacement work*. Until now, we have applied the basic laws to an elementary duct part that is stationary to the coordinate system in which we determine the flow quantities. The forces on the duct wall stand then still. No work is associated with them for a control volume positioned with its envelope on the duct wall. So we consider, when drawing up the energy balance, the pressure force work and the friction force work on the envelope as being zero. Neither is there any work of the friction force on the inlet and outlet sections (velocity perpendicular to these sections). The total work of the friction forces is thus zero.

Total work consists of displacement work and deformation work. The latter is the sum of volume change work and form change work. The volume change work of a friction force is zero, however (principally: friction force tangential to the surface). Thus, with a stationary duct part, the displacement work of a friction force on a moving fluid is equal in magnitude to the form change work. The latter is the source of friction heat (see Fluid Mechanics and see Chap. 2, Sect. 2.1.5).

Energy dissipation is reduction of the work capacity of the sum of the energies in a system. For a steady flow, the energy components are enthalpy, kinetic energy and gravitational potential energy (Eq. 1.7). The work capacity is expressed by the mechanical energy (Eq. 1.8 for constant density), i.e. the part of the energy that can be used for work generation. The mechanism that reduces the mechanical energy is the deformation work associated to friction forces, either friction inside the fluid or friction between the fluid and walls. The deformation work converts mechanical energy into heat. Technically, it is mostly said that there is *head loss*. In the turbo-machinery literature, it is even common to use the terms "loss" or "losses" as an abbreviation for head loss.

The dissipation by the friction force will be analysed in Chap. 2, Sect. 2.1.5. The effect of dissipation may be grasped already now by taking the difference between the *energy equation* (1.7) and the *work equation* (1.5):

$$dh - \frac{1}{\rho}dp - dq_{irr} = dq \quad \text{or} \quad T\,ds = dh - \frac{1}{\rho}dp = dq_{irr} + dq \qquad (1.9)$$

which is the expression of the second law of thermodynamics on the production of entropy (s) in an infinitesimal duct part. Equation (1.9) already gives confidence that the term dq_{irr} represents energy dissipation. The equation may also be written as

$$de = dq_{irr} + dq - p\,d\left(\frac{1}{\rho}\right).$$

This expression shows that friction causes heating of the fluid. The third term is *heating by compression* (volume change work of pressure). This term is zero for a constant-density fluid.

When expressing the energy balance, the envelope of the control volume has been put on the duct wall. Since the choice of a control volume is arbitrary, we may also opt to position its envelope just inside the fluid. Then, the envelope has velocity v and $(-dq_{irr})$ represents the total work of the friction force on the control volume. The new choice does not change the formulation of the momentum equation, but it changes the formulation of the energy equation. With the one-dimensional flow representation, velocity is uniform over a cross-section. This implies that we should consider the entire flow retardation due to wall friction as concentrated between the envelope of the control volume and the duct wall. Within this zone, dissipation by friction occurs. Thus, denoting the heat transferred through the duct wall towards the fluid by dq, the heat transferred through the envelope of the control volume equals $dq_{irr} + dq$. The energy law thus has $dW - dq_{irr} + dq_{irr} + dq$ as its right-hand part, with the second term representing the friction force work and the third term representing the heat generated by dissipation. The resulting energy equation is the same as the one with the first choice of the control volume.

The reasoning becomes more complex for a streamtube part within the fluid, without a control volume envelope coinciding with a material wall. The product of the friction force and the flow velocity then no longer corresponds to the work dissipated into heat. In other words, this work may contain an active part. In Chap. 2, we will demonstrate that the work equation in form (1.5) and the energy equation in form (1.7) stay valid for an arbitrary infinitesimal part of a streamline within a steady flow. But, the general validity of the energy equation (1.7) is obvious on the basis of the first law of thermodynamics, if one accepts that a part of the work dW may be by friction forces. However, quantifying that work and the heat dq transferred to the fluid originating from the dissipation on the nearby streamlines may be difficult. The general validity of the work equation (1.5) is, in the same way, obvious on the basis of the second law of thermodynamics, but again a problem may arise in quantifying the active and the dissipative parts of the work. This is further discussed in Chap. 2.

1.5 Basic Laws for Rotating Duct Parts

1.5.1 Work and Energy Equations in a Rotating Frame with Constant Angular Velocity

In a relative frame rotating at a constant angular velocity $\left(\vec{\Omega}\right)$ with respect to an absolute frame, the basic laws of mechanics and thermodynamics still hold, provided that two fictitious forces are introduced: centrifugal force and Coriolis force. These follow from the relation between absolute and relative velocities according to (Fig. 1.6):

$$\vec{v} = \vec{\Omega} \times \vec{r} + \vec{w}, \tag{1.10}$$

where $\vec{\Omega}$ is the angular speed vector, $\vec{r}$ the coordinate vector of the considered point P with respect to an origin on the rotation axis, and $\vec{w}$ the relative velocity.

An absolute displacement $d\vec{r}$ and a relative displacement $\delta\vec{r}$ are connected by

$$d\vec{r} = \vec{\Omega} \times \vec{r}\, dt + \delta\vec{r} \quad \text{or} \quad \frac{d\vec{r}}{dt} = \vec{\Omega} \times \vec{r} + \frac{\delta\vec{r}}{dt}. \tag{1.11}$$

Formula (1.11) also defines the relation between time differentiation in the relative frame and time differentiation in the absolute frame for any vector quantity as this may be considered being proportional to the difference of two coordinate vectors.

The differentiation rule applied to the velocity relation (1.10) results in:

$$\begin{aligned}
\vec{a} = \frac{d\vec{v}}{dt} &= \vec{\Omega} \times (\vec{\Omega} \times \vec{r}) + \vec{\Omega} \times \vec{w} + \vec{\Omega} \times \frac{\delta\vec{r}}{dt} + \frac{\delta\vec{w}}{dt} \\
&= \vec{\Omega} \times (\vec{\Omega} \times \vec{r}) + 2\vec{\Omega} \times \vec{w} + \vec{a}_{rel},
\end{aligned}$$

Fig. 1.6 Relative and absolute velocities for constant speed of rotation

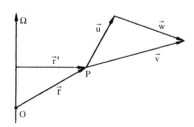

where $\vec{a}$ and $\vec{a}_{rel}$ respectively represent the absolute and the relative accelerations.

The basic laws may thus be applied in the relative frame, on condition of the introduction of (what is known as) the fictitious forces (per mass unit):

Centrifugal force:

$$\overrightarrow{Cf} = -\vec{\Omega} \times \left(\vec{\Omega} \times \vec{r}\right) = -\vec{\Omega} \times \left(\vec{\Omega} \times \vec{r}'\right) = \Omega^2 \vec{r}',$$

Coriolis force:

$$\overrightarrow{Co} = -2\vec{\Omega} \times \vec{w},$$

where $\vec{r}'$ represents the radial distance vector of point P with respect to the axis of rotation. From now on, this radial vector is denoted by $\vec{r}$, as in a cylindrical coordinate system.

An additional term comes in the right-hand part of the momentum equation:

$$\rho A dx \Omega^2 \vec{r}.\vec{1}_x = \rho A \Omega^2 r dr = \rho A \Omega^2 d\frac{r^2}{2} = \rho A d\frac{u^2}{2}.$$

Here, we applied $dx\vec{1}_x.\vec{1}_r = dr$ and $u = \Omega r$ represents the *blade speed*. The Coriolis force does not contribute to the work on the streamline.

From multiplication by the velocity and division by the mass flow rate it follows that the additional term in the right-hand part of the work equation makes $d\frac{1}{2}u^2$, with as result

$$d\frac{1}{2}w^2 + \frac{1}{\rho}dp + dU + dq_{irr} = dW + d\frac{1}{2}u^2. \tag{1.12}$$

There is thus a centrifugal force contribution to the work on a streamline. The same contribution has to be added in the energy equation.

In the relative frame, the work dW is zero on a rotor, since forces on the rotor perform no work within the frame that is turning with the rotor. Thus:

Work:

$$d\frac{1}{2}w^2 + \frac{1}{\rho}dp + dU + dq_{irr} = d\frac{1}{2}u^2, \tag{1.13}$$

Energy:

$$d\frac{1}{2}w^2 + dh + dU = d\frac{1}{2}u^2 + dq. \tag{1.14}$$

From now on, we use the notions *stagnation enthalpy* and *total enthalpy*. The stagnation enthalpy is the enthalpy that would be adopted by a fluid particle by locally bringing the flow to stagnation, i.e. zero velocity, adiabatically. With the subscript 0 denoting the stagnation, the stagnation enthalpies in the absolute and the relative frames are

$$h_0 = h + \frac{1}{2}v^2, \quad h_{0r} = h + \frac{1}{2}w^2.$$

With the gravitational potential energy added, the sum is called the total enthalpy. The total enthalpy is normally also denoted by the subscript 0, because it is always clear whether it is relevant to add the gravitational potential energy. With a gas, gravitational potential energy may mostly be ignored due to the low density. E.g., with 1.225 kg/m^3 (air in standard atmospheric conditions), a 1 m height difference represents 9.81 J/kg, which is the equivalent of a pressure difference of 12 Pa. With the density of water (1000 kg/m^3), the corresponding pressure difference is somewhat less than 10 kPa. Thus with a liquid, the gravitational potential energy is normally added.

Similarly, the *stagnation pressure* is the pressure that would be attained by bringing the flow locally to zero velocity in absence of heat exchange (adiabatic) and without losses (reversible). The *total pressure* is then the pressure that would be attained by additionally bringing (in mind) the fluid particle to the reference height for gravitational potential energy. Again, the subscript 0 is used for both the stagnation and the total pressure.

If necessary to make the difference with stagnation or total quantities, the basic quantities are called the static ones; so: *static pressure* and *static enthalpy*.

A remark is that with fundamental analyses of the flow through a rotor, very often, the effect of gravitational potential energy is ignored because of small height differences.

The energy equation in the relative frame is often expressed using a quantity $I = h_{0r} - \frac{1}{2}u^2$, called *rothalpy* (a short term for *rotating total enthalpy*). The centrifugal force is $\overrightarrow{Cf} = \Omega^2 \vec{r} = \nabla\left(\frac{1}{2}\Omega^2 r^2\right) = -\nabla\left(-\frac{1}{2}u^2\right)$. Thus, the term $-\frac{1}{2}u^2$ is the associated potential energy. The energy equation then becomes $dI = dq$. With an axial machine, $u =$ constant, and the energy equation simplifies to $dh_{0r} = dq$. We will not often use the term rothalpy. This means that the work term associated to the centrifugal force will mostly be written explicitly as in (1.13) and (1.14).

1.5.2   Moment of Momentum in the Absolute Frame: Rotor Work

Figure 1.7 sketches a meridional section of a streamtube enclosing the entire blade height of a mixed-flow pump rotor. A *meridional section* of a rotating object is a section containing the axis of rotation. This term is borrowed from geography. The *meridional component* of the absolute velocity, which is the velocity component in the meridional plane, is denoted by v_m. The meridional velocity component may be decomposed in an axial component v_a and a radial component v_r. The velocity

Fig. 1.7 Meridional
streamtube section: closed
and open rotors

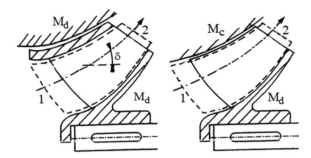

component in the circumferential (or tangential) direction is denoted by v_u. With the mean line flow representation, there is no velocity component perpendicular to the average circumferential streamsurface.

Figure 1.7 shows two possible rotor forms. The left-hand rotor has a *shroud*. This type is called *closed* or *shrouded*. The blades of the right-hand rotor fit directly in the casing with a clearance. This type is termed *open* or *unshrouded*. The part of the rotor that is fixed to the shaft is called the *hub*. The inner part of the rotor that carries the blades is mostly called the *rotor disc* (in general, the disc is curved), but the term hub is often used for the complete inner part. The flow is guided by the blade surfaces. With a closed rotor, the flow is guided by the shroud and the hub as well. The shroud and hub then constitute the *end walls*.

With a closed rotor, all guiding surfaces are in the relative frame. With an open type, the end wall at the hub side is in the relative frame and the end wall at the casing side is in the absolute frame. We first analyse the closed form.

Closed rotor (shrouded)

At the shroud and hub outsides, the surrounding fluid exerts friction forces. The moment of those friction forces around the shaft is usually termed *disc friction moment* M_d (here: disc and shroud together; we may call it *wheel friction moment*). By M_d we denote the moment exerted on the material rotor parts by the surrounding fluid. This moment is always braking: the sense of the moment is opposite to the rotation sense. Forces exerted at the flow side on the blade surfaces and the end walls (pressure forces and friction forces) form a moment around the shaft. The moment of these forces, calculated as exerted on the flow, is denoted with M.

We consider the moments M and M_d as positive in the running sense of the rotor. M_d is directed against the running sense and is negative. The flow moment M is positive with a driven machine (receiving shaft power) and negative with a driving machine (delivering shaft power). The sum of the reaction moment $(-M)$ and the disc friction moment (M_d) must be kept in balance by a moment exerted on the shaft (M_{shaft}). We also consider this shaft moment as positive in the running sense. So M_{shaft} is positive with a driven machine and negative with a driving one.

The moment balance of the material rotor parts is

$$M_{shaft} - M + M_d = 0.$$

With a driven machine we note

$$M_{shaft} = M - M_d. \tag{1.15}$$

The moment M_{shaft} exerted on the shaft by the external motor is then positive. M and $-M_d$ are both positive. The moment M_{shaft} is divided into two parts: the part M reaches the flow and the part $-M_d$ is absorbed by disc friction.

With a driving machine we note

$$-M = -M_{shaft} - M_d. \tag{1.16}$$

The flow moment M is negative, i.e. the flow exerts a positive moment $-M$ on the rotor. The shaft moment is negative as well, i.e. the machine exerts a positive moment $-M_{shaft}$ on the coupled load. The moment $-M$ is divided into two parts: the part $-M_{shaft}$ reaches the shaft and the part $-M_d$ is absorbed by disc friction.

We consider a control volume as sketched in Fig. 1.7. As usual, we indicate the rotor entry with subscript 1 and the exit with subscript 2. By taking the moment around the rotation axis of Newton's momentum law, we obtain that the change per time unit of the moment of momentum in the flow equals the moment of the forces exerted on the control volume:

$$\dot{m}(r_2 v_{2u} - r_1 v_{1u}) = M. \tag{1.17}$$

The moment M encompasses both contributions by pressure forces and by shear forces on the material surfaces. When drawing up of the former expression, it is assumed that shear stresses on the inlet and outlet faces of the control volume do not form a moment around the rotation axis. Actually, this is an approximation, which is justified if the inlet and outlet flows are not strongly distorted. This requires that the inlet and outlet surfaces of the control volume are at some distance from the entry and exit of the bladed part of the rotor. Clearly, there is no moment by the pressure forces on the inlet and outlet surfaces.

Multiplication of (1.17) by the speed of rotation Ω results in

$$\dot{m}(u_2 v_{2u} - u_1 v_{1u}) = P_{rot},$$

where P_{rot} is the power transferred from the rotor to the flow by the moment M.

After division by the mass flow rate, this is

$$u_2 v_{2u} - u_1 v_{1u} = \Delta W, \tag{1.18}$$

with ΔW the work done per mass unit on the flow by the moment M.

Multiplication of Eq. (1.15) by Ω results in

$$P_{shaft} = P_{rot} - P_d.$$

With a power receiving machine, P_{shaft} is the power supplied to the shaft by the driving motor. The term $-P_d$ represents the power associated to the dissipation by disc friction. After division by the mass flow rate, we note

$$\Delta W_{shaft} = \Delta W + q_{irr}^o. \qquad (1.19)$$

The term q_{irr}^o represents the part of the shaft work that is dissipated by disc friction. In (1.19) this term is represented with a positive sign, as it always is arithmetically positive in this equation.

Analogously, Eq. (1.16) for a power delivering machine results in

$$-\Delta W = -\Delta W_{shaft} + q_{irr}^o, \qquad (1.20)$$

where $-\Delta W$ is the work done by the flow on the rotor, $-\Delta W_{shaft}$ the part of $-\Delta W$ supplied to the shaft and q_{irr}^o the part of $-\Delta W$ dissipated by disc friction.

With a driven machine (1.19), ΔW is the work done by the rotor on the flow. With a driving machine (1.20), $-\Delta W$ is the work done by the flow on the rotor. For both, we use the term *rotor work*. With a driving machine, we note $-\Delta W$ as

$$-\Delta W = u_1 v_{1u} - u_2 v_{2u}. \qquad (1.21)$$

The work ΔW done on the flow with the driven machine is partly dissipated inside the flow by friction forces, as described by Bernoulli's equation (1.5). Analogously, internal dissipation occurs as well with a driving machine.

The effects of *dissipation during energy exchange* (q_{irr}^o) and *internal dissipation* (q_{irr}) are studied below.

Open rotor (unshrouded)

The foregoing derivations are similar with an open rotor. The difference is that the friction force on the casing directly contributes to the moment exerted on the flow. We again term M the resulting moment exerted on the flow by the forces on the blade surfaces and the end surfaces, including the casing. Equation (1.17) is then still valid. Equation (1.18) may formally be derived from it, but that $\Delta W = M\Omega$ is the rotor work is then not fully clear. Complete insight follows from the more detailed analysis hereafter.

The flow moment may be divided into two parts:

$$M = M_b + M_c.$$

M_b is the part originating from the blade surfaces and the hub, in other words, the material rotor surfaces and M_c is the contribution by the casing. Both moments are considered as being exerted on the flow. With a driven machine, M_b is positive and M_c negative. With a driving machine, M_b and M_c are both negative. M_c is, like M_d, always negative. The moment balance on the material rotor parts is

$$M_{shaft} - M_b + M_d = 0,$$

where M_d represents the disc friction moment of the hub.

It follows that

$$M_{shaft} = M_b - M_d = M - M_c - M_d. \tag{1.22}$$

After multiplication by Ω it follows that

$$P_{shaft} = M\Omega - M_c\Omega - M_d\Omega. \tag{1.23}$$

The term $-M_c\Omega$ is always positive and thus has the same nature as the term $-M_d\Omega$. After division by the mass flow rate we obtain

$$\Delta W_{shaft} = \Delta W + q_{irr}^o,$$

where $M\Omega = \dot{m}\Delta W$ and $-M_c\Omega - M_d\Omega = \dot{m}\, q_{irr}^o$.

The result for ΔW_{shaft} is then the same as (1.19).

Interpretation of the moment balance (1.22) is quite direct. For a driven machine, it means that the shaft moment M_{shaft} is divided into two parts: the part M_b reaches the rotor blades on the flow side and the part $-M_d$ is absorbed by disc friction. Further, the blade moment M_b is divided into two parts: the part M reaches the flow and the part $-M_c$ is absorbed by friction on the casing.

To interpret the power balance (1.23), we notice that the moment M_c equals the integral of the scalar product of the friction force $\vec{\tau}\,dS$ exerted on the flow and a unit vector along the blade speed $\vec{u}$, multiplied by the radius.

The term $M_c\Omega$ is thus the integral of the scalar product of the friction force $\vec{\tau}\,dS$ and the blade speed $\vec{u}$. With the relation between the absolute and the relative velocities, $\vec{v} = \vec{u} + \vec{w}$, it follows that

$$M_c\Omega = \int \vec{\tau}.\vec{u}\, dS = \int \vec{\tau}.\vec{v}\, dS - \int \vec{\tau}.\vec{w}\, dS \quad \text{or}$$

$$-M_c\Omega = \int -\vec{\tau}.\vec{v}\, dS - \int -\vec{\tau}.\vec{w}\, dS.$$

The term with the integrand $-\vec{\tau}.\vec{v}$ represents the dissipation by the friction force on the casing. The term with the integrand $-\vec{\tau}.\vec{w}$ is physically fictive, but may be considered as the dissipation by the friction force if it were acting in the relative frame. The difference between both terms may thus be considered as the disc friction dissipation associated to the friction force acting on a fictitious shroud of the rotor. The term with the integrand $-\vec{\tau}.\vec{w}$ has then to be considered as dissipation inside the flow on this fictitious shroud.

The interpretation of the power balance (1.23) for a driven machine is that the shaft power consists of three parts: the part $M\Omega$ reaches the flow, the part $-M_c\Omega$ is dissipated at the casing and the part $-M_d\Omega$ is dissipated at the rotor disc. The part $M\Omega$ is thus the rotor work. A part of the rotor work is dissipated internally in the flow, and this includes friction dissipation on the fictitious shroud.

The foregoing interpretations of the balances of moment and power are for work done by the rotor on the flow (driven machine). They are similar for work done in the opposite sense (driving machine).

1.5.3 Rotor Work in the Mean Line Representation of the Flow

We note, both for an open and a closed rotor, as equation for the rotor work:

$$\Delta W = u_2 v_{2u} - u_1 v_{1u}. \qquad (1.24)$$

This equation is termed *Euler's turbomachine equation* or *Euler's work equation*. It is considered to be the most important basic equation in turbomachinery theory. Equation (1.24), the same as (1.18), is written for work done by the rotor on the flow (driven machine). It is also valid for work done in the opposite sense (driving machine), but is then best written with inversed signs, thus as equation (1.21). Henceforth, we refer to Euler's work equation (1.18 or 1.21) as the *rotor work equation* and to Bernoulli's equation (1.5) as the *work equation*.

The work Eq. (1.5) and the energy Eq. (1.6) express the energy changes in the flow as a result of the work done on the flow or by the flow. The rotor work equation expresses the relation between the work done by the rotor or on the rotor and the change of angular momentum in the flow (change of angular momentum multiplied by angular speed). By the notation, it is assumed that the rotor work, expressed by Eqs. 1.18 and 1.21, derived from a moment of momentum balance, can be applied in the work and energy equations 1.5 and 1.6 for an average streamline through the rotor.

It is thus assumed that the active forces, exerted on the flow by the material parts, intervene on average as represented in Fig. 1.5. This means that the work done per time unit may be considered as the scalar product of the average force and the average flow velocity. This does not follow from the reasoning with the moment of momentum balance. There is no direct objection against the use of the work expressed by 1.24 in the work equation 1.5 and the energy equation 1.6, on the basis of the fundamental laws of thermodynamics. It will be verified in Chap. 2 (Sects. 2.2.5 and 2.2.7) that the representation of the work done in Fig. 1.5 is correct for an axial machine. This verification is less feasible for a more general machine. It requires three-dimensional formulation of the conservation laws.

The rotor work equation has been derived here based on a one-dimensional flow representation (mean line representation) but the formulation for a real three-dimensional flow follows from the same principles.

1.5.4 Moment of Momentum in the Relative Frame: Forces Intervening in the Rotor Work

From a velocity triangle (Fig. 1.8) follows geometrically: $\vec{v} = \vec{u} + \vec{w}$, or $v_u = u + w_u$.

The velocity triangle in Fig. 1.8 is drawn deliberately with negative w_u. From $v_u = u + w_u$, it follows that

$$u v_u = u^2 + u w_u.$$

So:

$$\Delta W = u_2^2 - u_1^2 + u_2 w_{2u} - u_1 w_{1u}. \tag{1.25}$$

Equation (1.25) also follows from a moment of momentum balance, applied in the relative frame. Fictitious forces to be introduced are

$$\overrightarrow{Co} = -2\vec{\Omega} \times \vec{w} \quad \text{and} \quad \overrightarrow{Cf} = \Omega^2 \vec{r}.$$

The relative velocity $\vec{w}$ has an axial component, a radial component and a tangential one. The axial component does not intervene in the Coriolis force. A tangential unit vector in the sense of the rotation may be noted as

$$\vec{1}_u = \vec{1}_\Omega \times \vec{1}_r,$$

so that

$$\overrightarrow{Co} = -2\vec{\Omega} \times \left(w_r \vec{1}_r + w_u \vec{1}_\Omega \times \vec{1}_r \right) = -2\Omega w_r \vec{1}_u + 2\Omega w_u \vec{1}_r.$$

According to Fig. 1.7, $w_r = w_m \vec{1}_m \cdot \vec{1}_r$, with $\vec{1}_m$ a unit vector in the meridional direction with positive axial and radial components. The centrifugal force does not contribute to the force moment. Neither does the radial component of the Coriolis force. The moment of momentum balance in the relative frame results in

Fig. 1.8 Velocity triangle

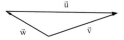

$$\dot{m}(r_2 w_{2u} - r_1 w_{1u}) = M + M_{co},\tag{1.26}$$

with

$$M_{co} = \int -2\Omega w_m \vec{1}_m . \vec{1}_r \rho \, dA \, dm \, r,$$

where dA is the cross-section area of an elementary annular streamtube around the average circumferential streamsurface in Fig. 1.7. The elementary length in the meridional plane is denoted by dm. So the term $\rho w_m \, dA$ represents the mass flow rate through the elementary streamtube. With the mean line representation, the result for the average streamsurface is considered to be representative for the entire flow.

We thus note

$$M_{co} = -\dot{m} \int 2\Omega r \vec{1}_m . \vec{1}_r \, dm.$$

With $dm \vec{1}_m . \vec{1}_r = dr$ it follows that

$$M_{co} = -\dot{m} \int 2\Omega r \, dr = -\dot{m}\Omega(r_2^2 - r_1^2).$$

After multiplication by Ω and division by $\dot{m}$, the moment of momentum equation in the relative frame (1.26) results in Eq. (1.25).

Equation (1.26) may be written as

$$M = -M_{co} - M_\ell,$$

where M_ℓ is the moment term associated to the turning of the relative flow (change of $r w_u$), thus by the lift. The expression demonstrates, for a driven machine, that the moment required to drive the rotor must balance the moment of two forces, namely *Coriolis force* and *lift force*. In Eq. (1.25) this means that the term $u_2^2 - u_1^2$ represents the work done against the Coriolis force and $u_2 w_{2u} - u_1 w_{1u}$ the work done against the lift force. The interpretation with a driving machine is with work done by the Coriolis force and work done by the lift force.

The above analysis implies that two forces intervene in the rotor work. The effect of the lift is the simplest to understand. For the axial pump and the axial turbine, analysed earlier (Figs. 1.2 and 1.3), it is obvious that turning of the flow within the rotor generates lift on the blades. Physically, lift is the result of a pressure difference between both sides of a blade or a rotor blade channel, generated by turning the flow. We will analyse the origin of this pressure difference in Chap. 2. Figures 1.2 and 1.3 demonstrate that lift intervenes in the rotor work. In these axial examples, $u_1 = u_2$, but, meanwhile, we understand that the moment of forces intervenes with work and that a change of radius affects the resulting moment.

The role of the Coriolis force in the rotor work only follows in a formal way from the above analysis. Obviously, rotation induces forces in the fluid, namely the Coriolis force and the centrifugal force. Only the Coriolis force has a component in the tangential direction. This component intervenes in the moment of momentum. The concrete way that rotation-induced forces intervene is by causing a pressure difference between the two sides of a blade or a rotor blade channel. Associated to the Coriolis force, there is a pressure difference, with a moment of the resulting rotor force around the rotation axis. This is analysed further for radial machines in Sect. 1.7 and in Chap. 3, Sect. 3.3.2.

Confusion may arise by the finding that there is work contribution by the centrifugal force in the work and energy equations on a streamline in a rotating frame; expressions (1.13) and (1.14). With the above analysis it is clear that the centrifugal force does not contribute to the energy exchange between the rotor and the fluid (the rotor work), because the centrifugal force has no moment around the rotation axis. It becomes obvious from expressions (1.13) and (1.14) that the centrifugal force intervenes in the change of kinetic energy and the change of pressure (pressure-associated energy) inside the flow.

The centrifugal force influences the distribution of the energy forms within the rotor flow. A potential energy is associated to the centrifugal force. A change in this potential energy due to a change in radius must thus be compensated by a change in another energy component. Distribution of energy forms, inside the flow, has to be distinguished from energy exchange between the flow and the rotating machine parts. The effect of the centrifugal force on the energy distribution is discussed in the next section.

1.5.5 Energy Component Changes Caused by the Rotor Work

From the cosine rule follows (Fig. 1.8):

$$w^2 = u^2 + v^2 - 2uv_u,$$

or

$$uv_u = \frac{1}{2}u^2 + \frac{1}{2}v^2 - \frac{1}{2}w^2.$$

Thus:

$$\Delta W = \frac{u_2^2 - u_1^2}{2} + \frac{v_2^2 - v_1^2}{2} + \frac{w_1^2 - w_2^2}{2}. \qquad (1.27)$$

The meaning of the rotor work equation in the form (1.27) becomes obvious by combining it with the energy equation. Ignoring the change of gravitational potential energy through the rotor, because of the small height difference, the energy equation in the absolute frame (1.7), in absence of heat transfer, is

$$\Delta W = \Delta h_0 = \Delta h + \Delta \frac{1}{2} v^2. \tag{1.28}$$

Combined with (1.27) it follows that

$$\Delta h = \frac{u_2^2 - u_1^2}{2} + \frac{w_1^2 - w_2^2}{2}. \tag{1.29}$$

Equations (1.28) and (1.29) demonstrate that work done becomes visible in two forms. For work done on the fluid, there is *kinetic energy increase* and *static enthalpy increase*. The kinetic energy increase is called the *action part* of the work, in other words, the directly visible effect. The static enthalpy increase is called the *reaction part*. The static enthalpy increase may be converted into kinetic energy by connecting a nozzle in which kinetic energy is generated (hence the term reaction).

The action-reaction terminology may sound peculiar with driven machines. It originates from turbines, as will be illustrated below. The concept applies to all turbomachines, however.

The meaning of the static enthalpy increase follows from the work equation and the energy equation in the relative frame. In this frame all forces on the rotor stand still and perform no work. So we write the work equation, infinitesimally (again, ignoring changes in gravitational potential energy), according to (1.13):

$$d\frac{1}{2}u^2 = d\frac{1}{2}w^2 + \frac{1}{\rho}dp + dq_{irr}. \tag{1.30}$$

The energy equation according to (1.14), in absence of heat transfer, is

$$d\frac{1}{2}u^2 = dh + d\frac{1}{2}w^2. \tag{1.31}$$

The energy equation is identical with Eq. (1.29). From the combination with the work equation it follows that

$$dh = \frac{1}{\rho}dp + dq_{irr}.$$

The latter equation demonstrates that the loss-free part of the enthalpy increase due to the work done corresponds to a pressure increase.

The term $(u_2^2 - u_1^2)/2$ is the enthalpy increase corresponding to the pressure increase by the centrifugal force. It is obvious that the centrifugal force generates this term by the work of the centrifugal force $d\frac{1}{2}u^2$ in the work equation (1.30). From the same equation it follows that the term $(w_1^2 - w_2^2)/2$, infinitesimally $-d\frac{1}{2}w^2$, is the enthalpy increase corresponding to the pressure increase due to flow deceleration.

For further clarification, we directly derive the pressure force, $-(1/\rho)\nabla p$, associated to the centrifugal force (equal and opposite) by

$$-\frac{1}{\rho}\nabla p = -\Omega^2 \vec{r} \quad \text{or} \quad \frac{1}{\rho}\frac{dp}{dr} = \Omega^2 r.$$

Thus:

$$\frac{1}{\rho}dp = \Omega^2 r\,dr = \Omega^2 d\frac{r^2}{2} = d\frac{u^2}{2}.$$

The enthalpy increase corresponding to the pressure increase by the centrifugal force constitutes a total differential and so may be integrated from the rotor entry to the rotor exit, along an arbitrary path. This means that the pressure increase by the centrifugal force is flow-independent and thus occurs without losses. The basic reason is, as already discussed, that a potential energy is associated to the centrifugal force. The pressure field associated to the centrifugal force is thus actually static in the rotating frame. It is completely similar to the pressure field associated to the gravity force. The loss term in the work equation (1.30) is associated to the friction in the flow and the conversion of kinetic energy into enthalpy, by flow deceleration. Conversion of kinetic energy into enthalpy, which is denominated *diffusion*, is a process liable to losses (see Chap. 2, Sect. 2.4).

1.6 Energy Analysis of Turbomachines

In an energy analysis, we write work or energy balances between successive stations in a machine and we define efficiencies.

1.6.1 Mechanical Efficiency and Internal Efficiency

For a work receiving machine, the work transfer from the shaft to the fluid side of the rotor is expressed by Eq. (1.19). The efficiency of this transfer is called the *mechanical efficiency*, defined by

$$\eta_m = \frac{\Delta W}{\Delta W_{shaft}}.$$

For a work delivering machine, the work transfer from the fluid side of the rotor to the shaft is expressed by Eq. (1.20). The efficiency of the transfer, also called the mechanical efficiency, is defined by

$$\eta_m = \frac{-\Delta W_{shaft}}{-\Delta W}.$$

Up to now, the difference between the shaft work and the rotor work is the dissipation by wheel friction (q_{irr}^o), outside the flow path (external dissipation). In practice, shaft work is measured at the flange of the driving motor or the driven load. This means that dissipation by friction in the bearings of the shaft and the seals at the passage of the shaft through the casing are included in the definition of the mechanical efficiency.

For a work receiving machine (pump), the relation between the rotor work and the increase of the mechanical energy in the fluid (the head) may be written as

$$\Delta W = \Delta E_m + q_{irr},$$

where q_{irr} is the dissipation inside the flow path (internal dissipation). The efficiency of the conversion of the rotor work (work done by the rotor on the flow) into mechanical energy increase in the fluid is called *internal efficiency*, defined by

$$\eta_i = \frac{\Delta E_m}{\Delta W}.$$

The work equation for a work delivering machine is

$$-\Delta W = -\Delta E_m - q_{irr} \quad \text{or} \quad -\Delta E_m = -\Delta W + q_{irr}.$$

The internal efficiency is the ratio of the rotor work (work done by the flow on the rotor) to the mechanical energy extracted from the flow, so that

$$\eta_i = \frac{-\Delta W}{-\Delta E_m} = \frac{|\Delta W|}{|\Delta E_m|}.$$

The terminology comes from an interpretation of the heating of the fluid by the dissipation mechanisms. In simplified turbomachine analyses, heat generated outside the flow path by wheel friction, friction in bearings and seals is supposed to be transferred to the surroundings. Heat generated inside the flow path by friction is supposed to stay in the fluid. This means that the flow, not the machine, is considered to be adiabatic (except when there is explicit heat exchange between the flow and the surroundings, as in a cooled turbine part of a gas turbine; see Chap. 11). From now on, this simplification is assumed. The energy equation (1.6) is then noted as

$$\Delta W = \Delta \left(e + \frac{1}{2}v^2 + \frac{p}{\rho} + U \right). \tag{1.32}$$

Because of the assumption of adiabatic flow, we consider q_{irr}^o as a *mechanical loss*. This means a loss with removal to the surroundings of the heat produced by the dissipation. A loss with the heat by dissipation absorbed by the fluid is called a *thermodynamic loss* or an *internal loss*.

We remark that the definitions of mechanical and internal efficiencies stay valid with another interpretation of heating due to dissipation. The interpretation only influences the writing of the energy equation (1.32).

1.6.2 Energy Analysis of an Axial Hydraulic Turbine

Figure 1.9 sketches mean streamlines in the flow through an axial hydraulic turbine.

Inlet: $0 \rightarrow 1$

Absolute frame (no work):

$$0 = d\frac{1}{2}v^2 + \frac{1}{\rho}dp + dU + dq_{irr}.$$

So

$$\frac{v_1^2}{2} - 0 + \frac{p_1 - p_a}{\rho} + gz_1 - gz_0 + q_{irr01} = 0,$$

or

$$g(z_0 - z_1) = \frac{v_1^2}{2} + \frac{p_1 - p_a}{\rho} + q_{irr01}.$$

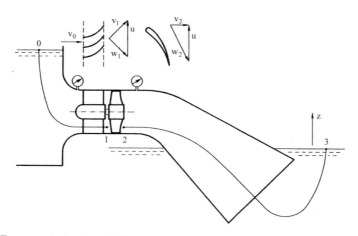

Fig. 1.9 Energy analysis of an axial hydraulic turbine

The gravitational potential energy consumed is converted into kinetic energy and pressure energy, with losses. The inlet supplies the kinetic energy and the pressure energy to the rotor. We remark that for the determination of the gravitational potential energy in the average flow, the points 1 and 2 in Fig. 1.9 have to be considered as being at the height of the machine shaft.

Rotor: $1 \rightarrow 2$

In the absolute frame:

$$dW = d\frac{1}{2}v^2 + \frac{1}{\rho}dp + dU + dq_{irr}.$$

For horizontal flow:

$$\Delta W = \frac{v_2^2 - v_1^2}{2} + \frac{p_2 - p_1}{\rho} + q_{irr12},$$

or

$$\frac{v_1^2 - v_2^2}{2} + \frac{p_1 - p_2}{\rho} = -\Delta W + q_{irr12}.$$

Kinetic energy and pressure energy (= mechanical energy) are consumed in the flow in order to generate mechanical energy on the rotor ($-\Delta W$), with losses. The kinetic energy consumption is visible in the velocity triangles ($v_1 > v_2$). The consumption of pressure energy follows from the work equation in the relative frame:

$$0 = d\frac{1}{2}w^2 + \frac{1}{\rho}dp + dq_{irr}.$$

So

$$\frac{w_2^2 - w_1^2}{2} + \frac{p_2 - p_1}{\rho} + q_{irr12} = 0,$$

or

$$\frac{p_1 - p_2}{\rho} = \frac{w_2^2 - w_1^2}{2} + q_{irr12}.$$

The relative flow accelerates ($w_2 > w_1$). Kinetic energy generation in the relative flow corresponds to pressure energy consumption.

Obviously, we also find

$$-\Delta W = \frac{v_1^2 - v_2^2}{2} + \frac{w_2^2 - w_1^2}{2}. \tag{1.33}$$

The work by the kinetic energy decrease in the absolute frame is termed the *action part*. The work by the acceleration of the relative flow is called the *reaction part*. The reaction part corresponds to the static enthalpy decrease: $h_1 - h_2$ ($h_{0r} = $ constant in adiabatic flow for $u_1 = u_2$). The work done on the rotor equals the total enthalpy decrease in the flow: $h_{01} - h_{02}$ (in adiabatic flow).

The *degree of reaction* is defined by

$$R = \frac{h_1 - h_2}{h_{01} - h_{02}}, \qquad (1.34)$$

with

$$h_1 - h_2 = \frac{w_2^2 - w_1^2}{2} \quad \text{and} \quad h_{01} - h_{02} = \frac{v_1^2 - v_2^2}{2} + \frac{w_2^2 - w_1^2}{2}. \qquad (1.35)$$

The action-reaction terminology results from the observation that a flow, in principle, may exert force on an object in two ways: by turning of the flow with constant relative velocity (*action*) and by acceleration of the flow (*reaction*), according to the sketches in Fig. 1.10. The right-hand sketch suggests the propulsion of a rocket, assuming that velocity w may be generated from an internal energy source.

The velocity triangles and blade shapes with pure action ($R = 0$) and with 50% degree of reaction ($R = 0.5$) are sketched in Fig. 1.11, as they occur in steam turbines.

It is obvious from Fig. 1.11 that the value of the degree of reaction has a strong influence on the shape of the velocity triangles and the shape of the rotor blades. In other words, the degree of reaction is a *kinematic parameter*. The effect of the choice of the degree of reaction will further be discussed at several occasions.

Draught tube: $2 \rightarrow 3$.

Absolute frame (no work):

$$0 = d\frac{1}{2}v^2 + \frac{1}{\rho}dp + dU + dq_{irr}.$$

So

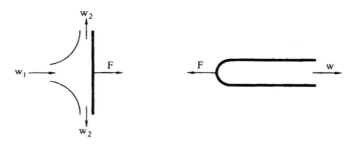

Fig. 1.10 Force by action ($w_1 = w_2$) and reaction ($\Delta w > 0$)

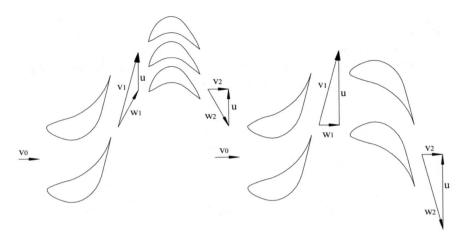

Fig. 1.11 Velocity triangles for $R = 0$ ($w_1 = w_2$) and $R = 0.5$ ($w_2 = v_1$; $v_2 = w_1$) in steam turbines

$$0 - \frac{v_2^2}{2} + \frac{p_a - p_2}{\rho} + g(z_3 - z_2) + q_{irr23} = 0,$$

or

$$\frac{v_2^2}{2} + g(z_2 - z_3) = \frac{p_a - p_2}{\rho} + q_{irr23}.$$

The kinetic energy and the gravitational potential energy available at the rotor exit are used to produce pressure decrease behind the rotor and thus increase of the pressure difference through the rotor.

The resulting three work equations are:

$$g(z_0 - z_1) = \frac{v_1^2}{2} + \frac{p_1 - p_a}{\rho} + q_{irr01},$$

$$\frac{v_1^2 - v_2^2}{2} + \frac{p_1 - p_2}{\rho} = |\Delta W| + q_{irr12},$$

$$\frac{v_2^2}{2} + g(z_2 - z_3) = \frac{p_a - p_2}{\rho} + q_{irr23}.$$

The sum is

$$g(z_0 - z_3) = |\Delta W| + q_{irr03}.$$

The internal efficiency of the whole turbine is

$$\eta_i = \frac{|\Delta W|}{g(z_0 - z_3)}.$$

Figure 1.9 is an example of a small turbine for low-head application. The head-water is close to the stator ring and the rotor. One can thus evaluate the whole installation as one system. But, with large head, there is a long supply duct to the entry of the vane ring. With such installations, it is not appropriate to include the losses in the supply duct in the definition of the turbine efficiency. The concept of *manometric head* is introduced for this purpose. Pressure is determined by a manometer at the turbine entry (see Fig. 1.9). The mechanical energy at the disposal of the turbine is then defined by

$$\frac{v_m^2}{2} + \frac{p_m - p_a}{\rho} + g(z_m - z_3) = g H_m.$$

with v_m the velocity at the position of the manometer, $p_m - p_a$ the measured pressure difference to the atmospheric pressure and $z_m - z_3$ the height difference of the position of the manometer to the tailwater. The *manometric head* H_m is then the mechanical energy expressed as a height. The manometric head H_m is lower than the geometric head $z_0 - z_3$ due to losses in the supply duct.

The losses in the supply duct follow from

$$\frac{v_m^2}{2} + \frac{p_m - p_a}{\rho} + q_{irr0m} = g(z_0 - z_m).$$

The losses in the draught tube may also be taken out by defining the mechanical energy with a manometer at the rotor exit (see Fig. 1.9). But, this is normally not done with a hydraulic turbine and the draught tube is considered as a turbine part.

1.6.3 Energy Analysis of an Axial Pump

Inlet: $0 \rightarrow 1$

  Absolute frame (no work):

$$0 = d\frac{1}{2}v^2 + \frac{1}{\rho}dp + dU + dq_{irr}.$$

So

$$\frac{v_1^2}{2} - 0 + \frac{p_1 - p_a}{\rho} + g(z_1 - z_0) + q_{irr01} = 0,$$

or

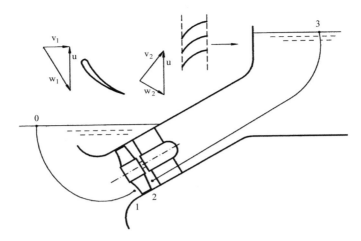

Fig. 1.12 Energy analysis of an axial pump

$$g(z_0 - z_1) = \frac{v_1^2}{2} + \frac{p_1 - p_a}{\rho} + q_{irr01}.$$

As with the turbine, the gravitational potential energy consumed is converted into kinetic energy and pressure energy, with losses.

Figure 1.12 shows a pump that is mounted below the suction level (submerged). The inlet is a suction mouth. If the pump is mounted above the suction level, with a suction pipe, $g(z_0 - z_1)$ is negative. The above expression is then written as

$$\frac{p_a - p_1}{\rho} = g(z_1 - z_0) + \frac{v_1^2}{2} + q_{irr01}.$$

The interpretation is then that the suction height is overcome by pressure lowering at the suction side of the pump (suction by the rotor) and that kinetic energy is generated, with losses.

Rotor: $1 \rightarrow 2$

We ignore the height difference through the rotor (see a remark afterwards). In the absolute frame:

$$dW = d\frac{1}{2}v^2 + \frac{1}{\rho}dp + dU + dq_{irr}.$$

So

$$\Delta W = \frac{v_2^2 - v_1^2}{2} + \frac{p_2 - p_1}{\rho} + q_{irr12}.$$

The work done on the fluid results in mechanical energy increase, with losses.

In the relative frame:

$$\frac{w_2^2 - w_1^2}{2} + \frac{p_2 - p_1}{\rho} + q_{irr12} = 0.$$

The kinetic energy increase in the absolute frame is visible in the velocity triangles. The pressure increase follows from the kinetic energy decrease in the relative frame. It follows as well that

$$\Delta W = \frac{v_2^2 - v_1^2}{2} + \frac{w_1^2 - w_2^2}{2}. \tag{1.36}$$

The first part of the expression is the *action part*, the second one the *reaction part*. The work done equals the total enthalpy increase (in adiabatic flow). The reaction part of the work equals the static enthalpy increase (in adiabatic flow).

The degree of reaction is

$$R = \frac{h_2 - h_1}{h_{02} - h_{01}}, \tag{1.37}$$

with

$$h_2 - h_1 = \frac{w_1^2 - w_2^2}{2} \quad \text{and} \quad h_{02} - h_{01} = \frac{v_2^2 - v_1^2}{2} + \frac{w_1^2 - w_2^2}{2}. \tag{1.38}$$

Diffuser ring and delivery pipe: $2 \rightarrow 3$
In the absolute frame:

$$0 = d\frac{1}{2}v^2 + \frac{1}{\rho}dp + dU + dq_{irr}.$$

So

$$0 = 0 - \frac{v_2^2}{2} + \frac{p_a - p_2}{\rho} + g(z_3 - z_2) + q_{irr23},$$

or

$$\frac{v_2^2}{2} + \frac{p_2 - p_a}{\rho} = g(z_3 - z_2) + q_{irr23}.$$

The kinetic energy and the pressure energy available downstream of the rotor are converted, with losses, into increase of gravitational potential energy.

Addition of the three work equations results in

$$\Delta W + g(z_0 - z_1) = g(z_3 - z_2) + q_{irr03},$$

or

$$\Delta W = g(z_3 - z_0) + q_{irr03}.$$

The internal efficiency of the whole installation is

$$\eta_i = \frac{g(z_3 - z_0)}{\Delta W}.$$

The losses in the pipes connected to the pump can be taken out of the definition by placing manometers at the entry and exit of the pump and by determining the manometric head. In the example of Fig. 1.12, the suction mouth is normally considered as part of the pump. If two manometers are mounted, the *manometric head of the pump* H_m is defined by

$$g H_m = \frac{v_d^2 - v_s^2}{2} + \frac{p_d - p_s}{\rho} + g(z_d - z_s).$$

The subscripts d and s refer to the *discharge*/delivery and *suction* sides of the pump. The term manometric head denotes the mechanical energy rise by the pump, expressed as a height. Mostly, the term discharge side is used for the side by which the pump delivers the fluid. Similarly, the term *discharge pipe* is used. We follow this terminology.

In the previous analysis of the axial pump and the axial turbine, the change of the gravitational potential energy through the rotor has been ignored (with the horizontal shaft of the axial turbine, there is no change of gravitational potential energy). This simplification has also been made in fundamental analyses in this chapter. But, it is not difficult to take the change into account. For constant density, it follows directly from the work and energy equations 1.5 and 1.7 (absolute system) and 1.13 and 1.14 (relative system) that gravitational potential energy (U) has to be added to the pressure potential energy (p/ρ) in a work equation and to the enthalpy in an energy equation, thus: $h = e + (p/\rho) + U$. Equations (1.33) and (1.36) follow directly from Euler's rotor work equation and do not change. In the relations (1.34), (1.35), (1.37) and (1.38), the gravitational potential energy has to be included in the enthalpy.

1.7   Examples of Radial Turbomachines

In order to gain insight in the working principle of radial machines, we discuss two examples, first a centrifugal pump. Figure 1.13 shows a meridional section of a pump and sketches the section with the mean circumferential streamsurface. Figure 1.14 sketches the velocity triangles at the rotor entry and exit. The rotor is mounted in overhung to the shaft (4). With this type, the rotor suction side is free, allowing axial water intake. Flow is turned to the radial direction within the rotor (1). There is a

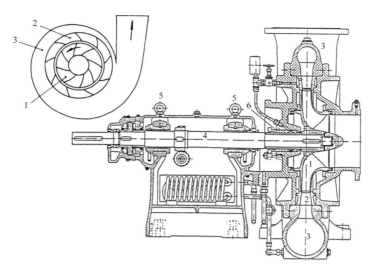

Fig. 1.13 Centrifugal pump (classical example)

Fig. 1.14 Radial pump rotor
with backward curved blades
(drawn is $w_1 = w_2$)

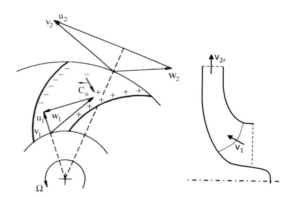

radial diffuser ring (2) and a collector (3). Further components are the bearings (5: here: sliding bearings) and the shaft seal (6: here: compression packing).

At the entry of the bladed part of the rotor, the absolute velocity v_1 has gained a component in the radial direction due to turning from the axial inflow direction towards the radial direction in the *suction eye* of the rotor. Theoretically, there is no component in the tangential direction, as no guide vanes have been passed through. In practice, the flow is somewhat entrained upstream of the rotor blades in the suction eye, which creates a small tangential component. This is termed *spontaneous pre-whirl*. This pre-whirl does not decrease the rotor work, as it actually is caused by the torque of the rotor, influencing the upstream flow. Therefore, we draw the velocity triangle at the rotor entry without pre-whirl. No axial velocity component is present anymore at the rotor exit. In the example drawn, the blades have a rather strong inclination with respect to the meridional plane at the rotor exit, backward compared to the rotation sense and with an angle nearly the same as at the rotor entry. This rotor blade shape is mostly described as *backward curved*.

Towards the rotor exit, cylindrical sections of the blades are straight. Figure 1.14 suggests blade filaments in the axial direction near the exit, but these may have an angle with respect to the meridional plane (the rake angle). The blade shape near the rotor exit is called *simply-curved*, meaning that the curvature is the same in every orthogonal section. Due to the diagonal flow direction at the entry of the bladed rotor part, there is blade curvature in the direction perpendicular to the mean streamsurface in the entry part. The blade shape near the rotor entry is called *doubly-curved*, meaning that there is curvature in orthogonal sections and cylindrical sections. But, in a mean line analysis, blade curvature (inlet part) and blade inclination (outlet part) perpendicular to the average streamsurface do not intervene.

The Coriolis force is $\vec{Co} = -2\vec{\Omega} \times \vec{w}$, with the sense as in Fig. 1.14. A pressure difference across a blade channel corresponds to the Coriolis force, as indicated. This generates a blade force with tangential component against the sense of rotation, thus causing energy transfer from the rotor to the flow. In the expression (1.25), the rotor work is $\Delta W = u_2^2 - u_1^2 + u_2 w_{2u} - u_1 w_{1u}$. The $u_2^2 - u_1^2$ part originates from the Coriolis force and is always positive for flow in centrifugal sense through the rotor ($u_2 > u_1$). The $u_2 w_{2u} - u_1 w_{1u}$ part is produced by turning of the flow, thus by lift, and may be positive or negative, depending on the blade shape. In the example drawn, the lift term is negative. The magnitude of $u_2 w_{2u}$ (the term is negative) is larger than the magnitude of $u_1 w_{1u}$ (the term is negative). The lift thus works in the adverse sense, as the work associated lies in the turbine sense.

To make the lift term less negative, the blades at the rotor exit should be inclined less backward. With a constant meridional velocity component, as drawn, this implies that deceleration then has to be built in ($w_2 < w_1$). It is impossible however to achieve strong deceleration in a diverging channel of limited length.

As will be discussed in Chap. 2, a velocity ratio down to about 0.7 may be reached with optimally-shaped stationary diffusers of limited length, i.e. about half of the kinetic energy at the entry may be converted into pressure energy. Due to curved blade channels, strong changes of the shape of the cross-section and to rotation effects, the velocity ratio w_2/w_1 in a rotor cannot be lower than about 0.8 to 0.9, with as a consequence that lift cannot intervene in the pump sense. The functioning is thus essentially based on Coriolis force. The rotor has to be considered as composed of rotating channels. The blades do not function as lifting objects.

The rotor work may also be written as

$$\Delta W = \frac{u_2^2 - u_1^2}{2} + \frac{v_2^2 - v_1^2}{2} + \frac{w_1^2 - w_2^2}{2}.$$

The term in the relative kinetic energies is not very significant in the sum. The centrifugal term corresponds to a pressure increase. As a pump is intended to increase pressure, the kinetic energy at the rotor exit $\frac{1}{2}v_2^2$ has to be converted into pressure

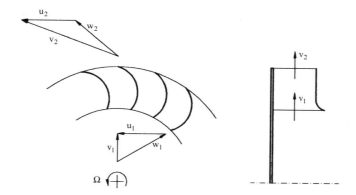

Fig. 1.15 Radial fan rotor with forward curved blades (drawn is $w_1 = w_2$)

energy by reducing it to the kinetic energy level at the entry $\frac{1}{2}v_1^2$. This may be achieved by a diffuser immediately downstream of the rotor. In the example of Fig. 1.13, this is an annular bladed diffuser, in which the velocity v_2 is forced towards the radial direction (tangential component decreases). Also the radial velocity component decreases due to the increase in radius and axial width. Downstream of the diffuser follows a collector with the shape of a spiral. Such a collector is mostly called a *volute*. At this stage, principal understanding of the role of the stator components downstream of the rotor is sufficient. These components will be discussed with more detail in the chapter on pumps (Chap. 8).

There is a means, despite the limitation on the rotor deceleration ($w_2/w_1 > 0.8$ to 0.9), to realise a positive work contribution by the lift, namely by curving the blades strongly forward at the rotor exit, as sketched in Fig. 1.15. The term $u_2 w_{2u} - u_1 w_{1u}$ then becomes highly positive.

With *backward curved* blades, work is $\Delta W = u_2 v_{2u} \approx \frac{1}{2}u_2^2$ (Fig. 1.14), whereas it is $\Delta W = u_2 v_{2u} \approx \frac{3}{2}u_2^2 (u_2 \approx 2u_1)$ to $2u_2^2$ ($u_2 \approx u_1$) with *forward curved* blades. Forward curved blades generate a high kinetic energy at the rotor exit. The consequence is that reduction of the velocity v_2 to the level of the velocity v_1 is not possible without significant energy dissipation. The machine, in principle, only makes sense if intended to generate velocity, in other words as a fan. This fan type will be discussed further in the chapter on fans (Chap. 3). A fan may also be designed in the way of a centrifugal pump, i.e. with backward curved blades. In that case, the machine mainly realises pressure increase and less kinetic energy increase. Such a fan is required if the application is feeding an extended ductwork, in which significant losses occur. There also exist rotor shapes in between those shown in Figs. 1.14 and 1.15 (see Chap. 3).

1.8 Performance Characteristics

By a *performance characteristic* is meant a functional relationship between two operation quantities, such as flow rate and manometric head, with constant other quantities. In principle, many such relationships can be defined. In practice however, only a few are relevant.

With a hydraulic turbine, normally, the head is constant. The same applies to the rotational speed, as the turbine typically drives a generator with a fixed rotational speed. With a constant geometry, this machine does not have any variable operation quantities. At a fixed head and rotational speed, the flow rate and power can only be varied by changing the rotor blade pitch angle (the angle of the blade chord with respect to the tangential direction) of an axial turbine (see Fig. 1.2). A characteristic may thus be: the power as a function of the rotor blade pitch angle at a constant rotational speed and a constant head.

With a pump, the rotational speed is set by the driving motor. The load is typically a pipework filling a reservoir under pressure (see Fig. 1.16). The pressure within the reservoir is variable, for instance due to the presence of an air-filled bag. With an installation for household use, it is normal to vary the reservoir pressure between 1.5 and 3 bar. A pressure sensor starts the pump when the pressure falls under the set minimum (1.5 bar) and stops the pump when the pressure exceeds the maximum (3 bar). The pump thus functions with a variable manometric head. The change of the manometric head affects the flow rate. With the characteristic of a pump is thus normally meant the relationship between the manometric head and the flow rate at a constant rotational speed.

The form of this relationship may easily be derived from the rotor work. As an example, we take the centrifugal pump in Fig. 1.14.

The velocity triangles drawn with full lines in Fig. 1.17 correspond to the design flow rate (velocity components with *). This means that w_1 is tangent to the entry side of the rotor blades and that v_2 is tangent to the entry side of the diffuser vanes. In that case, no inlet incidence is said to occur at the rotor and stator components. When the pump flow rate is decreased, e.g. by applying a constriction to the discharge pipe, the velocity triangles change as drawn with dashed lines in Fig. 1.17.

Fig. 1.16 Pump with reservoir under variable pressure

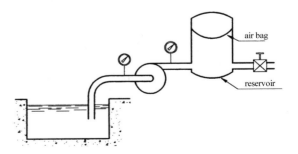

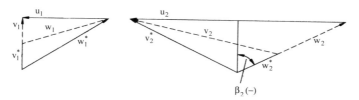

Fig. 1.17 Change of velocity triangles by flow rate decrease in a centrifugal pump; full line: design (*); dashed line: decreased flow rate

With decreased flow rate, the rotor work is still $u_2 v_{2u}$, with $v_{2u} = u_2 + w_{2u}$ and $\tan \beta_2 = w_{2u}/v_{2r}$ ($w_{2u} < 0, \beta_2 < 0$). The flow rate is proportional to v_{2r}. The relationship between work and flow rate thus has the form

$$\Delta W = u_2^2 + k(\tan \beta_2) Q, \qquad (1.39)$$

with k being a positive constant. The relation is a descending straight line. The mechanical energy rise in the fluid is smaller than the rotor work, due to friction losses (approximately proportional to Q^2) and incidence losses. At a flow rate different from the design flow rate Q^*, the relative entry velocity of the rotor (w_1) and the absolute entry velocity of the diffuser (v_2) are no longer tangent to the blades or vanes. This generates a loss proportional to $(Q - Q^*)^2$ (see Chap. 3). The manometric head may be derived from the rotor work, in principle, as sketched in Fig. 1.18. The characteristic is a curve with an approximate parabolic shape.

The correct determination of the characteristic is somewhat more complex than sketched in Fig. 1.18, because the flow does not exactly follow the blades at the rotor exit. A phenomenon called slip occurs. The relationship between work and flow rate therefore deviates somewhat from the purely geometrically determined relation (1.39). Slip will be discussed in Chap. 3. The above approximate reasoning should provisionally suffice to understand that there is a functional relationship between head and flow rate.

Fig. 1.18 Characteristic of a centrifugal pump (backward curved rotor blades); $\Omega =$ constant

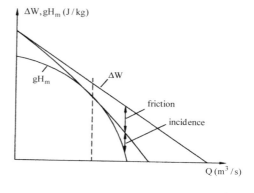

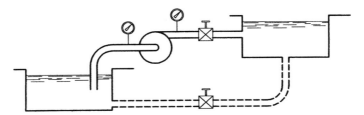

Fig. 1.19 Determination of a pump characteristic (used in Chap. 5)

It is common practice to consider flow rate as an independent quantity and head as a dependent one. This option is due to the common method for experimental determination of the characteristic (Fig. 1.19). The pump displaces fluid from a suction reservoir to a discharge reservoir. A valve is mounted on the discharge pipe. Manometers are placed at the suction and discharge sides of the pump. Operation of the valve at first suggests flow rate control. The manometer readings change as well, of course. The manometric head is determined from these readings. A flow rate meter may be mounted additionally in the discharge pipe for flow rate measurement. A more classical method is flow rate determination by the level increase in the discharge reservoir, with a closed return pipe.

It is typical, with a work receiving machine, that the relationship between the mechanical energy rise and the flow rate at a constant rotational speed is a descending curve. Figure 1.12 (axial pump) demonstrates that v_{2u} increases with flow rate decrease. The characteristic of the axial pump is thus similar to that of the centrifugal pump with backward curved rotor blades.

The dependence between head and flow rate of a hydraulic turbine may be derived in a similar way. Figure 1.20 shows the velocity triangles for the design flow rate

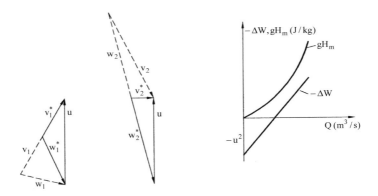

Fig. 1.20 Change of the velocity triangles with flow rate increase in an axial hydraulic turbine; full line: design (*); dashed line: increased flow rate; turbine characteristic: flow rate depending on head at constant rotational speed and constant geometry

and with flow rate increase (by increasing the head) for an axial hydraulic turbine as sketched in Fig. 1.9.

In Fig. 1.20 we notice (* indicates the design flow rate):

$$v_{1u} = v_{1u}^* \left(\frac{Q}{Q^*} \right); \quad v_{1u} > 0,$$

$$w_{2u} = w_{2u}^* \left(\frac{Q}{Q^*} \right); \quad w_{2u} < 0; \quad w_{2u}^* = -u; \quad v_{2u} = u + w_{2u}.$$

$$-\Delta W = (v_{1u} - v_{2u})u = \left[v_{1u}^* \left(\frac{Q}{Q^*} \right) - u + u \left(\frac{Q}{Q^*} \right) \right] u$$

$$= (v_{1u}^* + u)u \left(\frac{Q}{Q^*} \right) - u^2.$$

Figure 1.20 sketches the work and the head. The work increases linearly with the flow rate. Due to friction losses and incidence losses, the energy extracted from the flow exceeds the rotor work. The observation is that the head increases with the flow rate. A more natural reading of the graph is increasing flow rate with increasing head. So, normally, the characteristic is drawn with the flow rate (in the ordinate) depending on the head (in the abscissa). The figure is drawn with zero flow rate for zero head, but this is not necessarily exact in practice.

1.9 Exercises

1.9.1 (Force by action) The left-hand figure shows a liquid jet impacting perpendicularly a flat plate in an open space. Assume 2D flow (no variation perpendicular to the drawing). Denote by A the section area of the nozzle. Use a momentum balance on a control volume to determine the force on the plate in the direction of the oncoming flow. Assume that the flow, after turning, perfectly follows the plate. Assume steady flow and ignore friction. Consider gravity perpendicular to the jet, as indicated. Write a momentum balance in the vertical direction. Use the Bernoulli equation for determination of the exit velocities at the top and the bottom of the control volume. Assume that $2gh$ is small with respect to v^2 so that linearization is justified. Assume that the mass flow rates to the top and the bottom are equal (in reality, there is more mass flow rate to the bottom, increasing with increasing distance between the nozzle exit and the plate). Because the tangential component of the force on the plate is exactly zero in absence of friction, the vertical momentum balance results in an estimate of the weight of the liquid (G) inside the control volume. Express the weight G as a function of A and h.

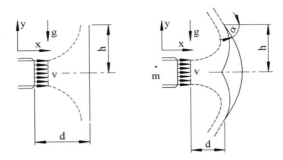

The right-hand figure shows a liquid jet impacting the blade (bucket) of a Pelton turbine (tangential hydraulic turbine, see Chap. 9). The bucket has the shape of a double spoon. The jet leaves the bucket with the indicated angle α. Determine first the force component in the direction of the jet when the bucket stands still, assuming that gravity is perpendicular to the figure, i.e. not as drawn in the figure. Determine then the force components in vertical and horizontal directions assuming that gravity is as indicated in the figure. Take similar assumptions as with the flat plate. Conclude that there is a downward force due to gravity for α sufficiently lower than 90°.

A: I: $F_x = \dot{m}v$, $G \approx \rho g Ah$,

II: $F_x = \dot{m}v(1 + \cos\alpha)$; $F_x \approx \dot{m}v(1 + \cos\alpha)$, $F_y \approx \rho g Ah(\sin\alpha) - G$.

1.9.2 (Force by reaction) The figure shows a trolley with a water tank. The trolley moves at 3 m/s. The driving force is generated by a pump taking water ($1000\ \mathrm{kg/m^3}$) from the tank and ejecting it with a 10 m/s speed relative to the trolley, the tank and the pump, the flow rate being $2\ \mathrm{m^3/s}$.

Determine the drag force on the trolley by the wheel and air friction. Reason first with a control volume attached to the trolley, i.e. in a relative frame. This is the easiest to do. Verify that the result is the same for a control volume in the absolute frame.

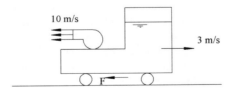

A: $F = 20$ kN.

1.9.3 This exercise is a further illustration of force by reaction and is a demonstration that the kinetic energy in the flow behind a propulsion jet constitutes a loss (residual kinetic energy). The figure is a sketch of a hydrofoil boat with speed $v = 20$ m/s. The mass of the boat is 100 tons (g = 9.81 m/s²). The foils that keep the vessel above the water surface have a lift/drag ratio $L/D = 20$. Water ($1000\ \mathrm{kg/m^3}$) is drawn in by a pump and ejected through a nozzle at 45 m/s relative to the boat.

Determine the mass flow rate to be handled by the pump, ignoring the air drag on the vessel. What power should (theoretically) be supplied to the pump (i.e. ignoring losses in the pump)? What is the useful propulsive power supplied to the vessel by the jet (force times speed of the vessel)? Observe that this power differs from the pump power. Explain the difference by the kinetic energy dissipated behind the vessel. Reason first with a control volume attached to the boat, i.e. in a relative frame. This is the easiest to do. Verify that the result is the same for a control volume in the absolute frame.

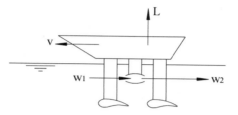

A: $\dot{m} = 1962$ kg/s, $P_{pump} = 1594$ kW, $P_{propulsion} = 981$ kW, $P_{residual} = 613$ kW.

1.9.4 This exercise is similar to the previous one, but for an open flow. The figure sketches the flow of air (1.225 kg/m^3) through an aircraft propeller. Assume that the flight speed is 75 m/s and that the propeller accelerates the air to 120 m/s relative to the aircraft.

Determine the thrust generated per m^2 frontal propeller area, assuming a uniform flow through the propeller with velocity equal to the arithmetic average of the upstream and downstream velocities (for justification, see a similar problem for a wind turbine in Chap. 10).

Determine, as in the previous exercise, the power to be supplied theoretically to the propeller, the useful power and the power dissipated behind the aircraft (all per m^2). Ignore post-whirl actually generated by the work done by the propeller on the air. Observe that there is no substantial difference with the findings concerning the jet-propelled hydrofoil ship in Exercise 1.9.3.

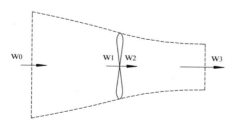

A: $P_{shaft} = 524.0$ kW, $P_{propulsion} = 403.1$ kW, $P_{residual} = 120.9$ kW.

1.9.5 The figure shows a lawn sprinkler with two arms. The length of an arm part in the radial direction (R) is 150 mm. The tip part is perpendicular to the radial part and is at an angle of 30° to a horizontal plane. The length of the tip part (L) is 30 mm. The internal diameter of the arm tubes is 4 mm. The water flow rate is 7.5 l/min. The rotational speed is 30 rpm.

Determine the friction torque exerted on the lawn sprinkler shaft. Reason first in the absolute frame and verify then the result in the relative frame. Notice that the moment of the Coriolis force on the flow through the arms has to be taken into account in the relative frame. Hint: reason with a projection on a horizontal plane. Determine the rotational speed if the friction torque were halved?

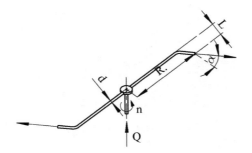

A: $M_{friction} = 71.66$ Nmm, $n_{new} = 148.1$ rpm.

1.9.6 The figure sketches an extraction fan with radial blades. The flow in the suction pipe gets pre-whirl by a guide-vane ring, so that the flow entering the rotor is tangent to the blades. Inlet and outlet width are such that the relative velocity within the rotor is constant. Determine the work done on the flow, the parts of kinetic energy increase and enthalpy increase, and thus the degree of reaction.

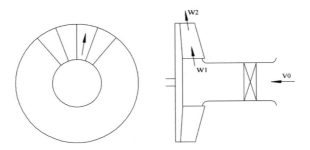

Ignore blade thickness and friction losses. Assume alignment of flow and blades. The fan is intended to remove air through the suction pipe. The velocity in the pipe generates a pressure drop. Sketch the pressure evolution through the suction pipe and the fan. Derive the required pressure rise by the fan from a pressure balance. Observe that there are two essential loss mechanisms with the considered application: one at the rotor entry and another at the rotor exit. Determine how the rotor shape can be made more efficient. Consider a rotor with constant meridional velocity and constant relative velocity (no deceleration).

A: The figure sketches the velocity triangles at the rotor entry and exit.

$$\Delta W = u_2 v_{2u} - u_1 v_{1u} = u_2^2 - u_1^2,$$

$$\Delta E_k = \frac{v_2^2}{2} - \frac{v_1^2}{2} = \frac{u_2^2}{2} - \frac{u_1^2}{2}, \quad \Delta E_p = \frac{u_2^2}{2} - \frac{u_1^2}{2} + \frac{w_1^2}{2} - \frac{w_2^2}{2} = \frac{u_2^2}{2} - \frac{u_1^2}{2}.$$

So:

$$\Delta E_p = \Delta E_k \text{ and } R = 0.5.$$

Pressure balance: A pressure drop $\rho v_0^2/2$ occurs due to the generation of the velocity in the suction pipe v_0. The acceleration in the guide vanes generates an additional pressure drop $\rho v_1^2/2 - \rho v_0^2/2$. The pressure rise by the fan is $\rho u_2^2/2 - \rho u_1^2/2$. The fan has to raise the pressure to the atmospheric pressure.
The pressure balance is:

$$\Delta E_p = \frac{u_2^2}{2} - \frac{u_1^2}{2} = \frac{v_0^2}{2} + \frac{v_1^2 - v_0^2}{2}.$$

So:

$$\Delta W = \Delta E_p + \Delta E_k = \frac{v_0^2}{2} + \frac{v_1^2 - v_0^2}{2} + \frac{v_2^2 - v_1^2}{2} \left(= \frac{v_2^2}{2} \right). \tag{1.40}$$

The work consists of three terms in kinetic energy with sum $v_2^2/2$. The first term is the useful term. It is the work required to compensate the pressure drop due to the velocity generation within the suction pipe (unavoidable). The second term is the kinetic energy rise within the stationary vane ring. This increase causes a second pressure drop to be compensated by the fan. This second term is not useful. The third term is the kinetic energy increase within the rotor. This term is also useless, as the fan ideally only should generate a pressure rise. All kinetic energy generated within the rotor causes a larger kinetic energy dissipated in the atmosphere downstream of the rotor. The useless terms in (1.40) may be eliminated or reduced by adapting the rotor entry and exit, as sketched below. The blades are inclined backward at the rotor entry, making the stator vane ring unnecessary. The blades are inclined backward at the rotor exit in order to decrease the exit velocity in the absolute frame. Constant meridional velocity and constant relative velocity are assumed (no deceleration).

$$\Delta W = u_2 v_{2u} - u_1 v_{1u} = u_2 v_{2u} = u_2(u_2 - u_1),$$

$$\Delta E_k = \frac{v_2^2}{2} - \frac{v_1^2}{2} = \frac{(u_2 - u_1)^2}{2}, \quad \Delta E_p = \frac{u_2^2 - u_1^2}{2} + \frac{w_1^2 - w_2^2}{2} = \frac{u_2^2 - u_1^2}{2}.$$

The pressure balance is now

$$\Delta E_p = \frac{u_2^2 - u_1^2}{2} = \frac{v_0^2}{2}.$$

So:

$$\Delta W = \Delta E_p + \Delta E_k = \frac{v_0^2}{2} + \frac{(u_2 - u_1)^2}{2} \left(= \frac{v_2^2}{2}\right). \tag{1.41}$$

The first term is the useful term once more in (1.41). The second term is the kinetic energy increase within the rotor. This term is not useful, but cannot be reduced to zero, due to the imposed conditions on constant meridional velocity and relative velocity. Expression (1.41) is obviously much more favourable than expression (1.40). The work is, again, equal to the kinetic energy at the rotor exit, but this is now much lower.

Work may further be reduced by pressure increase by deceleration in the rotor ($w_2 < w_1$) and by pressure increase by reduction of the kinetic energy in a diffuser downstream of the rotor. Limited deceleration in the rotor is easy and usually applied. A diffuser is a supplementary component and typically not added.

1.9.7 Consider the same problem as in the foregoing exercise, but with an axial fan, as sketched in the figure. Determine the power required for the application. Compare with the result from the foregoing exercise.

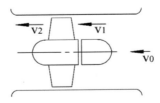

A: The figure sketches the velocity triangles at the rotor entry and exit. Notice that, with constant meridional velocity, there is deceleration in the rotor.

$$\Delta W = u v_{2u},$$

$$\Delta E_k = \frac{v_2^2}{2} - \frac{v_1^2}{2} = \frac{v_{2u}^2}{2}, \quad \Delta E_p = \frac{w_1^2 - w_2^2}{2} = \frac{u^2}{2} - \frac{(u - v_{2u})^2}{2} = u v_{2u} - \frac{v_{2u}^2}{2}.$$

The pressure balance is

$$u v_{2u} - \frac{v_{2u}^2}{2} = \frac{v_0^2}{2} + \frac{v_1^2 - v_0^2}{2}.$$

So:

$$\Delta W = u v_{2u} = \frac{v_0^2}{2} + \frac{v_1^2 - v_0^2}{2} + \frac{v_{2u}^2}{2} \left(= \frac{v_2^2}{2} \right). \tag{1.42}$$

The first term in (1.42) is useful. The second term expresses the flow acceleration by the hub. The third term is the kinetic energy increase within the rotor, useless but unavoidable. The expressions with the axial fan are not fundamentally different from those with the radial fan in the foregoing exercise.

1.9.8 The figure is a sketch of a meridional section of the rotor of a centrifugal pump and an orthogonal section of a blade at the mean entry of the bladed part of the rotor. At this mean entry, the diameter is $d_1 = 92$ mm (internal diameter) with blade angle $\beta_1 = -68°$ (with respect to a meridional plane) on the mean streamsurface. The external diameter of the rotor is $d_2 = 196$ mm and the blade angle is $\beta_2 = -60°$ at rotor exit. The rotor inflow is in the meridional direction in the absolute frame (no pre-whirl). The rotor channel width at the entry of the bladed part is $b_1 = 22$ mm (perpendicular to the mean streamsurface) and is $b_2 = 14$ mm at rotor exit.

The entry blade angle of the orthogonal blade section is somewhat larger than on the mean streamsurface due to projection of the meridional velocity component on an orthogonal plane: $\tan \beta_1' = \tan \beta_1 / \cos \delta_1$, with δ_1 the angle with respect to an orthogonal plane of the mean flow at the entry: $\beta_1' \approx -72°$. The camber line of the blade is a circle segment.

The pump runs at 2900 rpm and pumps water ($\rho = 1000$ kg/m^3).

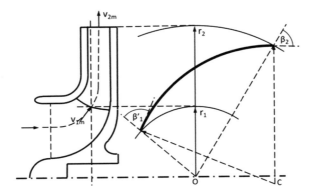

Determine the velocity triangles at rotor entry and exit. Ignore blade thickness and friction losses. Assume alignment of flow and blades. Determine the rotor work ΔW. Split the rotor work into the part by the Coriolis force and the part by the lift force. Determine the pressure energy rise and the kinetic energy rise in the flow through the rotor. Calculate the degree of reaction R. Determine the volume flow rate Q through the rotor and the rotor power P_{rot}.

In reality, the rotor flow rate is smaller (about 85% of the estimated value) due to obstruction by blade thickness and the rotor work is smaller (about 80% of the estimated value) due to slip of the flow at the rotor exit (the relative flow is more tangential than the blade orientation): see Exercise 8.9.2 in Chap. 8.

A: $v_1 = 5.64$ m/s, $w_1 = 15.07$ m/s, $v_{2m} = 4.16$ m/s, $v_{2u} = 22.55$ m/s, $w_2 = 8.33$ m/s, $\Delta W = 671.13$ J/kg, Coriolis part = 102.9%, pressure energy rise = 424.13 J/kg, kinetic energy rise = 247.00 J/kg, $R = 0.632$, $Q = 35.89$ l/s, $P_{rot} = 24.08$ kW.

1.9.9 Consider an axial hydraulic turbine according to Fig. 1.9. The outer diameter of the turbine and the supply duct diameter are 1 m. The hub diameter is 0.5 m. $n = 300$ rpm, $\rho = 1000$ kg/m^3. On the arithmetically average radius ($r = 0.375$ m), the stator deviates the flow by $+30°$ with respect to the axial direction, while the rotor outflow is at $-60°$ in the relative frame. The outflow of the rotor is axial in the absolute frame on the entire rotor radius. Stator vanes and rotor blades are designed for constant work and constant axial velocity along the radius. Assume that the head losses (dq_{irr}) in the turbine components may be expressed by a relevant kinetic energy multiplied with a loss coefficient. Assume for the loss coefficients: stator: $\xi = 0.05$ related to the exit kinetic energy $\frac{1}{2}v_1^2$; rotor: $\xi = 0.05$ related to the relative exit kinetic energy $\frac{1}{2}w_2^2$; diffuser $\xi = 0.30$ related to the entry kinetic energy $\frac{1}{2}v_2^2$ (all velocities on the mean radius).

Sketch the velocity triangles on the average radius (upstream and downstream of the rotor). Determine the axial velocity, the flow rate and the angle of the relative flow at rotor entry with respect to the axial direction on the average radius.

Determine the rotor power. Determine the pressure on the manometer (gauge pressure: relative to atmospheric pressure) on the supply duct, with centre of the manometer 3 m above the downward water level (tailwater). Determine the manometric head and the internal efficiency. Determine the stator outflow angle and the rotor outflow angle at the rotor periphery, measured to the axial direction.

A: $Q = 4.01$ m^3/s; $\beta_1 = -49.1°$; $P_{rot} = 185.4$ kW; $p - p_a = 16.93$ kPa; $gh_m = 59.37$ J/kg; $\eta_i = 0.779$; $\alpha_1 = 23.4°$; $\beta_2 = -66.6°$.

1.9.10 The figure shows a Poncelet-type waterwheel. This is an undershot waterwheel with curved blades. The water accelerates under the slide due to the height difference ($v_1^2/2 = gh$, ignoring losses). The direction of the water jet forms the angle α with the tangential direction at the periphery of the wheel.

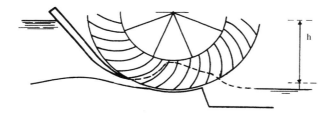

The wheel speed at the periphery (u) is allowed to vary. Magnitude and direction of the relative velocity at the wheel entry change by this. Assume that the entry angle β is adapted so that the jet enters the wheel aligned with the blade direction (with a given machine, this condition is only correct for one operating point; thus we consider a design optimisation with a blade angle not given a priori).

Assume that the water leaves the wheel at the same height as at the entry, so that $w_2 = w_1$. Derive a formula for the rotor work for fixed values of v_1 and α and a variable value of u. Express the internal efficiency of the wheel, ignoring all friction losses. Determine the peripheral speed for maximum efficiency. Determine the velocity triangle at the wheel exit together with a formula for the exit velocity v_2. Interpret the efficiency attained, in other words determine which losses occur. Conclude from the result that a small α is beneficial (in practice $\alpha = 15°$ is applied). Is there still another apparent loss mechanism?

A: The figure sketches the velocity triangles at the wheel entry and exit.

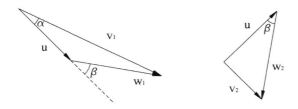

$$w_{1u} = v_{1u} - u = v_1 \cos \alpha - u;$$
$$w_{2u} = -w_{1u}; \Delta W = u(w_{1u} - w_{2u}) = 2u(v_1 \cos \alpha - u).$$

Maximum work is attained by

$$u = \frac{v_1 \cos \alpha}{2}.$$

Maximum work is

$$(\Delta W)_{max} = \frac{v_1^2}{2} \cos^2 \alpha \quad \text{with} \quad \frac{v_1^2}{2} = gh.$$

So:

$$\eta_i = \frac{\Delta W}{gh} \quad \text{and} \quad (\eta_i)_{max} = \cos^2 \alpha.$$

In the inlet triangle, then

$$w_{1u} = \frac{v_1 \cos \alpha}{2} = u.$$

In the outlet triangle, then

$$w_{2u} = -u.$$

The outlet triangle is orthogonal:

$$v_{2r} = v_{1r} = v_1 \sin \alpha.$$

The exit kinetic energy is

$$\frac{v_2^2}{2} = \frac{v_1^2}{2} \sin^2 \alpha = \frac{v_1^2}{2} \left(1 - \cos^2 \alpha\right) = \frac{v_1^2}{2} - \Delta W.$$

The only loss within the flow is the kinetic energy at the rotor exit. Therefore optimum operation corresponds to perpendicular outflow. Another loss is the downward head that is not used. The head used is somewhat lower than the geometric head.

As discussed in Sect. 1.2.1, hydraulic turbines have a draught tube which transforms the downstream head and part of the exit kinetic energy into pressure decrease at the rotor exit. Application of a draught tube is not possible with a Poncelet wheel, because the wheel runs in an open atmosphere. The relative flow in the wheel is particular because it comes to a halt and reverses direction. Apart from this aspect, the functioning is similar to that of a Pelton turbine (see Sect. 9.3.1 in Chap. 9).

Chapter 2
Basic Components

Abstract The blades of the axial hydraulic turbines and pumps, studied up to now, have profiles resembling aircraft wing profiles (*aerofoils*). In machines with large spacing between blades, this resemblance is strong. With axial compressors, gas and steam turbines, blades are closer to each other and cause larger flow turning. A circumferential section is then a row of blade profiles with tangential spacing comparable to, or smaller than, the largest profile dimension. We then use the term blade row or *blade cascade*. Radial machine rotor blades principally do not function as lifting objects. The blades constitute *channels*. The blade profiles have no resemblance to aerofoils. Channel flows may be accelerating (turbines) or decelerating (pumps, fans, compressors). With decelerating or diffusion flows, avoidance of separation between flow and geometry is difficult. As we have already learned, *diffusers* also can be stator components of turbines. Aerofoils, cascades, channels and diffusers constitute the basic components of turbomachines, which we study in this chapter.

2.1 Aerofoils

2.1.1 Force Generation

Figure 2.1 sketches the streamlines close to an aerofoil in an oncoming flow with uniform velocity v_∞ and fluid density ρ_∞ far upstream.

The streamlines are intended to curve due to camber of the profile (curvature in the longitudinal direction) and due to incidence of the flow (angle difference between the oncoming flow and the profile). To understand the effect of the curvature, the simplest is considering an observer moving along with a fluid particle. This observer then stands still relative to the flow. In a moving coordinate frame with origin on the observer, with x-axis along the flow and y-axis perpendicular to it, a centrifugal force must then be introduced with a value v^2/R, perpendicular to the streamline, so along the y-axis, away from the centre of curvature (R is the radius of curvature). In the flow-normal direction (y-axis) the momentum is zero. Conservation of momentum thus means a balance of forces. This means that the centrifugal force must be kept in balance by a pressure difference. The arrows perpendicular to the streamlines

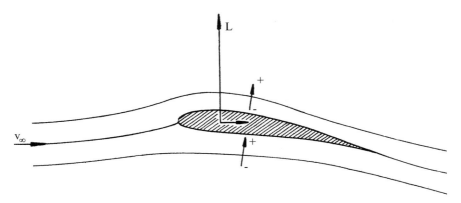

Fig. 2.1 Streamlines with flow over an aerofoil

in Fig. 2.1 represent the centrifugal force. The corresponding pressure difference is indicated with $+$ and $-$ symbols. The consequence is that the pressure is lower than the distant pressure at the top side of the aerofoil, while at the bottom side it is higher than the distant pressure. It is said that a *suction side* and a *pressure side* form. The pressure difference generates a force in the upward direction, approximately perpendicular to the oncoming flow.

In fluid mechanics, it is proved that, with a lossless flow, the force on an aerofoil stands exactly perpendicular to the oncoming flow (Kutta-Joukowski theorem). In a flow with losses there is also a component in the flow direction. The perpendicular component is termed *lift L*. The component in the flow direction is termed *drag D*. These forces are usually expressed per span unit of the aerofoil (N/m). From fundamental fluid mechanics we recall that forces exerted by a flow on objects (N), according to similitude theory, are proportional to $\rho_\infty v_\infty^2/2$ (= dynamic pressure, the difference between stagnation pressure and static pressure for constant density; see Sect. 2.1.3) and to a characteristic object area. We thus define lift coefficient and drag coefficient by

$$C_L = \frac{L}{\frac{1}{2}\rho_\infty v_\infty^2 c}, \quad C_D = \frac{D}{\frac{1}{2}\rho_\infty v_\infty^2 c}.$$

The symbol c denotes the chord length of the aerofoil, which is a measure of the largest aerofoil dimension. It approximately equals the largest distance between two points on the aerofoil surface (defined below).

An aerofoil profile (Fig. 2.2) is characterised by a camber line and a thickness distribution. The camber line is a longitudinal line, such that points on the aerofoil surface are at half the thickness set perpendicularly to both sides of it. The camber line is approximately obtained by connecting the midpoints of the inscribed circles within the aerofoil. The leading side (A) is rounded with most aerofoil shapes, causing the inscribed circle to have a finite radius there. In theoretical aerofoil representations the trailing side (B) mostly has no thickness (a practical aerofoil is obtained by truncating). The *leading edge* is the final point on the camber line at the leading

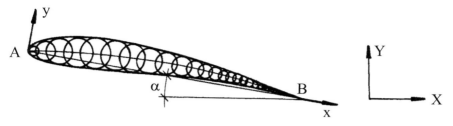

Fig. 2.2 Camber line, chord, thickness and angle of attack of an aerofoil

side and the *trailing edge* is the final point at the trailing side. The *chord* is the line segment between the leading and the trailing edges (AB).

The camber line is described by the distance to the chord (y) as a function of a coordinate (x) along the chord from the leading edge to the trailing edge. The local thickness is commonly specified as a function of the same coordinate. The angle between the chord and the oncoming flow is termed *chord angle* or *angle of attack*. Generally, chord angle means the angle between the chord and a geometric reference line, while angle of attack means the angle between the chord and the oncoming flow direction. The concepts coincide with a free-standing aerofoil, but become different for an aerofoil in a turbomachine.

Most aerofoils have camber so that there is lift with a zero angle of attack. The angle of attack must then be negative in order to obtain zero lift. The aerofoil position for zero-lift defines a direction termed *zero-lift line*. From fluid mechanics it follows that this line is found with a good approximation by connecting the trailing edge to the point of maximum camber on the camber line.

2.1.2 Performance Parameters

Figure 2.3 sketches, as an illustration, the lift coefficient as a function of the angle of attack for a NACA 4412 aerofoil [1]. This aerofoil was formerly (before the contemporary common use of computational fluid dynamics techniques) typically applied in axial pumps and axial hydraulic turbines. The aerofoil shape is sketched in the figure as well. The NACA denomination refers to the classification by the National Advisory Committee for Aeronautics, the forerunner of the NASA, National Aeronautics and Space Administration.

Also shown in Fig. 2.3 is the drag coefficient as a function of the lift coefficient. The tangent from the origin to the curve determines the angle of attack with maximum L/D ratio. Use of the aerofoil with this angle of attack realises maximum efficiency in most applications. The corresponding lift coefficient is about 1.00, the drag coefficient about 0.01. The lift increases with the angle of attack until an angle where the boundary layer on the suction side in the vicinity of the trailing edge separates from the surface. Beyond this angle, the lift coefficient decreases strongly and the drag

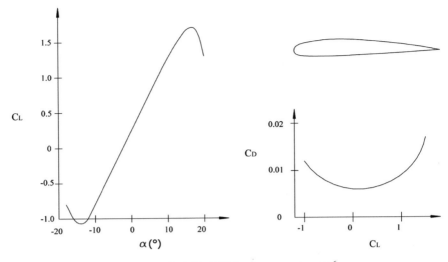

Fig. 2.3 Lift and drag coefficient of NACA 4412 for $Re = v_\infty \, c/\nu = 10^6$

coefficient increases strongly. This phenomenon is called *stall*, which literally means loss of proper functioning. An analogous stall phenomenon is observed with negative angles of attack, but aerofoils are commonly not designed to function well with highly negative angles of attack.

Boundary layer separation will be analysed in a following section, but one can acquire an intuitive understanding of it for a very large chord angle, in particular when the profile is positioned perpendicularly to the flow. A large recirculation zone is then created downstream of the aerofoil. At the upstream side of the aerofoil, the pressure is higher than in the oncoming flow due to flow stagnation. Within the recirculation zone the pressure approximates the static pressure of the oncoming flow. The contribution to the drag by the pressure difference is termed *pressure drag*. With a weaker separation, as sketched in Fig. 2.4, pressure drag is generated as well. This adds to the *friction drag*. The pressure drag is the resultant in the flow direction of the pressure on the surface. The friction drag is the resultant of the shear stress on the surface. Pressure drag is rather low with attached flow, but separation causes a strong increase. Moreover, the curvature of the streamlines at the suction side is then

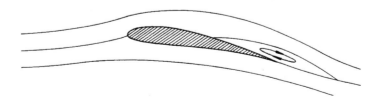

Fig. 2.4 Separated flow over an aerofoil

reduced. The consequence is that the pressure minimum at the suction side is weaker than with attached flow. This explains the decrease of the lift after separation.

2.1.3 Pressure Distribution

The pressure distribution for optimal operation of a NACA 4412 is sketched in Fig. 2.5. At the leading edge, the velocity is near to zero and the pressure is near to the *stagnation pressure* of the oncoming flow. The stagnation pressure (*total pressure* if the effect of gravity is taken into account; see Sect. 2.1.5) is the pressure obtained by bringing the flow to zero velocity in an adiabatic reversible way. This means that kinetic energy is converted into pressure energy. For constant density, the stagnation pressure is $p_0 = p_\infty + \frac{1}{2}\rho_\infty v_\infty^2$. The pressure is somewhat lower in the leading edge zone. The trailing edge pressure is typically slightly higher than the pressure of the oncoming flow.

The pressure coefficient in Fig. 2.5 is $C_p = (p - p_\infty)/(\frac{1}{2}\rho_\infty v_\infty^2)$. It is common practice to plot this coefficient with negative values to the upside in order to have the suction side at the upside of the figure. On the pressure side of the aerofoil, the pressure is uniformly decreasing. The boundary layer flow is accelerating everywhere. On the suction side, the flow accelerates at the leading edge, causing a pressure minimum. Already with a low aerofoil lift (small angle of attack), the minimum pressure is lower than the pressure of the oncoming flow (see Fig. 2.1). From the minimum pressure point to the trailing edge, the boundary layer on the suction side is subjected to an adverse pressure gradient.

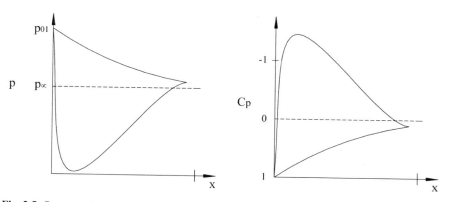

Fig. 2.5 Pressure distribution over an aerofoil with attached flow (NACA 4412)

2.1.4 Boundary Layer Separation

We can obtain more detailed understanding of the flow over an aerofoil by deriving
the momentum equations for an infinitesimal streamtube close to a wall. Figure 2.6
sketches such a streamtube at the suction side. We consider a flow that follows the
surface, in other words, that does not separate.

The axis in flow direction is denoted by x. This direction approximately follows
the surface. The direction perpendicular to the flow is denoted by y. Shear stress
on a face of the control volume is denoted by τ. The flow between the streamtube
considered and the wall has a braking effect on the streamtube. The flow between the
streamtube and the free flow has a driving effect. The momentum balance in the flow
direction for an infinitesimal part of the streamtube, on condition of small curvature,
is

$$\rho v (dy) dv = -(dy) dp - \rho (dy) dU + (dx) d\tau.$$

Thus:

$$\rho v \frac{dv}{dx} = -\frac{dp}{dx} - \rho \frac{dU}{dx} + \frac{d\tau}{dy}. \tag{2.1}$$

From now on, we assume constant density, but extension for variable density is
possible. With constant density, the effect of gravity on the pressure may be expressed
by adding ρU to the pressure. This is relevant for liquid flow. The effect of gravity
may simply be ignored for gas flow. We thus simplify (2.1) into

$$\rho v \frac{dv}{dx} = -\frac{dp}{dx} + \frac{d\tau}{dy}. \tag{2.2}$$

We first consider a flow with a zero pressure gradient. Far from the wall, the shear
stress is zero. There is a certain value on the wall. So: $d\tau/dy < 0$. The momentum
equation demonstrates that friction reduces the velocity within the streamtube in
the flow sense. The height of the streamtube thus increases in the flow sense. The
same applies to the entire zone close to the wall, where the flow is retarded by the
wall friction. The zone affected by the wall friction is called the *boundary layer*. Its

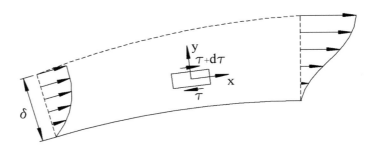

Fig. 2.6 Infinitesimal streamtube part close to a wall and velocity profile evolution with an adverse
pressure gradient

thickness is denoted by δ in Fig. 2.6. Note that this thickness cannot be determined precisely. The shear stress is partly of molecular nature due to the relation

$$\tau = \mu \frac{dv}{dy}, \tag{2.3}$$

where μ is the dynamic viscosity coefficient.

We recall from fluid mechanics that friction is generated between two adjacent fluid layers with different velocity by the chaotic motion of the microscopic fluid particles (molecules or atoms) around the average macroscopic motion. Fluid particles leap from the fast layer to the slower one, where they arrive with a higher average momentum and so push forward the slower layer. Conversely, fluid particles leaping from the slower layer to the faster one produce a breaking effect. The resulting effect is described macroscopically by the relation (2.3), with the viscosity coefficient μ as a positive quantity. According to the momentum equation (2.2), friction affects the velocity change with μ/ρ. We thus define the kinematic viscosity coefficient by

$$v = \mu/\rho. \tag{2.4}$$

The dimension of kinematic viscosity is m²/s. The value for water at sea-level atmospheric temperature is about 10^{-6} m²/s. The value for air at sea-level atmospheric temperature and pressure is about 15×10^{-6} m²/s. So water and air are not highly viscous fluids. This results in a very thin boundary layer. It is demonstrated in fluid mechanics that, on a smooth wall without a pressure gradient, the boundary layer thickness is approximately

$$\delta/x \approx 5/\sqrt{\frac{x v_\infty}{v}},$$

where v_∞ represents the velocity far away from the wall (outside the boundary layer zone) and x is the distance covered by the boundary layer. A boundary layer thickness with laminar flow of air or water maximally amounts to some thousandths of the covered length. It is very important to keep in mind the small thickness of a boundary layer in boundary layer analyses. When drawing a boundary layer, the thickness must necessarily be exaggerated, which creates a false impression.

Turbulent flow contains eddies (whirling flow patterns) as a result of the fragmentation due to stretching and bending of the vortices generated by shear zones. Eddies have a macroscopic size. Their motion is chaotic around an average flow, similar to, but on a much larger scale, than the microscopic fluid particles. The turbulent motion thus also generates shear stress on the average flow. This stress is expressed by an eddy viscosity μ_t according to

$$\tau = (\mu + \mu_t)\frac{dv}{dy}. \tag{2.5}$$

The eddy viscosity coefficient μ_t is no fluid characteristic and depends on the local flow. It is a positive quantity that may be up to 100 times larger than the molecular viscosity coefficient. So, due to turbulence, the boundary layer thickness increases significantly. The thickness stays small however compared to the covered length, namely some hundredths. From the positive values of μ and μ_t and the negative value of $d\tau/dy$, we understand that the velocity within the streamtube decreases in flow sense, due to friction, but that inversion of the flow sense is impossible. Separation thus cannot be caused by the viscosity effect only.

In order to understand the role of the pressure gradient, we must also analyse the pressure variation in the normal direction. The easiest way is with a relative frame attached to a fluid particle, as used in Sect. 2.1.1, leading to a balance between the normal pressure gradient and the centrifugal force caused by the curvature of the streamline:

$$\frac{1}{\rho}\frac{dp}{dy} = \frac{v^2}{R}. \tag{2.6}$$

R is the radius of curvature. It is assumed in (2.6) that friction forces do not contribute. As is clear in Fig. 2.6, there is also shear stress on the inlet and outlet faces of the infinitesimal streamtube. A stress change in the flow direction thus contributes to the force in the normal direction. As boundary layers are very thin, with changes in the normal direction being much larger than in the flow direction, the contribution of friction may be ignored.

A consequence of Eq. (2.6) is that the pressure distribution over an aerofoil, as sketched in Fig. 2.5, is only slightly dependent on the fluid viscosity, except with very viscous fluids. Another consequence is an almost identical pressure variation along the various streamlines in a boundary layer, as the radius of curvature R is very large compared to the boundary layer thickness. The decrease of kinetic energy due to an adverse pressure gradient is thus, according to Eq. (2.2), about the same on all streamlines. Consequently, the velocity decrease is relatively stronger on a streamline nearer to the surface. This causes, as sketched in Fig. 2.6, the velocity profile to become more concave as the flow evolves. With a sufficiently large adverse pressure gradient, the velocity close to the profile surface may attain a very low value, causing separation downstream of this position.

With the reasoning with streamtubes following the surface, as in Fig. 2.6, it cannot be analysed how the change from attached flow to separated flow exactly occurs, because the velocity within a streamtube cannot change sign. The reasoning however explains the origin of separation, which is sufficient here. We conclude that separation may occur on the suction side, if the adverse pressure gradient is sufficiently large. We also conclude that separation cannot occur on the pressure side if the flow is everywhere accelerating.

We further notice that a turbulent boundary layer has better resistance to separation than a laminar one. A turbulent boundary layer is thicker than a laminar one, but turbulence generation is most intense close to the wall. With a turbulent boundary layer, the near-wall velocity gradient is much larger and there is thus more momentum near the wall than with a laminar boundary layer.

2.1.5 Loss Mechanism Associated to Friction: Energy Dissipation

The loss mechanism associated to friction in a boundary layer may be understood by analysing the work by friction exerted on the streamtube shown in Fig. 2.6. The bottom streamline is a stationary wall (velocity equal to zero).

The overlying flow exerts positive work $(\tau + d\tau)(v + dv)$ and drives the streamtube. The underlying flow brakes by the negative work $(-\tau v)$. The resulting work leads to the energy balance, in absence of heat transfer:

$$\rho(dy)v\, d\left(h + \frac{1}{2}v^2\right) = (\tau + d\tau)(v + dv)dx - \tau v dx,$$

or

$$\rho v \frac{dh_0}{dx} = \frac{d}{dy}(\tau v). \tag{2.7}$$

Potential energy, when relevant, is included in the enthalpy. $h_0 = h + \frac{1}{2}v^2$ is thus total enthalpy. The shear stress τ varies in the y-direction from the wall value to zero in the main flow. The velocity varies from zero to the main flow value. The τv quantity thus goes from zero to zero through a positive maximum. Thus, friction work causes redistribution of total enthalpy within the boundary layer, but, with an adiabatic wall, the total enthalpy flux stays constant within the entire boundary layer. This follows from the integration of Eq. (2.7) in the y-direction.

The work equation associated to the momentum equation (2.2), with the effect of gravity, if relevant, included in the pressure, is

$$\rho v \frac{d}{dx}\left(\frac{1}{2}v^2\right) = -v\frac{dp}{dx} + v\frac{d\tau}{dy} \quad \text{or} \quad \rho v \left(\frac{d}{dx}\left(\frac{1}{2}v^2\right) + \frac{1}{\rho}\frac{dp}{dx}\right) = v\frac{d\tau}{dy}. \tag{2.8}$$

We still assume constant density. With ρU added to the pressure, the mechanical energy is $E_m = \frac{1}{2}v^2 + p/\rho = p_0/\rho$, with p_0 the total pressure. The work equation (2.8) then means that the mechanical energy decreases due to the *displacement work* of the friction force, since $d\tau/dy < 0$.

Subtracting the work equation (2.8) from the energy equation (2.7) results in

$$\rho v \frac{de}{dx} = \tau \frac{dv}{dy}. \tag{2.9}$$

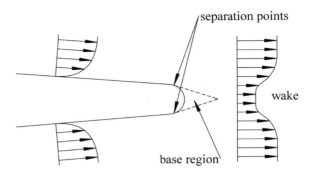

Fig. 2.7 Merging boundary layers at a trailing edge with generation of a wake

The part of the work $\tau \, (dv/dy)$ is the *deformation work*: total work minus displacement work. This work part is always positive in view of Eq. (2.5). Friction thus results in increase of internal energy (Eq. 2.9) and decrease of mechanical energy (Eq. 2.8) with constancy of the flux of the total enthalpy (Eq. 2.7), the sum of mechanical energy and internal energy. The conversion of mechanical energy into internal energy is called energy dissipation.

With the boundary layer of Fig. 2.6 there is no exchange of work and heat with the surroundings. Therefore, the flux of the total enthalpy is constant and the decrease of mechanical energy is exactly equal to the increase of internal energy. So, energy dissipation may be seen in both ways. This is not a general result, however. Strictly, energy dissipation is caused by the deformation work of the friction. This may be understood by adding work done by an external force on the boundary layer in Fig. 2.6. The same work term then adds to the balance of total enthalpy (2.7) and the balance of mechanical energy (2.8). In the difference of both balances, the work of the external force cancels.

This demonstrates that actually the balance of internal energy (2.9) determines the dissipation. This observation may cause confusion since in Chap. 1 the displacement work of the friction force resulting in mechanical energy decrease was denoted as dissipation. Strictly, this is incorrect, since the deformation work is the dissipative part of the friction work. The numerical value of the dissipation is correct, because the magnitude of the integral across the boundary layer of the displacement work equals the magnitude of the integral of the deformation work, since the integral of the total work equals zero with a shear flow on a stationary wall. It is essential for this result that the friction acts on a stationary wall.

From the analysis with the differential equations it follows as well that the energy dissipation is not exactly equal to the product of the shear stress on the wall and the average velocity in a channel, as obtained in a one-dimensional analysis. This product has the correct order of magnitude, but the exact result depends on the distribution of the shear stress and the velocity gradient across a channel section. With a trapezoidal velocity profile and with an infinitesimally thin transition from core to wall, it follows that the energy dissipation integral (2.9) equals the product of the velocity in the flow core and the shear stress on the wall.

The reasoning based on the energy equation (2.7) and the work equation (2.8) demonstrates that the wall friction affects the entire boundary layer. Energy dissipation by friction is thus no local phenomenon. Moreover, the dissipation continues downstream of the profile. Figure 2.7 illustrates how the boundary layers on the suction and pressure sides merge at the trailing edge of an aerofoil and how a wake is formed in which further energy dissipation occurs.

As explained above, the drag of an aerofoil is composed of friction drag and pressure drag. Energy dissipation associated to drag may be decomposed in two parts as well, termed *friction loss* within the boundary layers and *mixing loss* within the wake downstream of the aerofoil. The term mixing loss expresses the dissipation by the mixing of the flows from the suction side and the pressure side with each other and with the surrounding flow. Figure 2.7 makes clear that the expansion of the wake, in principle, continues up to an infinite distance from the aerofoil.

It is thus very difficult to estimate the total loss by an integral of the energy dissipation within the fluid. The result of the integration is known a priori, however. Let us consider, instead of a stationary aerofoil in a flow with oncoming velocity v_∞, the motion of an aerofoil with velocity v_∞ in a stationary atmosphere. For the last case, the totally dissipated energy per time unit is $\left|\vec{D}.\vec{v}_\infty\right|$, since all work is dissipated. Obviously, the amount must be the same for the stationary aerofoil in the steady flow. For flow over an aerofoil, the following must thus apply (where S is a surface enclosing the aerofoil at a very large distance):

$$\vec{D}.\vec{v}_\infty(>0) = \int_S (h_0 - E_m)\rho \vec{v}.\vec{n} dS = -\int_S E_m \rho \vec{v}.\vec{n} dS. \qquad (2.10)$$

2.1.6 Profile Shapes

The optimal profile shape strongly depends on the particular application. The NACA 4412 in Fig. 2.3, was, as already mentioned, frequently used in older designs of pumps and hydraulic turbines. The intended flow deflection is rather small, the Reynolds number moderately high to high, i.e. in the order of 1–5×10^6 and with a high level of turbulence within the main flow. There first is flow acceleration at the suction side, which keeps the boundary layer laminar (the laminar boundary layer becomes more stable in accelerating flow). Deceleration follows after that. With a high Reynolds number and high turbulence level in the main flow, the boundary layer quickly becomes turbulent when subjected to an adverse pressure gradient (the laminar boundary layer becomes less stable in decelerating flow). This is intended, as a turbulent boundary layer better resists separation than a laminar one. An aerofoil shape with a pressure distribution as sketched in Fig. 2.5 is therefore termed a *turbulent profile*, which means that the boundary layer at the suction side is turbulent on a large part of it.

The profile shape is not optimal for pumps and hydraulic turbines. The predominantly turbulent boundary layer at the suction side allows a large adverse pressure gradient. Consequently, the attainable lift coefficient is high. But the drag coefficient is rather high as well, due to the higher friction than with laminar flow. Another disadvantage is the strong pressure minimum at the suction side, which may cause cavitation (local water evaporation) with some applications. Modern profile shapes have a less deep suction pressure minimum. The laminar part of the leading edge boundary layer is larger, resulting in a lower drag coefficient. Nowadays, these profiles are designed with computational fluid dynamics techniques.

Figure 2.8 shows an aerofoil profile and the accompanying pressure distribution for the middle part of a wind turbine blade [2]. The pressure distribution is completely different from that of the NACA 4412. Flow features are here: low deflection, slight acceleration within the general flow, rather high Reynolds number, i.e. 3–5 10^6, and low turbulence level in the main flow. Strong acceleration occurs at the leading edge part of the suction side, followed by a zone of weak acceleration. The boundary layer stays laminar until the end of the weak acceleration zone.

This succeeds due to the low turbulence level in the oncoming flow. From the beginning of the deceleration phase, the boundary layer becomes turbulent. This succeeds due to the rather high Reynolds number. At first, acceleration occurs at the pressure side due to the high aerofoil thickness. Then follows a weak deceleration intended to increase the pressure difference between the pressure and the suction sides. The obtainable lift coefficient is moderately high ($\approx$1.00). The drag coefficient is low ($\approx$0.0075), because the boundary layer at the suction side is laminar over a rather significant part. The aerofoil shape is termed a *laminar profile*. The shape is similar to profiles applied in wings of aircraft.

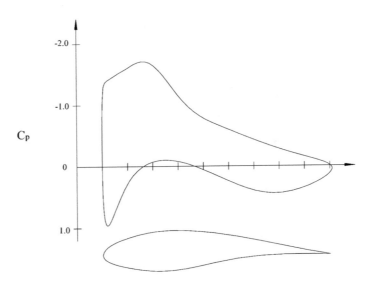

Fig. 2.8 Aerofoil profile and pressure distribution for a central part of a wind turbine blade

Present-day aerofoil profiles have the features of the profile of Fig. 2.8: strong acceleration followed by weak acceleration at the suction side leading edge, with a limitation of the minimum pressure; deceleration in the middle part of the pressure side in order to enlarge the lift. Gradients within the zones and the extensions of the zones vary strongly among different applications. The optimum is critically determined by the Reynolds number, the mean pressure gradient in the flow (globally accelerating or decelerating flow) and the turbulence level in the core of the flow. There is no universal optimum. Determination of the optimum profile requires advanced computational methods.

2.1.7 Blade Rows with Low Solidity

With the axial turbine or the axial pump discussed in Chap. 1, the blade profiles do not have the same performance as isolated blades, because the suction and pressure sides of neighbouring blades interact. This may increase or decrease the lift. The zero-lift direction changes and drag stays approximately the same. Figure 2.9 sketches a blade row for decelerating flow (pump, fan, compressor).

The tangential distance between the blades is called *spacing* (s) or pitch. The ratio of the chord (c) to the spacing is termed *solidity* ($\sigma = c/s$). With blade rows, the term solidity means the ratio of the blade area to the flow area. The blade area is typically defined by integration of the chord: $\int c \, dr$.

With a good approximation, the zero-lift line is found by connecting the trailing edge to the point of maximum deflection on the camber line, taking the maximum on a reduced chord between the trailing edge and a point obtained by projection in the axial direction of the trailing edge of the neighbouring profile [3].

Figure 2.10 demonstrates how the lift coefficient changes. It varies nearly linearly over a broad range of the angle of attack. This is the range without boundary layer separation. The slope of the lift curve changes with a factor k, plotted in the figure. Results originate from measurements on various profile shapes, but all with a relative thickness of 8% [3]. The angle in the diagram is between the zero-lift line of the isolated profile and the tangential direction. The diagram becomes unreliable for high solidity. It may be used confidently for $\sigma = c/s < 1.3$. With higher solidity,

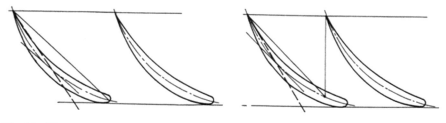

Fig. 2.9 Blade row; change of zero-lift line. Chord: full line, zero-lift line: dashed line

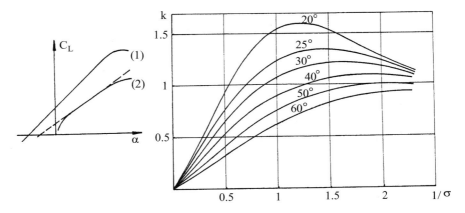

Fig. 2.10 Lift coefficient of a cascade. Isolated profile (1); profile in a row (2)

it is not possible to derive the blade row characteristics from the characteristics of isolated blades (see next section).

2.2   Linear Cascades

2.2.1   Relation with the Real Machine

The flow within the blade passages of an axial turbomachine is three-dimensional. A two-dimensional approximation is obtained by making a cylindrical section of the blades at the mean radius and by unrolling it to a plane. A linear cascade or blade row is built by setting up prismatic blades with the profiles obtained. In principle, an infinite number of blades are required in order to keep the periodicity of the blades in the original machine. The two-dimensional flow is representative for the real three-dimensional flow. The expressions of axial momentum are the same in both configurations. Moment of momentum with the three-dimensional flow corresponds to tangential momentum with the cascade (tangential = circumferential). Constant tangential momentum along the blade height with the linear cascade is equivalent to constant angular momentum ($v_u r$ = constant) along the radius in the real machine.

A three-dimensional whirling flow with constant angular momentum is called a *free-vortex flow*. In the pioneering time, axial turbomachines were built with free-vortex flow because of the theoretical correspondence with a two-dimensional flow. They show certain disadvantages, discussed later (Chap. 13: axial compressors; Chap. 15: axial and radial turbines). Present-day machines are designed with some vortex forcing, which means $v_u r \neq$ constant along the radius. The deviation from constant angular momentum is not very large, however.

With free-vortex blades in a cylindrical axial turbomachine, an exact transformation exists between the flow in the machine and the linear cascade (of course with top and bottom walls added). This does no longer apply with some vortex forcing. For instance, no cylindrical streamsurfaces exist anymore. The analogy between the three-dimensional flow within the machine and the two-dimensional flow within the linear cascade is then only met approximately. Nevertheless, studying the characteristics of linear cascades is very instructive as the flow within a linear cascade is simpler, but still representative for the flow in an average cylindrical section of a real machine.

2.2.2 Cascade Geometry

The cascade is determined by the shape of the blades, their position to the axial direction and their *spacing* (tangential distance between the blades). The term *pitch* is often used for spacing, but pitch may also mean, similarly to the term with screws, the axial distance covered by the flow when it makes a full turn of 360°. This applies in particular to propellers and wind turbines with adjustable rotor blades. Then often, also the term pitch angle is used. Results are also a function of the positioning of the blades relative to the flow.

Fig. 2.11 Cascade notation (example of decelerating flow)

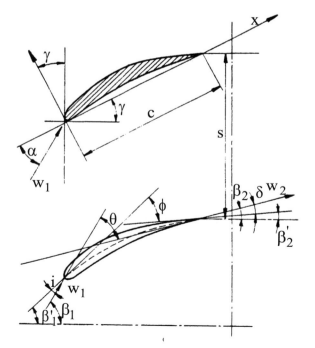

The profile shape is, as with aerofoils, determined by the camber line and the thickness distribution. The profile size is determined by the chord. *Solidity* σ means the ratio of the chord to the spacing. Some characteristics are shown in Fig. 2.11. The *camber angle* ϕ is the angle difference between the tangents to the camber line at the leading and trailing edges. The profile orientation relative to the axial direction is determined by the chord angle or *stagger angle* γ. The inlet velocity w_1 forms an angle of attack α with the chord, an angle β_1 with the axial direction and an *angle of incidence* i with the tangent to the camber line at the leading edge. The angle of incidence is counted positive when the flow turning increases by it. The outlet velocity w_2 forms with the inlet velocity w_1 an *angle of deflection* θ, with the tangent to the camber line at the trailing edge an *angle of deviation* δ and with the axial direction an angle β_2. The deviation angle is counted positive when it causes a decrease of θ.

2.2.3 Flow in Lossless Cascades: Force Components

Figure 2.12 sketches the flow through a rotor cascade with decelerating flow. We apply the fundamental laws (for constant density) to a cascade with profiles on spacing s. The flow velocities in the relative frame w have w_a in the axial direction and w_u in the tangential direction (= circumferential direction) as their components. Similarly, the force L exerted on the blades has L_a and L_u as components. The velocities w_1 and w_2 are considered sufficiently far from the cascade, so that velocity is assumed to be constant in sections 1 and 2.

We select a right-handed coordinate frame with the x-axis in the axial direction, positive in the through-flow sense, and the y-axis in the circumferential direction, positive in the running sense of the rotor. For the z-axis in the radial direction, with the outward sense as positive sense, the machine must be left-turning $\left(\vec{\Omega} = -\Omega \vec{1}_x \right)$. This is a minor complication. A right-handed coordinate frame with a right-turning machine is obtained by the x-axis in the running sense, the y-axis in the axial direction in the through-flow sense and the z-axis in the radial direction in the outward sense $\left(\vec{\Omega} = \Omega \vec{1}_y \right)$. With this last convention, often used in the past, angles are calculated with respect to the circumferential direction and vary from $0°$ to $180°$.

The last choice has the disadvantage that application of the goniometric tangent function is difficult. Due to the general use of electronic calculation devices, the first convention (axial convention) is preferred at present, with angles varying from $-90°$ to $+90°$. This convention has already been applied systematically in Chap. 1. In the entire book, we reason on left-turning machines. Both the left and right rotation senses are used in practice (some illustrations concern right-turning machines). More generally, we take a coordinate frame with first axis in the meridional direction (this is the tangent to the intersection of the average circumferential streamsurface and a

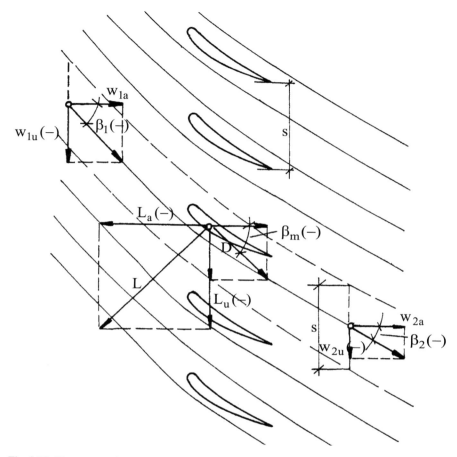

Fig. 2.12 Force exerted on a cascade blade (rotor pump)

meridional plane), positive in the through-flow sense; second axis in the circumferential direction, positive in the running sense; third axis perpendicular to the average streamsurface, positive in the outward sense.

The tangential velocity components are negative for the cascade in Fig. 2.12, with lift components L_a and L_u being negative as well. We reason in the relative frame. Because the centrifugal force and the Coriolis force lie in the radial direction with axial machines, these forces do not intervene with momentum relations in the axial and tangential directions. The derived relations thus automatically apply to stator cascades, as we always apply the momentum laws in an algebraically consistent way. The derived relations also apply to cascades with accelerating flow (this will be verified later on). The pump rotor cascade in Fig. 2.12 is the most complex one to analyse (w_u, L_a and L_u being negative). That is the reason why we take it as an example.

The mass conservation law is:

$$w_{1a}s = w_{2a}s = w_a s.$$

The work equation (relative frame, no losses: Eq. (2.8) with $\tau = 0$) is

$$p_{01r} - p_{02r} = 0 \quad \text{or} \quad p_1 + \frac{\rho}{2}w_1^2 = p_2 + \frac{\rho}{2}w_2^2,$$

from which

$$p_2 - p_1 = \frac{\rho}{2}\left(w_1^2 - w_2^2\right).$$

With

$$\Delta p = p_2 - p_1; \quad q_1 = \frac{\rho w_1^2}{2} \quad \text{and} \quad w_a = w_2 \cos \beta_2 = w_1 \cos \beta_1,$$

it follows that:

$$C_p = \frac{\Delta p}{q_1} = 1 - \frac{\cos^2 \beta_1}{\cos^2 \beta_2}. \tag{2.11}$$

The term $\Delta p/q_1$ is often denoted by C_p. A criterion for maximum possible flow deceleration is: $C_p = 0.5$ ($w_2/w_1 \approx 0.7$). This results in a relation between β_1 and β_2. E.g., corresponding values are $\beta_1 = -60°$ and $\beta_2 = -45°$, $\beta_1 = -55°$ and $\beta_2 = -35°$.

In order to apply the momentum conservation law, we consider the dashed-line contour in Fig. 2.12. It forms a control volume with two periodic streamsurfaces and two planes parallel to the cascade front. On the front and back surfaces, pressures are p_1 and p_2. Pressure forces on the periodic streamsurfaces counterbalance. The force by the blade on the flow is $-L$.

The momentum balance in the tangential direction is

$$-L_u = \rho s w_a(w_{2u} - w_{1u}), \quad \text{or} \quad L_u = \rho s w_a(w_{1u} - w_{2u}). \tag{2.12}$$

The momentum balance in the axial direction is

$$-L_a + sp_1 - sp_2 = 0, \quad \text{or} \quad L_a = s(p_1 - p_2) = s\frac{\rho}{2}\left(w_{2u}^2 - w_{1u}^2\right). \tag{2.13}$$

The circulation along the dashed-line contour in the positive sense, as periodic parts intervene with counterbalancing amounts, is

$$\Gamma = \int_C \vec{w}.\vec{dl} = s(-w_{1u} + w_{2u}) = s(w_{2u} - w_{1u}).$$

Combination results in

$$L_u = -\rho\Gamma w_a \quad \text{and} \quad L_a = \rho\Gamma \tfrac{1}{2}(w_{1u} + w_{2u}).$$ (2.14)

The result is that the force L is perpendicular to the velocity w_m, which has w_a and $(w_{1u} + w_{2u})/2$ as components and that the magnitude of L is

$$|L| = \rho|\Gamma|w_m.$$ (2.15)

This relation is termed the Kutta-Joukowski law for cascades. It is entirely analogous to that for aerofoils, being $|L| = \rho_\infty\,|\Gamma|\,w_\infty$. The expression for aerofoils may be derived from (2.15), by keeping Γ and allowing s to increase to infinity. Then $w_1 = w_2 = w_\infty$ becomes the velocity of the parallel flow.

2.2.4 Significance of Circulation

The work equation on a streamline through the cascade is Eq. (2.2), but with $d\tau/dy = 0$ in the core of the flow (negligible shear force), which we write for a rotor cascade as

$$\rho w \frac{dw}{dx} + \frac{dp}{dx} = 0.$$ (2.16)

The force balance in the direction perpendicular to the streamline is

$$\frac{1}{\rho}\frac{dp}{dy} = \frac{w^2}{R}.$$ (2.17)

Equation (2.16) may be integrated along a streamline to

$$\frac{1}{2}w^2 + \frac{1}{\rho}p = \text{constant}.$$ (2.18)

With the cosine rule on the velocity triangles follows

$$w^2 = u^2 + v^2 - 2uv_u.$$

Thus:

$$\frac{1}{2}v^2 + \frac{1}{\rho}p + \frac{1}{2}u^2 - uv_u = \text{constant}.$$ (2.19)

This means that, for constant mechanical energy of the incoming flow in the absolute frame and uniform v_u, the constant in Eq. (2.19) and thus in Eq. (2.18) is

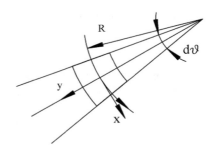

Fig. 2.13 Infinitesimal contour for evaluation of the rotation of the velocity vector

the same on all streamlines. Equation (2.18) then also implies

$$w\frac{dw}{dy} + \frac{1}{\rho}\frac{dp}{dy} = 0. \tag{2.20}$$

Combined with Eq. (2.17), this gives

$$w\frac{dw}{dy} + \frac{w^2}{R} = 0. \tag{2.21}$$

The significance of (2.21) may be understood by calculation of the circulation along an infinitesimal contour consisting of two pieces of neighbouring streamlines connected with straight segments as shown in Fig. 2.13 (x is streamline direction, y is normal direction), and by applying Stokes' circulation theorem, which is an integral theorem for the rotor of a vector quantity:

$$\int_C \vec{w}.\vec{dl} = \int_S (\nabla \times \vec{w}).\vec{n}\,dS,$$

where S is a surface spanned by the contour and $\vec{n}$ is the unit normal vector on the surface in the sense corresponding with the sense on the contour. On the infinitesimal contour of Fig. 2.13, this gives, with R the radius of curvature of the mean streamline and with ω the value of the rotor of the relative velocity vector in the z-direction:

$$\left(w + \frac{1}{2}\frac{dw}{dy}dy\right)\left(R + \frac{1}{2}dy\right)d\theta - \left(w - \frac{1}{2}\frac{dw}{dy}dy\right)\left(R - \frac{1}{2}dy\right)d\theta = \omega R d\theta dy,$$

or

$$\frac{dw}{dy}dy R d\theta + w\,dy\,d\theta = \omega R d\theta dy;$$

thus

$$\frac{dw}{dy} + \frac{w}{R} = \omega. \tag{2.22}$$

With (2.22), we understand that (2.21) means that the rotor of the velocity vector is zero everywhere in the core of the flow. Such a flow is called *irrotational*. The meaning of the term is that the rotational speed of each fluid particle is zero, since the rotational speed is equal to half of the rotor of the velocity vector (see fluid mechanics). The consequence is that the circulation is zero on every contour non-enclosing a blade and that the circulation on a contour enclosing a blade is the same for every contour around that blade. Strictly, this result is only valid for lossless flow, as the mechanical energy has to be the same for every streamline. In a flow with losses, the result remains approximately valid for every contour around a blade as long as this contour does not enter the boundary layers on the blade. The contour has to cut the wake, of course, which causes a small error on the strict equality of the circulation on each contour around a blade.

2.2.5 Flow in Lossless Cascades: Work

When the cascade moves in the tangential direction with velocity u, the absolute velocities v may be derived from u and w by

$$v_{1u} = u + w_{1u}; \quad v_{2u} = u + w_{2u}; \quad v_a = w_a.$$

Formula (2.12) then becomes

$$L_u = \rho s v_a (v_{1u} - v_{2u}).$$

The power transferred to the fluid becomes

$$P = -L_u u = \rho s v_a (v_{2u} - v_{1u}) u.$$

The power per mass flow rate unit, i.e. the work per mass unit, becomes

$$\Delta W = \frac{-L_u u}{\rho s v_a} = u(v_{2u} - v_{1u}).$$

This is Euler's formula, applied to an axial machine ($u_1 = u_2$).

It is informative to remark that the work also may be found from $-\vec{L}.\vec{v}_m$, since $-\vec{L}.\vec{v}_m = -\vec{L}.\vec{u} - \vec{L}.\vec{w}_m = -\vec{L}.\vec{u}$. So, the work done on the flow by an active force is the force multiplied with the displacement, as applied in Chap. 1. This seems self-evident. But it is not the case for a flow with losses, as we analyse below.

2.2.6 Flow in Cascades with Loss: Force Components

Loss results in a *total pressure drop* (Eq. 2.8), which we may express by

$$p_{01r} - p_{02r} = \left(p_1 + \frac{\rho w_1^2}{2}\right) - \left(p_2 + \frac{\rho w_2^2}{2}\right),$$

being a positive quantity. For further analysis, we assume equal loss on all streamlines. A *loss coefficient*, based on inlet dynamic pressure, is then

$$\xi_1 = \frac{p_{01r} - p_{02r}}{q_1}.$$

With F the force of the flow on the profile, then

$$F_u = \rho s w_a (w_{1u} - w_{2u}) = \rho s w_a^2 (\tan \beta_1 - \tan \beta_2),$$

$$F_a = s(p_1 - p_2) = \frac{\rho s}{2} w_a^2 \left(\tan^2 \beta_2 - \tan^2 \beta_1\right) + s(p_{01r} - p_{02r})$$

$$= \rho s w_a^2 \tan \beta_m (\tan \beta_2 - \tan \beta_1) + s(p_{01r} - p_{02r}),$$

With

$$\tan \beta_m = \frac{\tan \beta_1 + \tan \beta_2}{2}.$$

The drag follows from (see Fig. 2.12):

$$D = \left(F_u \vec{1}_u + F_a \vec{1}_a\right) \cdot \left(\sin \beta_m \vec{1}_u + \cos \beta_m \vec{1}_a\right)$$

$$= F_u \sin \beta_m + F_a \cos \beta_m = s(p_{01r} - p_{02r}) \cos \beta_m. \qquad (2.23)$$

The drag components are

$$D_a = s(p_{01r} - p_{02r}) \cos^2 \beta_m, \quad D_u = s(p_{01r} - p_{02r}) \sin \beta_m \cos \beta_m.$$

The lift components are

$$L_u = \rho s w_a^2 (\tan \beta_1 - \tan \beta_2) - s(p_{01r} - p_{02r}) \sin \beta_m \cos \beta_m,$$

$$L_a = \rho s w_a^2 \tan \beta_m (\tan \beta_2 - \tan \beta_1) + s(p_{01r} - p_{02r}) \sin^2 \beta_m.$$

The magnitude of the lift for the cascade in Fig. 2.12 ($\tan \beta_1 - \tan \beta_2 < 0$) is

$$L = \rho s \frac{w_a^2}{\cos \beta_m} (\tan \beta_2 - \tan \beta_1) + s(p_{01r} - p_{02r}) \sin \beta_m.$$

From this follows:

$$C_{L1} = \frac{L}{q_1 c} = \frac{2}{\sigma} \frac{\cos^2 \beta_1}{\cos \beta_m} (\tan \beta_2 - \tan \beta_1) + \frac{\xi_1}{\sigma} \sin \beta_m,$$

$$C_{D1} = \frac{D}{q_1 c} = \frac{\xi_1}{\sigma} \cos \beta_m.$$

Elimination of ξ_1 results in

$$C_{L1} = \frac{2}{\sigma} \frac{\cos^2 \beta_1}{\cos \beta_m} (\tan \beta_2 - \tan \beta_1) + C_{D1} \tan \beta_m.$$

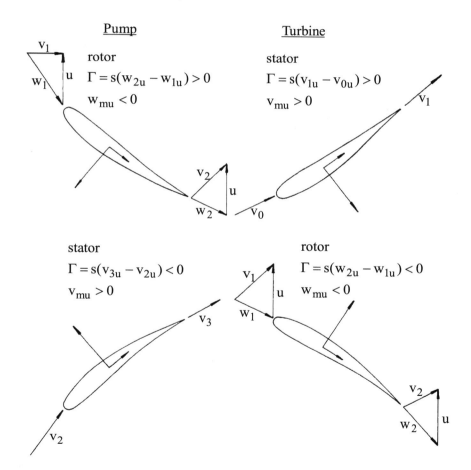

Fig. 2.14 Forces with various axial machine components

This relation implies that the lift coefficient does not depend solely on the cascade solidity and the flow angles, but on the drag coefficient as well.

With the above formulae it should be taken into account that the β-angles are negative with the cascade in Fig. 2.12. This means that lift decreases due to losses. When deriving the relations, sign conventions have been applied consistently. The expressions for F_u and F_a have general validity, also for a turbine cascade ($\tan \beta_1 - \tan \beta_2 > 0$, $F_u > 0$, $F_a > 0$). The expression for drag remains the same with a turbine. The expression for lift, with lift being a positive quantity, becomes ($\beta_m < 0$) with a turbine:

$$C_{L1} = \frac{2 \cos^2 \beta_1}{\sigma \cos \beta_m} (\tan \beta_1 - \tan \beta_2) - C_{D1} \tan \beta_m.$$

It is remarkable that, with a given deflection in a turbine, losses cause lift increase. The different behaviour compared to the pump can be understood by analysing the four configurations shown in Fig. 2.14. The four possible combinations of the sign of the circulation and the mean tangential velocity occur. First, the correspondence of the sign combinations with the expressions (2.14) may be verified. The tangential components of lift and drag have the same sense with pump cascades. For a given change of tangential velocity, this means given F_u, D_u decreases the magnitude of L_u. This is the opposite with a turbine cascade.

2.2.7   Flow in Cascades with Loss: Energy Dissipation and Work by Drag Force

Equation (2.23) for drag implies

$$D w_m = \rho s w_m \cos \beta_m \frac{p_{01r} - p_{02r}}{\rho} = \rho s w_a \frac{p_{01r} - p_{02r}}{\rho}.$$

Thus:

$$D w_m = \dot{m} q_{irr}. \tag{2.24}$$

Equation (2.24) is the same as the expression (2.10) for energy dissipation with an aerofoil. The right-hand part in (2.24) represents the difference between the mechanical energy fluxes through the inlet and outlet planes of the cascade. The term $D w_m$ is thus the total amount of energy dissipation. This result expresses that, with flow through a stationary cascade (we reason in the relative frame), the energy dissipation equals the sign-changed displacement work of the drag force exerted on the average flow. As explained before, the reason is that the total work is zero and thus the deformation work, which physically is the energy dissipation, has the same magnitude as the displacement work.

The finding that, for a given flow deflection, drag affects lift, is due to the contribution of the drag to the tangential velocity change. For a moving cascade, this means that the drag contributes to the work. The total work done by the drag force on the fluid is $-\vec{D}.\vec{u}$. Analogously to the lift, we can verify $-\vec{D}.\vec{v}_m$.

It follows that

$$-\vec{D}.\vec{u} = -\vec{D}.\vec{v}_m + \vec{D}.\vec{w}_m = -\vec{D}.\vec{v}_m + \dot{m}q_{irr}. \qquad (2.25)$$

With the configurations in Fig. 2.14, the total work by the drag force, $-\vec{D}.\vec{u}$, is always in the sense from material parts to the fluid. The energy dissipation during the work transfer (deformation work) is $\dot{m}q_{irr}$. The displacement work, $-\vec{D}.\vec{v}_m$, may be positive (work to the fluid) or negative. It is negative for the four configurations in Fig. 2.14. This is what we intuitively expect and is always correct for stator components. With rotor components, the displacement work of the drag force may be positive. This requires a high blade speed compared to the other velocity components (see Exercise 2.5.6). Equation (2.25) demonstrates that the concept of displacement work has to be used with care. Only in the case of a purely active force, without any deformation work associated, as a lift, does the displacement work equal the total work.

2.2.8 The Zweifel Tangential Force Coefficient

With an axial cascade, the component of the force on the blade useful for work is not the lift but the tangential component. It is therefore appropriate to define a coefficient on the basis of F_u as well. For a constant-density fluid, the obvious reference force is $\frac{1}{2}\rho w_2^2 c_a$, with c_a the *axial chord*, as illustrated in Fig. 2.15.

In loss-free flow, the pressure at the cascade outlet is lower than the total pressure at the inlet with the amount $\frac{1}{2}\rho w_2^2$. The area between the pressure curves on the pressure and suction sides is the magnitude of F_u. We see the relation with the reference force $(p_{01r} - p_2)c_a$. This quantity is replaced by the value for lossless constant-density flow: $\frac{1}{2}\rho_2 w_2^2 c_a$ (ρ_2 is outlet density for a compressible fluid).

This allows the definition of a tangential force coefficient, expressed solely as a function of velocity quantities and the axial solidity $\sigma_a = c_a/s$, by

$$C_{Fu} = \frac{|F_u|}{\frac{1}{2}\rho_2 w_2^2 c_a} = \frac{\rho_2 w_{2a} s |w_{1u} - w_{2u}|}{\frac{1}{2}\rho_2 w_2^2 c_a} = \frac{2|\Delta w_u| w_{2a}}{\sigma_a w_2^2}. \qquad (2.26)$$

The concept of the tangential force coefficient was introduced by Zweifel in the 1940s. He found that the coefficient was about 0.8 for cascades with good efficiency. The separation values were around 1.1–1.2. Good efficiency means small losses compared to the tangential force. With $C_{Fu} = 0.8$ follows an optimal value of the axial solidity $\sigma_a = c_a/s$ by

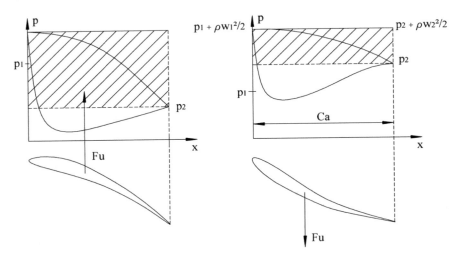

Fig. 2.15 Estimation of the tangential force capability; left: cascade with moderate flow acceleration (turbine); right: cascade with moderate flow deceleration (pump, fan)

$$\sigma_a = \frac{2}{0.8}\frac{w_a|w_{1u} - w_{2u}|}{w_2^2} = 2.5\frac{w_a^2|\tan\beta_1 - \tan\beta_2|}{w_2^2}$$

$$= 2.5\cos^2\beta_2|\tan\beta_1 - \tan\beta_2|.$$

For a compressor with $\beta_1 = -55°$ and $\beta_2 = -35°$, this formula results in
$\sigma_a = c_a/s = 1.22$. The chord solidity is $\sigma = \frac{c}{s} \approx \frac{c_a/s}{\cos\beta_m} = 1.78$. For a turbine with
$\beta_1 = 0°$ and $\beta_2 = -60°$ corresponds $\sigma_a = c_a/s = 1.08$ and $\sigma \approx 1.43$. Both solidity
values are above unity.

The tangential force coefficient is related to the lift coefficient. Ignoring the effect
of losses, the lift coefficient of a cascade is

$$L = \rho|\Gamma|w_m = \frac{1}{2}\rho w_1^2 C_{L1}c.$$

So:

$$C_{L1} = \frac{2|\Gamma|w_m}{w_1^2 c} = \frac{2s|w_{2u} - w_{1u}|w_m}{w_1^2 c} = \frac{2|\Delta w_u|w_m}{\sigma w_1^2}. \tag{2.27}$$

With $c_a/c \approx \cos\beta_m = w_a/w_m$ (for w_a constant), the tangential force coefficient
(2.26) is

$$C_{Fu} = \frac{2|\Delta w_u|w_a s}{w_2^2 c_a} \approx \frac{2|\Delta w_u|w_m s}{w_2^2 c} = \frac{2|\Delta w_u|w_m}{\sigma w_1^2}\left(\frac{w_1}{w_2}\right)^2. \tag{2.28}$$

From the comparison between the expressions (2.27) and (2.28) follows that the
tangential force coefficient is a lift coefficient, related to the outlet dynamic pressure.

Zweifel's reasoning thus demonstrates that the lift coefficient of a cascade has to be related to the outlet kinetic energy.

So, for an axial cascade:

$$C_{L2} = \frac{2|\Delta w_u|w_m}{\sigma w_2^2} \approx C_{Fu} = \frac{2|\Delta w_u|w_a}{\sigma_a w_2^2}. \tag{2.29}$$

The design value of the tangential force coefficient of present-day turbine cascades is about 1.0–1.2. The separation value is 1.4–1.6. These values are the same for constant density and variable density fluids and do not depend on the outlet Mach number for compressible fluids. The tangential force coefficient may very reliably be applied to turbine cascades. With pump, fan and compressor cascades, application of C_{Fu} is possible as well. The optimal value is about 1.0–1.2 as well, but values show a much larger spreading than with turbines (see next section).

2.2.9 The Lieblein Diffusion Factor

With strongly decelerating cascades, the lift potential is also determined by the inlet velocity level. Therefore, the lift coefficient C_{L2} is not universally applicable. Figure 2.16 shows velocity distributions near the blade surface (boundary layer edge) of a strongly accelerating turbine cascade and a strongly decelerating compressor cascade with large velocity difference between the suction and pressure sides, thus a large tangential force. It is often said that such cascades are highly loaded. The figure demonstrates that the tangential force with the turbine is strongly correlated to the outlet dynamic pressure. With the compressor cascade, the correlation is weak with both the dynamic pressure at inlet and at outlet. With the compressor, the cause is the strong accelerations downstream of the leading edge stagnation point, both on suction and pressure sides.

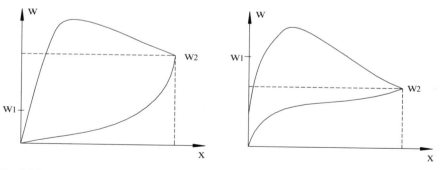

Fig. 2.16 Velocity distribution near the blade surface in a highly loaded turbine cascade (left) and a highly loaded compressor cascade (right)

The deceleration is moderate with fans and pumps (Fig. 2.15, right) and the tangential force coefficient (2.29) can be applied more reliably than with compressors. With compressor cascades, the concept of the *diffusion factor* is mostly used to determine the lift capacity of a cascade. The concept was introduced by Lieblein in the 1950s and there are some variants.

A diffusion factor is the ratio of the deceleration on the suction side ($w_{max} - w_2$) to a reference velocity, mostly w_1. The diffusion factor in this sense, denoted by DF, will be discussed in Chap. 13 (axial compressors). With axial and radial fans and pumps, the concept of *local diffusion factor* may be used as well. It is the ratio of ($w_{max} - w_2$) to w_{max}. A criterion for avoidance of separation is that this factor should be limited to about 0.5. This means that ($w_{max} - w_2$) must not exceed w_2. This criterion may also be applied to estimate the lift capacity of a cascade. We will use the local diffusion factor for radial cascades in Chap. 3. So:

$$DF = \frac{w_{\max} - w_2}{w_1} \quad \text{and} \quad D_{loc} = \frac{w_{\max} - w_2}{w_{\max}}.$$

2.2.10 Performance Parameters of Axial Cascades

The losses are expressed by the total pressure drop ($p_{01} - p_{02}$), relative to the inlet dynamic pressure q_1 with a decelerating cascade or to the outlet dynamic pressure q_2 with an accelerating cascade. The literature offers correlations to determine loss coefficients. Methods differ for decelerating and accelerating cascades. The deviation angle δ can be determined, as well as the optimal value of the angle of incidence i. We further use simple correlations. We refer to Dixon and Hall [4], Japikse and Baines [5] for more complete correlations concerning cascades.

2.3 Channels

2.3.1 Straight Channels

For a straight channel with constant cross-section area A and with fully developed flow (velocity profile does not change in flow direction), the relation between the pressure drop (denoted here by $-\Delta p$) and the shear stress on the wall is

$$A(-\Delta p) = \tau_w O L,$$

with O being the circumference and L the length of the part considered. Thus

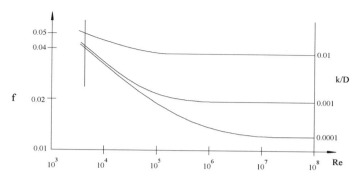

Fig. 2.17 Friction factor with circular ducts (Moody friction factor chart)

$$\frac{-\Delta p}{\rho} = \frac{\tau_w}{\rho} \frac{O}{A} L = \frac{\tau_w}{\frac{1}{2}\rho \bar{v}^2} 4 \frac{L}{D} \left(\frac{1}{2} \bar{v}^2 \right).$$

D is the hydraulic diameter ($D = 4A/O$) and $\bar{v}$ the average velocity. The term $-\Delta p/\rho$ represents the mechanical energy loss $-\Delta E_m$. A loss coefficient is defined by

$$\xi = \frac{-\Delta E_m}{\frac{1}{2}\bar{v}^2} = f \frac{L}{D}, \tag{2.30}$$

where f is the Darcy friction factor. The Fanning friction factor is used as well, defined by $\tau_w/\left(\frac{1}{2}\rho \bar{v}^2\right)$. There is a ratio 4 between both factors. Figure 2.17 is a schematic of the well-known Moody diagram for ducts with a circular section. The Darcy friction factor is represented. The full diagram is published in almost all books on basic fluid mechanics. It can also be found in many internet sites (look for Moody friction factor chart).

The Colebrook-White equation renders the friction factor for turbulent flows:

$$\frac{1}{\sqrt{f}} = -2 \log \left[\frac{2.51}{\text{Re}\sqrt{f}} + \frac{k}{3.7D} \right].$$

The Reynolds number is defined by $\text{Re} = \bar{v}D/v$, with v being the kinematic viscosity coefficient. Roughness is represented by the equivalent sand-grain roughness k. The implicit equation may be replaced with a very good approximation (better than 1.5%) by the explicit equation [6]:

$$\frac{1}{\sqrt{f}} = -1.8 \log \left[\frac{6.9}{\text{Re}} + \left(\frac{k}{3.7D} \right)^{1.11} \right].$$

Channels in turbomachines never have a constant cross section and a fully developed flow. It is customary however to estimate friction losses with the Moody-diagram on the basis of an average length and an average hydraulic diameter. Therefore, the applied method only gives an approximation.

2.3.2 Bends and Curved Channels

Flow in a bend changes pressure and velocity distributions, generates adverse pressure gradients and secondary flows and the curvature of the streamlines affects the turbulence.

Figure 2.18 sketches the velocity distribution with an ideal fluid (frictionless). The centrifugal force due to streamline curvature causes static pressure increase and velocity decrease at the bend outer part. There is an adverse pressure gradient at the entrance of the outer part and at the exit of the inner part of the bend.

Friction causes lower velocity near walls than within the flow core. The consequence is that the centrifugal force due to streamline curvature is higher within the flow core for equal curvature of the streamlines. The difference generates two vortex flows as sketched in Fig. 2.19. This transverse flow is termed *secondary flow*. In the view perpendicular to the bend, the curvature of the streamlines increases near the walls and decreases in the core flow. This realises pressure equilibrium in the direction perpendicular to the bend.

Low-momentum fluid migrates in wall vicinity from the outer part of the bend to the inner part. With 45° bends, with ratios of the radius of curvature of the bend to the diameter of 1–3, which are common values, it comes out that low-momentum fluid arrives in the adverse pressure gradient zone at the bend exit. There is then a separation risk. With larger bend angles, core fluid arrives in the adverse pressure gradient zone at the bend exit. There is thus a far lower separation risk with a 90° degree bend, which is a surprising observation.

A further phenomenon is that turbulence is affected by flow curvature. The magnitude of the centrifugal force due to curvature is proportional to the second power of

Fig. 2.18 Pressure and velocity distributions for flow of an ideal fluid in a bend

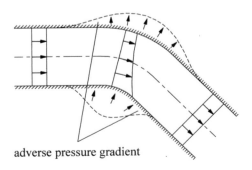

adverse pressure gradient

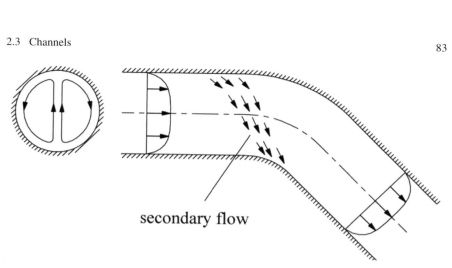

secondary flow

Fig. 2.19 Secondary flow; vortex motions in a cross section (left); secondary flow pattern near the wall (right)

the magnitude of the through-flow velocity. It points towards the wall at the outer part of the bend and away from the wall at the inner part.

A force with magnitude proportional to a positive power of the magnitude of the velocity enhances turbulence in a boundary layer when it points towards the wall and damps turbulence when it points away from the wall. The mechanism with the force pointing towards the wall is that a fluid parcel which leaps by the turbulent motion from a position nearer to the wall to a position further away from the wall experiences in its new position a smaller force than the fluid parcels at that position, due to its lower velocity in the mean streamwise direction (the parcel conserves approximately its momentum). The consequence is that its motion away from the wall is amplified. This way, the turbulent motion is enhanced near the wall. The effect is the inverse with a force pointing away from the wall. Thus, turbulence increases in the outer part of the bend and decreases in the inner part. A consequence is partial laminarization of the boundary layer in the inner part. This further increases the boundary layer separation risk in the exit region of the inner part.

Secondary flow phenomena are intense in centrifugal rotor channels. At rotor entry, the flow turns from the axial to the radial direction. The associated centrifugal force causes transversal migration of low-momentum fluid in the blade boundary layers from the hub side to the shroud side of the rotor. The radial component of the Coriolis force within the rotor channels (see Fig. 1.14) has a similar effect.

The tangential component of the Coriolis force causes transversal migration of low-momentum fluid in the hub and shroud boundary layers from the pressure side to the suction side. With a shrouded rotor, low-momentum fluid thus accumulates in the corner of the suction side of the blades and the shroud.

The magnitude of the Coriolis force is proportional to the first power of the magnitude of the through-flow velocity. Consequently, the Coriolis force has a similar effect on the turbulence as a centrifugal force. Turbulence is thus damped in the corner of the suction side of the blades and the shroud.

The combined effects of low-momentum accumulation and turbulence damping means that the risk for separation is the largest in the adverse pressure gradient zone of the boundary layers in this corner zone. But, even without flow separation, the outflow of a centrifugal rotor is very inhomogeneous, due to the accumulation of low-momentum fluid in one corner of the blade channels. The through-flow velocity is lowered there. The phenomenon is described as *jet-wake* flow. The term jet refers to the zone with higher through-flow velocity and the wake to the zone with lower through-flow velocity. With an unshrouded rotor, leakage flows influence the jet-wake flow, but the general phenomenon persists (see Chap. 14, Sect. 14.4.3).

A bend represents a supplementary loss, partly by increased friction within the bend, and partly by flow equilibration downstream of the bend. By equilibration is meant the evolution of the velocity profile towards an equilibrium profile adapted to the downstream channel. Velocity rearrangement of the fluid layers produces vortices that break down into turbulent eddies. These eddies interact and further break down to eddies of smaller size. The transfer of kinetic energy from larger to smaller structures is termed the *energy cascade*. The smaller the turbulent eddies, the larger is the impact of viscosity forces on their motion and the larger is the fraction of the energy dissipated into heat during the breakdown from larger to smaller structures. Finally, the kinetic energy of the smallest eddies is completely dissipated.

Figure 2.20 illustrates the dissipation process associated to the velocity profile evolution after a sudden expansion. The evolution after a bend is similar. With a bend of 90° and $r/d = 2$, the bend loss coefficient is about 0.16. About half of the loss occurs in the bend itself and about half is due to equilibration downstream of the bend. The loss by equilibration is called the *mixing loss*, referring to the mixing of fluid layers with different velocity. So, similarly as with aerofoils and cascades, a distinction is made between friction loss and mixing loss..

At the entry of a centrifugal rotor, there is no loss due to equilibration, because the inlet is not followed by a duct in which equilibration can occur. So, the loss coefficient due to the meridional flow turning is about 0.1, but the irregular inflow of the bladed part of the rotor reduces the rotor efficiency. In the same way, mixing

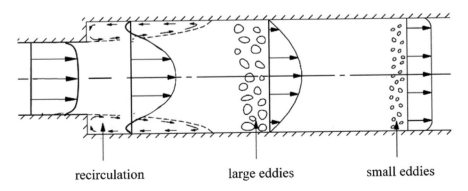

recirculation   large eddies   small eddies

Fig. 2.20 Deceleration by sudden expansion (dump diffusion)

loss downstream of a rotor is not complete, unless a large mixing space is placed around the rotor. But the efficiency of a diffuser downstream of the rotor is impaired due to inhomogeneous inflow. The effect of inhomogeneous inflow with a diffuser is discussed in the next section.

2.4 Diffusers

2.4.1 Dump Diffusers

Figure 2.20 sketches the flow in a sudden expansion of a channel. The flow is decelerated. A sudden expansion thus functions as a diffuser. The figure illustrates that the loss is mainly by flow equilibration after the expansion, thus mixing. The loss is the kinetic energy related to the velocity difference (see Exercise 2.5.4):

$$q_{irr} = \frac{(v_1 - v_2)^2}{2} = \left(1 - \frac{v_2}{v_1}\right)^2 \frac{v_1^2}{2} = \left(1 - \frac{A_1}{A_2}\right)^2 \frac{v_1^2}{2} = \xi \frac{v_1^2}{2}. \tag{2.31}$$

Deceleration by sudden expansion or dump diffusion is efficient with a moderately large area ratio. For instance, $A_2/A_1 = 1.25$ gives $\xi = 0.04$. A higher area ratio is useful as well. For instance, $A_2/A_1 = 2$ gives $\xi = 0.25$. A loss coefficient of 0.25 is not disadvantageous for a diffuser, as discussed below. That is why dump deceleration is applied in turbomachines, which is rather surprising at a first glance. The main advantage of dump deceleration is its realisation in a short distance.

2.4.2 Inlet Flow Distortion

Diffusers are channels in which the flow is decelerated and dynamic pressure is converted into static pressure. This process is called *pressure recovery*. The attainable pressure recovery depends strongly on the uniformity of the incoming flow. Figure 2.21 compares decelerations with uniform and non-uniform incoming flow

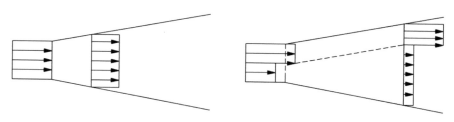

Fig. 2.21 Equal pressure recovery with uniform and non-uniform inlet flows

Fig. 2.22 Velocity profile
evolution within a diffuser
with limited length

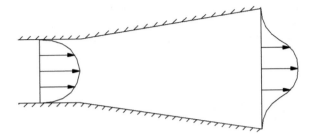

of an ideal fluid. The figure shows the position where a recovery equal to 50% of the kinetic energy associated to the average inlet velocity is obtained. With the uniform flow, the corresponding velocity is $\sqrt{1-0.5} \approx 0.71$ times the inflow velocity and the corresponding section area is 1.41 times the inflow section area. With a non-uniform inlet, a larger area ratio is required.

The required covered length is larger with a given opening angle, due to the much faster velocity decrease of the slower part of the flow. In the example, the velocities at inflow are 1.25 and 0.75 times the average velocity. The necessary reduced velocities are $\sqrt{(1.25)^2 - 0.5} \approx 1.03$ and $\sqrt{(0.75)^2 - 0.5} = 0.25$ times the average inflow velocity. The corresponding section area is 2.11 times the inflow section area. The conclusion is that irregularity of the inlet flow causes a reduction of the possible pressure recovery. There are almost always length limitations with turbomachines, causing an inherent limitation of the pressure recovery.

Within a real flow, the velocity on the walls equals zero. The slow flow near the walls can only participate in the diffusion process when there is momentum transfer from the core flow. This transfer is due to molecular viscosity and to turbulent mixing. Both processes are coupled with energy dissipation. In the first phase of a diffusion process, strong deceleration occurs near the walls. This results in a shear zone moving to the centre as the fluid advances through the diffuser. Turbulence produced by the shear enhances the momentum exchange. But, at the exit of a diffuser with a limited length, the core flow is nearly unaffected.

Figure 2.22 illustrates the resulting inhomogeneous velocity at the exit. The velocity has only decreased little within the flow core. Figure 2.22 thus shows that the loss mainly consists of mixing loss downstream of the diffuser. The exit velocity profile is only a little more regular than with dump diffusion (Fig. 2.20). Diffusion within a gradually widening channel so intrinsically is a process with large loss.

2.4.3 Flow Separation

Separation occurs when the divergence of a diffuser is too large. Pressure recovery within a diffuser with large separation is low and the exit flow is heavily deformed, with backflow at some places. Large-scale backflow is illustrated in Fig. 2.23 (left). It

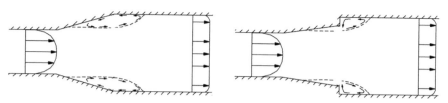

Fig. 2.23 Diffusion with separation; left: large separation zone; right: reduced separation zone by combination with sudden expansion

occurs with strong divergence and is nearly steady. The flow pattern is denominated *steady stall*. With less strong divergence, the separation is typically time dependent. Separation zones then build up repeatedly inside the diffuser and are moved out by convection. This growth and removal pattern is termed *transitory stall*. When diffusers function in transitory stall, large fluctuations in pressure recovery and exit velocity distribution occur.

Separation with strong backflow, as in Fig. 2.23 (left), causes large mixing loss downstream of the diffuser. It may then be more advantageous to combine a diffuser with a smaller opening angle with a dump diffuser, as shown in Fig. 2.23 (right).

2.4.4   Flow Improvement

Since the efficiency of a diffuser depends strongly on the uniformity of the velocity at the entry, efficient deceleration may be impossible when the entry velocity is very irregular. Figure 2.24 shows, as an example, a diffuser downstream of a bend. It is then better to postpone the deceleration in order to reduce the distortion of the velocity profile and to organise the deceleration at the high-momentum side.

The diffuser efficiency may be improved as illustrated in Fig. 2.25. Very efficient is removal of low-momentum flow near the wall by suction (Fig. 2.25 left, top wall) or injection of high-momentum flow in the boundary layers by tangential slots (Fig. 2.25

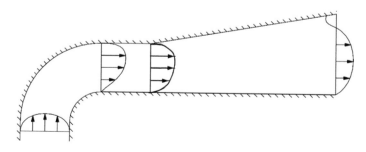

Fig. 2.24 Partial equilibration and asymmetric deceleration after a bend

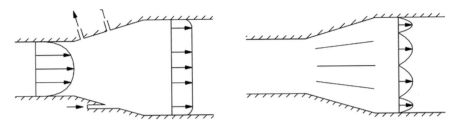

Fig. 2.25 Removal of low-momentum fluid (left; top wall); injection of high momentum fluid (left; bottom wall); application of guiding surfaces (right)

left, bottom wall). These techniques are applied with wind tunnel diffusers or when fluid is drained for secondary purposes. A simple method, but much less efficient, is mounting guiding surfaces in a diffuser with a large opening angle (see next section for dimensioning and see Exercise 2.5.10).

2.4.5 Representation of Diffuser Performance

Geometrical characteristics are shown in Fig. 2.26. These are the inlet and outlet areas A_1 and A_2, the length L and the hydraulic radius R of the inlet section (R_1 for a circle; H_1 for a rectangle with high ratio of width to height). A common way to represent pressure recovery is shown in Fig. 2.27 [7]. Lines of constant pressure recovery are plotted in a diagram with diffuser length to inlet radius ratio in the abscissa and the area ratio in the ordinate.

Pressure recovery is expressed with a coefficient C_p defined by

$$C_p = \frac{p_2 - p_1}{\frac{1}{2}\rho \bar{v}_1^2}.$$

The loss coefficient with connected upstream and downstream ducts is

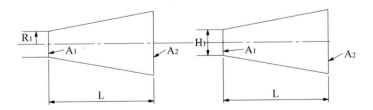

Fig. 2.26 Geometrical characteristics of diffusers

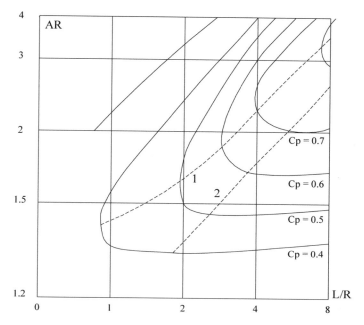

Fig. 2.27 Performance of conical diffusers with uniform inlet velocity and a free outlet; adapted from Miller [7]

$$\xi = \frac{p_{01} - p_{02}}{\frac{1}{2}\rho\bar{v}_1^2} = 1 - \left(\frac{\bar{v}_2}{\bar{v}_1}\right)^2 - C_p = C_{p,id} - C_p,$$

where $C_{p,id}$ is the ideal pressure recovery coefficient:

$$C_{p,id} = 1 - \left(\frac{A_1}{A_2}\right)^2.$$

For a diffuser with a free outlet, outlet kinetic energy is a loss as well, and so

$$\xi = \frac{p_{01} - p_2}{\frac{1}{2}\rho\bar{v}_1^2} = \frac{p_1 + \frac{1}{2}\rho\bar{v}_1^2 - p_2}{\frac{1}{2}\rho\bar{v}_1^2} = 1 - C_p.$$

The loss coefficient of a diffuser with a free outlet may be described again as $C_{p,id} - C_p$, with the ideal pressure recovery coefficient being unity.

The dashed lines 1 and 2 in the diagram of Fig. 2.27 separate different flow regimes. At the right of line 2, where the pressure recovery lines are nearly horizontal, the flow is stable. With a constant area ratio, pressure recovery decreases with larger length due to increase of the friction surface. Diffusers in this zone are unnecessarily long.

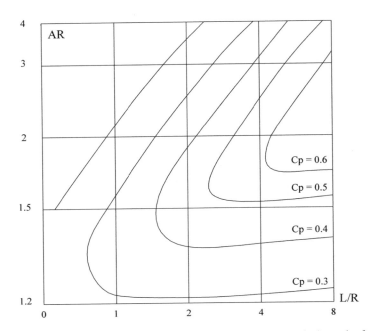

Fig. 2.28 Performance of conical diffusers with fully developed inlet velocity and a free outlet; adapted from Miller [7]

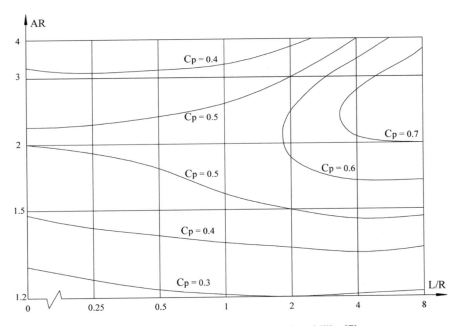

Fig. 2.29 Performance of conical diffusers in ducts; adapted from Miller [7]

Thus, for given area ratio, the most efficient diffusers with stable flow are just on line 2. Steady stall occurs above line 1. For a given length, pressure recovery decreases with increasing area ratio. So, it is useless to operate a diffuser in this zone. In the zone between the two lines, transitory stall occurs. Many diffusers in practice have geometries in this region, near to line 2.

Line 1 in Fig. 2.27 is approximately the area ratio with minimum total pressure loss for a given length ratio (vertical tangents to the contour lines of C_p). Line 2 is approximately the length ratio with minimum total pressure loss for a given area ratio (horizontal tangents to the contour lines of C_p). So, mostly, the lines separating the flow regimes are not drawn in a diffuser diagram and the user is supposed to know about the flow regimes. In turbomachinery applications, there are always length limitations. Therefore, in principle, the most efficient operation is for the area ratio corresponding to the minimum total pressure loss at a given length ratio, thus on line 1. But this operation implies heavy unsteady stall or large-scale steady stall. So, for limiting flow oscillations or pressure jumps, it is better to choose the geometry in the middle of the zone where the contour lines of pressure recovery change from horizontal to vertical direction.

Figure 2.27 shows the pressure recovery with a uniform inflow. The corresponding diagram for a fully developed inflow is shown in Fig. 2.28 [7]. Both diagrams are valid for Re = 10^6. A minor correction for the change of the Reynolds number is required. The performance is better with uniform incoming flow. E.g., with $L/R_1 = 2$ and $AR = 1.5$, the C_p with uniform flow at inlet is about 0.50, while it is about 0.45 with fully developed flow.

Figure 2.29 shows the performance of conical diffusers connected to circular ducts at inlet and outlet, with fully developed inlet flow. For large length, the contour lines are similar to these of the previous diagrams. For small dimensionless length, the performance is that of a sudden expansion.

The mixing downstream of a diffuser in a constant section pipe results in a static pressure increase, followed by a decrease due to friction. The length of the duct required to obtain maximum static pressure depends on the area ratio and the diffuser angle but is typically about 4 diameters. After the maximum pressure position, the friction factor remains somewhat higher than that of a fully developed flow during 20–50 diameters. The loss coefficient for this additional loss is about 0.1. The pressure recovery shown in Fig. 2.29 is the pressure recovery at the maximum pressure position.

Very often, the diffuser length available for a desired velocity reduction, thus an imposed area ratio, is too small for avoiding separated flow. Either a diffuser followed by a sudden expansion (Fig. 2.23) or a diffuser with guiding surfaces (Fig. 2.25) may then be applied. The number of guiding surfaces should be determined so that the L/R ratio becomes sufficiently large for each partial diffuser.

2.4.6 *Deceleration in a Bend*

All phenomena in bends and straight diffusers come together in a still more extreme form in a bent diffuser. By adding a bend immediately downstream of a diffuser, the adverse pressure gradient in the outward part of the bend entrance increases (see Fig. 2.18) and there is high risk for separation. Addition of a diffuser downstream of a bend steepens the pressure gradient in the inward part of the exit of the bend (see Fig. 2.18) and makes separation very likely. These examples make clear that diffusion within a bend is delicate. The phenomena discussed before, such as secondary flow and turbulence amplification or damping, strongly intervene. However, a diffuser in which bend phenomena and flow deceleration are well matched allows a pressure recovery that is far superior to that of a serially connected bend and a straight diffuser.

The crucial aspect is useful application of secondary flow in order to bring high-momentum fluid to the places with imminent separation. The optimum geometry strongly depends on the inflow conditions. The consequence is that there is no universal optimum geometry. For instance, bent diffusers are optimised for vertical shaft hydraulic turbines. Figure 2.30 shows an example (Francis turbine). A pressure recovery up to 75% may be attained. The optimisation requires computational techniques and turbulence models. At present, optimisation of stationary diffusers is well possible. It is more difficult for rotating decelerating channels (radial pumps and compressors), as complex secondary flow patterns intervene. Notice that a large pressure recovery requires a sufficient covered length (Fig. 2.30).

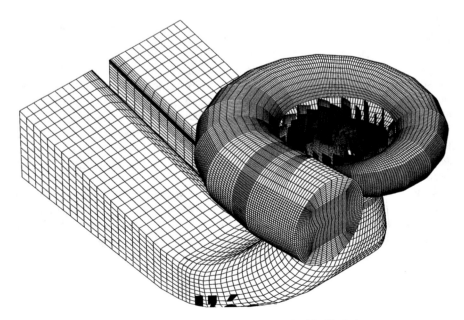

Fig. 2.30 Optimised diffuser of a Francis-turbine (*Courtesy* Andritz Hydro)

Most radial pumps, fans and compressors have limited rotor channel lengths. Hence, the earlier mentioned limit of velocity deceleration ratio w_2/w_1 of about 0.8–0.9 in order to avoid separation. In some radial compressors, rotor channels with larger length are applied, by installing an inducer (see Chap. 14).

Figure 2.24 is way of combining a bend with a diffuser without applying means of optimisation. It is advisable to place the bend first, and then a duct segment with constant section (length about one diameter) in which mixing occurs by secondary flow, followed by an asymmetric diffuser.

2.5 Exercises

2.5.1 The figure sketches the velocity profile in a laminar flow between a moving block and a stationary flat wall. The block moves parallel to the wall at velocity v. Assume that pressure is constant in the space between the block and the wall. Derive from the momentum equation (2.2) that the shear stress τ within the shear layer is constant and that the velocity profile is linear. Demonstrate that the dissipated work per surface unit and per time unit equals the displacement work $(v.\tau)$ exerted by the object on the shear flow. Demonstrate that this result remains valid with turbulent flow, but that then the velocity profile is not linear.

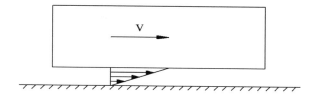

2.5.2 The left figure sketches the velocity profile in a 2D-channel with fully developed laminar flow of a constant-density fluid (ρ = constant, μ = constant).

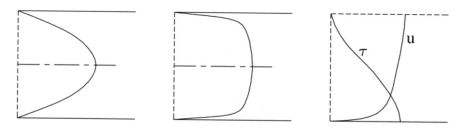

Fully developed flow means that the velocity profile is the same at all streamwise positions. Derive that the shear stress varies linearly from the wall value τ_w to zero in the channel centre. Demonstrate that the velocity profile is a parabola. Derive that the energy dissipated per wall surface unit and per time unit equals $U_m \tau_w$, with U_m being the average velocity in the channel. Note that this result only requires that the shear stress varies linearly across the channel. So argue that the result is the same with a turbulent profile (central figure) ($\mu_t \neq$ constant).

The right-hand figure sketches the velocity and shear stress profiles with turbulent boundary layer flow on a flat plate under a zero pressure gradient. The boundary layer grows. So the flow is not fully developed. The shear stress profile differs little from a linear one. So, the former result is still valid with a good approximation.

2.5.3 The figure sketches the mixing of two flows of a constant-density fluid with velocities $3/2v$ and $1/2v$ within a 2D-channel. Both flows occupy half the section. The velocity therefore is v after complete mixing. Determine the pressure increase by complete mixing. Determine the energy dissipation by mixing per mass flow rate unit, ignoring friction.

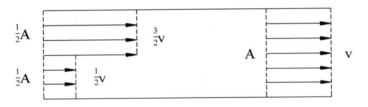

A: $\frac{\Delta p}{\rho} = \frac{1}{2}\frac{v^2}{2}$; $q_{irr} = \frac{1}{4}\frac{v^2}{2}$.

2.5.4 The figure sketches a flow of a constant-density fluid with sudden expansion (dump diffusion). Determine the pressure increase and the energy dissipation per mass flow rate unit. Assume that the pressure at the position of the expansion is the pressure of the oncoming flow in the whole of the expanded section.

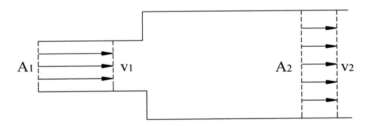

A: $\frac{\Delta p}{\rho} = v_1 v_2 - v_2^2 = 2\frac{A_1}{A_2}\left(1 - \frac{A_1}{A_2}\right)\frac{v_1^2}{2}$, $q_{irr} = \frac{1}{2}(v_1 - v_2)^2 = \left(1 - \frac{A_1}{A_2}\right)^2\frac{v_1^2}{2}$.

2.5.5 The figure is a sketch of a jet pump. Water is injected from an external source at velocity v_1 into a tube, driving water from the suction pipe to velocity v_2 at the injector exit position. Determine the pump characteristic as the mechanical energy increase of the driven flow $\Delta E_{m2} = (p_{03} - p_{02})/\rho$ as a function of its mass flow rate for given values of A_1, A and v_1. Define surface ratio $\alpha = A_1/A$ and velocity ratio $U = v_2/v_1$. Take $\alpha = 0.4$ as an example. Define a pressure coefficient as

$\psi = \Delta E_{m2}/(v_1^2/2)$ and consider U as a mass flow rate measure. Observe that the characteristic is decreasing ψ as a function of U.

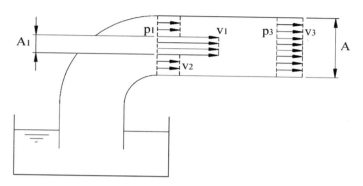

Study the energy exchange. An expression for the mechanical energy increase in the driven flow has already been formulated. Determine the mechanical energy decrease in the driving jet flow $(p_{01} - p_{03})/\rho$. In most applications, the driving water is tapped from the flow in the discharge pipe. Driving water is then produced with a pump increasing the total pressure p_{03} to the total pressure p_{01}. With an ideal pump, the (mechanical) energy increase required exactly equals the energy decrease in the driving jet between the injection position (1) and the jet pump exit (3). The efficiency η of the jet pump may be defined as the ratio of the product of the energy increase (J/kg) and the mass flow rate (kg/s) in the driven flow to the product of energy decrease and the mass flow rate in the driving flow. This efficiency definition also applies with driving water from another origin.

Define the total pressure ratio δ as the total pressure increase in the driven flow to the total pressure decrease in the driving flow. Define the mass flow ratio μ as the mass flow rate of the driven flow to the mass flow rate of the driving flow. With these definitions, the efficiency is $\eta = \mu\delta$. Determine ψ, δ, μ and η for varying α and U. Study the values of $\alpha = 0.2, 0.4, 0.6, 0.8$ combined with the values of $U = 0, 0.25, 0.50, 0.75, 1$. Observe that the efficiency is weakly dependent on α, but strongly dependent on U.

The calculated efficiency is unity with the velocity ratio U equal to unity. This is obvious, as the only loss considered is mixing loss. But for velocity ratio equal to unity, there is no energy increase in the driven flow ($\psi = 0$). For a useful energy increase, the velocity ratio must be much lower than unity, but for an acceptable efficiency, the velocity ratio must be sufficiently high. A compromise is a velocity ratio around 0.5 (theoretical efficiency about 0.7). This implies that the velocity difference between driving and driven flows at the mixing position must not be too large. The driven jet must thus be accelerated around the injector. The acceleration is illustrated by the geometry of a real jet pump as shown in the figure below.

Mostly, the velocity of the driven jet must be limited in order to avoid cavitation. For instance, to $v_2 = 10$ m/s corresponds a pressure decrease in the suction pipe of

$\rho v_2^2/2 = 50$ kPa. Such a strong pressure decrease implies that the geometric suction height must be much lower than 5 m.

Jet pumps are therefore mostly mounted below the water surface in a well (submerged). A diffuser downstream of the mixing chamber is required as well. With the example, the velocity exceeds 10 m/s after mixing, but the velocity in a pipe is typically at maximum about 2.5 m/s. So the diffuser generates a substantial loss. High efficiency is thus not possible.

Observe that, for comparable efficiency, as well a small mass flow rate together with a large pressure increase and a large mass flow rate together with a small pressure increase may be chosen. The choice of the combination is determined by the surface ratio α with only a weak effect of the velocity ratio U. Since the analysis takes neither the friction loss in the mixing chamber nor the diffuser loss into account, the actual efficiency is much lower than obtained by the analysis. With the velocity ratio about 0.5, the efficiency in the analysis is about 0.70, but is it only about 0.35 in practice.

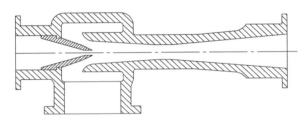

2.5.6 The figure sketches a moving linear cascade of cylindrical rods with spacing s. The oncoming flow has velocity v_1. The speed of the rods is u. The ratio of the blade speed to the oncoming flow velocity is termed the *speed ratio*, being here $\lambda = u/v_1 = 8$. The solidity of the cascade is $\sigma = d/s$, where d is the diameter of the rods. Take the solidity $\sigma = 1/8$ and the drag coefficient $C_D = 1$. Determine the flow velocity downstream of the rods. Assume that the flow velocity at the position of the cascade is the arithmetic average of the velocities far upstream and far downstream (for justification, see Chap. 10, Sects. 10.3.1 and 10.3.2).

With cascade analysis it is customary to express the flow velocity components at the cascade position relative to the oncoming flow velocity and the blade speed. The axial and tangential flow velocity components with a driven cascade are represented by $w_a = v_1(1 + a)$, $w_u = -u(1 - b)$. The values at large distance downstream of the cascade are then $w_{2a} = v_1(1 + 2a)$, $w_{2u} = -u(1 - 2b)$. The factors a and b are called *interference factors*. With a linear cascade, as sketched in the figure, $a = 0$.

Determine the work done on the fluid, the mechanical energy increase in the flow and the energy dissipated. Observe that there is increase of mechanical energy, which proves that energy transfer to a flow is possible by drag forces. The efficiency is very low, however.

A: $b = 0.173$, $\Delta W/u^2 = 2b = 0.346$, $\Delta E_m/u^2 = 0.053$, $q_{irr}/u^2 = 0.293$.

2.5.7 The figure is a sketch of an annular streamtube with an infinitesimal height through the rotor of an axial fan without stator parts. The rotor blade speed in the section of the streamtube is u. As speed ratio we choose $\lambda = u/v_a = 3$, being a typical value for a half radius section (tip value $\lambda_T = 6$). Applying an interference factor, as in the previous exercise, we set $w_{2u} = -u(1 - 2b)$. The solidity of the cascade is $\sigma = c/s = 2/3$. Take $C_L = 1$ as lift coefficient and ignore drag, so $C_D = 0$.

Determine the work done on the fluid. Determine the static pressure increase through the rotor and the degree of reaction. Observe that the efficiency of the work transfer is unity due to the absence of drag (compare to the previous exercise). Remark that by adding a downstream stator, the kinetic energy associated to the tangential velocity component can be converted into static pressure increase.

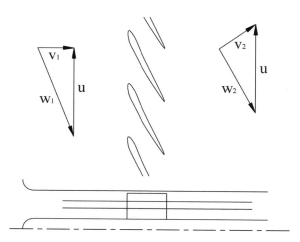

A: $b = 0.152$, $\Delta W/u^2 = 2b = 0.304$, $\Delta(p/\rho)/u^2 = 2b - 2b^2 = 0.258$, $\Delta(v^2/2)/u^2 = 2b^2 = 0.046$, $R = 0.848$.

2.5.8 Estimate the tangential force coefficients (Zweifel) of the steam turbine cascades shown in Fig. 1.11 of Chap. 1.

A: $R = 0$: stator $C_{Fu} \approx 0.50$ (low load), rotor $C_{Fu} \approx 0.90$ (moderate load); $R = 0.50$: stator and rotor: $C_{Fu} \approx 0.50$ (low load).

2.5.9 Estimate the local diffusion factor of the compressor cascade with velocity distribution on the right-hand side of Fig. 2.16. Observe that the deceleration on the suction side is the maximum possible one for avoiding flow separation.

A: $D_{loc} \approx 0.50$.

2.5.10 A cylindrical tube should undergo a diameter increase with a factor 2. For this increase, there is only a length of 1 inlet diameter available. Design a diffuser with the Fig. 2.29 diagram. What is the realisable pressure recovery?

A: The ideal pressure recovery coefficient is $C_{p,id} = 0.9375$. A conical diffuser between the two sections, with area ratio 4 and length to radius ratio 2, realises the very low recovery value $C_p = 0.375$, which is the same value as with a sudden expansion. A simple solution is a conical diffuser with an optimum opening angle (area ratio 2; $C_p = 0.615$), followed by a sudden expansion (area ratio 2). The loss coefficient of the conical diffuser is $C_{p,id} - C_p = 0.75 - 0.615 = 0.135$. The loss coefficient of the sudden expansion is 0.25 according to Eq. 2.31. In reality, it is somewhat higher due to inflow irregularity. The resulting loss coefficient based on inlet kinetic energy is $0.135 + 0.25/4 = 0.1975$. Thus: $C_p = 0.740$, and, in reality, probably about 0.70. Mounting guiding surfaces is not a practical solution with a very large area ratio. The area ratio does not change with guiding surfaces. By subdividing into 4 flow parts (3 surfaces), the equivalent inlet radius halves. This brings L/R onto 4, but the corresponding $C_p = 0.5$ is still not good.

References

1. Abott IH, Von Doenhoff AE (1959) Theory of wing sections. Dover Publications. ISBN 486-60586-8 (NACA-aerofoil classification)
2. Knill TJ (2005) The application of aeroelastic analysis output load distributions to finite element models of wind turbines. Wind Eng 29:153–168
3. Pfleiderer C (1961) Die Kreiselpumpen für Flüssigkeiten und Gase, 5th edn. Springer, No ISBN (cascade effect on aerofoil characteristics)
4. Dixon SL, Hall CA (2014) Fluid mechanics and thermodynamics of turbomachinery, 7th edn. Elsevier. ISBN 978-0-12-415954-9 (cascade theory)
5. Japikse D, Baines N (1994) Introduction to turbomachinery. Oxford University Press. ISBN 0-933283-06-7 (cascade theory)
6. Haaland S (1983) Simple and explicit formulas for the friction factor in turbulent flow. J Fluids Eng 103:89–90
7. Miller DS (1990) Internal flow systems. Gulf Publishing Company. ISBN 0-87201-020-1 (ducts and diffusers)

Chapter 3
Fans

Abstract Fans are machines that move air or a similar gas. With regard to types, there are axial, radial and mixed-flow fans. Principles for the analysis of axial machines have been elaborated in the foregoing chapters, but for radial machines, only a limited theory has been developed. The fan chapter is therefore also used to complete the theory for power receiving radial machines. It is appropriate to do this with fans. First, a fan is a simple turbomachine. Further, there are various rotor shapes with radial fans (forward curved blades, radial-end blades and backward curved blades), as opposed to radial pumps (only backward curved blades) and radial compressors (radial-end and backward curved blades). In this chapter, we also discuss the performance evaluation of radial fans, rotor design choices with radial fans and performance of axial and mixed-flow fans.

3.1 Fan Applications and Fan Types

3.1.1 Fan Applications

A fan moves air or a similar gas, producing a small pressure rise. A general criterion is that the total pressure ratio does not exceed 1.25 ($p_{02}/p_{01} < 1.25$). The fluid density then only changes a little. The fluid may be considered, with a good approximation, with constant density equal to the arithmetic average of the entry and exit values (a justification for the application of the average density is given in Chap. 4: compressible fluids). With total pressure ratios exceeding 2 ($p_{02}/p_{01} > 2$), the term *compressor* is used. It denominates a machine intended to increase the pressure of a compressible fluid. The term *blower* is typically used for intermediate machines ($1.25 < p_{02}/p_{01} < 2$), but means also a machine that blows air with rather large velocity (snow blower, leaf blower, hot air blower). So, the term blower may refer to a fan with a high work input or a compressor with a low work input, mainly producing kinetic energy. There is no specific theory for blowers. Depending on the case, a blower is analysed as a fan (constant density) or as a compressor (variable density).

E. Dick, *Fundamentals of Turbomachines*, Fluid Mechanics and Its Applications 130,
https://doi.org/10.1007/978-3-030-93578-8_3

Fans are primarily applied to exhaust polluted air from industrial spaces, office buildings and homes. Fresh air supply, possibly through an air conditioning system, is an analogous application. These fans exist in strongly different sizes.

Extreme examples are fans for ventilation of mines, with flow rates and pressure rises as large as 500 m^3/s and 5000 Pa (two-stage axial fans or double-suction centrifugal fans). Most applications are more modest, with flow rates up to 5 m^3/s and pressure rises up to 2000 Pa, i.e. power up to 10 kW (radial fans). Low-power applications are also found in this sector, as cooker hood fans or fans in cars. Fans for supply of combustion air or exhaust of combustion gas constitute a second group. There are various sizes, from large steam boiler fans to small fans in home heating systems. Fans for cooling by air, mainly fans in electronic devices and heat exchangers, constitute a third group. Some applications have a very low power, e.g. with a PC fan, it is typically 1 W. All fan types occur. High-power examples are fans for cooling air supply to surface condensers (axial fans) in electric power plants. The above list is not complete. We mention fans in vacuum cleaners and hear dryers.

3.1.2 Large Radial Fans

From about a 200 mm rotor diameter onwards, the entire machine is manufactured from steel plate, with most parts welded. Figure 3.1 shows an example. The main components are a volute and a rotor with backward curved or forward curved blades. The rotor may directly be driven by a motor, which is the simplest construction form. Figure 3.1 shows a belt-driven type. The fan itself has bearings in that case.

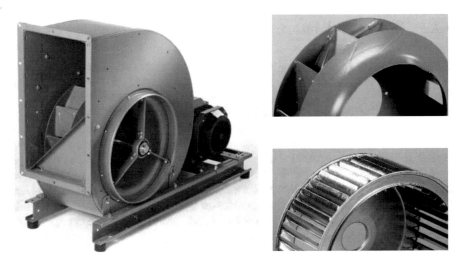

Fig. 3.1 Large radial fan with rotors with backward curved blades (top right) or forward curved blades (bottom right) (*Courtesy* FläktGroup)

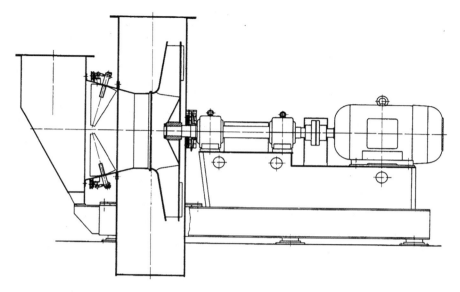

Fig. 3.2 Radial fan with inlet guide ring and directly driven shaft with bearings (*Courtesy* TLT-Turbo)

The fan shown in Fig. 3.1 is left-turning, but also a right-turning version is made. Rotors with backward curved blades and forward curved blades fit in the same volute. Figure 3.1, right top, is a left-turning rotor with backward curved blades and Fig. 3.1, right bottom, is a right-turning rotor with forward curved blades. We recall (Chap. 1) that backward curved blades are mainly intended for rising pressure, whereas forward curved blades aim at generation of kinetic energy. Larger machines sometimes have variable inlet guide vanes in order to realise a comparable pressure rise at a lower flow rate (see Sect. 3.8), as is the case with the fan in Fig. 3.2, which is a type with separate bearings.

3.1.3 Small Radial Fans

Rotor and stator manufacturing from welded steel plate is difficult with a rotor diameter under 200 mm. Rotors are sometimes made from punched steel plate, with parts joined by tabs and slots, but more common is mould injected plastic. The volute may be manufactured from aluminium by casting, from plastic by injection moulding or from steel plate. The rotor is generally mounted directly on the driving motor, often with the motor inside the hub (see Sect. 3.6).

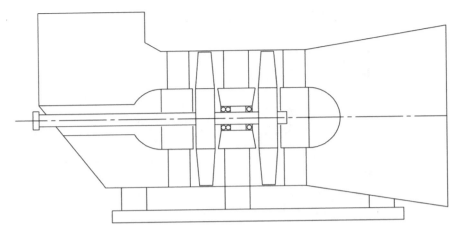

Fig. 3.3 Two-stage axial fan; flow from left to right; the normal build is struts with axial inflow and outflow, followed by two stages rotor–stator

3.1.4   Large Axial Fans

Large axial fans typically have welded hubs (sometimes forged). The rotor blades have aerofoil shaped profiles and are mostly cast (aluminium).

The normal configuration is a rotor with a downstream stator. The number of rotor blades depends on the pressure rise to be realised and may vary between 2 for a fan with a low pressure rise to the order of 20 for a fan with a high pressure rise. A fan with high pressure rise is normally also one with low flow rate and thus has a high ratio of hub diameter to tip diameter. A low pressure rise fan typically has a low diameter ratio. Figure 3.3 is a sketch of a meridional section of a two-stage axial fan. With some fans, the rotor blade angle may be changed at standstill, individually or collectively by means of a collar with lever arms (called adjustable blades) or during operation (called variable blades) for changing the flow rate and the pressure rise (see Sect. 3.8).

3.1.5   Small Axial Fans

Stator vanes are mostly omitted with small axial fans. The flow then contains post-whirl impairing the efficiency. The rotor is commonly mounted on the driving motor, often with the motor in the hub. The rotor is mostly manufactured from plastic by injection moulding.

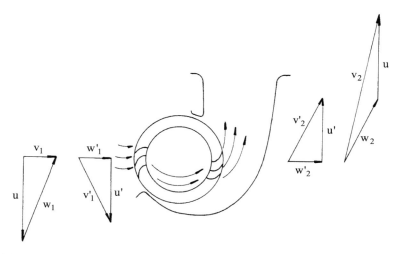

Fig. 3.4 Cross-flow fan

3.1.6 Cross-Flow Fans

Axial fans and radial fans with forward curved blades are often used for small-power applications, such as generation of an air flow with a small pressure rise. But, there is a third possible type. Figure 3.4 shows an orthogonal section of a cross-flow fan. The air passes twice through the rotor. The stator shape determines the suction and the discharge sides. The diameter ratio (inner diameter to outer diameter) is very high, analogous to a radial fan with forward curved blades.

Velocity triangles are drawn in Fig. 3.4. Blades are radial at the inner side. The rotor work and degree of reaction are comparable to that of a fan with forward curved blades. The dimensions in the axial direction are nearly unlimited with this fan type, which makes it appropriate for flow through a slot-shaped area. Applications are mostly cooling air in electric appliances (projector, copier) or a heated air jet (electric convector, hair dryer).

3.2 Idealised Mean Line Analysis of a Radial Fan

3.2.1 Idealised Flow Concept: Infinite Number of Blades

A mean line analysis is performed in a plane perpendicular to the rotor shaft, first with an infinite number of infinitely thin blades. Figure 3.5 represents a meridional section and an orthogonal section of a rotor. The term *infinite number of blades* means that the flow is supposed to follow perfectly the blade geometry, which would be

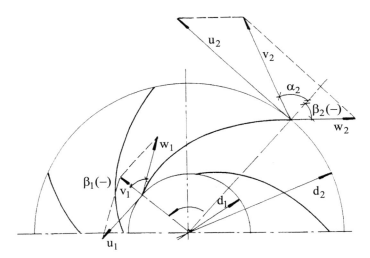

Fig. 3.5 Flow through the rotor of a radial fan; drawn for radial inflow; flow direction coinciding with blade orientation (in reality, the exit velocities w_2 and v_2 are much more tangential due to slip)

attained with an infinite number of infinitely thin blades. The relative velocity would then be tangent to the blades at each point of the flow. The flow at the rotor entry has no incidence and there is no friction of the fluid on the blades. An incidence-free entry means that the flow direction just upstream of the entry coincides with the blade orientation.

In a general case, the absolute inflow is not perpendicular to the inlet circle and has a pre-whirl velocity component v_{1u}. With most machines, no pre-whirl occurs, however. Analogous to the rotor, we assume an incidence-free entry to the volute and no friction in the volute.

Quantities related to the above defined flow get as subscripts t (theoretical flow = no losses) and ∞ (infinite number of blades). The assumptions regarding losses and flow direction coinciding with blade orientation will gradually be abandoned in further analysis steps.

3.2.2 Degree of Reaction

The degree of reaction with adiabatic flow is the ratio of the static enthalpy increase to the total enthalpy increase through the rotor. The total enthalpy increase through the rotor equals the total enthalpy increase though the machine. The static enthalpy increase is

$$h_2 - h_1 = \frac{p_2}{\rho_2} - \frac{p_1}{\rho_1} + e_2 - e_1.$$

The reversible part of this enthalpy increase is $\int \left(\frac{1}{\rho}\right) dp$.
This follows from the entropy definition:

$$T ds = dh - (1/\rho) \, dp \quad \text{or} \quad dh_s = (1/\rho) \, dp.$$

For a small pressure increase with a compressible fluid, the isentropic enthalpy increase corresponding to the pressure increase is well approximated by

$$(h_2 - h_1)_s \approx \frac{p_2 - p_1}{\bar{\rho}},$$

with $\bar{\rho}$ being the average density.

A strict justification of this approximation is presented in Chap. 13. The degree of reaction may thus be approximated by

$$R_p = \frac{\Delta p_{rot}}{\Delta p_0}, \tag{3.1}$$

which is the ratio of the static pressure rise through the rotor to the total pressure rise through the machine.

The degree of reaction according to (3.1) is applied in practice to fans. There is a small difference with the general definition. This is obvious, as both values of the degree of reaction coincide for lossless flow of a constant-density fluid. The degree of reaction according to the general definition is termed *kinematic degree of reaction* because this degree of reaction may be expressed with velocity components, as demonstrated in Chap. 1 and derived once more hereafter. There is no specific term in the fan literature for the degree of reaction according to definition (3.1). We use the term *pressure degree of reaction*. By the simple term degree of reaction, we mean the *kinematic degree of reaction*.

A definition of the degree of reaction, slightly deviating from the general definition is applied with other machines as well, for instance, with steam turbines (Chap. 6). Deviating definitions are always used for practical purposes. Pressure differences required to determinate the degree of reaction according to (3.1) may be measured easily. Reliable measurement of the enthalpy differences for the general definition is impossible, because temperature differences with fans are very small. For instance, a total pressure increase of 3600 Pa corresponds to an adiabatic temperature difference of only 3 K with a density 1.20 kg/m^3 and a specific heat capacity 1000 J/kgK. This difference cannot be measured reliably because of the effect of heat transfer.

3.2.3 Relation Between Rotor Blade Shape and Performance Parameters

The kinematic degree of reaction is

$$R_{t\infty} = \frac{h_2 - h_1}{h_{02} - h_{01}} = \frac{u_2^2 - u_1^2 + w_1^2 - w_2^2}{2(u_2 v_{2u} - u_1 v_{1u})}. \tag{3.2}$$

In the absence of pre-whirl,

$$v_{1u} = 0 \quad \text{and} \quad w_1^2 = u_1^2 + v_1^2.$$

Then, (3.2) becomes

$$R_{t\infty} = \frac{u_2^2 + v_1^2 - w_2^2}{2u_2 v_{2u}}.$$

In order to obtain a simple expression for the degree of reaction, we assume that the radial component of the velocity is equal at the rotor entry and exit. This is approximately met in real rotors and is achieved by decreasing the axial rotor width with increasing diameter.

With

$$v_1 = v_{2r} \quad \text{and} \quad w_2^2 - v_{2r}^2 = (u_2 - v_{2u})^2,$$

it follows, with $\zeta = v_{2u}/u_2$, that

$$R_{t\infty} = \frac{u_2^2 - (u_2 - v_{2u})^2}{2u_2 v_{2u}} = 1 - \frac{\zeta}{2}. \tag{3.3}$$

A total pressure coefficient is defined by

$$\psi_{0,t\infty} = \frac{\Delta p_{0,t\infty}}{\rho u_2^2} = \frac{u_2 v_{2u}}{u_2^2} = \frac{v_{2u}}{u_2} = \zeta. \tag{3.4}$$

A static pressure coefficient is similarly, with $\Delta p_{rot,t\infty}$ the static pressure rise in the rotor:

$$\psi_{r,t\infty} = \frac{\Delta p_{rot,t\infty}}{\rho u_2^2} = R_{t\infty} \psi_{0,t\infty} = \zeta \left(1 - \frac{\zeta}{2} \right). \tag{3.5}$$

Expressions (3.4) and (3.5) are only correct for lossless flow.

The dimensionless coefficients for the total pressure rise in the machine and for the static pressure rise in the rotor are termed *performance parameters*.

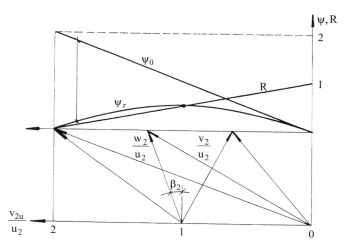

Fig. 3.6 Degree of reaction and pressure coefficients as functions of the blade angle at the rotor exit (β_2); the radial velocity component is drawn exaggerated for clarity (theoretical flow with an infinite number of blades)

Figure 3.6 shows the pressure coefficients and the degree of reaction as functions of v_{2u}/u_2. The positive sense of the abscissa is to the left in order to accord with the blade speed of a left-turning machine. Some dimensionless velocity triangles at the rotor exit are represented. The total pressure coefficient increases with increasing v_{2u}/u_2. The degree of reaction decreases. The degree of reaction is 1 for $\zeta = 0$. The total pressure rise is then zero and the rotor blades are strongly curved backward. No work is done by the rotor. A machine without any rotor work is useless, of course. So, a practical minimal value is $\zeta \approx 0.2$. The degree of reaction is 0.5 for $\zeta = 1$. Within the rotor, the increase of pressure energy then equals the increase of kinetic energy. The outflow part of the rotor blades is in the radial direction. For $\zeta = 2$, there is no static pressure rise, only kinetic energy rise, within the rotor. The blades are strongly curved forward and the total pressure rise is high. Most fan applications require some static pressure rise. Some deceleration is then required downstream of the rotor with $\zeta = 2$. But the fan efficiency decreases when kinetic energy at the rotor exit is to be converted into pressure energy. Therefore, it is advisable to realise some static pressure rise in the rotor itself. So, a practical maximum value is $\zeta \approx 1.8$.

Figure 3.7 shows schematically the blade shapes with *forward curved blades*, *radial-end blades* and *backward curved blades*. The blade orientation at rotor entry is always similar. The corresponding degrees of reaction are respectively low, 0.5 and high. So it is clear that the degree of reaction determines the rotor blade shape.

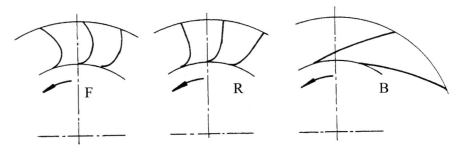

Fig. 3.7 Forward curved (F), radial-end (R) and backward curved (B) blades

3.2.4 Performance Characteristics with Idealised Flow

A first characteristic represents the total pressure rise by the fan as a function of the flow rate. Without pre-whirl, the relation resulting from Euler's equation is

$$\frac{\Delta p_{0,t\infty}}{\rho} = u_2 v_{2u} = u_2(u_2 + v_{2r} \tan \beta_2) = u_2\left(u_2 + \frac{Q}{\pi d_2 b_2} \tan \beta_2\right). \qquad (3.6)$$

The relation (3.6) represents a straight line in the $\left(\frac{1}{\rho}\Delta p_{0,t\infty}, Q\right)$-diagram (Fig. 3.8). With forward curved blades, the theoretical total pressure rise increases with the flow rate. The total pressure rise stays constant with radial-end blades and it is a decreasing function of the flow rate with backward curved blades. The theoretical power $P_{t\infty}$ follows from

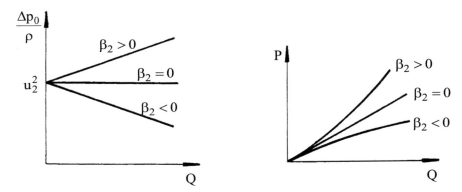

Fig. 3.8 Total pressure rise and absorbed power as functions of flow rate with forward curved ($\beta_2 > 0$), radial-end ($\beta_2 = 0$) and backward curved ($\beta_2 < 0$) blades (theoretical flow with an infinite number of blades)

$$P_{t\infty} = \Delta p_{0,t\infty} Q = \rho u_2^2 Q + \rho u_2 \frac{Q^2}{\pi d_2 b_2} \tan \beta_2. \tag{3.7}$$

The relation is linear for radial-end blades, quadratic and steeper than linear for forward curved blades, and quadratic and flatter than linear for backward curved blades (Fig. 3.8). The characteristics of real radial fans are discussed in Sects. 3.4 and 3.6.

3.3 Lossless Two-Dimensional Flow Through a Radial Rotor with a Finite Number of Blades

3.3.1 Relative Vortex in Blade Channels

We still consider the flow as occurring in a plane perpendicular to the shaft, but the effect of the spacing of the blades on the mean flow is taken into account. The flow within a blade channel of a rotor with a finite number of blades may be considered as the superposition of two flows. The first flow is the translation flow analysed until now. This flow follows the blade direction and has a uniform velocity at a given radius. The second flow is a circulation flow in the relative frame, as sketched in Fig. 3.9, showing the superposition of both flows as well.

A side forms with a higher velocity and a lower pressure (*suction side*) and another one with a lower velocity and a higher pressure (*pressure side*). The generation of suction and pressure sides has been explained in Chap. 1 as a consequence of the Coriolis force (rotation effect), but influenced by the lift force (streamline curvature). Herewith, we understand that a rotational motion is superimposed on the translational motion. Figure 3.10 (left) sketches the velocity gradient caused by the Coriolis force in an orthogonal plane. A velocity gradient is also due to the centrifugal force by streamline curvature. Figure 3.10 (right) shows the effect of curvature in the meridional plane, but the effect is similar in an orthogonal plane.

Fig. 3.9 Relative vortex motion (left) and superposition with the translational motion (right) in a blade channel of a radial rotor

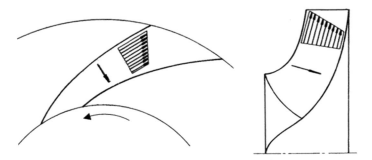

Fig. 3.10 Velocity gradient by Coriolis force (left); by centrifugal force (right)

3.3.2 Velocity Difference Across a Rotating Channel

We study the relative vortex motion by the momentum equations in the relative frame with Coriolis force and centrifugal force as intervening forces.

Figure 3.11 is a sketch of a streamline within an infinitesimal streamtube part through a radial rotor with backward curved blades. The streamline direction is indicated by x, the normal direction by y. The streamtube part is an infinitesimal control volume with dimension dx in flow direction and average dimension dy in the normal direction. The intervening body forces per mass unit are:

- centrifugal force by rotor rotation: $\overrightarrow{Cf} = \Omega^2 r \vec{1}_r$.
- Coriolis force by rotor rotation: $\overrightarrow{Co} = -2\vec{\Omega} \times \vec{w} = -2\Omega w \vec{1}_y$.
- centrifugal force by streamline curvature for an observer attached to a fluid particle (streamline radius of curvature is R): $\overrightarrow{Cu} = \frac{w^2}{R} \vec{1}_y$.

Pressure forces on the infinitesimal control volume in x and y directions are

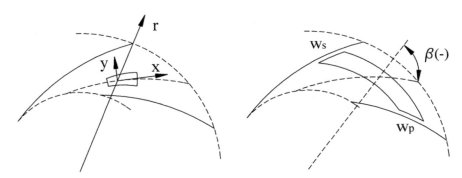

Fig. 3.11 Curved streamline and infinitesimal streamtube part (left); contour for determining the velocity difference across a blade channel (right)

$$-\frac{\partial p}{\partial x}dx\,dy\,b \quad \text{and} \quad -\frac{\partial p}{\partial y}dy\,dx\,b,$$

where b is the axial width of the rotor.

The x-oriented momentum balance for steady flow within the relative frame, for an observer attached to the rotor, is

$$\rho w\,dy\,b\,\frac{\partial w}{\partial x}dx = -\frac{\partial p}{\partial x}dx\,dy\,b + \Omega^2 r\vec{1}_r.\vec{1}_x\rho\,dx\,dy\,b.$$

The y-oriented force balance in the stagnant fluid seen by an observer attached to a fluid particle is

$$0 = -\frac{\partial p}{\partial y}dy\,dx\,b + \left(\Omega^2 r\vec{1}_r.\vec{1}_y - 2\Omega w + \frac{w^2}{R}\right)\rho\,dx\,dy\,b.$$

The changes of r with a change of x or y are given by

$$dr = \vec{1}_r.\vec{1}_x dx \quad \text{and} \quad dr = \vec{1}_r.\vec{1}_y dy.$$

The balances thus become

$$w\frac{\partial w}{\partial x} = -\frac{1}{\rho}\frac{\partial p}{\partial x} + \Omega^2 r\frac{\partial r}{\partial x}, \tag{3.8}$$

$$0 = -\frac{1}{\rho}\frac{\partial p}{\partial y} + \Omega^2 r\frac{\partial r}{\partial y} - 2\Omega w + \frac{w^2}{R}. \tag{3.9}$$

The momentum balance (3.8) is the Bernoulli equation on a streamline in the relative frame, as derived in the one-dimensional analysis in Chap. 1. A second momentum relation is now the force balance in the direction perpendicular to the flow (3.9). The Bernoulli equation may be integrated along streamlines if ρ depends only on p. This particularly applies to $\rho = $ constant. The result is ($u = \Omega\,r$):

$$\frac{w^2}{2} + \frac{p}{\rho} - \frac{u^2}{2} = \text{constant.} \tag{3.10}$$

From the velocity triangle follows: $w^2 = u^2 + v^2 - 2uv_u$. The constant is the same on all streamlines if the absolute flow at the rotor entry ($u = $ constant) is a free-vortex flow ($rv_u = $ constant) with a uniform total pressure $\left(\frac{p}{\rho} + \frac{v^2}{2} = \text{constant}\right)$. This condition is mostly met, especially in absence of inlet guide vanes. With the same integration constant in Eq. (3.10) on all streamlines, this relation implies

$$w\frac{\partial w}{\partial y} + \frac{1}{\rho}\frac{\partial p}{\partial y} - \Omega^2 r\frac{\partial r}{\partial y} = 0. \tag{3.11}$$

The same result is obtained for an isentropic flow of a compressible fluid (no entropy creation), the entropy definition being $T\nabla s = \nabla h - (1/\rho)\nabla p$. For constant entropy on a streamline, (3.8) may be integrated to

$$\frac{1}{2}w^2 + h - \frac{1}{2}u^2 = \text{constant.}$$

Under analogous conditions as for an incompressible fluid, constant total enthalpy $\left(h_0 = h + \frac{1}{2}v^2\right)$ and free-vortex flow ($rv_u = \text{constant}$) at the rotor entry, the integration constant is the same on all streamlines, which implies constant entropy within the entire flow inside the rotor (homentropic flow). So, Eq. (3.11) applies. Combination of Eqs. (3.9) and (3.11) results in

$$w\frac{\partial w}{\partial y} - 2\Omega w + \frac{w^2}{R} = 0 \quad \text{or} \quad \frac{\partial w}{\partial y} = 2\Omega - \frac{w}{R}. \tag{3.12}$$

Equation (3.12) demonstrates that the transversal velocity variation mainly originates from the Coriolis force, but that it decreases due to backward curving of the streamlines ($R > 0$).

The significance of (3.12) follows, as in Chap. 2, by calculating the circulation on the infinitesimal contour formed by the edges of the infinitesimal streamtube part of Fig. 3.11 (left) and by applying the Stokes circulation theorem. On the infinitesimal contour, run in clockwise sense (negative sense according to the z-axis), this gives, with ω the value of the rotor of the relative velocity vector in the z-direction and $d\theta$ the slope angle variation of the infinitesimal streamline part:

$$\left(w + \frac{1}{2}\frac{\partial w}{\partial y}dy\right)\left(R + \frac{1}{2}dy\right)d\theta - \left(w - \frac{1}{2}\frac{\partial w}{\partial y}dy\right)\left(R - \frac{1}{2}dy\right)d\theta = -\omega R d\theta dy.$$

Thus:

$$\frac{\partial w}{\partial y}dyRd\theta + wdyd\theta = -\omega Rd\theta dy \quad \text{or} \quad \frac{\partial w}{\partial y} + \frac{w}{R} = -\omega. \tag{3.13}$$

With Eq. (3.13), we understand that Eq. (3.12) means that the rotor of the relative velocity vector is equal to -2Ω everywhere in the flow. Since the angular speed of a fluid particle is equal to half of the rotor of the velocity vector (see Fluid Mechanics), this means that the relative vortex motion is such that the angular speed of a fluid particle seen in the relative frame is exactly equal in magnitude to the angular speed of the machine rotor. The sense of rotation of the fluid particle is opposite to the rotation sense of the machine rotor, so that the fluid particle rotation is zero in the absolute frame. This means that the relative vortex motion is an inertia reaction to the rotation of the machine rotor.

With the knowledge of the rotor of the relative velocity, the circulation may be calculated along a contour as sketched in Fig. 3.11 (right). This contour has two parts at constant radius with infinitesimal distance between each other. Calculation of the circulation allows determining the velocity difference between the suction and pressure sides of a blade channel at constant radius. We choose the clockwise sense on the contour. The corresponding rotation sense of the fluid particles in the relative frame is positive. The result is

$$w_s\left(\frac{dr}{\cos\beta}\right) - w_p\left(\frac{dr}{\cos\beta}\right) - (r+dr)\Delta\theta\, w_u(r+dr) + r\Delta\theta\, w_u(r) = 2\Omega r\,\Delta\theta dr.$$

In the expression above, w_u has to be seen as an average over the circular segment with angle variation $\Delta\theta$. Thus:

$$w_s - w_p = \Delta\theta \cos\beta\left[2\Omega r + \frac{d}{dr}(rw_u)\right]. \tag{3.14}$$

Equation (3.14) shows quantitatively that the local velocity difference across a blade channel has two causes: the Coriolis force and the lift force.

3.3.3 Pressure Difference Across a Rotating Channel

We assume that the Bernoulli constant (3.10) is the same on all streamlines.
    The pressure difference across the blade channel on a segment with constant radius (u = constant) is then

$$(p_p - p_s) = \frac{1}{2}\rho(w_s^2 - w_p^2) = \rho(w_s - w_p)w_{mean}. \tag{3.15}$$

The velocity w_{mean} is the average of the velocities at the suction and pressure sides and may be interpreted as an average across the blade channel so that the mass flow rate in the channel (Z is the number of blades) is

$$\dot{m}_{channel} = \frac{\dot{m}}{Z} = \rho w_{mean} r\,\Delta\theta\, b \cos\beta. \tag{3.16}$$

Combining (3.14), (3.15) and (3.16) results in

$$(p_p - p_s)r\,b = \Delta p\,r\,b = (w_s - w_p)\rho w_{mean}\,r\,b = \dot{m}_{channel}\left[2\Omega r + \frac{d}{dr}(rw_u)\right];$$

$$\Delta p\,r\,b = \frac{\dot{m}}{Z}\left[\Omega\frac{d}{dr}(r^2) + \frac{d}{dr}(rw_u)\right] = \frac{\dot{m}}{Z}\left[\frac{d}{dr}(ru) + \frac{d}{dr}(rw_u)\right]. \tag{3.17}$$

Integration of this expression results in the torque transferred by the rotor:

$$M = Z \int \Delta p \, r \, b \, dr = \dot{m}[r_2 u_2 - r_1 u_1 + r_2 w_{2u} - r_1 w_{1u}]. \qquad (3.18)$$

We recover the result already obtained in Chap. 1, that the rotor torque has contributions from the Coriolis force and from the lift force. Remark that Eq. (3.17) can be obtained directly by a moment of momentum balance in the absolute frame on a control volume formed by the contour shown in Fig. 3.11 (right) and that the expression (3.14) for the velocity difference can be obtained from it with the Bernoulli equation in the relative frame. So, Eq. (3.14) may be derived without explicit knowledge of the value of the rotor of the relative velocity.

3.3.4 Slip: Reduction of Rotor Work

Due to the relative vortex motion (Fig. 3.9), the average outflow of a blade channel deviates from the blade direction in the sense opposite to the rotation. This phenomenon is termed *slip*. Figure 3.12 shows the velocity triangle at the rotor exit with slip, compared to the velocity triangle without slip. Slip reduces the rotor work.

The effect of slip may be estimated in two ways, depending on whether the rotor can be considered as composed of channels (channel flow analysis according to Stodola) or rather as composed of individual blades with little overlap (blade loading analysis according to Pfleiderer).

The derivations are done hereafter with infinitely thin blades. Adaptation of the formulae for finite blade thickness is done later (see the later Eq. 3.35).

With the flow pattern of Fig. 3.9, Stodola (1924) approximates the average tangential slip velocity by half of the magnitude of the maximum velocity in the relative vortex motion at the rotor exit:

$$\delta v_u = \delta w_u = \frac{(w_s - w_{mean})_2}{2} = \frac{(w_{mean} - w_p)_2}{2} = \frac{(w_s - w_p)_2}{4}. \qquad (3.19)$$

By w_{mean} is meant the average of the velocities on the suction and pressure sides, which thus is w_2 in the mean line flow representation. Stodola further assumes that the contribution of the lift to the velocity difference in (3.14) may be ignored at the rotor exit. This means an assumption of constant angular momentum in the outflow. Then:

Fig. 3.12 Effect of slip on the velocity triangle at the rotor exit: full line with slip, dashed line without slip; the radial velocity is drawn exaggerated for clarity

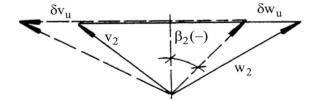

$$\delta v_u = \frac{1}{2} \Delta \theta \, \Omega \, r_2 \cos \beta_2 = \frac{\pi}{Z} \Omega \, r_2 \cos \beta_2 = \frac{\pi}{Z} u_2 \cos \beta_2.$$

From the velocity triangle of Fig. 3.12 it follows that

$$v_{2u} = u_2 + v_{2r} \tan \beta_2 - \delta v_u = \left(1 - \frac{\pi}{Z} \cos \beta_2 \right) u_2 + v_{2r} \tan \beta_2,$$

or

$$v_{2u} = \sigma \, u_2 + v_{2r} \tan \beta_2. \tag{3.20}$$

It appears as if not u_2 is transferred by the rotor motion, but a lower velocity, denoted by $\sigma \, u_2$, with σ being the *slip factor*:

$$\sigma = 1 - \frac{\pi}{Z} \cos \beta_2.$$

With constant σ, the characteristic $\Delta p_0 / \rho$ as a function of the flow rate is parallel to the theoretical characteristic with an infinite number of blades, shifted by $(1 - \sigma) u_2^2$. In reality the slip factor is not completely constant, but varies somewhat with varying flow rate.

Stodola's reasoning holds well when blade passages clearly form channels. This requirement is often marginally satisfied by pumps, but fans usually do not satisfy it. A formula by Wiesner (1967) is mostly employed with pumps. It is a corrected version of Stodola's formula for effects of curvature, friction and partial loss of blade overlap, with use of theoretical and experimental results.

Wiesner's formula is

$$\sigma = \left(1 - \frac{\sqrt{\cos(\beta_2^b)}}{Z^{0.7}}\right)\left(1 - \left[\frac{\max\left[(r_1/r_2) - a, 0\right]}{1 - a}\right]^3\right), \tag{3.21}$$

with $a = \exp(-8.16 \frac{\cos(\beta_2^b)}{Z})$, where β_2^b is the material blade angle.

From now on, we distinguish between the blade angle β_2^b and the flow angle β_2. The second factor in Eq. (3.21) describes the overlap effect. For $r_1/r_2 < a$, the factor is unity, which is commonly met by pumps. For example, $\beta_2^b = -60°$, $Z = 6$ makes $a = (r_1/r_2)_{lim} \approx 0.5$. Typical is $r_1/r_2 \approx 0.4$.

The expression of the velocity difference (3.14) is not well justified for small blade overlap. The reasoning presumes the presence of channels, so large enough solidity. The condition is noticed as the correction factor in Eq. (3.21) intervenes for a high value of r_1/r_2 and a low value of Z. Some fans with backward curved blades have only few blades and a high radius ratio (r_1/r_2), so that overlap of blades even does not exist.

In case of small overlap, Pfleiderer (1924) proposes to keep the average slip velocity estimated to one fourth of the difference between the suction side and the pressure side velocities at the rotor exit, but to calculate the velocity difference from the pressure difference on the blade, using the Bernoulli relation (3.15).

The following derivation is an adaptation of Pfleiderer's original reasoning by Eck [4], specifically for fans (the definition of the static moment differs). The average pressure difference on the blade is determined from the rotor torque by

$$M = Z \int_1^2 \Delta p \, b r \, dr = Z \, \overline{\Delta p} \int_1^2 b r \, dr = Z \, \overline{\Delta p} \, M_{st}, \quad \text{with} \quad M_{st} = \int_1^2 b r \, dr,$$

$$(3.22)$$

where M_{st} is the static moment of the meridional section around the rotation axis and b is the local width in axial direction. The torque also follows from the moment of momentum by

$$M = \dot{m}(r_2 v_{2u} - r_1 v_{1u}) = \dot{m} \frac{\Delta W}{\Omega} = \rho v_{2r} 2\pi \, r_2 b_2 \frac{\Delta W}{\Omega}.$$

The pressure difference between the pressure and suction sides of a blade follows from the Bernoulli equation and is given by Eq. (3.15). So, the velocity difference at the trailing edge between suction and pressure sides may be estimated, using the average pressure difference, by

$$\overline{\Delta p} = \rho w_2 (w_s - w_p)_2 = \frac{M}{Z \, M_{st}} = \frac{\rho v_{2r} 2\pi \, r_2 b_2}{Z \, M_{st}} \frac{\Delta W}{\Omega}, \qquad (3.23)$$

from which follows $(w_s - w_p)_2$.

The estimated slip velocity is

$$\delta v_u = \frac{1}{4}(w_s - w_p)_2 = \frac{1}{2} \frac{v_{2r}}{w_2} \frac{\pi \, r_2 b_2}{Z \, M_{st}} \frac{\Delta W}{\Omega} = \frac{\cos \beta_2}{2} \frac{\pi \, r_2^2 b_2}{Z \, M_{st}} \frac{\Delta W}{u_2}.$$

The rotor work is

$$\Delta W_t = u_2 v_{2u} - u_1 v_{1u} = \Delta W_{t\infty} - u_2 \delta v_u.$$

With Pfleiderer's reasoning, there is a relation between rotor work with and without slip according to

$$\Delta W_t = \frac{\Delta W_{t\infty}}{1 + Pf}, \quad \text{with the Pfleiderer factor: } Pf = \frac{\cos \beta_2}{2} \frac{\pi r_2^2 b_2}{Z \, M_{st}}.$$

A correction for using the mean value of the pressure difference by Eq. (3.22) is required. The pressure difference over a blade starts from zero at the leading edge and becomes again zero at the tailing edge. The average pressure difference thus underestimates the pressure difference near the trailing edge at the start of the zone where this difference begins to evolve towards zero. Also a correction for friction is required and, to be more practical, the blade angle, and not the flow angle, should be used. Therefore, Pfleiderer adapted the formula by comparing it to experimental results and determined the *work reduction factor* as

$$
\varepsilon = \frac{\Delta p_{0,t}}{\Delta p_{0,t\infty}}, \quad \text{with } \varepsilon = \frac{1}{1 + Pf} \quad \text{and} \quad Pf = \xi \frac{r_2^2 b_2}{Z \, M_{st}}, \tag{3.24}
$$

with:

$$
\xi = \lambda \left(2.5 + \frac{\beta_2^b}{60} \right). \tag{3.25}
$$

β_2^b is the blade angle at the rotor exit, expressed in degrees (negative with backward curved blades). The factor λ expresses the embedding of the rotor, with:

$\lambda = 0.65$ to 0.85 for a rotor surrounded by a volute;

$\lambda \approx 0.6$ for a rotor surrounded by a vaned diffuser;

$\lambda = 0.85$ to 1 for a rotor surrounded by a vaneless diffuser.

In practice, the integral for the determination of the static moment may be approximated by $M_{st} = \frac{1}{2}(b_2 r_2 + b_1 r_1)(r_2 - r_1)$.

With Pfleiderer's representation, the characteristic $\Delta p_0 / \rho$ as a function of the flow rate is a straight line that intersects the theoretical characteristic with an infinite number of blades on the abscissa (with $\beta_2 \neq 0$). In reality, the work reduction factor is not exactly constant, but varies somewhat with varying flow rate.

The turbomachinery literature contains many formulae for the slip factor or the work reduction factor. They all give approximate results, best workable near the design operating point. Wiesner's formula is best with high solidity rotors. Pfleiderer's formula is best with small blade overlap. However, no formula is perfect and, even nowadays, research on new formulae continues. The slip factor and the work reduction factor are flow rate dependent, but it is not possible to express this dependency in a simple way.

3.3.5 Number of Blades and Solidity: Pfleiderer Moment Coefficient

When a rotor has many blades, slip is small and rotor work is close to the maximum attainable. This is no optimum choice regarding efficiency, as a large blade surface implies a large friction surface. Work reduction due to slip is no loss in itself, but the effect of losses between the driving motor and the rotor (mechanical losses, see Sect. 3.5.1) becomes important with a very small rotor work. So, except for cases with small rotor work, a rather large slip is acceptable. This implies that the optimum number of blades for efficiency is mostly quite close to the minimum that is sufficient to realise an attached flow. Of course, other reasons may require a larger number. A common reason is margin against flow separation at a flow rate lower than the design flow rate (see next section).

From the flow pattern in Fig. 3.9 we learn that there are two criteria for avoiding separation at the rotor exit. The number of blades must be large enough to prevent the Coriolis force and lift force from generating reversed flow at the pressure side and the adverse pressure gradient near the trailing edge at the suction side must not become so high that flow separation occurs. The pressures at the suction and pressure sides become equal at a rotor blade trailing edge. So, there is always an adverse pressure gradient at the suction side trailing edge.

Flow reversal at the pressure side is prevented for

$$w_p > 0, \quad \text{thus} \quad \frac{1}{2}(w_s - w_p) < w_2. \tag{3.26}$$

Prevention of boundary layer separation at the suction side may be expressed by a local diffusion factor criterion as

$$\frac{w_s - w_2}{w_s} < D_{loc} \approx 0.5 \quad \text{or} \quad w_s - w_2 < \frac{D_{loc}}{1 - D_{loc}} w_2. \tag{3.27}$$

This criterion also gives the result (3.26) for $D_{loc} = 0.5$. This D_{loc} value is the critical value for separation with axial cascades. Deceleration within the boundary layer at the suction side is less favourable with a centrifugal rotor, due to the jet-wake flow with lower energy in the corner of the suction side and the shroud (see Sect. 2.3.2 in Chap. 2). Therefore, we further apply the criterion (3.26) for separation at the suction side, but with a factor in front of w_2 in the right-hand side set to 0.8.

We note:

$$C = \frac{(w_s - w_p)_2}{w_2} < C_{\lim} \approx 1.6. \tag{3.28}$$

With Eq. (3.14), the velocity difference at the rotor exit is

$$(w_s - w_p)_2 = \frac{2\pi}{Z} \cos \beta_2 \left[2u_2 + \frac{d}{dr}(rw_u)_2 \right] = \frac{4\pi}{Z} u_2 \cos \beta_2 \, f_R, \tag{3.29}$$

with

$$f_R = 1 + \frac{1}{2} \frac{\frac{d}{dr}(rw_u)_2}{u_2}.$$

The curvature factor, f_R, is the ratio of the effect of the Coriolis force and lift force (= curvature) together to the effect of the Coriolis force alone. With radial compressors with large rotor blade solidity, the curvature factor may be estimated as being somewhat lower than unity. We may then use Eq. (3.29) together with Eq. (3.28) and $f_R \approx 0.80$ (see Chap. 14: radial compressors).

For radial fans and pumps with little blade overlap, the curvature factor cannot easily be estimated. An alternative for the velocity difference by Eq. 3.14 is using the pressure difference estimate (3.22) by Pfleiderer and the corresponding velocity difference by Eq. (3.23). The criterion (3.28) then becomes:

$$C_M = \frac{v_{2r}}{w_2^2} \frac{2\pi \ r_2 b_2}{Z \ M_{st}} \frac{\Delta W}{\Omega} = \frac{2\pi \ r_2^2 b_2}{Z \ M_{st}} \frac{v_{2r}}{u_2} \frac{\Delta W}{w_2^2} < C_{M,\lim} \approx 1.4. \tag{3.30}$$

We assume a lower limit value than in Eq. (3.28) to take the underestimation of the velocity difference by Eq. (3.23) into account.

Equation (3.30) for a radial rotor is similar to the expression of the Zweifel coefficient (Eq. 2.29 in Chap. 2) for an axial cascade. The factor

$$\sigma_M = \frac{Z M_{st}}{2\pi r_2 b_2 r_2}, \tag{3.31}$$

may be considered as the *moment solidity* of the meridional section, so that Eq. (3.30) may be written as

$$C_M = \frac{v_{2r}}{u_2} \frac{\Delta W}{\sigma_M \ w_2^2} < C_{M,\lim} \approx 1.4. \tag{3.32}$$

We compare with the Zweifel tangential force coefficient:

$$C_{Fu} = 2\frac{v_{2a} \ \Delta w_u}{\sigma_a \ w_2^2} < C_{Fu,\lim} \approx 1.4,$$

with the axial solidity $\sigma_a = c_a/s$.

We call the factor C_M by Eq. (3.32) the *Pfleiderer moment coefficient* since it is based on the average pressure difference reasoning by Pfleiderer.

The criterion (3.32) is used in the next section for estimation of the necessary solidity for avoiding boundary layer separation in a centrifugal rotor in design conditions.

A warning is that this criterion is not justified for estimating the start of flow separation by flow rate decrease. An essential ingredient is the assumption of an approximately constant pressure difference distribution along a rotor blade by Eq. (3.22), which is only well justified for the design flow rate.

There are no generally applicable criteria in the turbomachinery literature for estimation of a suitable number of blades in centrifugal rotors. Nowadays, the number of blades is mostly determined by a CFD analysis and the target is not necessarily the best possible efficiency. Other features like realisation of the largest possible flow rate or the largest possible pressure rise may be chosen. But, a discussion of possible design options is outside the objectives of this course book.

3.3.6 Number of Blades: Examples

Figure 3.13 is a sketch of 4 centrifugal rotor blade shapes. The characteristics are:
Backward curved blades: $\beta_1^b \approx -60°$, $\beta_2^b \approx -60°$, $r_1/r_2 \approx 0.7$, $\beta_2 \approx -70°$;
Straight backswept blades: $\beta_1^b \approx -55°$, $\beta_2^b \approx -25°$, $r_1/r_2 \approx 0.4$, $\beta_2 \approx -40°$;
Radial-end blades: $\beta_1^b \approx -55°$, $\beta_2^b \approx 0°$, $r_1/r_2 \approx 0.4$, $\beta_2 \approx -20°$;
Forward curved blades: $\beta_1^b \approx -30°$, $\beta_2^b \approx +60°$, $r_1/r_2 \approx 0.8$, $\beta_2 \approx +45°$.
Radial-end blades are applied in fans, but the blade shape is mainly appropriate for high rotational speed, intended to produce large rotor work, in other words, compressor application. In the radial blade part, no bending stress by centrifugal force occurs. The inlet has to be adapted to reach the same goal there. An axial part is added, called an inducer (see Chap. 14: radial compressors).

We determine the minimum number of blades with Eq. (3.32) and $\Delta W = u_2 v_{2u}$. From the velocity triangle at the rotor exit in Fig. 3.5, we derive

$$v_{2u} = v_{2r} \tan \alpha_2 \quad \text{and} \quad v_{2r} = w_2 \cos \beta_2.$$

So:

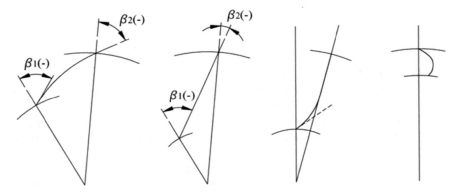

Fig. 3.13 Centrifugal rotor blade shapes: backward curved, straight backswept, radial-end (inducer not drawn), forward curved

Table 3.1 Number of blades for the rotor shapes shown in Fig. 3.13

β_2	r_1/r_2	ϕ	ψ	$\sigma_{M,calc}$	Z_{calc}	Z_{real}
$-70°$	0.7	0.182	0.500	0.230	4.82	6–7
$-40°$	0.4	0.279	0.766	1.152	12.06	14–16
$-20°$	0.4	0.321	0.883	1.733	18.15	20
$+45°$	0.8	0.571	1.572	0.981	30.82	36–40

$$C_M = \frac{v_{2r}\, v_{2u}}{\sigma_M w_2^2} = \frac{\tan \alpha_2 \cos^2 \beta_2}{\sigma_M}.$$

By taking $b_1 r_1 = b_2 r_2$, which means constant radial velocity component in the rotor, the moment solidity (3.31) may be estimated as

$$\sigma_M = \frac{Z}{2\pi}\left(1 - \frac{r_1}{r_2}\right).$$

From the velocity triangle at rotor exit in Fig. 3.5, we further derive

$$\tan \alpha_2 = \frac{v_{2u}}{v_{2r}}, \quad \tan \beta_2 = \frac{w_{2u}}{v_{2r}}, \quad v_{2u} = u_2 + w_{2u} \quad \text{or} \quad u_2 = (\tan \alpha_2 - \tan \beta_2)v_{2r}.$$

The flow coefficient and work coefficient are

$$\phi = \frac{v_{2r}}{u_2} = \frac{1}{\tan \alpha_2 - \tan \beta_2} \quad \text{and} \quad \psi = \frac{v_{2u}}{u_2} = \frac{\tan \alpha_2}{\tan \alpha_2 - \tan \beta_2}.$$

The results are listed in Table 3.1. The angle α_2 is set to $70°$ and C_M is set to 1.4. The numbers typically used in practice are in the last column. For the rotor with radial-end blades, the safety margin on the number of blades may seem small, but in reality the rotor has an axial part at the entry (the inducer), which contributes to the moment solidity (see Chap. 14: radial compressors). The number of blades might be larger than given in the last column, depending on the margin that the designer wants against separation at the suction side and flow reversal at the pressure side at reduced flow rate. Anyhow, these phenomena cannot be avoided at very low flow rate (flow pattern of Fig. 3.9, left).

3.4 Internal Losses with Radial Fans

This section is an analysis of internal losses with radial fans, but most results apply to radial pumps as well. With internal losses is meant the losses inside the flow from the machine entry to exit that passes through the rotor.

3.4.1 Turning Loss in the Rotor Eye

The fluid axially enters the eye of the rotor (entry part without blades; see Fig. 3.2) and turns 90° to the radial direction. The turning generates a loss analogous to loss in a bend. In energy measure, this loss is about 10% of the kinetic energy of the flow in the narrowest section of the rotor eye.

3.4.2 Incidence Loss at the Rotor Entry

Figure 3.14 is a sketch of the inflow of a cascade (radial, axial or mixed flow), with infinitely thin blades. The blades are drawn straight, but this is not crucial for the result. The flow just upstream of the cascade is not aligned with the blades, but makes an *incidence angle* δ with them. The flow deflects from the inlet velocity w_1 to the velocity w_2 inside the blade passage with a deflection angle $\delta = \beta_1 - \beta_2$.

The mass flow rate per unit span through a blade passage with spacing s is:

$$\dot{m} = \rho s \, w_1 \cos \beta_1 = \rho s \, w_2 \cos \beta_2.$$

The meridional component of the velocity is constant. The deflection velocity w_{def} is in the tangential direction.

The momentum conservation equation, projected on the β_2-direction is:

$$\dot{m} \, (w_2 - w_1 \cos \delta) = (p_1 - p_2)s \cos \beta_2.$$

Thus:

$$p_1 - p_2 = \rho w_2(w_2 - w_1 \cos \delta).$$

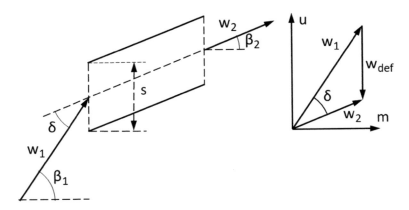

Fig. 3.14 Incidence and deflection at rotor entry

The total pressure drop $-\Delta p_{0,\,def}$ due to tangential deflection is

$$\frac{-\Delta p_{0,\,def}}{\rho} = \frac{p_1 - p_2}{\rho} + \frac{w_1^2}{2} - \frac{w_2^2}{2} = \frac{1}{2}\left(w_1^2 + w_2^2 - 2w_1 w_2 \cos\delta\right) = \frac{1}{2}w_{def}^2.$$

In energy measure, the loss equals the kinetic energy associated to the deflection velocity. This result is similar to that of dump diffusion. With a real cascade, the incidence loss is overestimated with the formula found, as the deflection is not sudden, but is spread over a finite length. Incidence loss follows from

$$-\Delta p_{0,\,def} = \mu_{def}\,\rho\,\frac{w_{def}^2}{2}, \tag{3.33}$$

with μ_{def} being a coefficient of about 0.7–0.8.

3.4.3 Displacement by Blade Thickness

The blades constitute, due to their thickness t, an obstruction at the rotor entry (Fig. 3.15). The flow passage width in tangential direction $s = \pi d_1/Z$ decreases to $s\text{-}t/\cos\beta_1^b$. We may express the displacement effect by an *obstruction factor*:

$$\tau_1 = (s\text{-}t/\cos\beta_1^b)/s. \tag{3.34}$$

Continuity of the mass flow implies, with subscript 1 indicating the station upstream of the entry and 1^b the station downstream of the entry:

$$v_{1r}^b = v_{1r}/\tau_1.$$

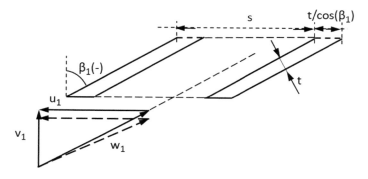

Fig. 3.15 Blade thickness obstruction at rotor entry; dashed line: before entry; full line: after entry

The direction of the relative flow changes through the rotor entry. For flow without incidence, the blades must be tangent to w_1^b instead of w_1.

From now on, we distinguish, if necessary, between the velocity triangles drawn immediately upstream of the rotor entry (station 1) and immediately downstream of it (station 1^b). The tangential velocity component in the relative frame is the same in these stations while the radial velocity component changes due to the blade thickness. A similar displacement effect occurs at the rotor exit, with a change in radial velocity component between positions immediately upstream (station 2^b) and downstream of the exit (station 2).

We remark that for taking into account the blade thickness at the rotor exit, the slip formulae of Stodola (3.20, 3.21) and Pfleiderer (3.24, 3.25) for the relation between the tangential velocity components at the rotor exit, with and without slip, have to be adapted. Without slip, the tangential velocity components are w_{2u}^b and v_{2u}^b, with w_{2u}^b calculated inside the rotor, applying the blade angle β_2^b, and $v_{2u}^b = u_2 + w_{2u}^b$. The adapted expressions are:

$$v_{2u} = \sigma \ u_2 + w_{2u}^b \quad \text{and} \quad u_2 v_{2u} - u_1 v_{1u} = \varepsilon(u_2 v_{2u}^b - u_1 v_{1u}). \tag{3.35}$$

3.4.4   Rotor Friction Loss and Rotor Diffusion Loss

With a rotor that can be considered as composed of blade channels, the friction loss may be estimated with friction coefficient formulae for ducts, applying an average hydraulic diameter. This is mostly appropriate for pumps. For fans, formulae for flow over plates are more appropriate. For a flow over a plate, the friction force and associated total pressure drop follow from

$$F = c_f A \rho \frac{w^2}{2}, \quad \frac{-\Delta p_0}{\rho} = \frac{F w}{\dot{m}}$$

where the friction coefficient c_f is around 0.005.

A diffusion loss is predominantly due to mixing of the layers with different velocity at the exit of a diffuser. Therefore, no diffusion loss is counted inside a rotor. The effect of diffusion is taken into account in the mixing loss downstream of the rotor (see Sect. 3.4.7).

3.4.5   Dump Diffusion Loss at the Rotor Exit

A rotor with backward curved blades is customarily combined with a volute which is much wider than the rotor exit (see Figs. 3.1, 3.2 and 3.16). A width ratio of

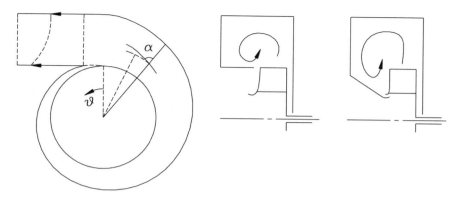

Fig. 3.16 Volute shape; left: flow with assumed uniform distribution over the width; middle: vortex motion in a meridional section; right: cross section in reality for useful application of the leakage flow

2–3 is typical and is intended to reduce the radial velocity component at the entry to the volute. This reduction is necessary for limiting the radial dimensions of the volute, as follows from the reasoning hereafter. We denote with subscript $2'$ the entrance conditions of the volute at radius r_2 (the rotor exit is station 2), thus after the reduction of the radial velocity component by the width leap.

When ignoring friction in the volute, the flow satisfies constant angular momentum:

$$v_u\, r = v'_{2u}\, r_2.$$

The volute is almost always manufactured with a constant width. The flow rate equation then reads

$$Q = 2\pi r\, b\, v_r \quad \text{or} \quad v_r\, r = v'_{2r}\, r_2.$$

In a volute with constant width, both the tangential and the radial velocity components vary inversely proportional to the radius. A streamline thus forms, in each of its points, the same angle with the radial direction. The equation of a streamline in a polar coordinate system is (Fig. 3.16):

$$\tan \alpha = \frac{r\, d\theta}{dr} = \text{constant}, \quad \text{or} \quad dr/r = d\theta/\tan \alpha, \quad \text{thus} \quad \ln(r/r_2) = \theta/\tan \alpha.$$

The shape of the streamline is termed a logarithmic spiral. In order to exert no force on the flow, the shape of the external wall of the volute has to be a logarithmic spiral.

The radius ratio when passing an entire circumference is

$$\ln (r_3/r_2) = \frac{2\pi}{\tan \alpha}. \tag{3.36}$$

From Fig. 3.12 it follows that the angle, by which the flow in the absolute frame leaves the rotor, may be around $70°$ (the radial velocity is exaggerated in the figure). The corresponding radius ratio is about 10, which is enormous. So, a more tangential outflow is wanted. It is thus necessary to reduce the radial velocity component before volute entrance. With a width ratio 3 and a rotor outlet angle $\alpha_2 = 70°$, the flow angle after the width leap is $\alpha'_2 = 83°$. The corresponding radius ratio is 2.16, which is still somewhat too large to be practical (see next section).

The sudden widening may be seen as dump diffusion and its loss estimated by

$$-\Delta p_0 = \rho \frac{1}{2} \left(v_{2r}^b - v'_{2r} \right)^2, \tag{3.37}$$

with v_{2r}^b and v'_{2r} being the radial velocity components upstream and downstream of the width leap. With e.g. $v'_{2r} = \frac{1}{3} v_{2r}^b$ it follows that

$$-\Delta p_0 = \mu_{dump} \, \rho \frac{1}{2} (v_{2r}^b)^2, \quad \text{with } \mu_{dump} = \frac{4}{9} \approx 0.444.$$

With (3.37), we include the effect of the blade thickness at the rotor exit in the dump loss. Notwithstanding the large loss coefficient, the loss by width leap is never extremely large, because the radial velocity component is not very large.

The loss estimation with a sudden leap overestimates the loss at the volute entrance itself, since the flow does not suddenly occupy the available space. Figure 3.16 (middle) demonstrates that the flow enters the volute with a vortex motion. A width leap around 2.5 is ideal to give sufficient room to the swirl. The resulting swirl velocity (velocity component in the meridional plane) is much larger than the velocity calculated as v'_{2r} in Eq. (3.37), but the associated kinetic energy is almost completely dissipated by the time that the flow reaches the volute exit.

This means that the dump loss formula (3.37) is not a good estimate for the loss associated to the rotor exit itself, but that it becomes a good estimate if the mixing loss associated to the swirl motion in the whole of the volute is included. The loss expression (3.37) produces thus realistic values for fans [4]. Similar, but more complex, dump loss models are also used for more streamlined volutes as in pumps and compressors [2, 6].

A real volute cross section is sketched in Fig. 3.16 (right). The leakage flow is injected into the rotor such that the boundary layer on the rotor shroud is energised. This prevents or, at least, reduces the separation of the inlet flow, which has to turn $90°$. The boundary layer that would separate without the injection is sucked to the wall by momentum transfer from the leakage flow. This is termed *Coandă-effect*.

Figure 3.16 (left) is a sketch of a volute with an external wall of logarithmic spiral shape. A clearance between the spiral and the rotor creates a tongue. The tongue clearance is typically about 5% of the rotor diameter.

3.4.6 Deceleration Loss in the Rotor Eye

Entry velocities of a centrifugal fan may be up to 30 m/s for a fan drawing air directly from a large space. Within an air transporting duct in an industrial plant, the velocity may be as high as 20 m/s and even higher for short ducts (the economic velocity follows from a balance between duct cost and energy dissipation cost). Fan entry velocities of such high level usually require flow deceleration between the rotor eye entry (plane perpendicular to the shaft) and the actual rotor entry (entry to the bladed part) and large enough blade speed at rotor entry.

The surface ratio (eye entry diameter $d_0 \approx$ rotor entry diameter d_1) is approximately $A_1/A_0 = 4b_1/d_1$. With backward curved rotor blades, this ratio may attain until 1.70 without occurrence of separation. The corresponding velocity reduction is about 0.60. The Coandă-effect enables this strong deceleration (the Coandă-effect cannot be used with forward curved blades; see section hereafter). The diffusion loss in the decelerating flow in the rotor eye is approximately balanced by the energising effect by the leakage flow. So, formally, deceleration loss in the rotor eye may be ignored.

With $\beta_1 \approx -60°$ is $v_{1m} \approx 0.6\ u_1$. Thus, with the strongest possible velocity reduction in the rotor eye, the blade speed at rotor entry has to be approximately equal to the entry velocity of the rotor eye. With many fans, this is easily realisable and very strong velocity reduction may even not be necessary.

With house ventilation, duct velocities are mostly limited to about 5 m/s because of noise limitation. It is then not possible to match the low velocity in the suction duct with the necessary much higher velocity at the fan entry. A typical arrangement is then a fan inside a box. The fan takes air from the box through a suction eye (a double entry is also possible) with a much smaller diameter than the suction duct connected to the box. The fan delivers air by dump diffusion into a second part of the box or directly into the delivery duct with a much larger section area than the fan outlet (an example with dump diffusion into the delivery duct is Fig. 3.1). The box arrangement is possible with backward and forward curved blades.

3.4.7 Flow Separation at Rotor Entry and Rotor Exit

Centrifugal fans with straight, slightly backswept, or radial-end blades (Fig. 3.13, second and third blade shapes), typically have separated flow at the rotor exit, even for the design flow rate. The reason is that the deceleration ratio in the rotor, the velocity ratio w_2/w_1, cannot be lower than about 0.7 for avoiding flow separation.

For actual calculations, we assume here that this limit ratio applies between the inlet flow after rotor entrance (w_1^b) and the outlet flow before rotor exit, so before slip (w_2^b). With w_2 near to the radial direction, this limit is rapidly obtained (see the design example 7.6 in Chap. 7).

The one-dimensional calculation procedure can still be used for separated flow, with a flow representation of jet-wake type at the rotor exit, which is a core flow with relative velocity equal to the limit value (called the jet) adjacent to a zone with very low velocity (called the wake). Physically, the wake is in the corner of the blade suction side and the shroud. In the simplest flow representation, the net through-flow velocity in the wake is set to zero (stagnant wake). The one-dimensional calculation is then applied to the jet flow through the rotor channels (see Exercise 7.7.5 in Chap. 7). With the simplest methods, slip formulae and loss formulae are used as for full through-flow (we follow this approach in Exercise 7.7.5), but corrections to the formulae are sometimes applied. With jet-wake rotor outflow, the dump diffusion at the entrance to the volute is to be considered as between the core flow in the rotor channels and a uniform flow immediately downstream of the rotor exit, filling the full width of the volute. Due to the partial filling of the rotor channels, the dump diffusion loss is then larger than with full through-flow.

Centrifugal fans with forward curved blades (Fig. 3.13, fourth blade form) have recirculation flow in the rotor, also for the design flow rate. These machines have a very low degree of reaction. Therefore, they can be built with a diameter ratio d_1/d_2 close to 1 (0.85–0.9 are typical) since the pressure increase by the centrifugal force, $(u_2^2 - u_1^2)/2$, does not have to be large. With a large inlet diameter, the flow rate can be maximised for a given rotor diameter. Since the work coefficient is also large, centrifugal fans with forward curved blades are the most compact ones for a given duty of flow rate and total pressure rise. For this reason, this type of fan is commonly applied when space is limited. An example is car ventilation, but also home ventilation (see further discussion in Sect. 3.6).

Figure 3.17 is a sketch of the flow pattern in a meridional section of a radial fan with forward curved blades, at the design flow rate.

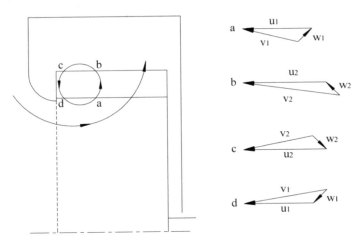

Fig. 3.17 Recirculation flow in a rotor with forward curved blades

Since the static pressure rise in the rotor is small, the stabilising effect of the Coandă-flow is weak and separation of the entrance flow of the rotor cannot be avoided at the shroud. The choice is then typically for a constant rotor width, much larger than follows from equal through-flow areas of the entries of the rotor eye and the bladed rotor part ($b_1/d_1 = 0.25$). The through-flow of the rotor can then take the space as needed, and the recirculation flow then also gets room.

Figure 3.17 explains how the recirculation flow is energised in the outward radial motion (a to b: $u_2 v_{2u} - u_1 v_{1u} > 0$), how the velocity is reduced during the turning in the volute (b to c: $|v_2|_b > |v_2|_c$) with energy transfer to the through-flow, how it returns to the rotor eye without energy exchange with the rotor (c to d: $u_2 v_{2u} - u_1 v_{1u} = 0$), how the velocity is reduced during the turning at the rotor entrance (d to a: $|v_1|_d > |v_1|_a$) with energy transfer to the through-flow.

The energy transfer from the rotor to the through-flow by the intermediate action of the recirculation flow cannot be derived from basic conservation laws with the one-dimensional flow representation and there do not seem to be models for this effect in the fan literature. For illustration of the flow in this type of fan, we refer to [1, 7]. These studies reveal that the size of the recirculation zone is not constant along the periphery of the rotor and that the size of the recirculation zone gets larger at lower delivered flow rate. For zero net flow, the recirculation flow is the strongest and creates significant energy transfer to the fluid in the volute. For some choices of the blade angles, it is even higher than for the design flow rate.

3.4.8 Incidence Loss at the Volute Entry

The radius ratio of the volute, adapted to the flow after the width leap at the volute entry, is mostly still somewhat too large to attain acceptable volute dimensions. To a radius ratio 2 corresponds a slope angle $\alpha_s = 83.7°$ (the subscript s stands for scroll). A radius ratio 2 already yields workable dimensions, but many manufacturers opt for a somewhat smaller radius ratio. Typically, the opening angle of the volute (α_s, but calculated tangentially) is about 5°. The corresponding radius ratio is about 1.75 (see Fig. 3.1 and the later Fig. 3.25).

Figure 3.18 is a sketch of the velocity triangles at the rotor exit (absolute velocity

Fig. 3.18 Top: velocity triangles at rotor exit before and after slip (dashed line without slip); Bottom: velocity triangles at volute entry after width leap, before and after tangential velocity deflection due to incidence; the radial velocity is drawn exaggerated for clarity

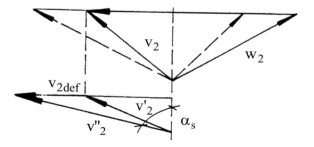

v_2, taking slip into account), after the width leap (velocity v'_2) and after the tangential deflection enforced by the volute (velocity v''_2).

The incidence loss at the volute entry may be estimated with a formula like (3.33) with deflection velocity $v''_{2u} - v'_{2u}$. The tangential velocity imposed by the volute, v''_{2u}, follows from the flow rate, assuming equal angular momentum at the volute entry and exit [2, 6], $(rv)_3 = r_2 v''_{2u} = C$, so that $Q = C \int \frac{bdr}{r}$ at the volute exit (Fig. 3.16, left).

By making the volute more tangential, the volute is adapted to a lower flow rate than the design flow rate of the rotor. The rotor and the volute thus become mismatched. The above demonstrates that, mostly, some mismatch is unavoidable. The usual mismatch causes acceleration in the volute: $v''_{2u} - v'_{2u}$. The spontaneous deceleration in the volute (Fig. 3.16, left) may completely be undone by it.

Another consequence of the mismatch is enforcement of a non-uniform pressure distribution to the exit of the rotor, namely pressure decreasing in the flow sense. A pressure leap at the tongue is thus generated. The non-uniform pressure distribution causes increase of losses at the rotor exit.

3.4.9 Friction Loss Within the Volute

The friction loss may be estimated analogously to the friction loss in the rotor. The entire friction surface must be determined.

3.4.10 Applicability of the Loss Models

Loss models, in the style as described above, are used with hand calculations of the performance of centrifugal fans, pumps and compressors. Modern loss correlations are more complex than the formulae given above, but are inspired by the same principles [2, 4–6]. More detailed expressions of the loss coefficients are employed, while the constant loss coefficients used here, give only an approximate value, like $\mu_i = 0.7$ or $c_f = 0.005$.

Here, we only aim at a principal discussion. For more details, we refer to the literature cited and to an overview paper on loss correlations for fans and blowers by Kim et al. [8]. In particular, this paper shows that the dump loss model and the incidence loss model for the volute entrance are still used nowadays, giving reasonable loss estimates.

Performance evaluation with a one-dimensional flow representation, using the slip formulae of Stodola or Pfleiderer and with the loss formulae is only well justified if the flow through the rotor channels is sufficiently homogeneous in a cross section. This requirement is satisfied for centrifugal machines with backward curved blades operating not extremely far away from the design condition. For very low flow rate, the flow in the rotor becomes a recirculation flow as shown on Fig. 3.9 (left). Such a flow cannot be analysed as a one-dimensional flow.

3.4.11   Optimisation of the Rotor Entry of a Centrifugal Fan

In the design of a power receiving turbomachine with a constant-density fluid, the primary targets are the flow rate (volume flow rate Q) and the mechanical energy rise ($\Delta p_0/\rho$). The machine parameters to be determined in a first design phase are the rotor diameter (d_2) and the angular speed (Ω). These follow from similitude quantities derived by a global optimisation (see Chap. 7) and considerations about the application of the machine. In essence, the application type determines the degree of reaction, thus the division of the rotor work into enthalpy increase and kinetic energy increase in the rotor (see Sect. 3.6). Once the main parameters of the machine are determined (d and Ω), other parameters typically follow from local optimisation considerations. These differ somewhat between a fan and a pump.

As a further design step, it is generally assumed that optimum efficiency of a centrifugal fan is approximately obtained with absence of pre-whirl at rotor entry and by minimising the relative velocity at the entrance to the bladed part of the rotor. The most important losses occur at the rotor exit by the sudden expansion and at the entrance to the volute by the incidence loss due to tangential mismatch. The expansion loss is minimised by minimum relative velocity at rotor exit (w_2). This requires minimum relative entry velocity (w_1) and strongest possible deceleration in the rotor. The incidence loss at the entrance to the volute is mostly the lowest for the rotor outflow as near as possible to the tangential direction in the absolute frame (v_2), which again, implies minimum w_2.

The deceleration in the rotor eye may be expressed by a velocity factor

$$\zeta = \frac{v_{1r}^b}{v_0} \quad \text{with } v_{1r}^b = \frac{Q_{rotor}}{\pi d_1 b_1 \tau_1} \quad \text{and} \quad v_0 = \frac{4Q}{\pi d_0^2}. \tag{3.38}$$

Thus:

$$\zeta = \frac{Q_{rotor}}{Q} \frac{d_0^2}{4 d_1 b_1 \tau_1} = \frac{1}{\eta_V \tau_1} \left(\frac{d_0}{d_1}\right)^2 \frac{1}{4} \frac{d_1}{b_1}.$$

The diameter of the eye inlet d_0 is somewhat smaller than the inlet diameter of the bladed rotor part d_1, say $d_0 \approx 0.9\, d_1$.

With $\eta_V = Q/Q_{rotor} \approx 0.9$ (volumetric efficiency); $\tau_1 \approx 0.9$ (obstruction factor), the result is

$$\zeta \approx \frac{1}{4} \frac{d_1}{b_1} \quad \text{or} \quad \frac{b_1}{d_1} \approx \frac{1}{4\zeta}.$$

The strongest possible deceleration in the rotor eye is about 0.60, so that $b_1/d_1 = 0.25$ (no deceleration) to 0.40 (maximum deceleration). Then:

$$v^b_{1r} = \frac{Q}{k\pi d_1 b_1} = 4\frac{\zeta}{k\pi}\frac{Q}{d_1^2}, \text{ with } k = \eta_V \tau_1 \approx (0.9)^2 \text{ and } u_1 = \frac{\Omega d_1}{2}.$$

Thus, in absence of pre-whirl:

$$(w^b_1)^2 = (v^b_{1r})^2 + u_1^2 = a^2(d_1^2)^{-2} + b^2(d_1^2) \text{ with } a = 4\left(\frac{\zeta}{k\pi}\right)Q \text{ and } b = \frac{\Omega}{2}.$$

With the minimum of w_1 corresponds an optimum value of d_1:

$$(d_1)_o = \left(\frac{2a^2}{b^2}\right)^{1/6} = 2^{1/6} 2\left(\frac{\zeta}{k\pi}\right)^{1/3}\left(\frac{Q}{\Omega}\right)^{1/3}. \tag{3.39}$$

The corresponding velocity components are:

$$(v^b_{1r})_o = 2^{-1/3}\left(\frac{\zeta}{k\pi}\right)^{1/3}Q^{1/3}\Omega^{2/3} \text{ and } (u_1)_o = 2^{1/6}\left(\frac{\zeta}{k\pi}\right)^{1/3}Q^{1/3}\Omega^{2/3}. \tag{3.40}$$

The magnitude of the optimum relative velocity at the rotor entry is proportional to the factor $(\zeta/k\pi)^{1/3}$. This means, as expected, that it is advantageous to set the velocity factor to the lowest possible value. The second observation is that the blade angle corresponding to the optimum diameter does not depend on the velocity factor. The minimum of w_1 is obtained for $tg\beta^b_1 = -\sqrt{2}$, thus $\beta^b_1 \approx -55°$. In principle, this is a universal result. The corresponding inlet diameter (3.39) is, for $\zeta = 0.625(b_1/d_1 = 0.40)$:

$$(d_1)_o \approx \sqrt{2}\left(\frac{Q}{\Omega}\right)^{1/3}. \tag{3.41}$$

The expressions of the velocity components then become

$$(v^b_{1r})_o \approx \frac{1}{2}Q^{1/3}\Omega^{2/3} \text{ and } (u_1)_o \approx \frac{\sqrt{2}}{2}Q^{1/3}\Omega^{2/3}. \tag{3.42}$$

The minimum of w^b_1 as a function of the diameter is quite flat, however. For a diameter increase of 20% or a diameter decrease of 15%, the increase of the magnitude of w^b_1 is about 6%, which means about 12% in rotor-related losses.

The corresponding blade angles are −68 and −41°. $\beta^b_1 = -55°$ is thus not a strongly universal value.

In practice, the blade angle and the rotor entry diameter may deviate from the theoretical optimum values with the purpose to match the fan to the suction duct. With $|\beta^b_1| = 55°$, the corresponding velocity in the suction duct may be too high (see Sect. 3.4.6). Matching is then obtained by a more tangential flow at the rotor entry. Therefore, according to Eck [4], $|\beta^b_1| = 55°$ should be considered as a minimum value. Moreover, there may be other reasons for deviating from the theoretical optimum at rotor entry. Possible reasons are enlarging the flow rate or enlarging the pressure rise.

A further aspect concerns pre-whirl. With positive pre-whirl ($v_{1u} > 0$), the magnitude of w_1 is reduced. This suggests that it is advantageous to use positive pre-whirl. However, as w_1 diminishes, more solidity is necessary in the rotor. With the Pfleiderer moment coefficient (3.30), the necessary solidity is inversely proportional to the kinetic energy at the rotor exit. Thus, with pre-whirl, the kinetic energy multiplied by the blade surface stays approximately the same and thus also the friction loss in the rotor. But, as the rotor work diminishes by positive pre-whirl ($u_1 v_{1u} > 0$), a larger v_{2u} is necessary (for already chosen d_2 and Ω). The necessary larger v_{2u} is almost spontaneously attained by the lower w_2 and the lower slip due to larger solidity. With lower w_2, the expansion loss at the rotor exit is lower, but with larger v_{2u}, the incidence loss at the volute entry is larger. Further, the inlet guide vanes cause friction loss. The result is that there is no efficiency gain by using pre-whirl. Therefore, inlet guide vanes are not applied, unless they function for flow rate variation (see Sect. 3.8).

3.4.12 Characteristics Taking Losses into Account

Some losses are proportional to the flow rate squared: turning loss at rotor entrance, rotor friction loss, rotor diffusion loss, loss by dump at volute entrance, friction loss in the volute. Incidence losses are zero with an adapted flow rate Q* and change proportionally to $(Q - Q^*)^2$. This becomes clear from Fig. 3.19 showing the incidence at the rotor entry. The direction indicated with β_1^* is the direction for incidence-free entry, obtained from the blade orientation, but taking the displacement due to the blade thickness into account.

The deflection flow velocity follows from triangle similarity. The deflection velocity and corresponding incidence loss are:

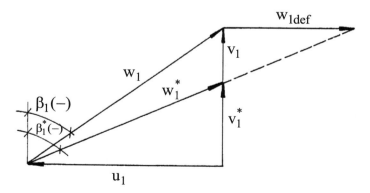

Fig. 3.19 Incidence at rotor entry; radial velocity is exaggerated for clarity

$$\frac{w_{1def}}{u_1} = \frac{v_1 - v_1^*}{v_1^*} = \frac{Q}{Q_r^*} - 1, \quad -\Delta p_{0, def} = \mu_{def}\rho\frac{w_{1def}^2}{2}. \tag{3.43}$$

An analogous reasoning applies to the volute. The velocity triangles at the volute entrance are sketched in Fig. 3.20, using Stodola's slip representation. From Fig. 3.20 it follows that

$$\frac{v_{2def}}{\sigma u_2} = \frac{v'_{2r} - v'^{*}_{2r}}{v'^{*}_{2r}} = \frac{Q}{Q_s^*} - 1. \tag{3.44}$$

In Fig. 3.20, the direction AC is fixed and determined by the rotor exit angle. The direction AD is fixed as well, as a result of the scaling of the radial components by the surface ratio at the width leap. The direction BE is imposed by the volute. The radial velocity component v'^{*}_{2r} has thus a fixed value, allowing to calculate the flow rate Q_s^*. With Pfleiderer's slip representation, the reasoning is similar, but with a different value of Q_s^*.

Figure 3.21 (same figure as Fig. 1.18) draws the characteristic Δp_0 as a function of the flow rate, theoretically, but with slip (both Stodola's and Pfleiderer's reasoning lead to a linear characteristic) and obtained by subtracting friction losses (a common term for all losses proportional to the flow rate squared) and incidence losses. The representation is simplified by assuming that the volute is matched to the rotor, so that a unique design flow rate $Q^* = Q_r^* = Q_s^*$ exists. With this simplification, the resulting characteristic becomes a parabola, assuming constant loss coefficients and a constant work reduction factor.

At zero flow rate, $\Delta p_0 \approx \rho u_2^2/2$ with backward curved blades. The average absolute velocity at the rotor exit equals then approximately u_2, since the rotor flow rate is small because there are only leakage flows. The rotor work is approximately u_2^2. There is nearly standstill within the volute, with the pressure loss by incidence about

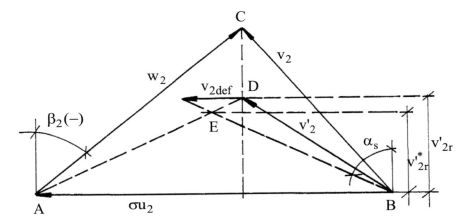

Fig. 3.20 Incidence at volute entry; radial velocity is exaggerated for clarity

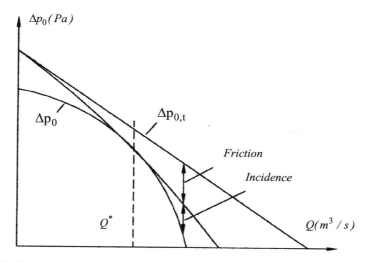

Fig. 3.21 Characteristic of a radial fan with backward curved blades

$\rho u_2^2/2$. With forward curved blades, Δp_0 is much larger for zero net flow rate, due to the energy transfer by the recirculation flow within the rotor (Fig. 3.17). The resulting characteristic then typically has a maximum and a minimum (see Sect. 3.6).

The *internal efficiency* is

$$\eta_i = \frac{\Delta p_0}{\Delta p_{0,\,t}}. \tag{3.45}$$

The internal efficiency attains its maximum at a flow rate that is lower than the design flow rate, due to part of the losses being proportional to the square of the flow rate (friction losses and diffusion losses). This phenomenon persists, even if the machine has a unique design flow rate (operating point without incidence losses in rotor and stator).

The flow rate at the operating point with maximum efficiency is often considered as the design flow rate. In principle, this is not correct. Considering the operating point with incidence-free rotor entry as design operating point is the correct attitude.

3.5   Overall Performance Evaluation

In this section, we derive the performance definitions of turbomachines, with the example of radial fans. The definitions are generalised for machines with constant-density fluids.

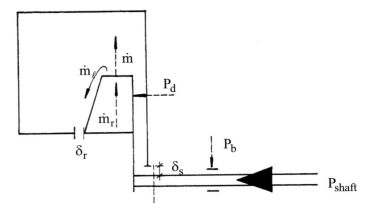

Fig. 3.22 Power receiving machine

3.5.1 Mechanical Loss

Figure 3.22 sketches a meridional section of a radial fan. The power transferred to the shaft by the driving motor is termed *shaft power* (P_{shaft}). The shaft power is not entirely transferred to the rotor, as there is mechanical loss in the bearings. The power dissipated in the bearings into heat, is conveyed to the surroundings. The outsides of the rotor shroud and rotor disc entrain surrounding fluid by friction. The energy required for this motion is taken from the shaft and dissipated into heat. Here, we consider the heat by disc friction as conveyed to the surroundings (see Chap. 1, Sect. 1.6.1), so that the disc friction loss becomes of the same nature as the bearing loss. Both are gathered in the mechanical loss, with power $P_m = P_b + P_d$. The power exchanged with the fluid within the rotor is termed *rotor power* (P_{rot}). The *mechanical efficiency* is defined by

$$\eta_m = \frac{P_{rot}}{P_{shaft}} = \frac{P_{rot}}{P_{rot} + P_m}. \tag{3.46}$$

3.5.2 Leakage Loss

The clearance of the shaft passage is δ_s (leakage to the outside) in Fig. 3.22. The volute fits to the rotor with clearance δ_r (leakage flow circulating through the rotor). Due to the clearances, the flow rate through the rotor ($\dot{m}_r$) exceeds the flow rate delivered by the machine ($\dot{m}$). The *volumetric efficiency* is defined by

$$\eta_v = \frac{\dot{m}}{\dot{m}_r} = \frac{\dot{m}}{\dot{m} + \dot{m}_\ell}. \tag{3.47}$$

In pumps, the gaps depicted in Fig. 3.22 are almost completely eliminated by seals. The shaft seal is typically a mechanical seal (almost without leakage) or a compression packing (with a small leakage flow). The rotor sealing is commonly realised with wear rings. These are rings in hard materials, one on the rotor side and one on the stator side (sometimes only on the stator side; see Chap. 8: pumps), with a small gap in between them (in the order of 2 ‰ of the diameter) and a quite large axial width. As these rings are subjected to some wear due to the high shear stress in the fluid in the narrow gap, they have to be replaced after a large number of duty hours, when the gap becomes too large.

Sealing is possible with pumps, because both the rotor and the stator have high rigidity. Sealing is impossible with fans, due to the low rigidity of plate materials. Leakage through the shaft gap is almost negligible. Leakage through the rotor gap is quite significant. The leakage flow rate depends on the gap width and on the static pressure increase within the rotor (degree of reaction). With a relative gap width of 1% of the diameter, the leakage flow rate, with backward curved blades, may be up to 10% of the delivered flow rate. A design according to Fig. 3.16 (right) is thus highly recommended, so that the rotor leakage flow may get a useful function (energising the boundary layer on the shroud).

3.5.3 Overall Efficiency with Power Receiving Machines

The following relations apply:

$$P_{shaft} = P_{rot} + P_m, \quad P_{rot} = (\dot{m} + \dot{m}_\ell)\Delta W, \quad \Delta W = \Delta E_m + q_{irr},$$

$$\eta_g = \frac{\dot{m}\,\Delta E_m}{P_{shaft}} = \frac{P_{rot}}{P_{rot} + P_m}\frac{\dot{m}}{\dot{m} + \dot{m}_\ell}\frac{\Delta E_m}{\Delta W} = \eta_m \eta_v \eta_i.$$

The overall efficiency (global efficiency) may thus be written as the product of three partial efficiencies.

3.5.4 Overall Efficiency with Power Delivering Machines

Figure 3.23 shows the intervening quantities.

The following relations apply:

$$P_{shaft} = P_{rot} - P_m, \quad P_{rot} = (\dot{m} - \dot{m}_\ell)(-\Delta W), \quad -\Delta E_m = -\Delta W + q_{irr},$$

$$\eta_g = \frac{P_{shaft}}{\dot{m}(-\Delta E_m)} = \frac{P_{rot} - P_m}{P_{rot}}\frac{\dot{m} - \dot{m}_\ell}{\dot{m}}\frac{-\Delta W}{-\Delta E_m} = \eta_m \eta_v \eta_i.$$

Fig. 3.23 Power delivering
machine

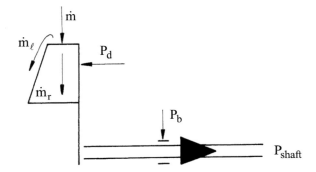

Just as with power receiving machines, the overall efficiency may be written as the product of three partial efficiencies.

3.5.5 Detailed Efficiency Analyses

The foregoing efficiency definitions are based on the mean line description of the flow. More detailed analyses are often used, but there is no universal methodology. Distinction may be made between the friction loss in the bearings and seals of the shaft and the disk friction loss. Often, especially with pumps and hydraulic turbines, the power on the shaft, but inside the machine, is then called the internal power. The meaning is the internal shaft power. For a power receiving machine, the internal efficiency may then be defined as the ratio of the useful power to this internal shaft power and the mechanical efficiency as the ratio of the internal shaft power to the external shaft power.

In some literature sources, the term internal power is used for the rotor power. For a power receiving machine, the internal efficiency may then be defined as the ratio of the useful power to the rotor power [9].

In this course book, we do not use the term internal efficiency in the previous two senses. Moreover, due to the ambiguity, we avoid the term internal power.

The rotor power may be split into several terms, associated to different flow paths in the rotor. For a power receiving machine, distinction may be made between the main flow from the entry to the exit of the machine (or machine stage) that passes through the rotor, recirculating flows at the rotor entry and exit, a leakage flow circulating through the clearances between the rotor and the casing and a leakage flow to the outside.

With a constant-density fluid, the term hydraulic efficiency is then often used for the comparison between the mechanical energy change and the rotor work in the flow from the entry to the exit of the machine (or the stage) through the rotor. But, there is no universally applied term for this efficiency with compressible fluids. In this course book, we use the term *internal efficiency* with the meaning that it concerns losses inside the flow that goes from entry to exit of the machine (or a stage) and that passes through the rotor [10]. A warning is that the term is not used in this sense in many literature sources, but many sources are not precise.

With the rotor work split into several parts, the global efficiency of a power receiving machine may be written as the product of several partial efficiencies, being a mechanical efficiency and as many terms as there are flow paths considered, with as last term the internal efficiency in the sense as used here. The decomposition of the global efficiency in the previous sections is simpler, however. With the mean-line flow representation, the effects of recirculating flows at rotor entry and exit are ignored. With a power receiving machine, the work transferred per unit of mass to the leakage flows is supposed to be the same as that to the flow that reaches the machine exit, and the work to the leakage flows is supposed to be fully dissipated. Further, energy exchanges between the different flow paths are ignored. These simplifications allow writing the global efficiency as the product of three partial efficiencies. The description is similar with power delivering machines. Clearly, for fundamental analyses, the simplified representation is sufficient. It captures the three main loss categories: dissipation outside the flow paths, dissipation in parasitic flow paths and dissipation inside the main flow path.

3.5.6 Total-to-Total and Total-to-Static Efficiencies

The internal efficiency used up to now is called the *total-to-total efficiency*. It compares the rotor work in the flow from the entry to the exit of the machine (or a stage) to the change of the total mechanical energy in this flow. With the total mechanical energy is meant that kinetic energy is included (total state or stagnation state). In some applications, however, the kinetic energy at the machine exit is not useful. An example is an exhaust fan, which is a fan that removes polluted air from a room or a building and expels it to the outside. The kinetic energy in the exhaust flow is then dissipated and is not useful. For such machines, the useful mechanical energy at the exit is the static mechanical energy, thus without the kinetic energy.

The mechanical energy rise may then be calculated as total-to-static (lower than total-to-total), and the corresponding efficiency is the *total-to-static efficiency*. Also with turbines, the kinetic energy at the exit may not be useful. Likewise, a total-to-static energy drop (larger than total-to-total) and a total-to-static efficiency (lower than total-to-total) are then defined.

The total-to-static efficiency is always smaller than the total-to-total efficiency. It is used because it is this form of efficiency that has to be optimised in cases where the exit kinetic energy cannot be useful.

3.6 Rotor Shape Choices with Radial Fans

Figure 3.24 shows the characteristic curves of total pressure rise at several rotational speeds for rotors with backward curved and forward curved blades, fitting in the same volute (fan of Fig. 3.1). The 040 in the machine number means that the rotor diameter is 40 cm. The rotor blade shape (forward or backward curved) determines the shape of the characteristic Δp_0 as a function of Q, the value of the total pressure rise Δp_0 and the way in which Δp_0, at rotor level, is composed from pressure energy increase and kinetic energy increase (degree of reaction). The different rotor types are appropriate for different applications.

With backward curved blades, a characteristic features a weak maximum, with the design operation point value of the total pressure rise coefficient lower than 0.5 ($\psi_0 = \Delta p_0/\rho u_2^2 \approx 0.43$ on the maximum efficiency line in Fig. 3.24a; 0.30–0.45 being typical). For a machine with less backward leaning (straight blades, Fig. 3.13b), the maximum is sharper with $\psi_0 \approx 0.6$–0.7 in the design operating point. In Fig. 3.24b, the total pressure rise at constant rotational speed with forward curved blades is almost independent of the flow rate, in the range of flow rate shown. This is quite particular. With forward curved blades, the characteristic typically has a minimum and a maximum. The minimum at a flow rate lower than the design value is due to the high value of ψ_0 for zero flow rate caused by the intense recirculation flow inside the rotor, as explained earlier (Fig. 3.17). With forward curved blades, ψ_0 exceeds 1 in the design operating point ($\psi_0 \approx 1.20$ on the maximum efficiency line in Fig. 3.24b).

The shape of the characteristics is very different with both rotor forms, but the shape is mostly not critical with fans. The fan load usually consists of a ductwork or a space. The total pressure rise by the fan compensates friction losses, bend losses or diffusion losses, which are all proportional to the flow rate squared. The intersection of the fan characteristic and the load characteristic is mostly about perpendicular and the operating point is stable (unstable operating points are discussed in Chap. 8: pumps). But there are applications where the shape of the characteristic is crucial. Fans supplying combustion air to boilers with chain grates (coal or household waste) should feature a steep characteristic, as flow resistance of the fuel layer may vary strongly. It is advisable that this does not affect the flow rate much, which makes a fan with strongly backward curved blades suitable. A load characteristic of required total pressure rise proportional to the square of the flow rate is approximately coincident with a line of constant efficiency for varying rotational speed (the total pressure rise is on a logarithmic scale in Fig. 3.24). This allows efficient flow rate control by variable rotational speed (see Sect. 3.8).

The needed total pressure rise is mostly not critical. The peripheral speed of a welded rotor may easily be 50 m/s (about 2400 rpm for diameter 0.40 m). The corresponding $\rho u_2^2/2$ is about 1500 Pa. With many applications, the required total pressure rise is much lower. Typical air velocities in ducts in industrial plants are 10–20 m/s. At 16 m/s, the dynamic pressure ($\rho v^2/2$) is about 150 Pa. The total loss coefficient of the load thus must thus be about 10 in order to attain 1500 Pa. But a larger peripheral speed is not a big problem. E.g., the GTHB version in Fig. 3.24

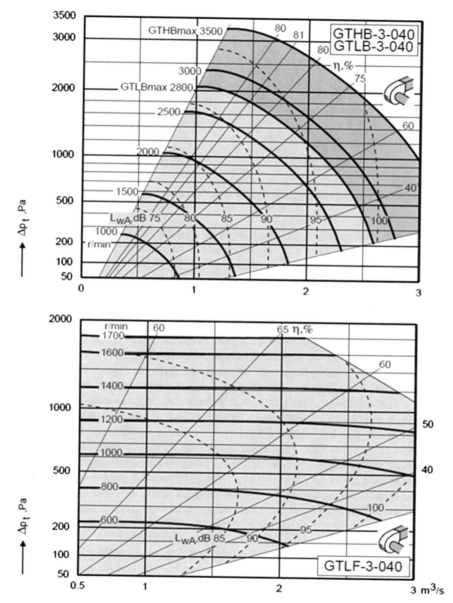

Fig. 3.24 Characteristics with a rotor with backward curved blades (top) and a rotor with forward curved blades (bottom) in the same volute; rotors as in Fig. 3.1, rotor diameter 400 mm (*Courtesy* FläktGroup)

Fig. 3.25 Rotors with motor incorporated in the hub (*Courtesy* ebm-papst)

has reinforced blades (Fig. 3.1). The peripheral speed is about 73 m/s at 3500 rpm. A peripheral speed of 80 m/s is attainable with straight backswept blades (almost no bending stress). Such a fan thus reaches ($\psi_0 \approx 0.6$) 4500 Pa total pressure rise. Peripheral speeds until 120 m/s are possible by thicker steel plates (10–20 mm thickness) and a large number of blades (until 16).

Fan noise increases with increasing rotational speed. For the fans in Fig. 3.24, with similar flow rate and total pressure rise, the noise with backward curved blades is only a little higher than with forward curved blades, notwithstanding the higher rotational speed. The noise is partly broadband due to turbulence and vortices and partly tonal (blade passing frequency and harmonics) due to impact of the jet-wake patterns on the volute tongue. A rotor with forward curved blades, with high flow turning and recirculation zones, is rather unfavourable with respect to broadband noise. The rotors with backward curved blades of Fig. 3.1 are specifically designed for low noise: e.g., a well-rounded inlet eye and a large number of blades, namely 11. This is much more than strictly necessary (the forward curved blade rotor has 42 blades). Due to the high number of blades, pressure differences between pressure and suction sides are reduced, which reduces tonal noise. An alternative method for noise reduction is use of hollow aerodynamically shaped blades (see Fig. 3.25, left).

With an equal diameter, the rotor with forward curved blades produces the same pressure rise as the rotor with backward curved blades for a rotor speed that is about a factor $0.575 \approx 1/\sqrt{3}$ lower (ψ about 1.2 compared to about 0.4). E.g., compare pressure rise 1800 Pa and flow rate 1.8 m³/s: 1700 rpm and 3000 rpm. Flow rates are comparable, with the inflow and outflow velocities of the fan being about the same. The rotor with forward curved blades is wider than the rotor with backward curved blades, but the through-flow does not fill the whole width, as explained before (Fig. 3.17). The outflow velocity of the rotor in the absolute frame is larger with the forward curved blades than with the backward curved blades. The ratio is about 1.5. With forward curved blades, a strong velocity reduction occurs at the volute entrance due to tangential dump diffusion with rather high energy dissipation. Therefore, the efficiency is significantly lower with forward curved blades than with backward curved ones. In Fig. 3.24, the maximum total-to-total efficiency with backward curved blades is 81%, compared to 65% with forward curved blades.

A practical consequence of the lower rotational speed with forward curved blades for the same pressure rise and flow rate is that the requirements concerning strength and vibrations are much lower. This allows a lighter and cheaper construction. Lower efficiency is less important for small-power applications. Fans with forward curved blades are therefore preferred for house ventilation, usually as a fan inside a box (see Sect. 3.4.6).

The degree of reaction is a main parameter adapted to the application. It is, as discussed in Sect. 3.2, determined by the blade shape. From Fig. 3.6 it follows that the total pressure rise of a pure reaction machine ($R = 1$) equals zero. No work is done by the blades on the fluid, which moves purely in radial direction in the absolute frame, if no pre-whirl is applied. The shape of the workless blade follows from $v_u r = $ constant, i.e. no change of angular momentum. If the blade is still more backward curved, the machine becomes a turbine. A centrifugal form of a radial turbine is not advisable, however, since the centrifugal force causes pressure rise in the flow sense. Because a turbine is intended to utilise a pressure drop, principally, a radial turbine should be a centripetal machine.

With β_2 between 0 and the value for $R = 1$, the rotor has backward curved blades. Rotor work is rather small, but a significant part of it is used for pressure increase in the rotor. This type is appropriate if mainly intended to build up pressure. Conversion of kinetic energy into pressure by a downstream diffuser generates large losses. Large fans, for which efficiency is important, therefore exclusively have backward curved blades. With moderately backswept blades, the relative velocity in the rotor channels is lower for the same flow rate. Solid particles in the gas flow cause then less blade erosion. This fan type is thus appropriate for gases containing solid particles, as e.g. flue gases. It is also applied for pneumatic transport. Fans with forward curved blades realise the required total pressure rise at a lower peripheral speed. But, the total pressure rise at the rotor level is mainly available as kinetic energy. Efficiency is impaired if a large part of the kinetic energy at the rotor exit must be converted into pressure. Such a fan is not suitable to deliver air through a large ductwork with high energy dissipation, if a low efficiency is not acceptable.

Figure 3.25 shows rotors with backward and forward curved blades, and a motor incorporated in the hub. This is a motor with an external rotor: the rotor runs externally around a stator coil instead of running within a stator coil, as is more usual. Such motors are commonly used with fans of rather small size.

3.7 Axial and Mixed-Flow Fans

3.7.1 Degree of Reaction with Axial Fans

Figure 3.26 sketches two customary axial fan types. Figure 3.27 shows the corre-

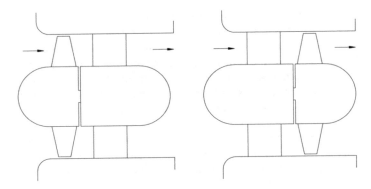

Fig. 3.26 Axial fans with high degree of reaction

sponding velocity triangles. The first type has an outlet guide vane ring with decelerating flow, downstream of the rotor. The second type has an inlet guide vane ring with accelerating flow, upstream of the rotor.

Both vane rings aim at axial flow at the machine entry and exit. The degree of reaction is very high, near to unity, with both types, being

$$R = \frac{\frac{1}{2}w_1^2 - \frac{1}{2}w_2^2}{u\,\Delta w_u} = \frac{\frac{1}{2}\left(w_{1u}^2 - w_{2u}^2\right)}{u(w_{2u} - w_{1u})} = -\frac{w_{mu}}{u}. \tag{3.48}$$

The meaning of w_{mu} is the tangential component of the mean relative velocity. With the machine with downstream vane ring $R < 1$ (80–90%); with upstream vane ring $R > 1$ (110–120%). Most fan applications aim at pressure increase and not at velocity increase (see Exercises 1.9.6 and 1.9.7 in Chap. 1). A high degree of reaction is thus beneficial. The velocities downstream and upstream of the fan are about equal (Fig. 3.26). There are also types with inlet guide vanes and outlet guide vanes and degree of reaction equal to unity (see Exercise 3.9.9.)

Differences in the performance of axial fans compared to radial ones concern the total pressure rise and the flow rate, relative to the peripheral speed and the diameter of the rotor, expressed by the total pressure coefficient $\psi_0 = \Delta p_0/\rho u_2^2$ and by the flow factor $\Phi = Q/(u_2\pi d_2^2/4)$. Figure 3.27 allows deriving that the pressure coefficient of an axial fan is about 0.15. For radial fans, the value is 0.30 (high degree of reaction) to 1.20 (low degree of reaction). The flow factor of an axial fan is of the 0.40 order, being 0.05 (backward curved blades with very narrow rotor) to 0.30 (forward curved blades with wide rotor) for radial fans. Axial fans are thus suitable for lower pressure rise and higher flow rate.

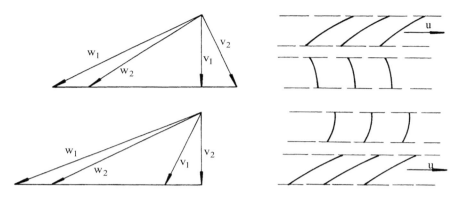

Fig. 3.27 Velocity triangles with axial fans; downstream vane ring (top) and upstream vane ring (bottom)

3.7.2 Free-Vortex and Non-free-vortex Types

Axial fans delivering to a duct or mounted in a duct are normally designed with (approximately) constant work along the radius. With free-vortex flow, $v_u r =$ constant, upstream and downstream of the rotor, a constant work distribution is obtained. Constant work is also an obvious choice with two-stage fans (Fig. 3.3). In practise, the free-vortex flow and constant work distribution are often not exactly satisfied. Figure 3.28 (left) shows a fan without stator with blades that satisfy approximately free-vortex flow and constant work distribution along the span.

With a single-stage axial fan, it may be advantageous to deviate strongly from free-vortex flow, with the objective to increase the work at larger radius. This potential is not used with free-vortex flow. Free-vortex blade sections near the hub have a higher lift coefficient and a higher chord than near the rotor periphery.

Not decreasing the lift coefficient with increasing radius and increasing the chord, the work by the fan may be strongly increased. Figure 3.28 (right) is an example. The

Fig. 3.28 Free-vortex axial fan (left, *Courtesy* Howden Group; right-turning) and non-free-vortex axial fan (right, *Courtesy* ebm-papst; left-turning)

work input varies then strongly along the radius, but this is no principal drawback for many applications of a single-stage fan. The efficiency decreases due to the non-constant work distribution, but forward sweep of the blade tip reduces the efficiency penalty.

Figure 3.29 (top, left) sketches the calculation of the circulation on a contour downstream of a rotor. When work increases with the radius, the angular momentum in the flow downstream of the rotor increases with the radius. The circulation on the contour drawn then does not equal zero:

$$\Gamma = \int_C \vec{v}.\, d\vec{s} = \int_S \vec{\omega}.\vec{n}\, dS.$$

With the Stokes circulation theorem follows that rotation $\vec{\omega}$ in the axial direction is created downstream of the rotor. Streamwise rotation is normally zero with constant work. The flow downstream of the rotor always features whirl due to the presence of the tangential velocity component, but, with a free-vortex flow, the rotation (ω) locally equals zero. This is the meaning of the term free-vortex flow: fluid particles nearer to the shaft have higher tangential velocity, keeping a direction linked to the flow parallel to itself during the vortex motion.

With non-constant work, the consequence is that a streamsurface that is cylindrical upstream of the rotor, does not remain so downstream.

Figure 3.29 (top, right) shows the twisted streamsurface where fluid at the pressure side migrates to the tip and fluid at the suction side migrates to the hub. Local vortex

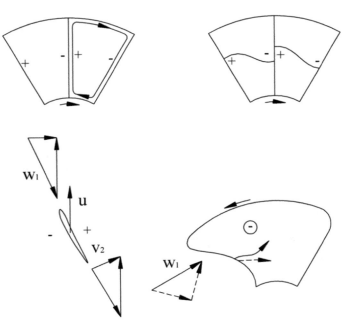

Fig. 3.29 Circulation on a contour downstream of the rotor; twist of the streamsurfaces; compensating twist by leading edge advancement at the tip

motions created in the blade wake take a part of the kinetic energy from the core flow and dissipate it downstream of the rotor. This generates a loss mechanism associated to the twist of the streamsurfaces.

Figure 3.29 (bottom) demonstrates that streamsurface twist may be compensated by advancing the blade leading edge at the tip. This is called forward sweep. The relative velocity at the leading edge has a component parallel to the leading edge and a component perpendicular to it. The parallel component remains almost unaffected with flow through the rotor. The perpendicular component increases at the suction side and decreases at the pressure side. As a consequence, the relative streamlines deviate as drawn. This generates twist on the streamsurfaces, which compensates the twist by the inhomogeneous work distribution.

3.7.3 Axial Fan Characteristics

Figure 3.30 (bottom left) is a sketch of the total pressure rise as a function of the flow rate of an axial fan with only a rotor. Flow patterns are sketched for a number of points on the characteristic. The characteristic curve has a maximum and a minimum. This feature remains when a guide vane ring is added.

Figure 3.27 demonstrates that, with flow rate decrease (with same direction of v_1), the angle of attack of the flow entering the rotor increases. This increases the rotor flow deflection and so the rotor work. The angle of attack on the stator, if downstream, increases as well (with same direction of w_2). The decrease of the flow rate thus increases the rotor work.

But boundary layer separation occurs below a certain flow rate, within the rotor or the stator. This diminishes the total pressure rise. The total pressure rise is higher

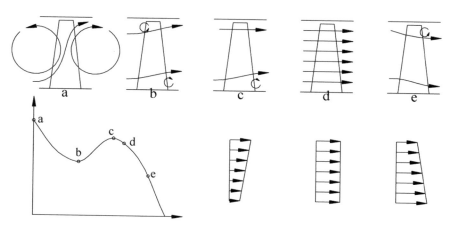

Fig. 3.30 Flow patterns through an axial fan and consequences for the shape of the characteristic curve; adapted from Pfleiderer [9]

however at a very low flow rate. Recirculation flows are generated, causing a significant radius increase for the through-flow (Fig. 3.30a). Due to the centrifugal force, the pressure rise increases in the flow.

From velocity triangles as on Fig. 3.27, at the rotor tip and hub, one sees that by flow rate reduction with respect to the design flow rate (Fig. 3.30c), separation first occurs at the hub (higher lift coefficient at the hub). By strong flow reduction, the flow pattern (a) is obtained. By flow rate increase with respect to the design flow rate (Fig. 3.30e), separation first occurs at the tip (negative angle of attack).

3.7.4 *Mixed-Flow Fans*

Some forms closely approach the axial form and others closely approach the radial form. Figure 3.31 shows the meridional section of a fan close to an axial type, with small radius change of the average flow. The corresponding velocity triangles show the axial velocity increase through the rotor by the decreasing blade height. The average blade speed increases as well, which increases the rotor work. Such a machine is, strictly spoken, a mixed-flow machine. But, because the blade shape remains the same as that with a purely axial machine, the machine is commonly termed as axial.

In Fig. 3.31, the pressure energy rise due to the centrifugal force $(u_2^2 - u_1^2)/2$ and the pressure energy drop due to the flow acceleration $(w_2^2 - w_1^2)/2$ are comparable. Consequently, the static pressure increase in the rotor is very low. The degree of reaction of the machine is thus very low. With the example, we understand that by appropriate blade height variation in the rotor, any degree of reaction may be obtained.

With a high degree of reaction fan, the total pressure coefficient related to the tip speed, amounts to about 0.15 (Fig. 3.27). With a low degree of reaction fan, this coefficient may be up to about 0.30 (Fig. 3.31), being equal to the lowest values obtained by radial fans with backward curved blades.

Advantages and disadvantages with the choice of the degree of reaction are the same with axial machines and radial ones: a lower degree of reaction implies a higher

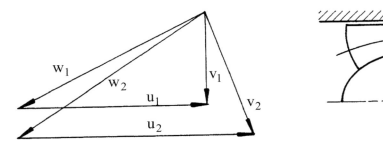

Fig. 3.31 Mixed-flow fan close to the axial form: low degree of reaction

Fig. 3.32 Mixed-flow fan
close to the radial form for
mounting in a duct; adapted
from Bleier [3]

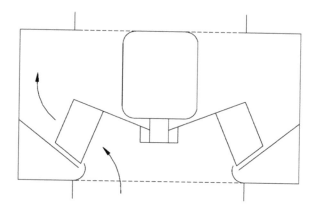

rotor work, but a lower efficiency if the kinetic energy downstream of the rotor has to
be converted into pressure rise. Many of these mixed-flow fans have a vaned diffuser
downstream of the rotor.

Figure 3.32 is a sketch of a mixed-flow fan, with a shape close to a radial fan.
This shape is sometimes chosen for fans mounted in a duct. The characteristics of
such a mixed-flow fan only deviate little from these of a centrifugal fan.

3.8 Flow Rate Control of Fans

The easiest is flow rate control by variable rotational speed. With Fig. 3.5 for a
centrifugal fan with backward-curved blades, one understands that the shape of the
velocity triangles is kept, if all velocity components vary proportional to the speed of
rotation. Incidence angles stay then the same. Thus, similarity of the velocity triangles
is obtained for flow rate proportional to rotational speed and energy rise proportional
to rotational speed squared. This similarity is thus realised on parabolas through
the origin in a map of characteristic curves. Further, losses stay proportional to the
rotor work and internal efficiency is thus maintained, provided that loss coefficients
are constant. The volumetric efficiency is also maintained, provided that contraction coefficients are constant. So, the global efficiency is approximately constant
on similarity parabolas. This may be verified on the characteristics for several rotational speeds shown in Fig. 3.24. Clearly, the result is the same for centrifugal fans
with forward-curved blades, axial fans and mixed-flow fans (velocity triangles on
Figs. 3.27 and 3.31).

With a fan moving gas in a ductwork, the load characteristic of required pressure rise as a function of flow rate is approximately parabolic, because friction and
dump losses are approximately proportional to the square of the flow velocity. The

load characteristic thus coincides approximately with a line of constant efficiency at varying rotational speed in a map of characteristic curves of the fan. So, the efficiency is approximately maintained when varying the flow rate by variation of the rotational speed. This form of flow rate control is therefore preferred and realised by variable frequency feeding of the electrical motor. But, for large power (above about 150 kW), the cost of the frequency controller becomes quite high. Mechanical means may then then be used.

The flow rate of a centrifugal fan with backward curved blades may be changed by adjustable inlet guide vanes. The velocity triangles at rotor entry on Fig. 3.33 show that pre-whirl with a tangential component of the absolute velocity in the rotation sense (closing the guide vanes) may be combined with flow rate reduction so that the direction of the relative velocity at the rotor entry remains the same. The rotor work stays then approximately the same due to the changed velocity triangle at the rotor exit (larger v_{2u}). The incidence at the volute entrance changes, but whether or not this decreases efficiency depends on the operating point. The conclusion is that with variable inlet guide vanes, characteristics may be shifted to lower or higher flow rates at approximately the same total pressure rise. In practice, there is some decrease of the efficiency due to incidences.

Figure 3.34 shows how by turning the rotor blades of an axial fan to a more tangential position (closing the rotor) the fan is adapted to operation at a lower flow rate. Also by changing the blade orientation, the angle of attack of the rotor blades may be lowered for obtaining a lower pressure rise. The characteristic thus shifts to a lower flow rate and a lower pressure rise. Similarly, by opening the rotor,

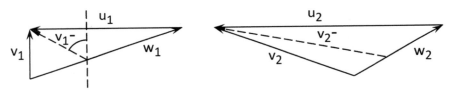

Fig. 3.33 Reduced flow rate by closing inlet guide vanes of a radial fan; corresponding velocity change at rotor exit for backward curved blades

Fig. 3.34 Reduction of flow rate and pressure rise by turning rotor blades of an axial fan to a more tangential position; top: vane ring downstream of rotor; bottom: vane ring upstream of rotor

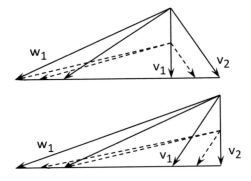

characteristics can be shifted to a higher flow rate and a higher pressure rise, but limited due to separation at too large angles of attack.

The figure suggests that, with the stator flow aligned with the vanes, the pressure rise and the flow rate change in the same proportion by turning the rotor blades, but this is not fully correct in practice due changes of incidences and deviations. These changes also cause an efficiency reduction, which increases with larger turning of the blades away from the design orientation.

With some larger industrial axial fans, the rotor blade angles can be changed at standstill, individually or collectively by means of a collar with lever arms. The rotor blades are then said to be adjustable. With some fans, the blade angles can be changed during operation, which is called variable blade angles.

3.9 Exercises

3.9.1 Large double-suction centrifugal fans are used for ventilation of mines. These are machines with rotor shape as on Fig. 3.2, but with two rotors mounted back-to-back and suction from two sides. The peripheral speed is set to a very high value for realisation of a large flow rate and a large pressure rise. The dimension data of (half) a rotor in the largest machines are in the order of $d_1 = 2750$ mm, $b_1 = 800$ mm, $d_2 = 5250$ mm, $b_2 = 450$ mm, together with $\beta_1^b = -55°$, $\beta_2^b = -45°$, and $n = 420$ rpm. Assume absence of pre-whirl by inlet guide vanes. Take the number of blades $Z = 10$ and blade thickness $t = 15$ mm.

- Draw the velocity triangle at rotor entry, just inside the rotor (station 1^b), for flow aligned with the blades. Calculate the rotor flow rate, taking into account the blade thickness (design flow rate).
- Draw the velocity triangle for the design flow rate, just upstream of the rotor exit (station 2^b). Assume flow aligned with the blades.
- Construct the velocity triangle just downstream of the rotor exit (station 2), taking blade thickness and slip into account. Estimate the slip by Wiesner's formula (3.21, 3.35) and by Pfleiderer's formula with $\lambda = 0.75$ (3.24, 3.25, 3.35). Observe the small difference between the two estimates of the tangential component of the rotor exit velocity v_{2u}. Take the average value as result.
- Calculate the rotor work and estimate the total pressure rise from an internal efficiency of 85% and air density equal to 1.22 kg/m³.
- Estimate the flow rate of the double-suction fan, assuming 95% volumetric efficiency.
- Estimate the shaft power, assuming 98% mechanical efficiency.
- Verify the deceleration ratio w_2^b/w_1^b and the moment solidity of the rotor (3.31).

A: $Q_{rot} = 283.81$ m³/s; $v_{2u} = 57.35$ m/s or 59.68 m/s; $\Delta W = 6755$ J/Kg; $\Delta p_0 \approx 7005$ Pa; $Q \approx 539$ m³/s; $P_{shaft} \approx 4770$ kW; $w_2^b/w_1^b = 0.74$; $\sigma_M = 0.73$.

3.9.2 A centrifugal fan has rotor internal and external diameters $d_1 = 500$ mm and $d_2 = 750$ mm. The diameter of the eye entry is $d_0 = 450$ mm. The rotor has

10 blades with thickness $t = 4$ mm. The blades are simply curved with width $b_1 = 180$ mm at the entry and $b_2 = 150$ mm at the exit. There is no pre-whirl. Blade angles are $\beta_1^b = -60°$ and $\beta_2^b = -45°$. The fan runs at 1450 rpm and delivers the volume flow rate $Q = 4$ m³/s ($\rho_{air} = 1.2$ kg/m³). Assume that the volumetric efficiency is $\eta_v = 0.95$ and the incidence loss coefficient at rotor inlet is $\mu_{def} = 0.80$. Take the finite number of blades into account. Apply Pfleiderer's formulae (3.24, 3.25, 3.35) to estimate the work reduction, with $\lambda = 0.75$.

- Determine the velocity triangles just upstream of the rotor entry (station 1) and just downstream of the rotor entry (station 1^b).
- Determine the deceleration ratio in the rotor eye: $\zeta = v_{1r}^b / v_0$ (acceleration by blade obstruction included).
- Determine the tangential deflection at the rotor entry and the corresponding incidence loss.
- Calculate the design flow rate of the rotor. Compare with $Q = 4$ m³/s.
- Determine the velocity triangles just upstream of the rotor exit (station 2^b) and just downstream of the rotor exit (station 2).
- Calculate the work transferred to the fluid and the degree of reaction.
- Determine the total pressure rise by the rotor ignoring all losses in the flow, except the incidence loss at the rotor entrance.
- Verify with the Pfleiderer moment coefficient (3.32) that the number of blades is appropriate. Verify for the design flow rate. Observe that the number of blades is larger than the minimum necessary for avoiding flow separation.

A: $\zeta = 0.624$; $q_{irr,inc1} = 46.5$ J/kg; $Q^* = 5.59$ m³/s; $\Delta W = 1772.0$ J/kg; $R = 0.749$; $\Delta p_{0,rot} = 2070.6$ Pa; $(C_M)^* = 0.86$.

3.9.3 Add a volute around the rotor of the previous exercise with width $b_3 = 375$ mm and height at the outlet $h_3 = 375$ mm. Assume a logarithmic spiral for the outer wall shape. Remark that the exit area of the volute is 88.4% of the suction inlet area of the fan ($d_0 = 450$ mm). With a diffuser downstream of the volute with an area ratio 1.131 the velocity at the discharge side of the fan thus becomes the same as at the suction side.

- Calculate the loss due to the radial dump at the entrance to the volute. Take into account that the complete rotor flow rate enters the volute during the dump (leakage flows included).
- Calculate the loss due to tangential deflection at the entrance to the volute. Assume as attenuation coefficient of this loss $\mu_{def} = 0.80$.
- Determine the rise of the total pressure through the resulting fan taking into account the dump loss at volute entrance and the incidence losses at rotor entry and at volute entry, ignoring all other losses in the flow.

A: $q_{irr,dump} = 27.68$ J/kg; $q_{irr,inc2} = 39.34$ J/kg; $\Delta p_0 = 1990.2$ Pa.

3.9.4 Assume with the fan of the two previous exercises a 3 mm gap between the volute and the rotor entry. Assume a contraction coefficient 0.90 for the gap flow. Calculate the leakage flow rate at the specified operating point (with the results of the previous exercises) and verify the estimated volumetric efficiency (ignore leakage to the outside). Assume that the total pressure in front of the gap on the volute side is the static pressure at volute entrance after the radial dump (station 2′) and that the static pressure on the side of the rotor is the static pressure just before entry to the rotor (station 1). **A:** $Q_{leak} = 0.22 \text{ m}^3/\text{s}$; $\eta_v = 0.95$.

3.9.5 The rotor of an axial fan has hub diameter $d_h = 0.24$ m and tip diameter $d_t = 0.48$ m. There is no pre-whirl. The inlet flow is axial on the quadratic average radius r_m ($2r_m^2 = r_t^2 + r_h^2$) and the relative velocity forms the angle $\beta_1 = -55°$ to the axial direction at the rotor entry and the angle $\beta_2 = -35°$ at the exit. The rotational speed is 2500 rpm and the air density is $\rho = 1.2 \text{ kg/m}^3$. Assume idealised flow, i.e. no internal losses. Assume as well that the flow on the average radius is characteristic for the entire machine.

- Determine the velocity triangles on the average radius.
- Determine the rotor work ΔW, the static pressure rise by the rotor Δp_{rot} and the degree of reaction R.
- Determine the volume flow rate Q and the rotor power P_{rot}.

A: $\Delta W = 1258 \text{ J/kg}$; $\Delta p_{rot} = 1125 \text{ Pa}$; $R = 0.745$; $Q = 4.72 \text{ m}^3/\text{s}$; $P_{rot} = 7124 \text{ W}$.

3.9.6 Assume a rotor with constant work and constant axial velocity along the radius (free-vortex flow) with the axial fan from the previous exercise.

- Determine the velocity triangles at the hub and the tip.
- Determine the degree of reaction and the work coefficient at hub and tip.
- Verify the realizability of the fan blades by calculating the deceleration ratios w_2/w_1 and the axial chord solidities, assuming a Zweifel coefficient of unity, at hub and tip. Assume constant axial velocity in the streamtubes at hub and tip. Observe that the necessary solidity at the hub is quite large.
- Perform the calculation once more, but with the work ΔW proportional to the radius (non-free vortex). Assume again constant axial velocity along the radius upstream and downstream of the rotor (in reality there is some variation: see Chap. 13). Compare with constant work distribution. Observe the much lower necessary solidity at the hub.

A: I: $R_{hub} = 0.36$; $\psi_{hub} = 1.27$; $R_{tip} = 0.84$; $\psi_{tip} = 0.32$; $(w_2/w_1)_{hub} = 0.77$; $(w_2/w_1)_{tip} = 0.77$; $(\sigma_a)_{hub} = 2.17$; $(\sigma_a)_{tip} = 0.46$; II: $R_{hub} = 0.60$; $\psi_{hub} = 0.81$; $R_{tip} = 0.80$; $\psi_{tip} = 0.40$; $(w_2/w_1)_{hub} = 0.75$; $(w_2/w_1)_{tip} = 0.71$, $(\sigma_a)_{hub} = 1.41$; $(\sigma_a)_{tip} = 0.67$.

3.9.7 Consider a two-stage axial fan according to Fig. 3.3, but with a stator-rotor pair followed by a rotor–stator pair, with the intermediate vane ring guiding the flow in the axial direction. The velocity triangles are as in Fig. 3.27 (but in reversed order). Two-stage axial fans are used for rather high flow rates (above 500 m³/s), combined with a rather high pressure rise (above 4000 Pa). Consider flow rate $Q = 660 \text{ m}^3/\text{s}$ and total pressure rise $\Delta p_0 = 6500 \text{ Pa}$, for rotor dimensions $d_{hub} = 2.10$ m and $d_{tip} = 4.20$ m at 590 rpm. Take as air density 1.25 kg/m³ and assume equal work done by both rotors. Estimate the internal efficiency $\eta_i = 0.885$ and ignore leakage flow ($\eta_v = 1$) and mechanical loss ($\eta_m = 1$).

- Determine the velocity triangles on the quadratic mean radius r_m.
- Determine the degree of reaction and the work coefficient on the radius r_m.
- Assume that the fan is realised with free-vortex flow. Estimate the necessary solidity at the hub of the rotor for a Zweifel lift coefficient of unity. Verify the deceleration ratio w_2/w_1. Determine the number of blades for an axial chord equal to 20% of the blade height.

A: I: $w_{2u} = -102.58$ m/s; $R = 1.14$; $(\sigma_a)_{hub} = 0.70$; $(w_2/w_1)_{hub} = 0.714$; $Z = 22$;
II: $w_{2u} = -73.93$ m/s; $R = 0.86$, $(\sigma_a)_{hub} = 1.30$; $(w_2/w_1)_{hub} = 0.732$; $Z = 41$.

3.9.8 There are several variants of the two-stage fan of the previous exercise. More common is 2 equal stages, either stator-rotor or rotor–stator. A special configuration is rotor-stator-rotor, which is a contraction of the rotor–stator and a stator-rotor of Fig. 3.26. The machine is kinematically equivalent with the fan in the previous exercise, but with the order of the stages reversed. Another variant is a two-stage fan with contra-rotating rotors. This variant is kinematically equivalent to the previous build. There are no stator vanes, but the machine requires two shafts rotating in opposite senses. The contra-rotating version is sometimes used with smaller fans where the rotors are mounted directly on the shafts of the driving motors. It is normally not used for larger fans with external motors.

The results of the previous exercise show that the stator-rotor combination is the most favourable concerning number of blades, which is due to the larger relative velocity. But, for the same reason, the rotor–stator combination is the best concerning efficiency. Estimate the friction force on the rotor blades by the kinetic energy on the mean position on the mean radius (velocity w_m) multiplied with the blade surface (and a friction force coefficient). Observe that the dissipation by friction, per pitch, is proportional to $w_m^3 \, c_m$, with c_m the average blade chord. The average blade chord may be estimated by $c_m/c_a \approx w_m/v_a$, with the axial chord c_a determined from a Zweifel lift coefficient, thus proportional to $s \, v_a \Delta w_u/w_2^2$. Conclude thus that the ratio of the dissipation by friction on the rotor blades to the rotor work is approximately proportional to $\zeta = w_m^4/(u_m v_a w_2^2)$. Prove by comparison of the ζ-coefficients that the rotor–stator combination is the most efficient (the losses in the stator rows are not estimated, but these are much smaller).

A: Rotor–stator: $\zeta = 2.26$; **Stator-rotor:** $\zeta = 3.30$.

3.9.9 Intermediate between the stator-rotor and rotor–stator stages of the previous exercises is two identical stator-rotor combinations with degree of reaction equal to unity (symmetry between v_1 and v_2) followed by a stator. Estimate the necessary number of rotor blades with this configuration for a Zweifel lift coefficient of unity at the hub. Derive that the rotor loss coefficient ζ is approximately the mean of the previous ones. **A:** $Z = 31$, $\zeta = 2.75$.

3.9.10 The number of rotor blades is quite large in the previous two exercises. This number may be reduced by larger blade speed. Estimate the necessary number of rotor blades for the configuration of the previous exercise for rotational speed multiplied by 1.25. Derive by the rotor loss coefficient ζ that dissipation by friction is about 6% larger (efficiency is about 1.5% lower). Observe that the reduction of the number of blades is significant. **A:** $Z = 18$, $\zeta = 2.91$.

References

1. Adachi T, Sugita N, Yamada Y (2004) Study on the performance of a Sirocco fan. Int J Rotating Machine 10:415–424
2. Aungier RH (2000) Centrifugal compressors: a strategy for aerodynamic design and analysis. ASME Press. ISBN 0-7918-0093-8
3. Bleier FP (1998) Fan handbook. McGraw-Hill. ISBN 0-07-005933-0
4. Eck B (1972) Ventilatoren, 5th edn. Springer. ISBN 3-540-05600-9
5. Gülich JF (2020) Centrifugal pumps, 4th edn. Springer. ISBN 978-3-030-14787-7
6. Japikse D, Marscher WD, Furst RB (1997) Centrifugal pump design and performance. Concepts ETI. ISBN 0-933283-09-1
7. Kim KY, Seo SJ (2004) Shape optimization of the forward-curved-blade centrifugal fan with Navier-Stokes analysis. J Fluids Eng 126:735–742
8. Kim S, Park J, Choi B, Baek J (2012) Flow analysis and assessment of loss models in the symmetric volute of a turbo-blower. J Fluids Engineering 134:011101
9. Pfleiderer C (1961) Die Kreiselpumpen für Flüssigkeiten und Gase, 5th edn. Springer, no ISBN
10. Wilson DG, Korakianitis (2014) The design of high-efficiency turbomachinery and gas turbines, 2nd edn. MIT Press. ISBN 978-0-262-52668-5

Chapter 4
Compressible Fluids

Abstract In further chapters we study machines that function with compressible fluids, first steam turbines (Chap. 6), then gas turbines and compressors (Chaps. 11–15). For analysis of these machines, knowledge of fundamentals of compressible fluid flow is necessary. We study compressible fluid flow in the present chapter, for one-dimensional steady flows. This term indicates a flow whose properties change in a single spatial direction, namely an average streamline, and that is uniform in the spatial directions perpendicular to this streamline and constant in time. The streamline need not be straight. The discussion is limited to what is strictly necessary for fundamental analysis of turbomachines. We justify the use of constant density for a compressible fluid with a limited change of the pressure, as done in the analysis of fans in the previous chapter. We refer to books on fluid mechanics for more in-depth study (Fox et al. in Fox and McDonald's introduction to fluid mechanics. Wiley, 2019; Gerhart et al. in Munson, Young and Okiishi's fundamentals of fluid mechanics. Wiley, 2016).

4.1 Basic Laws

The fundamental equations for one-dimensional flows were derived in Chap. 1. We apply them here to compressible fluids. The one-dimensional flow is an approximation of the flow in a blade passage of a turbomachine. This approximation is not very accurate. It is primarily intended to acquire fundamental insight. Multi-dimensional effects occur in real machines. We will study them introductorily in the chapter on steam turbines. More in-depth study of multi-dimensional effects is dealt with in the gas turbine and compressor chapters. Within most components of a turbine, the flow is accelerating. A stationary channel between two spaces in which a flow accelerates is called a *nozzle*, as shown in Fig. 4.1. Nozzle flow allows analysis of the functioning of steam turbines. In the present chapter, relations are derived for accelerating flow, but most apply to decelerating flow as well.

The mass conservation equation is $\rho v A =$ constant, where A is the cross-section area of the channel. Logarithmic differentiation results in

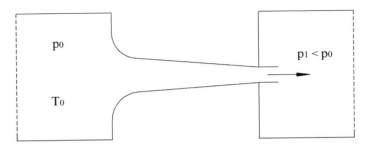

Fig. 4.1 Nozzle between two spaces (0 and 1)

$$\frac{d\rho}{\rho} + \frac{dv}{v} + \frac{dA}{A} = 0. \tag{4.1}$$

The work equation is

$$dW = d\frac{1}{2}v^2 + \frac{1}{\rho}dp + dU + dq_{irr}.$$

We consider a flow without work exchange ($dW = 0$) and as yet without losses ($dq_{irr} = 0$). Gravitational potential energy may always be ignored for a compressible fluid in a flow with significant pressure variation. This follows from the typical low density value; e.g. $\rho \approx 1.20$ kg/m^3 for air at sea-level atmospheric conditions. Example: a 1 m height change is a change of gravitational potential energy of about 10 J/kg. The corresponding pressure change is $dp = \rho dU \approx 12$ Pa, which is very small. Therefore, we simplify the work equation to

$$d\frac{1}{2}v^2 + \frac{dp}{\rho} = 0. \tag{4.2}$$

The energy equation in absence of work exchange and heat exchange is

$$0 = dh + d\frac{1}{2}v^2 + dU.$$

Gravitational potential energy may be ignored again. Example: a 10 J/kg enthalpy change represents a temperature change of $dT \approx 0.01$ K ($c_p = 1005$ J/kgK for air at 288 K). We simplify the energy equation to

$$dh + d\frac{1}{2}v^2 = 0. \tag{4.3}$$

A total differential is formed in the energy equation. We therefore use the concept of *total enthalpy* (equal to stagnation enthalpy when ignoring gravitational potential energy; see Chap. 1, Sect. 1.5.1), defined by

$$h_0 = h + \frac{1}{2}v^2. \tag{4.4}$$

The subscript 0 refers to the total value. The total enthalpy is constant in the nozzle: $h_0 = $ constant.

Combination of the work Eq. (4.2) and the energy Eq. (4.3) results in

$$T\,ds = dh - \frac{1}{\rho}dp = 0. \tag{4.5}$$

As expected, we find that the entropy is constant in a reversible adiabatic flow (= without heat exchange and without losses). In order to express this second constant, we use the concept of *total state*. The *stagnation state* is the state obtained by bringing to stagnation the flow in an adiabatic, reversible way. When the effect of gravitational potential energy may be ignored, the total state is equal to the stagnation state. The total state is denoted by the subscript 0.

From Eq. (4.5) follows

$$c_p dT = \frac{dp}{\rho},$$

and with the ideal gas law $p = \rho\,RT$:

$$c_p dT = RT\frac{dp}{p} \quad \text{or} \quad \frac{dp}{p} = \frac{c_p}{R}\frac{dT}{T}.$$

With $R = c_p - c_v$ and $\gamma = c_p/c_v$ this also results in

$$\frac{dp}{p} = \frac{\gamma}{\gamma - 1}\frac{dT}{T} \quad \text{or} \quad p \sim T^{\frac{\gamma}{\gamma-1}},$$

and

$$\rho T \sim T^{\frac{\gamma}{\gamma-1}} \quad \text{or} \quad \rho \sim T^{\frac{1}{\gamma-1}} \quad \text{and} \quad p \sim \rho^{\gamma}.$$

The process is thus polytropic with exponent γ.

For a given state p, ρ, T and v, the total state follows from

$$c_p T_0 = c_p T + \frac{1}{2}v^2, \quad \frac{p}{p_0} = \left(\frac{T}{T_0}\right)^{\frac{\gamma}{\gamma-1}} \quad \frac{\rho}{\rho_0} = \left(\frac{T}{T_0}\right)^{\frac{1}{\gamma-1}}. \tag{4.6}$$

The total state p_0, ρ_0, T_0 is constant within the nozzle and corresponds in Fig. 4.1 to the state within the first space, with an assumed stagnant fluid there.

With the formulae (4.6), it is assumed that the gas, apart from being *ideal* ($p = \rho RT$ and h only dependent on T), is *perfect* as well, i.e. that specific heat capacities c_p and

c_v are constant. In a real gas, these coefficients depend somewhat on temperature. For example, for a velocity change of $v = 0$ to $v = 300$ m/s, the kinetic energy change is 45 kJ/kg. The corresponding temperature change with air is about 40 K. The change of c_p over such a temperature difference is about 2–4 ‰. So, the approximation with constant coefficients is very accurate for air. A vapour cannot be represented accurately by an ideal and perfect gas, but some formulae for an ideal and perfect gas can still be used (see Exercise 4.10.5).

For adiabatic reversible flow, the energy equation in (4.6) may be written as

$$\frac{1}{2}v^2 = c_p T_0 \left[1 - \left(\frac{T}{T_0} \right) \right] = \frac{c_p}{R} \frac{p_0}{\rho_0} \left[1 - \left(\frac{T}{T_0} \right) \right], \tag{4.7}$$

or:

$$\frac{1}{2}v^2 = \frac{\gamma}{\gamma - 1} \frac{p_0}{\rho_0} \left[1 - \left(\frac{p}{p_0} \right)^{\frac{\gamma-1}{\gamma}} \right]. \tag{4.8}$$

This velocity–pressure relationship is termed the Barré de Saint Venant equation. Below (Sect. 4.7), we derive a similar equation for a flow with losses. The Saint–Venant equation is for a compressible fluid the equivalent of the Bernoulli equation for an incompressible (constant-density) fluid.

4.2   Compressibility and Velocity of Sound

A small pressure perturbation in a compressible fluid propagates. A pressure wave forms. The wave velocity is a function of the compressibility and may characterise it. Small pressure perturbations are called sound. The corresponding velocity is termed the *speed of sound*. Pressure waves with a large pressure change are called shock waves. Their propagation speed exceeds the speed of sound (see Fluid Mechanics).

Figure 4.2 represents a small pressure perturbation, propagating in a duct with a constant section, in which there is a stagnant fluid with state quantities p and ρ.

Fig. 4.2 Sound wave propagation in a duct

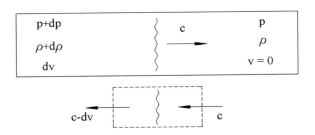

The propagation velocity is denoted by c. After passage of the sound wave, pressure, density and velocity become $p + dp$, $\rho + d\rho$ and dv.

We further consider a relative frame, moving together with the sound wave. The flow is steady within this frame and the laws formulated in the previous section apply. Before the wave, velocity equals c, after the wave c-dv.

Mass:

$$\frac{d\rho}{\rho} - \frac{dv}{c} = 0. \tag{4.9}$$

Work:

$$-cdv + \frac{dp}{\rho} = 0. \tag{4.10}$$

Elimination of dv from (4.9) and (4.10) gives

$$c^2\frac{d\rho}{\rho} = \frac{dp}{\rho} \quad \text{or} \quad c^2 = \frac{dp}{d\rho}. \tag{4.11}$$

Isentropy (reversible adiabatic flow) means

$$p \sim \rho^\gamma \quad \text{or} \quad \frac{dp}{p} = \gamma\frac{d\rho}{\rho}.$$

Thus:

$$c^2 = \gamma\frac{p}{\rho} \quad \text{or} \quad c = \sqrt{\frac{\gamma\,p}{\rho}}. \tag{4.12}$$

The compressibility of matter is characterised by the relative volume change caused by a pressure change. The ratio of the infinitesimal pressure increase to the corresponding infinitesimal relative volume decrease is called the bulk modulus or the volumetric elasticity modulus E. For a given mass m applies $m = \rho V$, so that $d\rho/\rho + dV/V = 0$. The compressibility coefficient β, being the inverse of the bulk modulus, is

$$\beta = \frac{1}{E} = -\frac{dV/V}{dp} = \frac{d\rho/\rho}{dp}.$$

For isentropic compression it follows that

$$E = \rho\frac{dp}{d\rho} = \rho c^2.$$

Thus, velocity of sound may be used as a measure of compressibility. Low velocity of sound corresponds to high compressibility. From Eq. (4.12) and the ideal gas law, it follows that

$$c = \sqrt{\gamma RT}. \tag{4.13}$$

With an ideal gas, the velocity of sound only depends on temperature. For air at 288 K ($\gamma = 1.4$; $R = 287$ J/kgK) it follows that $c = 340$ m/s.

According to the compressibility definition, an incompressible fluid (density not dependent on pressure) has an infinitely large elasticity modulus or an infinitely large speed of sound. No fluid is strictly incompressible, however. A liquid is compressible, but much less than a gas. For instance, water under sea level atmospheric conditions ($p = 1$ bar, $T = 288$ K) has a bulk modulus $E \approx 2 \cdot 10^9$ Pa. The corresponding velocity of sound is $c \approx 1400$ m/s, being larger than with air, but with a ratio under one order of magnitude.

4.3 Compressibility Effect on the Velocity–Pressure Relation

For a constant-density fluid the relation between velocity and pressure follows from the work equation or Bernoulli's Eq. (4.2), by

$$\frac{1}{2}v^2 + \frac{p}{\rho} = cst = \frac{p_0}{\rho} \quad \text{or} \quad \frac{1}{2}v^2 = \frac{p_0 - p}{\rho}. \tag{4.14}$$

For a compressible fluid, the relationship is more complex and given by the Saint–Venant formula (4.8). An important question for practice is how strictly necessary it is to use the compressible fluid Eq. (4.8) for a gas like air. The answer lies in the comparison of the term $\frac{1}{2}v^2$ to the term $\frac{\gamma}{\gamma-1}\frac{p_0}{\rho_0} = c_p T_0$. So, we may compare the flow velocity v to $a_0 = \sqrt{2c_p T_0}$, which is a velocity proportional to the velocity of sound $c_0 = \sqrt{\gamma R T_0}$ in stagnation conditions. For instance, for $T_0 = 288$ K, with $R = 287$ J/kgK and $\gamma = 1.4$ (air), $a_0 = 760$ m/s ($c_0 = 340$ m/s). In many flows, the flow velocity v is much smaller than a_0, which means that the term between the square brackets in Eq. (4.8) is small and thus p and p_0 do not differ much. We may then expand Eq. (4.8) as

$$\frac{1}{2}v^2 = \frac{\gamma}{\gamma-1}\frac{p_0}{\rho_0}\left[1 - \left(1 - \frac{p_0 - p}{p_0}\right)^{\frac{\gamma-1}{\gamma}}\right]$$
$$\approx \frac{\gamma}{\gamma-1}\frac{p_0}{\rho_0}\left[1 - \left(1 - \frac{\gamma-1}{\gamma}\frac{p_0-p}{p_0} + \cdots\right)\right],$$

Table 4.1 Comparison between results of the Saint–Venant and the Bernoulli equations for $T_0 = 288$ K, $R = 287$ J/kgK, $\gamma = 1.4$

p/p_0	0.999	0.99	0.95	0.8	0.4
v_{SV} (m/s)	12.860	40.732	91.747	189.042	365.060
v_{B,ρ_0} (m/s)	12.857	40.659	90.915	181.831	314.940
v_{B,ρ_m} (m/s)	12.860	40.731	91.744	188.923	361.296

and we recover the Bernoulli equation. This shows that for velocities that are low with respect to a_0, the Bernoulli equation is an accurate approximation of the Saint–Venant equation. In order to illustrate the quality of the approximation, Table 4.1 compares the results of both equations with the density in the Bernoulli equation set to ρ_0 or to $\rho_m = \frac{1}{2}(\rho + \rho_0)$. Therefore, Eq. (4.14) is written as

$$\frac{1}{2}v^2 = \frac{p_0}{\rho_0}\left[\frac{1 - (p/p_0)}{(\rho/\rho_0)}\right].$$

The differences between the formulae stay lower than 1% up to velocities of about 100 m/s. The Bernoulli equation with the mean value of the density stays accurate with an error less than 1‰ for velocities up to 200 m/s and less than 1% for velocities as high as the velocity of sound. So, in practice, very often, the Bernoulli equation with constant density may be used, even for a gas. The condition is that the flow velocity has to stay lower than about a third of the velocity of sound. This condition is certainly met in the analysis of fans in Chap. 3.

The ratio of the flow velocity to the velocity of sound is called the *Mach number*. The local Mach number is $M = v/c$. Up to a Mach number around unity, the Bernoulli equation with variable density, equal to the mean value, is accurate up to 1%. Substitution of the Saint–Venant equation by the Bernoulli equation with the mean density is very convenient for fundamental analysis of compressible fluid flow and is used in the chapter on axial compressors (Chap. 13).

With a liquid, the conclusion is also that compressibility may be ignored provided that the flow velocity stays low with respect to the velocity of sound. In water, the flow velocity typically is at maximum of the order of 10 m/s, while the velocity of sound is about 1400 m/s. So, constant density may be assumed with a very good approximation.

With high Mach number flows of an ideal and perfect gas, the compressible fluid relations (4.6)–(4.8) have to be used. It is then often more convenient to write these as functions of the Mach number. The energy equation in (4.6) may be written, with the use of the Mach number, as

$$\frac{T_0}{T} = 1 + \frac{v^2}{2c_p T} = 1 + \frac{\gamma R}{2c_p}\frac{v^2}{\gamma RT} = 1 + \frac{\gamma - 1}{2}M^2.$$

So

$$\frac{p_0}{p} = \left(1 + \frac{\gamma - 1}{2}M^2\right)^{\frac{\gamma}{\gamma-1}} \quad \text{and} \quad \frac{\rho_0}{\rho} = \left(1 + \frac{\gamma - 1}{2}M^2\right)^{\frac{1}{\gamma-1}}. \qquad (4.15)$$

4.4 Shape of a Nozzle

The cross-section area evolution follows from the mass Eq. (4.1):

$$\frac{dA}{A} = -\frac{d\rho}{\rho} - \frac{dv}{v}.$$

The density change is linked to the pressure change by isentropy (for adiabatic reversible flow):

$$\frac{d\rho}{\rho} = \frac{1}{\gamma}\frac{dp}{p}.$$

The velocity change is linked to the pressure change by the work Eq. (4.2):

$$-v dv = \frac{1}{\rho}dp \quad \text{or} \quad -\frac{dv}{v} = \frac{dp}{\rho v^2} = \frac{p}{\rho v^2}\frac{dp}{p} = \frac{1}{\gamma M^2}\frac{dp}{p}. \qquad (4.16)$$

The section change is

$$\frac{dA}{A} = \frac{1}{\gamma}\frac{dp}{p}\left(-1 + \frac{1}{M^2}\right) = \frac{1 - M^2}{\gamma M^2}\frac{dp}{p}. \qquad (4.17)$$

Further is

$$2\frac{dc}{c} = \frac{dp}{p} - \frac{d\rho}{\rho} = \left(1 - \frac{1}{\gamma}\right)\frac{dp}{p},$$

and

$$\frac{dM}{M} = \frac{dv}{v} - \frac{dc}{c} = -\left(\frac{1}{\gamma M^2} + \frac{\gamma - 1}{2\gamma}\right)\frac{dp}{p}. \qquad (4.18)$$

From Eq. (4.16) it follows that expansion ($dp < 0$) always causes velocity increase ($dv > 0$) and that compression ($dp > 0$) reversely causes velocity decrease ($dv < 0$). This follows also from the Saint–Venant formula (4.8). Equation (4.18) expresses that the Mach number increases by expansion and decreases by compression. From (4.17) it follows that expansion ($dp < 0$), within a subsonic flow ($M < 1$), requires a converging channel ($dA < 0$), while compression ($dp > 0$) requires a diverging channel

($dA > 0$). The opposite applies to a supersonic flow: expansion ($dp < 0$) requires a diverging channel ($dA < 0$) and compression ($dp > 0$) requires a converging one ($dA > 0$). The relations are visualised in Fig. 4.3.

We first consider the flow in a converging channel starting from a plenum as sketched in Fig. 4.1 and repeated in Fig. 4.4 (left). There is no flow when the backpressure at the outlet equals the pressure in the plenum. Flow is started when the backpressure is lowered. The Mach number then increases in the flow sense. With further lowering of the backpressure, the Mach number level in the channel increases. However, the Mach number cannot exceed unity at any place. Assuming that this occurs somewhere, the sketch in Fig. 4.3 (right bottom) demonstrates that the Mach number decreases then immediately. So, $M = 1$ can maximally be attained in the exit section of the converging channel.

The pressure in the exit section, corresponding to $M = 1$ is termed the *critical pressure* (p_*). The corresponding state is called *critical state*. The relevant state

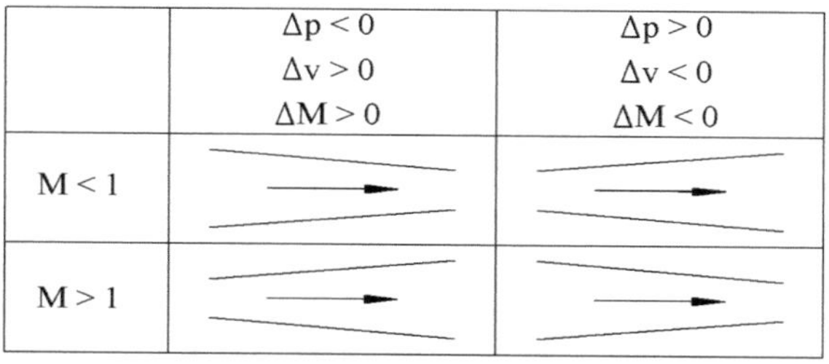

	$\Delta p < 0$ $\Delta v > 0$ $\Delta M > 0$	$\Delta p > 0$ $\Delta v < 0$ $\Delta M < 0$
$M < 1$		
$M > 1$		

Fig. 4.3 Evolution of flow quantities in subsonic and supersonic flow

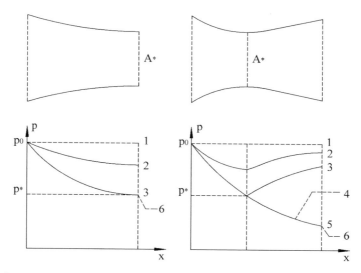

Fig. 4.4 Pressure evolution in a converging and converging–diverging nozzle

quantities follow from (4.15) for $M = 1$, being

$$\frac{T_*}{T_0} = \frac{2}{\gamma + 1}, \quad \frac{\rho_*}{\rho_0} = \left(\frac{2}{\gamma + 1}\right)^{\frac{1}{\gamma - 1}}, \quad \frac{p_*}{p_0} = \left(\frac{2}{\gamma + 1}\right)^{\frac{\gamma}{\gamma - 1}}. \quad (4.19)$$

For $\gamma = 1.4$ the corresponding numerical values are

$$\frac{T_*}{T_0} = 0.833, \quad \frac{\rho_*}{\rho_0} = 0.634, \quad \frac{p_*}{p_0} = 0.528.$$

As long as the backpressure of a converging nozzle exceeds the critical pressure ($p_3 = p_*$), the velocity in the exit section increases when the backpressure decreases. With the backpressure equal to the critical pressure $p_3 = p_*$, a sonic state is attained in the exit section. An additional decrease of the backpressure (p_6) cannot change the state within the nozzle compared to the state attained with $p_3 = p_*$, as the Mach number cannot exceed $M = 1$. This means that an expansion from p_3 to p_6 must occur within the space downstream of the nozzle. When the backpressure decreases, staying above p_3 however, the mass flow rate increases. It is blocked at the value corresponding to p_3 when the backpressure is lower than p. Such a state is termed *choked*. The choking mass flow rate equals $\rho_* v_* A_*$.

We consider now a converging–diverging channel, as shown in Fig. 4.4 (right). From the relations depicted in Fig. 4.3 we learn that, with a subsonic flow within a converging–diverging channel, the flow accelerates within the converging part and decelerates within the diverging part. Velocity and Mach number become maximal in the throat section. The channel is said to function as a subsonic Venturi tube. We notice that a similar flow, but everywhere supersonic, is possible when the inflow of the channel is supersonic. Velocity and Mach number attain then a minimum in the throat section. This is termed a supersonic Venturi tube.

From the relations in Fig. 4.3 follows that subsonic flow in the converging part may turn into supersonic flow in the diverging part. The critical state is then attained in the throat section. The backpressure p_3 in Fig. 4.4 (right) corresponds to a subsonic evolution with $M = 1$ being attained in the throat section. Pressure p_5 corresponds to a subsonic-supersonic evolution. For a backpressure between p_3 and p_5, the corresponding pressure evolution is no longer continuous everywhere. The supersonic branch of the pressure evolution in the diverging part is followed partially or completely, followed by a discontinuous compression, called a shock wave. Depending on the backpressure value p_4, the shock wave may occur within the diverging part or downstream of the exit section (see Fluid Mechanics).

With backpressure p_6 lower than p_5, there is an expansion downstream of the nozzle. In that case, the pressure evolution corresponding to p_5 is followed within the nozzle (see Fluid Mechanics).

With a converging–diverging nozzle, the mass flow is blocked at the choking value $\rho_* v_* A_*$, when the backpressure drops below p_3, as the critical state is then attained in the throat. The density ρ_3 is linked to the outlet pressure by an isentropic relation. The outlet velocity v_3 is linked to the outlet pressure by the Saint–Venant formula. Expressing constant mass flow rate, this implies a non-linear equation in outlet pressure, with two solutions, namely p_3 and p_5.

4.5 Expansion and Compression

In the foregoing analyses, the term expansion is used as equivalent for pressure decrease and the term compression for pressure increase. This terminology is common in texts on fundamental fluid mechanics and is inspired by the associated decrease or increase of density. The turbomachinery terminology is somewhat different, however. The term expansion is used for any process with significant pressure drop. The pressure decrease may be by increase of kinetic energy or by work extraction. The term compression refers solely to pressure rise by work supply. Pressure increase by decrease of kinetic energy is called *pressure recovery* or *diffusion*. From now on, we adhere to the turbomachinery terminology.

4.6 Nozzle with Initial Velocity

Up to now, we assumed that the velocity equals zero in the space at the nozzle entry, with an entry state being equal to the total state (state with subscript 0). From now on, we indicate the entry section with the subscript 0 (state $\neq$ total state) and take into account a velocity $v_0 \neq 0$. The above equations so must be adapted somewhat.

The energy equation is now

$$\frac{1}{2}v^2 + c_p T = \frac{1}{2}v_0^2 + c_p T_0.$$

With an isentropic relation between (p, T) and (p_0, T_0) follows

$$\frac{1}{2}v^2 = \frac{1}{2}v_0^2 + \frac{\gamma}{\gamma - 1}\frac{p_0}{\rho_0}\left[1 - \left(\frac{p}{p_0}\right)^{\frac{\gamma-1}{\gamma}}\right].$$

So an additional term appears in the Saint–Venant equation. But it is much more convenient to use the total state. We define the total state at station 0 (subscript double 0). The energy equation is then

$$\frac{1}{2}v^2 + c_p T = \frac{1}{2}v_0^2 + c_p T_0 = c_p T_{00}.$$

There is also an isentropic relation between (p, T) and (p_{00}, T_{00}). So it follows that

$$\frac{1}{2}v^2 = \frac{\gamma}{\gamma - 1}\frac{p_{00}}{\rho_{00}}\left[1 - \left(\frac{p}{p_{00}}\right)^{\frac{\gamma-1}{\gamma}}\right].$$ (4.20)

4.7   Nozzle with Losses: Infinitesimal Efficiency

In a flow with losses ($dq_{irr} > 0$), but without work and heat exchange, the total temperature is still a constant of the flow. The total pressure and total density are no longer constant. For $dq_{irr} > 0$ is $T\,ds > 0$.

From the first equality in (4.5) it follows that

$$T\,ds = c_p\,dT - RT\frac{dp}{p} \quad \text{or} \quad \frac{ds}{R} = \frac{\gamma}{\gamma - 1}\frac{dT}{T} - \frac{dp}{p}.$$

From (4.6) it follows that ($T_0 = cst$)

$$\frac{dp}{p} - \frac{dp_0}{p_0} = \frac{\gamma}{\gamma - 1}\frac{dT}{T} \quad \text{so that} \quad \frac{ds}{R} = -\frac{dp_0}{p_0}.$$

A decrease of total pressure corresponds to an increase of entropy.

Figure 4.5 (left) is the h–s diagram of the flow acceleration in a nozzle with losses between stations 0 and 1. With a quantity as p_{01}, the first subscript refers to the total state and the second to the location. Total enthalpy is constant ($h_{00} = h_{01}$). The entropy increases due to losses, causing a decrease of total pressure ($p_{01} < p_{00}$). Figure 4.5 (right) shows similarly the h–s diagram of flow deceleration with losses between station 1 and station 2. The total pressure decreases here as well.

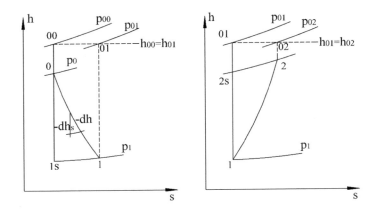

Fig. 4.5 Acceleration and deceleration with losses: decrease of total pressure

We consider an infinitesimal part of the expansion path in the h–s diagram in Fig. 4.5 (left). We define the efficiency of this infinitesimal process (subscript ∞), using the concept of isentropic efficiency (comparison of enthalpy decrease to isentropic enthalpy decrease; see thermodynamics) by

$$\eta_\infty = \frac{-dh}{-dh_s}.$$

With

$$T\,ds = dh - \frac{1}{\rho}dp, \text{ it follows that } dh_s = \frac{1}{\rho}dp \text{ and } \eta_\infty = \frac{\rho\,dh}{dp}.$$

For an ideal and perfect gas:

$$\eta_\infty = \frac{\rho c_p\,dT}{dp} = \frac{p}{RT}\frac{c_p\,dT}{dp} = \frac{c_p}{R}\frac{dT/T}{dp/p}.$$

Thus:

$$\eta_\infty \frac{dp}{p} = \frac{\gamma}{\gamma - 1}\frac{dT}{T}.$$

Assuming that $\eta_\infty = cst$ during the whole expansion, the process is polytropic. The efficiency η_∞ is called the *infinitesimal efficiency* or *polytropic efficiency*. The exponent of the process is denoted by n and is called the polytopic exponent.
From

$$p \sim \rho^n \text{ follows } p \sim T^{\frac{n}{n-1}} \text{ or } \frac{dp}{p} = \frac{n}{n-1}\frac{dT}{T}.$$

Thus:

$$\eta_\infty \frac{n}{n-1} = \frac{\gamma}{\gamma - 1}. \tag{4.21}$$

From

$$\eta_\infty n(\gamma - 1) = \gamma(n - 1) \text{ it follows that}$$

$$n = \frac{\gamma}{\gamma - \eta_\infty(\gamma - 1)} = \frac{\gamma}{\eta_\infty + (1 - \eta_\infty)\gamma}.$$

The denominator exceeds 1, since $\gamma > 1$, so it follows that $n < \gamma$.
Examples:

$$\gamma = 1.4; \ \eta_\infty = 0.9; \ n = 1.346,$$

$$\gamma = 1.3; \; \eta_\infty = 0.9; \; n = 1.262.$$

Taking the initial velocity into account, we obtain.

$$\frac{v_0^2}{2} + h_0 = h_{00} \quad \text{and} \quad \frac{v^2}{2} + h = h_{00}.$$

Polytropic:

$$\frac{h}{h_0} = \left(\frac{p}{p_0}\right)^{\frac{n-1}{n}}. \tag{4.22}$$

Isentropic:

$$\frac{h_0}{h_{00}} = \left(\frac{p_0}{p_{00}}\right)^{\frac{\gamma-1}{\gamma}}. \tag{4.23}$$

So (4.22) × (4.23) gives

$$\frac{h}{h_{00}} = \left(\frac{p_0}{p_{00}}\right)^{\frac{\gamma-1}{\gamma}} \left(\frac{p}{p_{00}}\right)^{\frac{n-1}{n}} \left(\frac{p_{00}}{p_0}\right)^{\frac{n-1}{n}}$$

$$= \left[\left(\frac{p_0}{p_{00}}\right)^{\frac{\gamma-1}{\gamma} - \frac{n-1}{n}}\right] \left(\frac{p}{p_{00}}\right)^{\frac{n-1}{n}} = C \left(\frac{p}{p_{00}}\right)^{\frac{n-1}{n}},$$

and

$$h_{00} = c_p T_{00} = \frac{c_p}{R} R T_{00} = \frac{\gamma}{\gamma - 1} \frac{p_{00}}{\rho_{00}}.$$

The factor C is very close to 1.

$$C = \left(\frac{h_0}{h_{00}}\right)^{1 - \left(\frac{n-1}{n}\right)/\left(\frac{\gamma-1}{\gamma}\right)} = \left(\frac{h_0}{h_{00}}\right)^{1-\eta_\infty} = \left(\frac{T_0}{T_{00}}\right)^{1-\eta_\infty} \quad \text{and}$$

$$\frac{T_{00}}{T_0} = 1 + \frac{\gamma - 1}{2} M_0^2.$$

Examples ($\eta_\infty = 0.9$):

$$\gamma = 1.4; \; M_0 = 0.3 : C = 0.998; \; M_0 = 0.5 : C = 0.995,$$
$$\gamma = 1.3; \; M_0 = 0.3 : C = 0.999; \; M_0 = 0.5 : C = 0.996.$$

The factor C may be set to 1 with a very good approximation.
With a very good approximation, the Saint–Venant formula thus becomes

$$\frac{v^2}{2} = \frac{\gamma}{\gamma - 1} \frac{p_{00}}{\rho_{00}} \left[1 - \left(\frac{p}{p_{00}} \right)^{\frac{n-1}{n}} \right]. \tag{4.24}$$

This approximation means considering a polytropic process between states 00 and 1. The approximation is obviously not accurate when the inlet kinetic energy is large with respect to the enthalpy drop in the nozzle.

Similarly to (4.16–4.18), the flow evolution follows now from

$$\frac{d\rho}{\rho} = \frac{1}{n} \frac{dp}{p},$$

$$d\frac{1}{2} v^2 = -dh = \eta_\infty \left(-\frac{dp}{\rho} \right) \quad \text{or} \quad \frac{dv}{v} = \eta_\infty \left(-\frac{dp}{\rho v^2} \right) = \eta_\infty \frac{-1}{\gamma M^2} \frac{dp}{p}.$$

$$\frac{dA}{A} = -\frac{d\rho}{\rho} - \frac{dv}{v} = \left(\frac{\eta_\infty}{\gamma M^2} - \frac{1}{n} \right) \frac{dp}{p},$$

$$\frac{dA}{A} = \frac{1}{n} \left(\frac{M_{th}^2}{M^2} - 1 \right) \frac{dp}{p} \quad \text{with} \quad M_{th}^2 = \eta_\infty \frac{n}{\gamma} = \frac{n-1}{\gamma - 1}. \tag{4.25}$$

$$2\frac{dc}{c} = \frac{dp}{p} - \frac{d\rho}{\rho} = \left(1 - \frac{1}{n} \right) \frac{dp}{p},$$

$$\frac{dM}{M} = \frac{dv}{v} - \frac{dc}{c} = -\left(\frac{\eta_\infty}{\gamma M^2} + \frac{n-1}{2n} \right) \frac{dp}{p}.$$

The last equation demonstrates that the changes of pressure and Mach number always have opposite signs. With Venturi flow, the minima or maxima of pressure and Mach number are still attained at the throat (Eq. 4.25), as in a nozzle without losses. The throat Mach number, M_{th}, now plays the role of critical Mach number. The distinction is now between subcritical flow ($M < M_{th}$) and supercritical flow ($M > M_{th}$). The choking mass flow rate is $\rho_{th} v_{th} A_{th}$. We notice that the throat Mach number is lower than 1 at choking.

State parameters in the throat at choking are

$$\frac{T_{th}}{T_{00}} = \frac{2}{n+1}, \quad \frac{\rho_{th}}{\rho_{00}} = \left(\frac{2}{n+1} \right)^{\frac{1}{n-1}}, \quad \frac{p_{th}}{p_{00}} = \left(\frac{2}{n+1} \right)^{\frac{n}{n-1}}.$$

These formulae replace the formulae (4.19). The corresponding throat velocity follows with (4.25) from

$$v_{th} = c_{th} M_{th} = c_{00} \sqrt{\frac{2}{n+1}} \sqrt{\frac{n-1}{\gamma - 1}}. \tag{4.26}$$

4.8   Isentropic and Polytropic Efficiencies

With isentropic flow acceleration in a nozzle as sketched in Fig. 4.5 (left), the exit velocity follows from

$$\frac{v_{1s}^2}{2} = h_{00} - h_{1s}.$$

With losses, the exit velocity is

$$\frac{v_1^2}{2} = h_{00} - h_1.$$

An often used definition of the isentropic efficiency of the expansion is then

$$\eta_s = \frac{v_1^2/2}{v_{1s}^2/2} = \frac{h_{01} - h_1}{h_{00} - h_{1s}}. \tag{4.27}$$

Such an *isentropic efficiency* is common, but the concept is somewhat misleading. Strictly, efficiency is the ratio of the useful part of an energy conversion process to the total amount of energy processed. Expression (4.27) does not satisfy this definition as it is the generated kinetic energy compared to the kinetic energy that would be generated with a loss-free process (isentropic). The results of two different processes are thus compared.

The work equation for the nozzle is

$$dW = 0 = d\frac{1}{2}v^2 + \frac{1}{\rho}dp + dq_{irr}.$$

The $d\frac{1}{2}v^2$ term is the useful part of the energy conversion. This term may be integrated to $\frac{1}{2}v_1^2$, if the total state (00) is considered as the initial state. The pressure energy used to generate the kinetic energy is

$$-\Delta h_r = -\int_{00}^{1} \frac{1}{\rho}dp. \tag{4.28}$$

The integral can only be determined if the expansion path is known. A fundamental problem arises since the internal details of a thermodynamic process are mostly unknown. Moreover, we wish to make a statement about the efficiency of a thermodynamic process, exclusively based on the initial and final states. From the above, we learn that this actually is impossible for a process with a compressible fluid.

Both the *isentropic efficiency* and *infinitesimal efficiency*, also called *polytropic efficiency*, are efficiency assessments, for which a process path has been agreed on for the calculation of (4.28).

The isentropic efficiency corresponds to the path $00 \rightarrow 1_s \rightarrow 1$(Fig. 4.5, left). To the $00 \rightarrow 1_s$ part applies

$$dW = 0 = d\frac{1}{2}v^2 + \frac{1}{\rho}dp.$$

The energy used is $\frac{1}{2}v_{1s}^2$. To the $1_s \rightarrow 1$ part applies

$$dW = 0 = d\frac{1}{2}v^2 + \frac{1}{\rho}dp + dq_{irr}.$$

As this path is isobaric, there is no contribution to (4.28) and integration is possible with result

$$q_{irr} = \frac{1}{2}v_{1s}^2 - \frac{1}{2}v_1^2. \tag{4.29}$$

The efficiency definition (4.27) corresponds to this interpretation. Dissipation during the process is considered as the enthalpy difference between states 1 and 1 s:

$$q_{irr} = h_1 - h_{1s}.$$

With the infinitesimal efficiency, the path is determined by defining the efficiency of an infinitesimal part, which is indisputably possible, and by assuming a constant infinitesimal efficiency on the total path. With the approximations introduced in the preceding section, the infinitesimal efficiency follows from expressions (4.24) and (4.21).

The generated kinetic energy (4.24) is

$$\frac{v_1^2}{2} = \frac{\gamma}{\gamma - 1}\frac{p_{00}}{\rho_{00}}\left[1 - \left(\frac{p_1}{p_{00}}\right)^{\frac{n-1}{n}}\right]. \tag{4.30}$$

The generated kinetic energy following the isentropic path is similarly

$$\frac{v_{1s}^2}{2} = \frac{\gamma}{\gamma - 1}\frac{p_{00}}{\rho_{00}}\left[1 - \left(\frac{p_1}{p_{00}}\right)^{\frac{\gamma-1}{\gamma}}\right]. \tag{4.31}$$

The integral (4.28) on a polytropic path with exponent n results in

$$-\Delta h_r = \frac{n}{n - 1}\frac{p_{00}}{\rho_{00}}\left[1 - \left(\frac{p_1}{p_{00}}\right)^{\frac{n-1}{n}}\right]. \tag{4.32}$$

For $n < \gamma$, it is verified that

Fig. 4.6 Approximation of
an expansion by isentropic
and isobaric paths

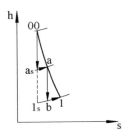

$$\frac{v_1^2}{2} < \frac{v_{1s}^2}{2} < |-\Delta h_r|.$$

The isentropic efficiency (ratio of $\frac{1}{2} v_1^2$ to $\frac{1}{2} v_{1s}^2$) is thus larger than the polytropic
or infinitesimal efficiency (ratio of $\frac{1}{2} v_1^2$ to $|-\Delta h_r|$).

The cause of the difference becomes obvious by considering a process consisting
of two isentropic and two isobaric paths, as shown in Fig. 4.6.

The efficiency for the $00 \rightarrow a_s \rightarrow a \rightarrow b \rightarrow 1$ path is

$$\eta = \frac{h_{00} - h_1}{(h_{00} - h_{as}) + (h_a - h_b)}.$$

The slope of an isobar in a h–s diagram is $\left(\frac{dh}{ds}\right)_p = T$. So, the isobars diverge in
the sense that the isentropic enthalpy difference between two isobars increases with
increasing entropy. It follows that $h_a - h_b > h_{as} - h_{1s}$. This inequality expresses
that a part of the loss on the $a_s \rightarrow a$ path, causing heating of the gas, is recovered
during the following expansion, because the corresponding isentropic enthalpy drop
increases. This effect is termed *reheat effect*, meaning heating due to internal loss.

Further breakdown of the $00 \rightarrow 1$ path in isentropic and isobaric parts results in a
polytropic representation as a limit case. This demonstrates that, with the polytropic
representation, the energy (4.28) used to generate the kinetic energy is larger than
with the isentropic-isobaric process.

We also understand that losses intervene in two ways: the useful result $\frac{1}{2} v_1^2$ is
smaller than with the isentropic process $\frac{1}{2} v_{1s}^2$ and the energy converted to achieve
this result $|-\Delta h_r|$ is larger than with the isentropic process $\frac{1}{2} v_{1s}^2$.

The infinitesimal representation describes the real process better than the
isentropic-isobaric representation. Both representations constitute models however,
as a correct assessment of efficiency is impossible without full details of the process
path. The infinitesimal efficiency has the advantage of an unchanged value when two
processes are connected in series, as demonstrated in Fig. 4.6. The series arrangement
of two expansions with the same isentropic efficiency has a larger overall isentropic
efficiency.

The different behaviour of isentropic efficiency and polytropic efficiency is similar
for expansions and compressions with work (see Chap. 11). Because of the preserva-
tion of efficiency with connection in series, use of polytropic efficiency is advisable

with cycle studies. Both definitions may be used when assessing a single machine component. Isentropic efficiency is the most visual, as the representation of losses in the h–s diagram is then clear. We therefore apply isentropic efficiency with the fundamental study of turbomachine components, as with the study of steam turbines in Chap. 6. The Saint–Venant formula (4.24) is used to calculate expansion processes within steam turbine components and infinitesimal efficiency indirectly intervenes when determining the polytropic exponent.

For completeness, we add two remarks. The first concerns alternative definitions of the isentropic efficiency. For the flow acceleration shown in Fig. 4.5 (left), the isentropic efficiency may also be defined by

$$\eta_s = \frac{h_0 - h_1}{h_0 - h_{1s}}. \tag{4.33}$$

The point of view is then to define the efficiency on the increase of the kinetic energy. A similar definition on the decrease of the kinetic energy in the flow deceleration shown in Fig. 4.5 (right) is

$$\eta_s = \frac{h_{2s} - h_1}{h_2 - h_1}. \tag{4.34}$$

The definitions (4.27), (4.33) and (4.34) are all definitions by which the loss is quantified by an enthalpy difference on the isobar through the static end point, thus $h_1 - h_{1s}$ or $h_2 - h_{2s}$. This is the most common way of conventionally defining the loss. But, looking at Fig. 4.5, there are obviously some more ways of expressing isentropic efficiency by a ratio smaller than unity (see Chaps. 13 and 14).

The second remark is that there is no difference between the isentropic and polytropic efficiencies with a constant-density fluid. Also with constant density, the definition of the infinitesimsal efficiency of an expansion is

$$\eta_\infty = \frac{-dh}{-dh_s} \quad \text{with} \quad dq_{irr} = T ds = dh - \frac{1}{\rho} dp,$$

so that

$$dh_s = \frac{1}{\rho} dp \quad \text{and} \quad \eta_\infty = \frac{-(1/\rho)dp - dq_{irr}}{-(1/\rho)dp}. \tag{4.35}$$

Further:

$$\Delta h_s = \frac{1}{\rho} \Delta p \quad \text{and} \quad q_{irr} = \Delta h - \frac{1}{\rho} \Delta p,$$

so that

$$\eta_s = \frac{-(1/\rho)\Delta p - q_{irr}}{-(1/\rho)\Delta p}. \tag{4.36}$$

For constant ρ and constant η_∞, the infinitesimal efficiency expression (4.35) can be integrated leading to the same result as the isentropic efficiency expression (4.36).

4.9 Effect of Heat Transfer

In the basic analysis of turbomachine components, adiabatic flow is assumed, but a real flow is seldom perfectly adiabatic. The definitions of isentropic efficiency and polytropic efficiency are based on adiabatic flow. One may thus doubt about the use of these notions in practice. Therefore, we analyse the influence of heat transfer on the example of nozzle flow of an ideal gas with p_{00} and T_{00} as inlet conditions and p_1 as backpressure. The h–s diagram for adiabatic flow is Fig. 4.5, left.

We assume distributed heat transfer proportional to the enthalpy drop according to $dq = \kappa(-dh)$, with constant κ, positive for heat addition to the nozzle and negative for heat removal. The equations for work and energy are (work is zero):

$$dW = d\tfrac{1}{2}v^2 + \tfrac{1}{\rho}dp + dq_{irr}, \quad \text{or} \quad d\tfrac{1}{2}v^2 = -\tfrac{1}{\rho}dp - dq_{irr},$$

$$dW + dq = dh + d\tfrac{1}{2}v^2, \quad \text{or} \quad d\tfrac{1}{2}v^2 = -dh + dq = (1+\kappa)(-dh).$$

The work equation allows the definition of the infinitesimal efficiency in the same way as with adiabatic flow: $d\tfrac{1}{2}v^2 = \eta_\infty(-\tfrac{1}{\rho}dp)$. With this relation substituted in the energy equation, it follows, for ideal gas and constant values of heat capacity, η_∞ and κ, that the expansion process is polytropic with exponent

$$\frac{n}{n-1} = \frac{1+\kappa}{\eta_\infty}\frac{\gamma}{\gamma-1}.$$

With the polytropic relations, we can derive the formulae for generated kinetic energy $\tfrac{1}{2}v_1^2$, converted pressure energy, $-\Delta h_r$, and enthalpy drop, $-\Delta h$, analogous to the formulae (4.30)–(4.32) for adiabatic flow. With constant coefficients, the following relations hold:

$$\tfrac{1}{2}v_1^2 = \eta_\infty(-\Delta h_r), \quad \tfrac{1}{2}v_1^2 = (1+\kappa)(-\Delta h).$$

The formula for $-\Delta h_r = -\int_{00}^{1}\left(\tfrac{1}{\rho}\right)dp$ is Eq. 4.32. It is the same for adiabatic and non-adiabatic flow, but with a different value of the exponent n. Table 4.2 lists results of $\tfrac{1}{2}v_1^2/h_{00}$ for $\eta_\infty = 0.90$ and $\gamma = 1.4$, for pressure ratios $p_{00}/p_1 = 1.2$, 1.5 and 2 and several values of κ. The generated kinetic energy is not much changed

Table 4.2 Effect of heat transfer on $\frac{1}{2}v_1^2/h_{00}$ for nozzle flow; $\eta_\infty = 0.9$, $\gamma = 1.4$

p_{00}/p_1	$\kappa = 0$	$\kappa = 0.1$	$\kappa = -0.1$	$\kappa = 0.2$	$\kappa = -0.2$
1.2	0.0458	0.0459	0.0457	0.0460	0.0455
1.5	0.0990	0.0995	0.0984	0.0999	0.0978
2.0	0.1633	0.1645	0.1617	0.1656	0.1598

by the heat transfer. But, by the relation $\frac{1}{2}v_1^2 = (1 + \kappa)(-\Delta h)$, the enthalpy drop $(-\Delta h)$, and thus the temperature drop, is significantly changed.

The conclusion is that, up to moderate levels of heat transfer, the performance of the nozzle can be evaluated by adiabatic formulae based on measured total pressure and total temperature at the nozzle entry and measured static pressure and total pressure at the nozzle exit. A measured exit temperature should not be used.

4.10 Exercises

4.10.1 Air flows through a pipe with a built-in nozzle. $D = 0.40$ m, $d = 0.20$ m. The inflow conditions are $p_1 = 150$ kPa, $T_1 = 300$ K. $\Delta p = p_1 - p_2 = 50$ kPa. Determine the mass flow rate, ignoring the losses within the nozzle (they are extremely small in practice) and assuming that no contraction occurs within the jet leaving the nozzle (contraction is small in practice). What is the flow rate error if the calculation is made assuming constant density determined from the upstream conditions? What is the flow rate error if the calculation is made assuming constant density determined by the average value in points 1 and 2? The calculations require derivation of mass flow rate formulae for constant density and for ideal gas with specified static flow conditions at inflow (see hereafter).

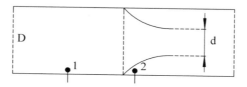

A: 10.71 kg/s; +26.5%; +18.3%.

The results obtained with the constant density formulae differ strongly from those for an ideal gas. We analyse the origin of these differences.

The relations for constant density are:

Bernoulli:

$$\frac{p_1}{\rho} + \frac{v_1^2}{2} = \frac{p_2}{\rho} + \frac{v_2^2}{2}.$$

Mass:

$$v_1 D^2 = v_2 d^2 \quad \text{or} \quad v_1 = v_2 \beta^2.$$

Thus

$$\frac{p_1 - p_2}{\rho} = \frac{v_2^2}{2}(1 - \beta^4) \quad \text{and} \quad \dot{m} = \rho A_{th} \frac{1}{\sqrt{1 - \beta^4}} \sqrt{\frac{2\Delta p}{\rho}}. \tag{4.37}$$

The relations for ideal gas are:
Conservation of energy:

$$c_p T_1 + \tfrac{1}{2} v_1^2 = c_p T_2 + \tfrac{1}{2} v_2^2.$$

Thus

$$\frac{v_2^2}{2} - \frac{v_1^2}{2} = c_p T_1 \left(1 - \frac{T_2}{T_1}\right).$$

Isentropy:

$$\frac{T_2}{T_1} = \left(\frac{p_2}{p_1}\right)^{\frac{\gamma-1}{\gamma}} \quad \text{and} \quad \frac{\rho_2}{\rho_1} = \left(\frac{p_2}{p_1}\right)^{\frac{1}{\gamma}}. \tag{4.38}$$

Mass:

$$\rho_1 v_1 = \rho_2 v_2 \beta^2.$$

So

$$\frac{v_2^2}{2}[1 - \beta^4(\rho_2/\rho_1)^2] = \frac{\gamma}{\gamma - 1} \frac{p_1}{\rho_1} \left[1 - \left(\frac{p_2}{p_1}\right)^{\frac{\gamma-1}{\gamma}}\right],$$

and

$$\dot{m} = \rho_2 A_{th} \frac{1}{\sqrt{1 - \beta^4(\rho_2/\rho_1)^2}} \sqrt{2 \frac{\gamma}{\gamma - 1} \frac{p_1}{\rho_1} \left[1 - (\frac{p_2}{p_1})^{\frac{\gamma-1}{\gamma}}\right]}. \tag{4.39}$$

The expression (4.39) can be brought into the form (4.37) with very good approximation as

$$\dot{m} = \rho_2 A_{th} \frac{1}{\sqrt{1 - \beta^4(\rho_2/\rho_1)^2}} \sqrt{\frac{2\Delta p}{\rho_m}}. \tag{4.40}$$

With (4.40), the mass flow result becomes very good. The last terms in (4.39) and (4.40) differ somewhat. The velocities by the formulae of De Saint Venant and Bernoulli) are 256.78 m/s (exact) and 256.24 m/s (approximate).

The large error by the formula (4.37) for constant density is not due to the approximation of the velocity–pressure relation, but comes from the large influence of the variable density in the mass flux balance between the flow in the pipe and the flow in the nozzle exit. So, one has to be careful with the approximation of a fluid of variable density by a fluid of constant density with density equal to a mean value. The approximation is accurate for the velocity–pressure relation, but, obviously, constant density should not be used in mass flux expressions.

4.10.2 Air flows through a pipe with diameter $D = 0.40$ m connected to a converging nozzle with exit diameter $d = 0.20$ m (similar geometry as in previous exercise). The inflow conditions are $p_1 = 150$ kPa, $T_1 = 300$ K. The flow ends in the atmosphere with $p_a = 100$ kPa ($T_a = 288$ K).

Calculate the mass flow rate (same question as in the previous exercise, losses and contraction to be ignored). Verify that the flow state is subcritical. What is the mass flow rate for $p_1 = 200$ kPa (slightly supercritical)? What is the mass flow rate for $p_1 = 500$ kPa (strongly supercritical)? What is the mass flow rate for $\eta_\infty = 0.95$ in that last case?

Remark that the pressure ratio through the nozzle for choking can be obtained by expressing that the mass flow rate attains a maximum. A hint is to introduce the pressure ratio to the power $\gamma/(\gamma - 1)$ as a new variable. The condition for maximum can easily be expressed because $2/(\gamma - 1) = 5$ for $\gamma = 1.4$. The value of the introduced variable can be obtained by natural iteration because the diameter ratio to the fourth power is a rather small number. The starting value can be the solution for diameter ratio equal to zero. With the expansion exponent different from 1.4 (last question), a similar procedure can be followed, but the expressions are more complicated.

A: 10.71 kg/s ($M_{th} = 0.78$); 14.85 kg/s ($M_{th} = 1$, $p_{th} = 107.25$ kPa); 37.13 kg/s ($2.5 \times$ foregoing); 35.94 kg/s ($M_{th} = 0.965$, $p_{th} = 270.48$ kPa).

4.10.3 The figure is a sketch of a rocket nozzle. The converging and the diverging parts are designed such that the exit flow is uniform, which is obtained by absence of shock waves in the diverging part. The section ratio exit to throat is 25. Determine the exit Mach number and the pressure ratio for the nozzle with isentropic flow of air ($\gamma = 1.4$). There are two solutions for the Mach number. The target is the Mach number for supersonic outflow. Observe that the ratio of exit pressure to inlet pressure is very small.

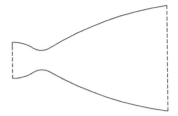

A: $M = 5$, $p/p_{00} \approx 0.00189$.

4.10.4 The figure shows the section at the mean radius of the stator and the rotor of an axial turbine with zero degree of reaction. The nozzle exit flow is supersonic. The rotor entry flow is near to sonic ($M \sim 1$). Turbines of this type are applied to drive fuel pumps in rocket engines. The fluid is gaseous hydrogen, obtained by heating liquid hydrogen by a coil that is wound around the diverging part of the rocket nozzle (see previous exercise).

The shaded zones indicate separated flow in the rotor at the suction side trailing edge. Due to flow separation, the rotor efficiency is rather low (for a definition of rotor efficiency see Chap. 6). For turbines with a very large work output, it is impossible however to attain an advantageous rotor efficiency. A low efficiency is no problem as the energy by an ideal expansion of the hydrogen is much larger than the energy required for driving the pumps. There is no separation in the turbine nozzles so that these may be calculated as lossless with a good approximation. The nozzle outlet angle is 72°. On the figure, we read for the nozzle solidity:$\sigma_a \approx 0.8$, and for the area ratio of the diverging part:$A_1/A_{th} \approx 1.60$.

Verify the correctness of the nozzle design by calculating the tangential force coefficient of Zweifel (Eq. 2.26 in Chap. 2). Assume a test with air ($\gamma = 1.40$). Determine the nozzle pressure ratio for design flow in the nozzle, thus without shock waves and without expansion waves downstream. Solve for the pressure ratio to the power $\gamma/(\gamma - 1)$. There are two solutions for this variable. The target is the pressure ratio for supersonic outflow.

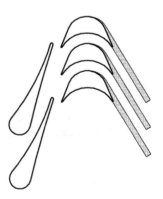

A: $C_{Fu} \approx 0.735$, thus enough solidity; $p_1/p_{00} \approx 0.141$.

4.10.5 Superheated steam can be represented as an ideal gas with a specific heat ratio γ around 1.30 for states far away from the saturation line. Wet steam deviates quite strongly from an ideal gas. But, remarkably, the polytropic relation between density and pressure, $\rho/\rho_{00} = (p/p_{00})^{1/\gamma}$, and the Saint–Venant enthalpy-pressure relation, $h_{00} - h = \frac{\gamma}{\gamma-1} \frac{p_{00}}{\rho_{00}} \left[1 - \left(\frac{p}{p_{00}} \right)^{\frac{\gamma-1}{\gamma}} \right]$, obtained for an ideal and perfect gas, can be fitted accurately to an isentropic expansion, even in the wet steam area. What thus has to be avoided is the use of the ideal gas law, $p = \rho R T$.

Verify first the density-pressure relation. Take as an example an isentropic expansion starting from saturated vapour at 6 bar until 0.5 bar. Calculate density for the pressure levels 3.5 bar, 2 bar, 1 bar and 0.5 bar.

Use steam tables from a book on thermodynamics or from a website. A basic Mollier-diagram for water-steam can be found on: http://www.engineeringtoolbox.com/mollier-diagram-water-d_308.html. Steam properties may be calculated with: https://beta.spiraxsarco.com/Resources-and-design-tools. The starting point of the expansion defines the entropy. Values of specific volume (inverse of the density), entropy and enthalpy are obtained by linear interpolation as a function of the dryness factor (also called steam quality: ratio of the mass fraction of vapour to the mass of the mixture of saturated liquid and saturated vapour).

Verify then that the enthalpy-pressure relation is accurately satisfied with the obtained value of the specific heat ratio. Observe that the ideal gas law, $p = \rho R T$, is not well satisfied (R is not a constant), but remark that the ideal gas law can be avoided in the calculation of the expansion through a nozzle.

A: The specific heat ratio γ is 1.1334 for a fitting to the 5 data couples.

References

1. Fox RW, McDonald AT, Mitchell JW (2019) Fox and McDonald's introduction to fluid mechanics, 10th edn. Wiley. ISBN 978-1-119-60376-4
2. Gerhart PM, Gerhart AL, Hochstein JI (2016) Munson, Young and Okiishi's fundamentals of fluid mechanics, 8th edn. Wiley. ISBN 978-1-119-08070-1

Chapter 5
Performance Measurement

Abstract Experimental performance analysis of turbomachines requires pressure measurement, temperature measurement and flow rate measurement at the flow side and torque and rotational speed measurement at the shaft side. The present chapter deals with the principles of the most fundamental measurement techniques for these variables. Three laboratory rigs for performance measurement are described: a hydraulic turbine, a fan and a pump. Measurement results are discussed. We demonstrate that performance evaluation of turbomachines with constant-density fluids is possible with the topics treated up to now.

5.1 Pressure Measurement

5.1.1 Metal Manometer

Figure 5.1 is a sketch. A hollow curved tube with an elliptical cross-section is connected with the fluid (liquid or gas). The tube bends outwards due to positive pressure difference with respect to the atmospheric pressure and inwards due to negative pressure difference. The displacement of the end point of the tube is transferred to a measuring needle. The linkage allows zero and scale adjustment. The measured pressure is called *gauge pressure*, which means that it is relative to the atmospheric pressure: above (+) or below (−). The absolute pressure is determined by adding the atmospheric pressure read from a barometer. Metal manometers are not suitable for values changing quickly in time.

5.1.2 Pressure Transducer

A pressure transducer is a device that converts an applied pressure into a measurable electrical signal. Figure 5.2 is a sketch. Two diaphragms bend by the pressure difference with respect to the atmosphere. In the example, the bending is detected by a strain gauge with variable electrical resistance, mounted in a Wheatstone bridge. The

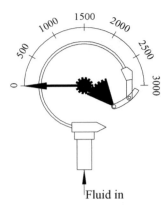

Fluid in

Fig. 5.1 The metal manometer

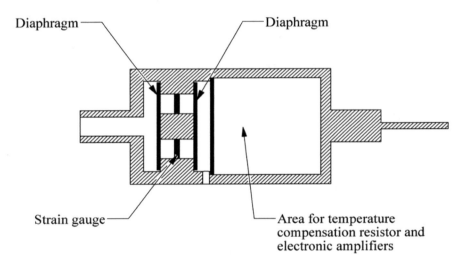

Diaphragm   Diaphragm

Strain gauge   Area for temperature
                                                     compensation resistor and
                                                     electronic amplifiers

Fig. 5.2 The pressure transducer

voltage output of the bridge is a measure of the pressure. The electronic circuit of the transducer includes a compensator for the temperature-related resistance change of the strain gauge. Similar pressure transducers exist with diaphragm movement detected by the change of an electrical capacitor (1 capacitor plate moves along with the diaphragm) or by generating a piezoelectric voltage (voltage within a crystal due to strain). Capacitive measuring is mainly suitable for low pressure, piezoelectric for high pressure. There are transducers for various pressure ranges and frequency responses. The voltage signal is read into a computer through an electronic circuit board.

5.1.3 Digital Manometer

A digital manometer functions similarly to a pressure transducer, but reading out is on a digital display. There are handheld devices with built-in transducer and display. With a handheld device, connections must be made between the device and the pressure measurement point. This is only workable for pressure measurement in air.

5.1.4 Calibration of Pressure Meters

Laboratory calibration of meters of pressure above atmospheric pressure is generally performed with a device in which calibrated weights are positioned onto a piston in an oil-containing cylinder (usually called a dead weight tester).

   Such a device is not suitable for small positive values and it cannot generate pressure below atmospheric pressure. U-shaped liquid columns are used as reference meters for low positive or for negative pressures. Pressure above or below atmospheric pressure may be generated in a closed container by means of a compressor. The pressure is derived from the height difference between the two levels of the liquid column, with devices for precise reading of the height difference. Mostly water is used as a liquid, but also lower-density liquids. For negative or positive pressure ranges until 1 bar, a mercury column is used, as in a barometer. In a laboratory, liquid columns are also used directly for pressure measurement in air. In industry, calibration is mostly done by comparing with calibration manometers on a reservoir in which positive or negative pressure is made. The calibration manometers are regularly checked in a calibration laboratory.

5.2 Temperature Measurement

5.2.1 Glass Thermometer

The principle is the expansion of a liquid, mostly mercury or an alcohol, within a spherical reservoir, by the influence of temperature. Glass thermometers do not require periodical calibration. Once the scale is determined, it is definitive. Glass thermometers are vulnerable, but there are types with metal protective tubes.

5.2.2 Temperature Transducer

A temperature transducer is mostly based on the temperature-dependence of an electrical resistance or functions by the voltage by a thermocouple. The variable resistance is part of a Wheatstone bridge, together with three fixed resistances. A thermocouple detects a temperature difference, typically with respect to the atmospheric temperature. This last one is then determined by a resistance measuring device. The electric signal is read into a computer. The difference between both types is the more limited range of a resistance temperature transducer compared to that of a thermocouple temperature transducer, but with a higher accuracy.

5.2.3   Digital Thermometer

The measuring principle is the same as with the transducer, but the device features a digital display. The display may be a separate handheld device or may be mounted directly to the probe.

5.3   Flow Rate Measurement

5.3.1   Reservoir

A direct method for measuring the volume flow rate of a liquid is filling of a reservoir calibrated in volume units or placed on a balance.

5.3.2   Flow Over a Weir

When a liquid flows over a weir, the flow rate is determined by the height of the liquid surface above the deepest point of the weir, as is sketched in Fig. 5.3. Rectangular and triangular weirs are applied. Formulae for volume flow rate are

$$Q = (2/3)C_R w \sqrt{2g(H + \Delta)^3} \text{ and } Q = (8/15)C_T tg\alpha \sqrt{2g(H + \Delta)^5},$$

Fig. 5.3 Flow over a weir

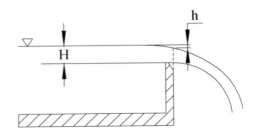

where w is the width of the rectangular weir, α half the opening angle of the triangular weir and H the height of the liquid level. The coefficients C_R, C_T and Δ are known for standardised weir forms.

5.3.3   Pressure Drop Devices

Figure 5.4 shows a nozzle mounted within a cylindrical duct or at a duct inlet. The flow of a liquid or a gas accelerates in the nozzle, causing pressure to drop. The pressure difference through the nozzle is a measure of the flow rate.

The flow rate is expressed by Eq. (4.40) (Chap. 4, Exercise 4.9.5):

$$\dot{m} = C_Q \rho_2 A_{th} \frac{1}{\sqrt{1 - \beta^4 (\rho_2/\rho_1)^2}} \sqrt{\frac{4 \Delta p}{\rho_1 + \rho_2}} \tag{5.1}$$

For a constant-density fluid $\rho_1 = \rho_2$. For a compressible fluid, the formula is accurate up to a throat Mach number of about 0.75. The parameter β is the ratio of the throat diameter to the tube diameter. C_Q is a coefficient taking into account the obstruction by boundary layers (see Sect. 5.7.1).

For a compressible fluid, determination of ρ_1 requires measurement of p_1 and T_1. Velocity determination further requires measurement of $\Delta p = p_1 - p_2$. The density ρ_2 is not determined by measuring p_2 and T_2, but by the isentropic relation between the density ratio and the pressure ratio: $\rho_2 = \rho_1 (p_2/p_1)^{1/\gamma}$.

Fig. 5.4 Flow rate measurement with a nozzle; left: at the inlet; right: mounted within a duct

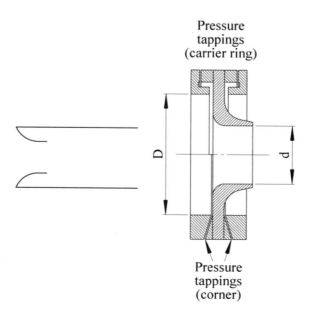

Pressure tappings (carrier ring)

D

d

Pressure tappings (corner)

An orifice plate (flat plate with a cylindrical orifice) or a Venturi (converging–diverging channel) may be used instead of a nozzle. The formulae for flow rate are similar. For a nozzle at a duct entry (Fig. 5.4, left), the formula is (5.1) with $\beta = 0$ and p_1 equal to atmospheric pressure $(=p_{01})$. All formulae have a flow rate coefficient C_Q. Expressions for C_Q are known for standardised forms. This coefficient is always somewhat lower than unity and depends on the Reynolds number of the approaching flow.

5.3.4 Industrial Flow Rate Meters

Industry mostly uses flow rate meters with a voltage signal read into a computer or a display. The signal may be obtained by several physical phenomena. Turbine meters, acoustic meters, magnetic meters and vortex meters are commonly used. With a turbine meter, a turbine wheel is mounted in a casing. The rotational speed is a measure for the flow rate. The rotational speed is mostly read out magnetically (see Sects. 5.5.2 and 5.5.3).

A Doppler-acoustic meter is possible with a fluid containing solid particles or bubbles larger than about 30 μm. An acoustic pulse is sent out by a piezoelectric transducer. The pulse is reflected by the particles or the bubbles and detected by a second piezoelectric transducer. The Doppler-effect modifies the frequency of the acoustic signal. Magnetic flow rate measurement is possible if the fluid is electro-conductive. The voltage generated within a conductive fluid flowing perpendicularly through a magnetic field is proportional to the flow velocity. An object is positioned in the flow with a vortex meter. The flow rate determination is based on the frequency of the vortices shed.

5.3.5 Positioning of Flow Rate Meters in Ducts

In principle, the approaching flow to a pressure drop device or a flow rate meter in a duct must be fully developed. This means a velocity profile that does not change in the flow direction. This is extremely difficult to realise in practice. Standards require approach lengths of 20–30 times the duct diameter. The approach length depends on the Reynolds number and the kind of upstream flow perturbations (e.g. the number of bends passed).

Smaller lengths apply when flow regulating devices, called flow conditioners, are placed upstream. These are plate systems that parallelise the flow and generate shear zones with intense turbulent mixing. Even with conditioners, the required approach lengths remain large. Flow rate measurement by an inlet nozzle, as in Fig. 5.4 (left), may be more convenient. The suction must be from a sufficiently large space, so that $\beta = 0$ applies.

5.4   Torque Measurement

5.4.1   Swinging Suspended Motor or Brake

With a driven machine, the driving motor may be suspended in ball or roller bearings, allowing it to swing. The driving motor torque on the machine then generates a reaction torque on the motor housing equal to the driving torque, apart from the bearing friction, which is very small. The reaction torque is balanced with a lever arm and a force. Force determination may be by a balance or a force transducer. A force transducer functions like a pressure transducer.

With a turbine, a brake is driven and similarly the reaction torque on the brake is measured. The brake principle may take various forms. Most obvious is an electric generator with power supply to the grid or with dissipation by resistors. Eddy current brakes are frequently used as well. Power is then dissipated by Foucault currents and the dissipation heat removed by cooling. Older braking systems use friction on a drum (Prony brake) or a water ring impacting on a bladed shell (Froude brake). These older systems are less precise and difficult to adjust.

5.4.2   Calibrated Motor

Swinging motor suspension cannot be applied on a driven turbomachine with integrated driving motor, as often used with smaller fans and pumps. A solution may consist in first taking off the electric motor and mounting it on a test rig with a swinging suspended brake, as described above. The characteristics of the motor, together with its controller may then be determined. A frequency-controlled asynchronous motor is often used. For set frequency, torque and rotational speed are determined as functions of output power. An example is detailed in Sect. 5.7.1.

5.4.3   Torque Transducer

This device is mounted between the shafts of the driving and the driven machine. An internal bar gets twisted by the torque transferred. The torsion angle is converted into a measurable electrical signal. The problem with a torque transducer is the necessity to measure the torsion in a rotating space. There are systems with strain gauges, so resistive determination, with capacitive determination and with inductive determination. The latter features four cored coils. Due to torsion, cores get deeper in two coils and less deep in the other two. The coils are mounted in a Wheatstone bridge. The measurement is similar with resistive and capacitive determination. The power supply of the bridge may be by AC, transferred at high frequency by a rotary transformer. The generated voltage is sent outside by a second rotary transformer.

This type has the advantage of being contactless. Slip ring systems exist as well. With those, power supply and read-out may be by DC with resistive determination.

5.5   Rotational Speed Measurement

5.5.1   Optical Tachometer

A tachometer is a general name of an instrument for measuring the rotational speed of a shaft. An optical pulse tachometer uses a disc with holes on the rotating shaft. The light of a lamp reaches a photocell. The resistance of the photocell, placed in a Wheatstone bridge, varies with incidence or absence of light. A pulse counter transfers its result to a digital display or it is read into a computer. Handheld devices function with a laser beam. A reflective plate is fixed to the shaft. The reflected beam reaches a photocell.

5.5.2   Electrical Tachometer

An electrical tachometer is mostly a permanent magnet DC generator producing a voltage proportional to the rotational speed or a permanent magnet AC generator producing a voltage with frequency and amplitude proportional to the rotational speed.

5.5.3   Rotational Speed Transducer

A rotational speed transducer or rotational speed sensor is an instrument that detects the motion of a toothed wheel in ferromagnetic material by the variable inductance of a permanent magnet near the wheel caused by the presence of a tooth or a gap. The change of the magnetic field strength is converted into a periodic electrical signal with frequency proportional to the rotational speed.

There are two typical ways of generation of the signal. A first is by a Hall-cell, which is a square plate in semiconductor material with four connections at the sides and a constant current through two opposing connections. When exposed to a magnetic field perpendicular to the plate, a voltage difference is created between the two other connections, proportional to the magnetic induction. The other method, called magnetic-inductive, is by a coil around the permanent magnet.

5.6   Laboratory Test of a Pelton Turbine

5.6.1   Test Rig

The Pelton turbine has one injector and a rotor with 110 mm mean diameter (diameter of the circle tangent to the centre line of the impacting jet), provided with 16 buckets (see Fig. 9.1 in Chap. 9). The rotor is made of accurately cast bronze and fixed on a stainless steel shaft, running in covered ball bearings. The injector has a 20 mm outlet orifice and a manually operated regulating needle. The inner diameter of the supply duct at the manometer just upstream of the injector is 40.0 mm. The shaft power is about 300 W at 1600 rpm, with a 21 m head and a 2.15 l/s flow rate.

The water is supplied by a single-stage centrifugal pump. The maximum head and flow rate of the pump are 24 m and 4 l/s at 2900 rpm. Pump pressure, and so turbine inlet pressure, is controlled by means of a manual valve. The flow rate is measured by a triangular weir between two reservoirs. Waves on the water surfaces are damped with bulkheads.

The turbine drives an asynchronous machine connected through an adjustable frequency convertor to the electricity grid and an electric resistor. The electric machine can function as a brake or as a motor. The rotational speed is controlled with a potentiometer. A torque sensor is mounted between the turbine and the asynchronous machine. For avoiding damage to this sensor, starting and stopping of the installation should only be done by an authorized operator. This has to be done gradually. e.g., the turbine has to rotate before the water jet impacts on it.

5.6.2   Measurements

The hydraulic input power of the turbine follows from

- The mass flow rate read out on the gauge of the highest reservoir, calibrated in flow rate units (Q in l/s).
- The static pressure in the supply duct to the injector, just upstream of the turbine, read out on a metal manometer (H in m water column).
- The dynamic pressure calculated from the velocity (v) in the supply duct to the injector (the duct diameter being 40.0 mm).

The manometric head in m water column is $H_m = H + v^2/2g$ and the hydraulic input power is $P = \rho Q g H_m$ (the height difference between the injector nozzle exit and the tailwater level is not taken into account).

The mechanical output power of the turbine follows from

- The rotational speed read on the display of the frequency converter (in rpm), calculated from the set frequency, ignoring the effect of slip of the asynchronous machine (about 2–5%). The rotational speed is thus somewhat underestimated.
- The brake torque read on the display of the torque sensor (in Nm).

5.6.3 Measurement Procedure

- Set the flow rate by opening the regulating needle from the fully closed position: 2, 4, 6, 8 or 10 rotations (4, 6 or 8 rotations yield the best efficiency, with flow rate between 1.5 and 2.5 l/s).
- Vary the rotational speed by adjusting the potentiometer from 500 rpm to the maximum attainable speed. Select eight operating points approximately evenly distributed over the rotational speed range from minimum to maximum. Read at each operating point: rotational speed, torque, flow rate (constant), head (constant).

5.6.4 Calculations

- Determine for each operating point: shaft power, mean blade speed, speed ratio and global efficiency (see hereafter for the definition of speed ratio).
- Plot torque T (Nm) as a function of rotational speed n (rpm) and plot global efficiency as a function of speed ratio (defined hereafter). Observe that the torque decreases with increasing speed with an approximately parabolic dependence, stronger than linear. Deduce from Exercise 1.9.1 in Chap. 1 that the theoretical dependence is linear. The theoretical velocity of the jet from the injector, called spouting velocity, is $v_0 = \sqrt{2g\,H_m}$. The actual velocity is somewhat smaller due to friction in the injector. In absence of friction on the rotor wheel, the force on a bucket is maximum for standstill of the bucket and becomes zero when the running speed of the bucket equals the jet velocity. This implies a linear descending dependence of the torque as a function of the rotational speed. In reality, there is friction on the wheel, mainly by windage. The friction torque is approximately a quadratic function of the rotational speed, which explains that the actual torque descends more rapidly than linear as a function of the wheel speed.
- Observe that, theoretically, the rotor power is maximum for the blade speed equal to half of the jet velocity. The actual maximum is at the ratio of the blade speed to the spouting velocity, called speed ratio, somewhat smaller than 0.5, due to the wheel friction. Determine the speed ratio at maximum efficiency.

Table 5.1 Test results for the Pelton turbine

Operating point	1	2	3	4	5	6	7	8
Rotational speed (rpm)	500	700	900	1100	1300	1500	1700	1900
Torque (Nm)	2.40	2.25	2.05	1.80	1.40	0.95	0.50	0.05

5.6.5 Measurement Example

- Injector needle position: 8 revolutions open.
- Static injector pressure: 11.2 m water column; flow rate: 2.5 l/s. (results are in Table 5.1)

A: The maximum efficiency is 0.74 at speed ratio 0.42. The maximum efficiency is rather low due to the small size of the turbine (losses make a relatively larger fraction of the work with a small machine size). For the same reason, the optimum speed ratio is lower than the value for large turbines, which is nearer to 0.5 (Sect. 9.3.1 in Chap. 9). Moreover, the rotational speed of the turbine is somewhat underestimated, which causes some underestimation of the efficiency and the speed ratio (see next test rig for a correct speed measurement).

5.7 Laboratory Test of a Centrifugal Fan

5.7.1 Test Rig

The design data of the fan are: $Q = 0.8$ m^3/s, $\Delta p_0 = 3500$ Pa, $n = 2900$ rpm. The rotor dimensions are: $d_1 = 200$ mm, $b_1 = 80$ mm, $d_2 = 450$ mm, $b_2 = 60$ mm. The rotor has 16 straight blades, with angles $\beta_1 = -57°$ and $\beta_2 = -22°$ to a meridional plane. The dimensions of the volute are: width $b_3 = 150$ mm, outlet height $h_4 = 170$ mm ($b_4 = 150$ mm). A description of the fan is given in Chap. 7, Sect. 7.6.

The fan is driven by a frequency-controlled asynchronous motor. The chosen frequency sets the rotational speed at zero-load. The motor slips when loaded. So, the characteristic of the fan is not measured with constant rotational speed. A prior mechanical measurement of the motor was done on a brake test rig. The torque and the rotational speed of the motor were measured under variable load together with the electric power supplied to the controller of the motor. The results for zero-load rotational speed 2750 rpm are shown as an example in Fig. 5.5. The curves indicate the torque on the shaft and the rotational speed as functions of the measured electric input power. The curves are linear with a very good approximation. Characteristics of torque and rotational speed as functions of electric input power are given in Table 5.2 by two data couples each time.

Figure 5.6 is a sketch of the set-up and the positions of the temperature and pressure taps for the measurement of the energy rise in the flow and the flow rate.

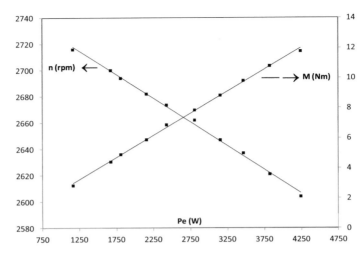

Fig. 5.5 Torque and rotational speed as functions of electric input power for zero-load rotational speed 2750 rpm

Table 5.2 Torque and rotational speed as functions of electric input power; chosen frequency (zero-load rpm); data couples: torque (Nm)—power (W) and speed (rpm)—power (W)

Zero-load rpm	Torque (Nm), power (W)		Speed (rpm), power (W)	
2000	(0, 65)	(12, 3100)	(2000, 300)	(1880, 2640)
2400	(0, 120)	(12, 3700)	(2400, 360)	(2280, 3280)
2750	(0, 165)	(12, 4225)	(2750, 400)	(2630, 3620)
3000	(0, 200)	(12, 4600)	(3000, 430)	(2880, 3900)
3200	(0, 225)	(12, 4900)	(3200, 500)	(3080, 4400)

Fig. 5.6 Fan set-up; measurement of energy rise and flow rate

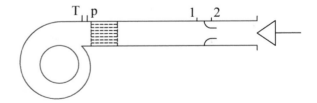

The increase of the mechanical energy in the flow is determined from the pressure and temperature p and T at the fan outlet and the pressure and temperature at the fan inlet p_a and T_a, being atmospheric values. The atmospheric pressure is read on the laboratory barometer (kPa) and the atmospheric temperature on the glass thermometer hanging next to the barometer (°C). The static outlet pressure p is read with a handheld digital manometer. The measured pressure is relative to the atmospheric pressure. The laboratory has devices with Pa, kPa and mBar read-out (1 mBar $= 100$ Pa). The temperature T at the fan outlet is read with a resistive digital temperature meter with a probe-mounted display (°C).

Flow rate measurement is done with a nozzle of a standardized design according to ISO-5167, type ISA 1932, with $D = 182.9$ mm, $d = 114$ mm, $\beta = d/D = 0.6233$. The mass flow rate follows from Eq. (5.1). The pressure p_1 upstream of the nozzle and the pressure difference over the nozzle $\Delta p = p_1 - p_2$ are measured with the same handheld digital manometer as used for p. For better accuracy, it is advisable to measure Δp and not to derive it from separate measurements of p_1 and p_2. The density ρ_1 and the density ratio ρ_2/ρ_1 are determined from the pressures p_1 and p_2 and the temperature T_1 (see Sect. 5.3.3). The temperature at the fan outlet is taken as an approximation for T_1. The approximation is appropriate as the outlet temperature of the fan just exceeds the atmospheric temperature a little and the temperature drop between the fan outlet and the measuring nozzle is very small.

The mass flow rate formula (5.1) contains a discharge coefficient, taking into account the boundary layer obstruction in the nozzle throat:

$$C_Q = 0.9900 - 0.2262\beta^{4.1} - (0.00175\beta^2 - 0.0033\beta^{4.15})(10^6/\text{Re}_D)^{1.15}.$$

The Reynolds number is $\text{Re}_D = (v_1 D\rho_1)/\mu_1 = 4\dot{m}/(\pi D\mu_1)$. The air viscosity is $\mu_1 = (17.177 + 0.0510\,T_1)\cdot10^{-6}$ Pas, with T_1 the temperature in °C (-10 °C $< T_1$ < 30 °C). The Reynolds number depends on the mass flow rate. As a consequence, determination of C_Q requires iteration. Start with $\text{Re}_D = 5\cdot10^5$ and one iteration is sufficient as the dependence of C_Q on Re_D is very weak.

5.7.2 Measurements

The mechanical input power of the fan follows from

- The shaft torque and the rotational speed of the fan determined from the characteristics of the driving motor and its controller by reading the electric input power.

The pneumatic output power of the fan follows from

- The mass flow rate determined with the measuring nozzle.
- The mechanical energy rise in the flow determined from the pressure and temperature at the fan inlet and outlet. The mechanical energy rise is calculated with the Bernoulli formula for density equal to the average value between fan inlet and outlet. The kinetic energy at the fan inlet is zero (atmospheric plenum). The kinetic energy at the fan outlet follows from the velocity, which is calculated from the mass flow rate and the density at the fan outlet, determined by p and T.

5.7.3   Measurement Procedure

- Select a rotational speed of 2000, 2400, 2750, 3000 or 3200 (idling rotational speed) on the frequency controller of the driving motor. Read atmospheric pressure and atmospheric temperature.
- Turn the flow control cone at the end of the measuring tube completely open. The flow rate through the fan and the electric input power are then maximal. The frequency controller has a power limiter. With an idling rotational speed of 3000 or 3200 rpm, the power limiter switches the frequency controller off with a completely open cone. Open the cone at these rotational speeds such that the power limiter just does not intervene. The power limiter does not intervene at lower rotational speeds. Read the maximum power input. Close then the flow control cone completely, however without forcing it, and read the corresponding input power.
- Vary the mass flow rate by opening or closing the cone. The fan has straight blades with only moderate backward leaning. So, the energy rise is almost constant for varying flow rate and thus is the rotor power approximately proportional to the flow rate. Select eight operating points approximately evenly distributed over the electric input power range from minimum (closed cone) to maximum.
- Measure at each operating point: P_{input}, T, p, p_1, $\Delta p = p_1 - p_2$.

5.7.4   Calculations

- Determine the rotational speed and the torque from the electric input power for every operating point, using the data in Table 5.2.
- Calculate the mass flow rate $\dot{m}$ for each operating point, using Eq. (5.1).
- Calculate for each operating point the mechanical energy rise $\Delta p_0/\rho$, the shaft power and the overall (total-to-total) efficiency of the fan, with the shaft power as input power. Use the average value of the density between the fan inlet and outlet.
- Observe from the Figs. 3.5 and 3.12 in Chap. 3, that the shape of the velocity triangles does not change when the rotational speed is varied together with a change of the flow rate proportional to the rotational speed ($Q \sim n$). This necessitates that the slip factor (Eq. 3.21) or the work reduction factor (Eq. 3.24) stay the same, which is assured by the formulae. The rotor work then varies proportional to the square of the rotational speed. This is also valid for the rise of the total pressure, under the assumption that losses due to incidence, due to dump and due to friction are proportional to the square of the flow rate. Observe that these dependencies are correct for constant loss coefficients. Observe that leakage flow rates scale with the square root of the pressure. The internal efficiency and the volumetric efficiency thus stay constant under varying rotational speed when the flow rate is varied simultaneously proportional to the rotational speed (in reality, this is not fully exact). Under the assumption of constant mechanical efficiency, the global

efficiency is then also constant. The observations form the basis of the theory of similitude of flows, which we elaborate in the later Chap. 7.

- Transform by similitude, thus $Q \sim n$, $\Delta p_0 \sim n^2$, $\eta =$ constant, for each operating point, the mass flow rate and the mechanical energy rise to the idling rotational speed. This way, the characteristic at constant rotational speed is obtained. The overall efficiency is assumed constant, but, strictly, a very small correction for the small change of the Reynolds number is necessary (see the later Chap. 7).
- Determine the work coefficient $\psi = \Delta W / u_2^2$ for maximum efficiency. Determine the rotor work from the shaft power and the flow rate, assuming that the mechanical and volumetric efficiencies equal unity. With a radial fan, the mechanical efficiency is near to unity (bearing friction and wheel friction are very small), but the volumetric efficiency is not near to unity (the gap between the volute and the rotor in the suction eye is rather large). This way, we overestimate the work coefficient (by about 10%). Evaluate the value of the work coefficient.

5.7.5 Measurement Example

- Rotational speed 2750 rpm (idling value).
- Atmospheric conditions: $p_a = 100.4$ kPa, $T_a = 21.0\,°C$ (results are in Table 5.3).

A: Calculated to constant rotational speed, the maximum mechanical energy rise in the flow is 2870 J/kg with mass flow rate 0.550 kg/s (weak maximum) and the maximum efficiency is 0.74 with mass flow rate 0.825 kg/s and energy rise 2690 J/kg (weak maximum). At maximum efficiency, the energy rise coefficient is 0.64 and the calculated work coefficient is 0.865. The real value is presumably about 10% lower, about 0.78. This value is little higher than the value of 0.74 assumed for the design flow rate (but the design flow rate is somewhat smaller than that with maximum efficiency; see Sect. 7.6).

Table 5.3 Test results for the fan

Operating point	1	2	3	4	5	6	7	8
P_{input} (W)	1200	1480	1840	2240	2640	3000	3400	3760
T (°C)	27.4	25.7	24.6	24.0	23.7	23.6	23.2	22.5
p (kPa)	2.92	3.06	3.13	3.13	3.08	2.92	2.60	2.23
p_1 (kPa)	2.92	3.04	3.09	3.06	2.96	2.76	2.38	1.95
Δp (kPa)	0.00	0.07	0.27	0.60	1.05	1.60	2.38	3.25

5.8 Laboratory Test of a Centrifugal Pump

5.8.1 Test Rig

The pump is a single-stage centrifugal pump, meant to be used in a milk collecting truck. Figure 5.7 shows a meridional section and a view on the rotor. The open rotor (1) is accurately cast in stainless steel. It has 5 blades with a simple backward curved form. The external diameter is 180 mm. The rotor disc is not full by removed material in between the blades (called scalloping). The clearances between the rotor and the casing are rather large and there are no wear rings for limiting leakage flows. These construction options are taken for avoiding damage to the pumped liquid and for allowing easy cleaning after operation, but they diminish the pump efficiency.

The axial force on the rotor is compensated by a space (2) at the back of the rotor, where the processed liquid circulates around the bellows mechanical seal (3). The casing of the pump (4) is made of pressed stainless steel.

The shaft of the pump (5) is connected to a DC motor with swinging suspended stator with lever arm connected to a balance where the torque provided by the motor can be read. The pump can run at a rotational speed up to 3000 rpm (being also the nominal rotational speed) and deliver a maximal total head around 42 m. Its maximum flow rate is approximately 28 l/s.

The pump is connected to a suction reservoir by a pipe with internal diameter 82.8 mm, and to a delivery reservoir by a pipe with internal diameter 70.3 mm. A flexible connector is positioned on the suction pipe to compensate for axial dilatations. The delivery pipe is provided with a valve to control the flow rate. The pumped

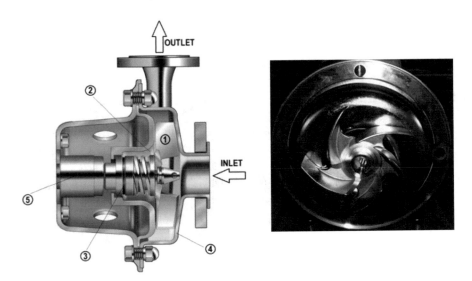

Fig. 5.7 Centrifugal pump used in the laboratory test (Courtesy Packo Inox Ltd.)

fluid is water. The water is spread over the delivery reservoir by a pipe with multiple openings in order to obtain a water surface that is as flat as possible. Upstream of this delivery reservoir there is a bypass pipe which leads directly to the suction reservoir (bypassing the delivery reservoir). Whether water is pumped to the delivery reservoir or to the suction reservoir may be chosen using valves. The rise of the water level in the delivery reservoir is used to determine the mass flow rate. After measurement, the delivery reservoir is emptied by opening a valve in the bottom which returns the water to the suction reservoir. During this emptying, the pump should be connected to the bypass pipe.

5.8.2 Measurements

The mechanical input power follows from

- The rotational speed measured with an optical pulse tachometer (disc with holes) in rpm.
- The torque by the driving motor with swinging suspended stator. The scale of the balance is expressed in kgfm (1 kgf $= 9.81$ N).

  The hydraulic output power follows from

- The suction pressure measured by a manometer on the suction pipe in kgf/cm^2 (1 kgf/cm$^2 = 98.1$ kPa), read as value below the atmospheric pressure.
- The delivery pressure measured by a manometer on the delivery pipe in kgf/cm^2, read as value above the atmospheric pressure.
- The height difference between the manometers at the suction and delivery sides. The manometers are provided with purge valves. Purging the delivery manometer guarantees that the connecting tube is filled with water. Purging the suction manometer guarantees that the connecting tube is filled with air. The delivery manometer is placed above the tapping point and thus measures the pressure in the delivery pipe at the tapping point minus the pressure corresponding to the height difference between the centre of the manometer and the tapping point. The suction manometer measures the pressure at the tapping point. The height difference to be taken into account between the manometers is the distance from the tapping point of the suction manometer to the centre of the delivery manometer. This height difference is 1.25 m.
- The flow rate by means of the calibrated delivery reservoir. The volume between the gauge glass marks is 100 L. Choose the number of measured marks in proportion to the flow rate to be measured (normally around 5 marks).

5.8.3 Measurement Procedure

- Select a rotational speed between 1500 and 2500 rpm. The driving motor is fed by a direct current controller (right part of the cabin). The rotational speed is adjusted with a knob on the controller and is read out digitally. After adjustment, the controller keeps the rotational speed constant, independent of the motor load.
- Select six to eight operating points on the pump characteristic by constricting the delivery pipe. Set the constriction valve on the suction side completely open. Start the test by closing the valve on the delivering pipe. This provides maximum delivery pressure. Then open the valve completely. This provides minimum pressure. Distribute the operating points approximately equally over the pressure range. Perform the initial actions with the pump supplying to the suction reservoir.
- When switching between the bypass pipe and the pipe connected to the delivery reservoir, first open both switching valves. The position of the handles allows one single person to do this. The pump is not damaged by closing both valves suddenly, but water hammer may be generated within the delivery pipe, and this is best avoided. After a flow rate measurement, switch to the bypass pipe and empty the delivery reservoir (but leave some water) by opening the connecting valve between the reservoirs. The delivery reservoir is sufficiently large, making it impossible to overflow it. The suction pipe inlet in the suction reservoir comes out of the water with a completely filled delivery reservoir.

5.8.4 Calculations

- Determine for each operating point: flow rate, delivery pressure, suction pressure and torque.
- Calculate for each operating point: the mechanical input power delivered to the pump, the velocity in the suction and delivery pipes, the mechanical energy rise ΔE_m in J/kg, the manometric head H_m in m and the overall efficiency.
- Plot manometric head, mechanical input power and overall efficiency as functions of flow rate.
- Determine the work coefficient for the operating point with maximum efficiency (interpolate between the results found, if necessary). When calculating the work coefficient, determine the rotor work from the shaft power and the flow rate, assuming that the mechanical and volumetric efficiencies equal unity. With most centrifugal pumps, the volumetric efficiency is close to unity (wear rings, labyrinth seals and shaft seals are applied to seal the rotor from the stator, both for internal and external leakage), but the mechanical efficiency may be significantly lower than unity (friction within the seals and disc friction are not negligible). With the pump tested, the mechanical efficiency is quite high, but the volumetric efficiency is low (no wear rings and large clearances). We thus overestimate the work coefficient. Evaluate the value of the work coefficient.

Table 5.4 Test results for the pump

Operating point	1	2	3	4	5	6	7	8
p_p (kgf/cm^2)	2.1	1.8	1.6	1.4	1.2	1.05	0.9	0.8
p_s (kgf/cm^2)	0	−0.01	−0.04	−0.08	−0.13	−0.17	−0.22	−0.26
T (kgfm)	0.75	1.75	2.1	2.3	2.45	2.55	2.65	2.75
Q (l/s)	0	10.1	13.9	15.9	17.9	19.7	21.3	22.2

5.8.5 Measurement Example

- Rotational speed 2250 rpm (results are in Table 5.4).

A: At 2250 rpm, the maximum mechanical energy rise is 218.3 J/kg with a zero mass flow rate and the maximum global efficiency is 0.505 (thus quite low) with a 13.9 kg/s mass flow rate. At maximum efficiency, the energy rise is 176.2 J/kg, the energy rise coefficient is 0.39 and the calculated work coefficient is 0.78. The real value of the work coefficient is certainly much lower, presumably about 20% lower, so about 0.62. This value is much lower than that found for the fan in the previous test, but the exit flow of the rotor is much more backwards with the pump.

Chapter 6
Steam Turbines

Abstract The working principles and the construction forms of steam turbines are discussed, starting from the two historical axial types. These are the impulse type and the reaction type. The chapter is also intended to formulate the general theory of axial turbines. In particular, the role of the degree of reaction is discussed. Typical construction forms of large steam turbines for electric power stations and small turbines for industrial applications are illustrated. The chapter ends with a discussion of blade and vane shapes.

6.1   Applications of Steam Turbines

A steam turbine produces shaft power from an enthalpy drop of steam. A special feature of steam turbines is the ability for very high power, due to the large enthalpy drop per mass unit of steam. Supply conditions in the most advanced (2021) coal-fired electric power stations are 330 bar (33 MPa), 650 °C (ultra-supercritical). Steam expansion to a 5 kPa vacuum provides about 2000 kJ/kg when applying a single reheat (in practice to 670 °C). This enables power in the 1200 MW order.

The basics of steam turbines are known since ancient times, but practical realisations were only achieved at the end of the nineteenth century, by Gustaf de Laval (Sweden) in 1883 and by Charles Parsons (UK) in 1884. Broad industrial application started around 1920. The development of steam turbines was mainly promoted by the use of electricity as an energy carrier. From 1920 onwards, electricity production was concentrated in power stations, requiring large driving machines. Steam turbine power strongly increased in the course of time. A 1 MW turbine was a large unit in 1920. At present, units with 1000 MW power per shaft (coal) are common, whereas the largest turbines yield a power of 1800 MW per shaft (nuclear) (see web sites of large manufacturers: General Electric Power, Siemens Energy, Mitsubishi Power; smaller manufacturers are Ansaldo Energia and MAN Energy Solutions).

Worldwide, the largest part of the electricity is produced by steam turbines. Steam is obtained by water evaporation under pressure, with solid, liquid, gaseous or nuclear fuels. Due to the low cost of nuclear fuel but the high investment cost of nuclear reactors, nuclear power stations are used for base load. Typically, the technically

E. Dick, *Fundamentals of Turbomachines*, Fluid Mechanics and Its Applications 130,
https://doi.org/10.1007/978-3-030-93578-8_6

highest possible turbine power (1500–1800 MW) is opted for. Due to the high power, the boiler of nuclear units has high thermal inertia, precluding these from following daily load variations.

Fossil fuel steam turbine power stations, without gas turbines, use nearly exclusively coal, but many can use oil or gas as back-up fuel. Coal units function as base load or as mid load. In the last case, it means that they partially follow the daily power consumption variation: full load during high power consumption hours, part load during low consumption hours. The need to limit boiler thermal inertia precludes from building such units with very large power. Mid-load units are mostly limited to about 400 MW. Coal units of 1000 MW or more are base-load units. Steam turbines are inadequate as peak units, mainly because of the thermal inertia of the steam generator. Fast-starting machines for peak load are hydraulic turbines and simple-cycle gas turbines.

Natural gas is not commonly used to fire directly a steam generator, but more typically in a combined-cycle power station. The gas is then burnt in one or two gas turbines with the exhaust flow led to a recovery steam generator (no burning of fuel), which feeds a steam turbine. At present, such power stations yield the highest net electrical efficiency (plant output to fuel), up to 64%. An ultra-critical coal-fired steam turbine power station yields about 48% net electrical efficiency. A nuclear power station reaches about 35%.

Combined steam and gas turbine stations are used for base load, due to their very high efficiency, but they are also used for mid load because the thermal inertia of the gas turbine is small. The rated power of the largest gas turbines is currently (2021) close to 600 MW. Two gas turbines combined with one steam turbine, with coupled shafts, allow an electric power above 1600 MW (e.g., GE Power: 2×570 MW $+ 540$ MW). But such a combination actually consists of five coupled machines, because the steam turbine is composed of 3 separate modules. A coal-fired steam turbine of 1200 MW consists also of 5 coupled modules (see Sect. 6.8.1).

The exhaust gas of a gas turbine in a combined-cycle plant must be sufficiently hot to produce steam at useful conditions. The exhaust temperature currently amounts to about 630 °C, due to the high inlet temperature of the turbine part, at present typically around 1500 °C, enabled by progress in metallurgy and blade cooling techniques. This allows steam generation at about 600 °C. Before the advent of combined units, gas turbines were exclusively applied as aircraft engines or as industrial turbines. These applications have limited power, at most 20–40 MW.

Steam turbines are also used as industrial machines, although decreasingly. Due to improved reliability of electric power generation and increase in available power, industrial drives are preferably electric nowadays. Steam turbines are only applied to drive machines requiring high power and turning at high rotational speed. A typical application is drive of turbo-compressors. Drive of turbo-pumps exists too, e.g., in electric power stations. Opting for a steam turbine as a motor is attractive if the industrial application also needs process heat. Cogeneration of electrical power and process heat is similar. Steam is then produced at high pressure and temperature and fed to the driving steam turbine. The process heat is supplied by extraction steam or backpressure steam from the steam turbine. The generated electric power is consumed locally or fed into the electricity grid.

Industrial steam turbines operate with lower values of inlet pressure and inlet temperature than large steam turbines in power stations, up to about 100 bar, 500 °C. The reasons are cost limitation and the need for quick adaptability of the machine load (no big thermal inertia). Dependent on the application, industrial steam turbines may differ strongly in power, with the largest around 250 MW. The largest machines also serve in combined-cycle plants. In current cogeneration applications, steam turbines experience strong competition from gas turbines. Gas turbines burn natural gas and process steam is produced in a recovery steam generator by the turbine exhaust gas. Investment costs of gas turbines are considerably lower than these of steam turbines. Waste gas from the process may also be burnt in a recovery steam generator (co-firing).

Steam turbines have had an important role in ship propulsion, but this application is nowadays extremely exceptional due to the power available by modern diesel engines, up to 80 MW. Large low-speed diesel engines also have a higher efficiency (45%). Pressure and temperature conditions of ship propulsion steam turbines are similar to those of industrial steam turbines. The efficiency amounts to about 38%. At present, ships are mainly propelled by diesel engines, but gas turbines are applied for high power when available space is limited: high-speed battle ships, coast guard patrol boats, fast ferries and fast container ships. Some ships with gas turbines use both gas turbines and diesel engines for propulsion. Use of the diesel engine at low navigation speed and the gas turbine or both the gas turbine and the diesel engine at high speed is typical.

6.2 Working Principles of Steam Turbines

Steam turbines are almost exclusively built in the axial form. The functioning is analogous with that of the axial hydraulic turbine discussed as an example in Chap. 1. A flow is generated in stator components by converting static enthalpy into kinetic energy. Work is produced by change of flow direction in the downstream rotor, i.e. by using the kinetic energy, but static enthalpy may be converted into kinetic energy during work in the rotor. A degree of reaction is then present.

Even though the principle of a steam turbine is the same as that of a hydraulic one, it has totally different appearance due to the nature of the fluid. The available enthalpy drop with steam is very high: of the 1000–2000 kJ/kg order. Even under extreme circumstances, a hydraulic turbine ranks two orders lower: a 1000 m height difference yields about 10 kJ/kg. Hydraulic turbines are always of the single-stage type for that reason. Almost all steam turbines are multistage machines. Further, the low steam density allows very high flow speeds. The flow velocity at the stator exit is typically about 400 m/s, even up to 600 m/s. The corresponding kinetic energy is of the 100 kJ/kg order. This value also represents the order of magnitude of the enthalpy drop of a stage.

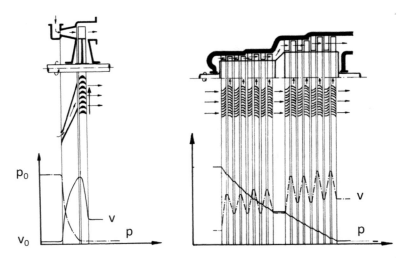

Fig. 6.1 Left: impulse turbine (single-stage example); right: reaction turbine (multistage example); sketches after the historical machines of Laval and Parsons

There are two historical axial steam turbine types: the impulse turbine and the reaction turbine. The impulse turbine was introduced by G. de Laval in 1883. It is a machine with no pressure drop in the rotor. The entire pressure drop and the accompanying enthalpy drop are used in the stator to produce kinetic energy. Principally, the degree of reaction is thus zero (correct for lossless flow; the degree of reaction is discussed in Sect. 6.7.1). Figure 6.1 (left) illustrates the functioning of a single-stage impulse turbine (sketch after Laval's original turbine: the stator is composed of separate nozzles).

The enthalpy relations in stator and rotor are:

Stator:

$$h_{00} = h_1 + \frac{v_1^2}{2}, \tag{6.1}$$

Rotor:

$$h_{0r} = h_1 + \frac{w_1^2}{2} = h_2 + \frac{w_2^2}{2}; \ h_1 - h_2 = \frac{w_2^2 - w_1^2}{2}. \tag{6.2}$$

We used here absence of work and heat transfer in the stator in the absolute frame and absence of work and heat transfer in the rotor in the relative frame, together with $u_1 = u_2$ (Eq. 1.14 of Chap. 1). With a provisional neglect of losses, the enthalpy-entropy diagram of Fig. 6.2 (left) shows that with $p_1 = p_2$, then follows $h_1 = h_2$ and thus $w_1 = w_2$. Thus rotor work is produced only by the momentum (impulse) made available by the nozzles.

The reaction turbine was introduced by Ch. Parsons in 1884. This is a machine with a pressure drop in the rotor. Parsons' original machine had a degree of reaction of

Fig. 6.2 Enthalpy-entropy diagrams and velocity triangles, drawn for lossless flow and axial outflow; left: zero degree of reaction; right: degree of reaction 0.5

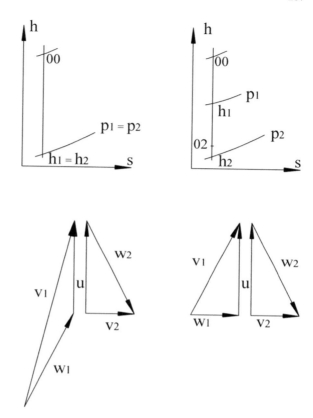

exactly 50%. This is no requirement for reaction turbines, but it was a reasoned choice by Parsons. Figure 6.1 (right) illustrates the functioning of a multistage reaction turbine (sketch after Parsons' original machine). The thermodynamic relations are again (6.1) and (6.2).

The total enthalpy drop is

$$h_{00} - h_{02} = h_1 + \frac{v_1^2}{2} - h_2 - \frac{v_2^2}{2}.$$

From $R = \frac{h_1 - h_2}{h_{00} - h_{02}} = 0.5$ follows

$$h_1 - h_2 = \frac{v_1^2 - v_2^2}{2} = \frac{w_2^2 - w_1^2}{2}.$$

For a stage with equal entry and exit velocities, $\vec{v}_0 = \vec{v}_2$, rotor and stator accelerations are equal for $R = 0.5$. This means that the stator and rotor blade shapes may be equal, but placed symmetrically. The application of a same blade shape for both the rotor and the stator motivated Parsons to choose $R = 0.5$. In Parsons' original machine all blades are prismatic and of the same cross section.

Comparison of the velocity triangles with $R = 0$ and $R = 0.5$ in Fig. 6.2 demonstrates that, with an equal blade speed (u), the work ($\Delta W = u\Delta v_u$) with $R = 0$ is double that with $R = 0.5$. This is a general rule: turbines with a low degree of reaction can more easily produce a large work per stage. This was one of the reasons why Laval opted for $R = 0$.

Parsons' machine has stator blades around the entire circumference. The stator part upstream of the rotor of Laval's machine consists of a number of nozzles which do not cover the entire circumference of the rotor. In other words, there is *partial admission* or, more in general, *partial flow*. The rotor blades only produce work when passing a nozzle. A partial-admission turbine can only function adequately with constant pressure in the rotor. With built-in pressure difference, the rotor blade channels not in front of a nozzle, would take flow from the nozzles. This flow cannot enter the rotor at the appropriate angle. Furthermore, each nozzle flow deviation causes incidence losses at the entry to rotor blade channels in front of a nozzle. Partial flow thus necessitates, in principle, constant pressure in the rotor. This was Laval's second reason to opt for $R = 0$. A partial-flow machine can be built for a fraction of the power it would produce with full flow.

Single-stage impulse turbines with partial admission are still built nowadays. They are mainly meant for mechanical drive. The degree of reaction is not exactly zero, but typically around $R = 0.10$. This is intended to compensate rotor losses. A partial admission impulse stage is also sometimes used as first stage in a multistage turbine (see Sect. 6.8.4).

6.3   The Steam Cycle

Figure 6.3 represents a steam cycle with loss-free components in the T-s diagram (ideal Rankine cycle). The cycle drawn is without reheat and at subcritical pressure. The parts are: isentropic pressure increase with slight temperature increase in the liquid phase (1–2) (condenser extraction pump and boiler feed pump), heating under constant pressure, consisting of heating in the liquid phase, evaporation and super-heating in the vapour phase (2–3) (boiler), isentropic expansion in the vapour phase (3–4) (turbine) and condensation at constant pressure (4–1) (condenser).

The *thermal efficiency* of the cycle is the ratio of the produced work to the supplied heat (both on the fluid side): $[(h_{03} - h_{04}) - (h_{02} - h_{01})]/(h_{03} - h_{02})$. We remind that the pump work is only 2–3% of the turbine work (see Thermodynamics). The internal turbine efficiency (equal to 1 in the figure) is the ratio of the produced work per unit of mass (on the fluid side) $(h_{03} - h_{04})$ and the isentropic total enthalpy drop $(h_{03} - h_{04s})$ supplied by the cycle. The best turbines yield about 94%.

Figure 6.3 shows the thermal efficiency of the simple cycle with ideal components. The maximum value is about 46% for 250 bar, 550 °C (just supercritical). This

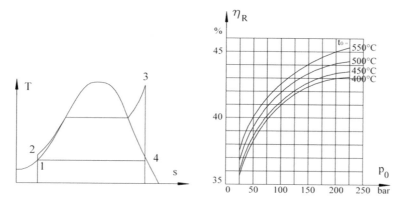

Fig. 6.3 Ideal Rankine cycle in the T-s diagram and corresponding thermal efficiency (subcritical cycle)

efficiency can be increased by higher inlet pressure and higher inlet temperature, and by adding reheat and regenerative feed water heating (see Thermodynamics).

At present (2021), a gross thermal efficiency (ratio of turbine shaft output to fuel heat input) of 53–54% is reached for inlet conditions 330 bar, 650 °C and one reheat to 670 °C or 330 bar, 620 °C and two reheats to 630 °C. The corresponding gross electrical efficiency (ratio of alternator output to fuel heat input) is about 52% and the net electrical efficiency (ratio of plant output to fuel heat input; power consumption by the pumps and other auxiliaries taken into account) is 47–48%.

6.4 The Single Impulse Stage or Laval Stage

For didactic reasons, we discuss a single-stage impulse turbine at first, its analysis being the simplest one. This should not convey the impression that this type prevails. Typical current designs will be discussed later on.

6.4.1 Velocity Triangles

Figure 6.4 sketches a meridional section of a stage and a cylindrical section of a rotor blade. The velocity triangles are drawn with proportions that are typical for impulse turbines. Angles are counted with respect to the axial direction, positive in the running sense. The coordinate frame is right-handed, with x-axis in axial direction, positive in the through-flow sense; y-axis in tangential direction, positive in the running sense; z-axis in radial direction positive away from the shaft. A machine with positive blade speed is *left-turning*: $\vec{\Omega} = -\Omega \vec{1}_x$. Velocity triangles are typically drawn with a vertical axial direction and the flow in the downward sense. We follow this convention.

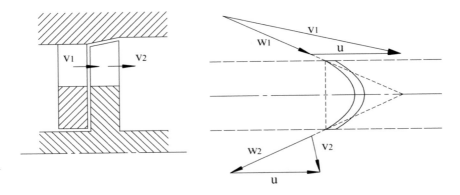

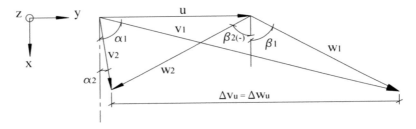

Fig. 6.4 Velocity triangles with impulse turbines: $R_s = 0$, $\psi = 1.95$ (work coefficient for optimum total-to-static efficiency)

6.4.2   *Work and Energy Relations*

The primary relations are ($dq = 0$, $dU = 0$):
   Energy:

$$dW = dh + \frac{1}{2}dv^2 = dh_0,$$

Work:

$$dW = \frac{1}{\rho}dp + d\frac{1}{2}v^2 + dq_{irr},$$

Rotor work (Euler):

$$\Delta W = u(v_{2u} - v_{1u})$$

From the velocity triangles it follows that

$$w_1^2 = u^2 + v_1^2 - 2uv_{1u}$$
$$w_2^2 = u^2 + v_2^2 - 2uv_{2u}$$
$$w_1^2 - w_2^2 = v_1^2 - v_2^2 - 2u(v_{1u} - v_{2u})$$

$$-\Delta W = u(v_{1u} - v_{2u}) = \frac{v_1^2}{2} - \frac{v_2^2}{2} - \left(\frac{w_1^2}{2} - \frac{w_2^2}{2}\right).$$

For a turbine, we conventionally consider delivered work as positive and write:

$$\Delta W = u(v_{1u} - v_{2u}) = \frac{v_1^2}{2} - \frac{v_2^2}{2} - \left(\frac{w_1^2}{2} - \frac{w_2^2}{2}\right). \tag{6.3}$$

With the energy equation, work is also:

$$\Delta W = \Delta h_0 = h_{01} - h_{02},$$

where Δh_0 represents the total enthalpy drop. From combination of the rotor work equation (6.3) and the energy equation, it follows that

$$h_1 + \frac{v_1^2}{2} - h_2 - \frac{v_2^2}{2} = \frac{v_1^2}{2} - \frac{v_2^2}{2} - \left(\frac{w_1^2}{2} - \frac{w_2^2}{2}\right),$$

or

$$h_1 + \frac{w_1^2}{2} = h_2 + \frac{w_2^2}{2} = h_{0r} = \text{constant}.$$

We recover that the total relative enthalpy is constant within the rotor. This result only applies to an axial machine ($u = \text{cst}$). Within the stator $\Delta W = 0$, thus

$$h + \frac{1}{2}v^2 = h_0 = \text{constant}.$$

Figure 6.5 sketches the processes with an impulse turbine in the h–s diagram, taking the obtained relations into account. The diagram is drawn for constant pressure in the rotor. So, we use the strict definition of an impulse turbine.

We first consider the nozzles (stator). The work equation within the stator cannot be integrated without determining the details of the expansion (see Chap. 4). Efficiency is therefore defined by comparing the result of the real expansion to that of a loss-free expansion. As isentropic nozzle efficiency we define:

Fig. 6.5 *h–s* diagram of an
impulse turbine stage; I:
nozzle loss; II: rotor loss; III:
outlet kinetic energy

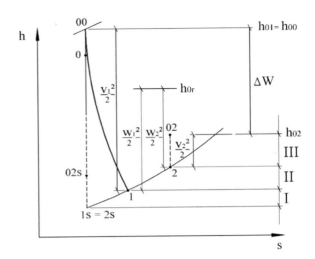

$$\eta_{ss} = \tfrac{1}{2}v_1^2 / \tfrac{1}{2}v_{1s}^2, \quad \text{with } \tfrac{1}{2}v_1^2 = h_{00} - h_1; \quad \tfrac{1}{2}v_{1s}^2 = h_{00} - h_{1s}. \tag{6.4}$$

Conventionally, the difference $\tfrac{1}{2}v_{1s}^2 - \tfrac{1}{2}v_1^2 = h_1 - h_{1s}$ is considered as nozzle
loss. This difference does not correspond exactly to the integral of dq_{irr} in the work
equation. We recall that it has been reasoned in Chap. 4 that the conventional loss
underestimates somewhat the real loss. As the real loss is practically undeterminable,
because it depends on the details of the expansion path, the conventional nozzle loss
is applied in the further analysis.

We define a *velocity coefficient* ϕ_s with

$$v_1 = \phi_s v_{1s}, \text{ so that } \eta_{ss} = \phi_s^2.$$

In the rotor of an impulse turbine, $dp = 0$. So the work equation can be integrated.
It follows that

$$0 = 0 + d\frac{1}{2}w^2 + dq_{irr}, \text{ or } q_{irr} = \frac{w_1^2}{2} - \frac{w_2^2}{2}.$$

The decrease of the kinetic energy in the relative system is thus the loss. Rotor
work (6.3) may be written as

$$\Delta W = \frac{v_1^2}{2} - \left(\frac{w_1^2}{2} - \frac{w_2^2}{2}\right) - \frac{v_2^2}{2}. \tag{6.5}$$

The interpretation is:

$\frac{1}{2}v_1^2$: kinetic energy supplied to the rotor,

$\frac{1}{2}(w_1^2 - w_2^2)$: rotor loss,

$\frac{1}{2}v_2^2$: outlet kinetic energy.

It is thus appropriate to define rotor efficiency by

$$\eta_r = \frac{\Delta W}{\frac{1}{2}v_1^2}. \tag{6.6}$$

Such a definition is only meaningful with an impulse turbine. As there is no pressure drop in the rotor, the kinetic energy supplied to the rotor is the only source of work. We define a *rotor velocity coefficient* by

$$w_2 = \phi_r w_{2s} = \phi_r w_1.$$

6.4.3 Stage Efficiency Definitions

In order to define stage efficiency, we must determine the usefulness of the kinetic energy at the stage exit. Rotor work is first written as

$$\Delta W = \frac{v_{1s}^2}{2} - \left(\frac{v_{1s}^2}{2} - \frac{v_1^2}{2}\right) - \left(\frac{w_1^2}{2} - \frac{w_2^2}{2}\right) - \frac{v_2^2}{2}. \tag{6.7}$$

If the outlet kinetic energy is not useful and thus has to be considered as a loss, it may be concluded from Eq. 6.7 together with Fig. 6.5, that the isentropic enthalpy drop supplied to the stage for work production is the *total-to-static* difference $\Delta h_s = h_{00} - h_{1s} = \frac{1}{2}v_{1s}^2$. The point of view is then that an ideal expansion is one with zero kinetic energy at the stage exit and zero losses in stator and rotor. The reference isentropic enthalpy drop is from the total state at the stage entry to a static state at the stage exit. The corresponding internal efficiency is called the *total-to-static efficiency*:

$$\eta_i = \frac{\Delta W}{\Delta h_s} = \frac{\Delta W}{h_{00} - h_{1s}} = \frac{\Delta W}{h_{00} - h_{2s}} = \eta_{ss}\eta_r. \tag{6.8}$$

It is quite straightforward to optimise this efficiency (see Sect. 6.4.6).

If the outlet kinetic energy is useful, e.g. because it is supplied to a next stage, with Eq. 6.7 and Fig. 6.5, the enthalpy drop available to the stage for work production may be defined by $\frac{1}{2}v_{1s}^2 - \frac{1}{2}v_2^2$. The point of view is then that the kinetic energy at the stage exit is not a loss and that the ideal expansion is one with zero losses in stator and rotor. The reference isentropic enthalpy difference may be visualised in the h–s diagram by the *total-to-total* difference $h_{00} - h_{02s}$, where the total state h_{02s} is obtained by adding the outlet kinetic energy to the state $1_s = 2_s$. The reference isentropic enthalpy drop is from the total state at the stage entry to a total state at the stage exit. The corresponding internal efficiency is called the *total-to-total efficiency*:

$$\eta_i = \frac{\Delta W}{h_{00} - h_{02s}}. \tag{6.9}$$

This efficiency cannot be expressed elementary. However, if the stage is supposed to supply the kinetic energy $\frac{1}{2}v_2^2$ to the next stage, it seems obvious to suppose that the stage receives the same kinetic energy: $\frac{1}{2}v_0^2 = \frac{1}{2}v_2^2$. Especially, this is satisfied when the stage inflow and outflow velocities are equal in magnitude and direction: $\vec{v}_0 = \vec{v}_2$. Such a stage is called a *repeating stage*. It is a model for a stage of a multistage machine. With a repeating stage, the *static-to-static* isentropic enthalpy difference can be used as reference and Eq. (6.9) changed into

$$\eta_i = \frac{\Delta W}{h_0 - h_{2s}} = \frac{\Delta W}{h_0 - h_{1s}}. \tag{6.10}$$

With impulse turbines, the outlet kinetic energy is completely dissipated or almost completely dissipated in some applications.

With partial admission, rotor blades not exposed to steam produce a windage flow. The fluid between the blades is driven to the periphery by centrifugal effect, which generates a circulating flow. The windage flow consumes wheel power (= windage loss). It also perturbs the outflow from steam-exposed blade channels and partially dissipates the outlet kinetic energy. An impulse stage with partial admission is sometimes used as a first stage in a multistage turbine. There is then typically a diameter reduction after this stage, intended to diminish the through-flow area in order to obtain full admission in the second and further stages. In such a case, the outlet kinetic energy in the first stage is completely dissipated (such a first stage will be discussed in Sect. 6.8.4).

With a single-stage impulse turbine as illustrated in Fig. 6.1, outlet kinetic energy is largely a loss, as the post-connected diffuser can only recover a small part of it. In other applications, outlet kinetic energy is fully useful, as in full admission stages of multistage turbines. Both efficiency definitions (6.8) of *total-to-static efficiency* and (6.9) of *total-to-total efficiency* are thus relevant. With a repeating stage the total-to-total efficiency equals a static-to-static efficiency (Eq. 6.10), but the latter name is not generally used.

6.4.4 Blade Profile Shape

Figure 6.6 represents the velocity triangles with a given nozzle exit angle α_1 and a given blade speed u. Without rotor losses, the path by which w_1 can be turned

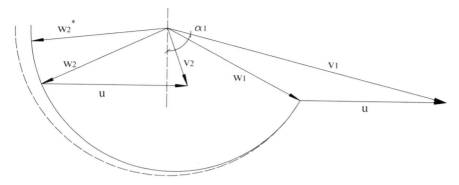

Fig. 6.6 Flow turning in the rotor of an impulse turbine

to w_2 would be the dashed line (circle). The theoretically maximum turning, and thus maximum rotor work, is then reached for a purely tangential outflow. This is practically unachievable, as there would be no axial velocity. With losses, the possible turning path becomes a curve, as drawn with a full line, with the radius decreasing as the turning proceeds. The maximum turning is attained for w_2^*, but the corresponding axial component is still too small for practical realisation. This follows from the details of the rotor blade force generation, as reasoned hereafter.

Figure 6.7 shows the traditional way of composing the profile of an impulse blade. The figure is drawn for a symmetric blade ($\beta_2 = -\beta_1$), but the same principle applies to an asymmetric blade. The pressure side of the rotor blade is a circular arc AB with centre O. The part CD of the suction side is an arc with centre O' shifted over the pitch s with respect to O, so that the pressure side of the adjacent blade A'B' has the same centre O'. The suction side parts EC and DF are drawn straight in the figure, parallel with the inflow and outflow directions. The flow in the channel part A'B'CD has circular streamlines with centre O'. Thus, in absence of losses, the flow satisfies

Fig. 6.7 Traditional shape of an impulse blade; the blade may be rounded at the leading edge with low Mach number

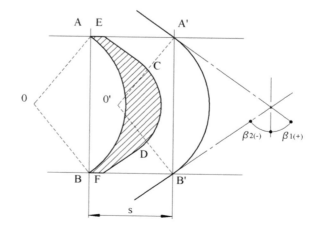

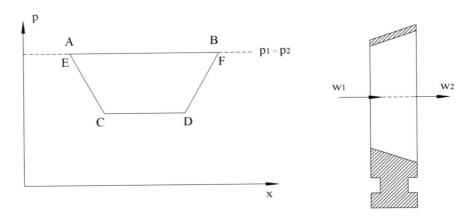

Fig. 6.8 Pressure distribution with a traditional impulse blade in a cylindrical section (left); meridional section through a blade passage (right)

the free-vortex law with constant angular momentum around the centre O'. This means that the streamline A'B' is at constant pressure, namely entry pressure and exit pressure. The CD part is at constant pressure as well, but at a lower value than the common entry and exit pressure. The pressure decreases in the flow sense in the EC part and it increases in the DF part.

The pressure distribution at the suction and pressure sides is drawn in Fig. 6.8. The transition parts EC and DF are drawn straight, but the pressure variation on these parts is normally not exactly linear. The exact form is determined by the geometrical form of parts EC and DF in Fig. 6.7, but the possibility to deviate from linear geometrical forms and linear pressure evolution is limited.

As is typical, the force generation capacity of the blade profile is determined by the adverse pressure gradient on the suction side near the trailing edge. The pressure diagram in Fig. 6.8 is very unfavourable with regard to force generation. For a given pressure gradient DF, the tangential blade force (surface of the pressure diagram) increases if a stagnation point can be created at the blade leading edge AE by rounding the edge. The Mach number of the velocity w_1 must then to be sufficiently lower than unity for avoiding shock waves at rotor entry. The nozzle exit velocity v_1 is usually supersonic, but the rotor entry velocity w_1 is lower. So, in many applications, leading edge rounding is possible (see the later Fig. 6.34).

For sufficiently high inflow Mach number, but this does not occur in steam turbines, some deceleration is possible on the blade pressure side at the blade passage entry (starting at point A) and then a compensating acceleration at the blade passage exit (ending in point B). This feature also increases the pressure difference between the two blade sides.

Even with a symmetric blade form, as represented in Fig. 6.7, the axial velocity decreases from rotor entry to exit ($w_{1a} > w_{2a}$) due to losses. Further, the fluid density decreases somewhat due to heating at constant pressure by losses ($\rho_1 > \rho_2$). Both effects cause the need for a diverging meridional shape of the blade passage (Fig. 6.8, right). At the periphery and the hub, the section area increase causes flow curvature in the meridional plane. The associated pressure decrease on the concave side of the streamsurfaces adds adverse pressure gradient to the boundary layer at the suction side trailing edge. The consequence is, with a symmetric blade, that the divergence of the meridional section is close to the possible maximum for avoiding flow separation.

A symmetric or nearly-symmetric blade profile, thus $\beta_2 \approx -\beta_1$, is mostly chosen in practise. At most a few degrees of deviation from symmetry are possible. Henceforth we take $\beta_2 = -\beta_1$ for the further analysis.

6.4.5 Loss Representation

Loss coefficients are defined with compressible fluids in the same style as for constant-density fluids in Chaps. 2 and 3. Two definitions are common. A loss coefficient associated to the loss representation of the nozzles in Fig. 6.5 is

$$\xi = \frac{\frac{1}{2}v_{1s}^2 - \frac{1}{2}v_1^2}{\frac{1}{2}v_1^2} = \frac{h_1 - h_{1s}}{h_{01} - h_1}. \tag{6.11}$$

The loss may also be expressed by the decrease of the total pressure (see Sect. 4.7 in Chap. 4) as

$$\omega = \frac{p_{00} - p_{01}}{p_{01} - p_1}. \tag{6.12}$$

The coefficient (6.11) is called the *energy loss coefficient* or the *enthalpy loss coefficient* and (6.12) the *pressure loss coefficient* or *total pressure loss coefficient*. For constant density, there is no difference between the coefficients because the difference between total pressure and static pressure is then density multiplied with kinetic energy. For a compressible fluid, there is a difference which increases with Mach number (see Exercise 6.10.1). The pressure loss coefficient is most convenient in experiments. For fundamental analysis of machine components, the energy loss coefficient is the most convenient. Variant forms of the definitions are often used. The denominator in Eq. (6.11) may be replaced by $\frac{1}{2}v_{1s}^2 = h_{00} - h_{1s}$. The denominator in Eq. (6.12) may also be $p_{00} - p_1$.

For the rotor in Fig. 6.5, the enthalpy loss coefficient is

$$\xi = \frac{h_2 - h_1}{h_{0r} - h_2} = \frac{h_2 - h_1}{\frac{1}{2}w_2^2}, \tag{6.13}$$

with the outlet kinetic energy as reference term.

Enthalpy loss coefficients may be estimated with Soderberg's loss correlation (1949), adapted by later researchers [1, 4]. We use here the simplified version by Hawthorne [4]. It represents the losses in an accelerating cascade with optimal solidity (Zweifel's formula 2.26 in Chap. 2) as a fraction ξ of the outlet kinetic energy, with

$$\xi = \xi_1 + \xi_2 \text{ with } \xi_1 = 0.025\left(1 + \left(\frac{\delta^0}{90}\right)^2\right) \text{ and } \xi_2 = 3.2\left(\frac{c_a}{h}\right)\xi_1, \qquad (6.14)$$

where δ^0 is the flow turning in degrees. The coefficient ξ_1 determines friction losses on the blades, with the factor 0.025 for a Reynolds number equal to 10^5, based on hydraulic diameter and exit velocity. With a lower Reynolds number, the factor is somewhat higher. Coefficient ξ_2 determines the friction losses in the end wall boundary layers and the losses by secondary flows (discussion of secondary flows in Sect. 6.9.1). The aspect ratio is the ratio of blade height h to axial width c_a. Henceforth we will take aspect ratio $= 4$ as an example.

The loss formula is intended for subsonic flow. If need be, losses by shock waves have to be added. Clearance losses are not included either. We will use Soderberg's formula for losses in the rotor too, although the formula is meant for cascades with a general acceleration. We do this for reasons of simplicity as the objective of the discussions hereafter on optimisation of the efficiency is only to derive global tendencies. For more complete, but much more complex, loss correlations, we refer to Dixon and Hall [1], Korpela [3] and Lewis [4].

6.4.6 Optimisation of Total-to-Static Efficiency

According to Eq. 6.8, optimisation of the total-to-static efficiency means, with given $h_{00} - h_{1s}$, the maximisation of the rotor work $\Delta W = u(w_{1u} - w_{2u})$. The tangential velocity change, with given u, increases with increasing nozzle exit angle α_1 (Fig. 6.6). But nozzles cannot be made with α_1 near to 90°. The value of α_1 that is practically realisable is at maximum about 75°. We take $\alpha_1 = 75°$.

With symmetric blades:

$$w_{2u} = -\phi_r w_{1u}, \quad w_{2a} = \phi_r w_{1a}.$$

Then:

$$\Delta W = u(w_{1u} - w_{2u}) = u(1 + \phi_r)w_{1u}.$$

With

$$w_{1u} = v_{1u} - u = \phi_s v_{1s} \sin\alpha_1 - u$$

then:

$$\Delta W = (1 + \phi_r)u(v_{1u} - u)$$

For constant ϕ_r, work is a parabolic function of u, with a maximum for

$$u = u_o = \frac{v_{1u}}{2} = \frac{\phi_s v_{1s} \sin \alpha_1}{2}.$$

The value of the maximum work is

$$(\Delta W)_o = (1 + \phi_r)u_o^2 = \left(\frac{1 + \phi_r}{2}\right)\phi_s^2 \sin^2 \alpha_1 \frac{v_{1s}^2}{2}.$$

The maximum efficiency value is

$$(\eta_i)_o = \frac{(\Delta W)_o}{\Delta h_s} = \left(\frac{1 + \phi_r}{2}\right)\phi_s^2 \sin^2 \alpha_1.$$

This result is somewhat approximate because the rotor velocity coefficient varies slightly with varying blade speed.

Applied to the nozzle for $\delta = \alpha_1 = 75°$, Soderberg's formula results in $\xi = 0.0763$, for aspect ratio $= 4$.

With

$$\xi_s \frac{v_1^2}{2} = \frac{v_{1s}^2}{2} - \frac{v_1^2}{2} : \eta_{ss} = \frac{1}{1 + \xi_s} = 0.929 \text{ and } \phi_s = \frac{1}{\sqrt{1 + \xi_s}} = 0.964.$$

The velocity obtained by loss-free conversion into kinetic energy of the relevant isentropic enthalpy drop of a turbine stage is called the *spouting velocity* and is taken as *reference velocity* of the stage. For an impulse turbine (constant pressure in the rotor) with loss of exit kinetic energy, the total-to-static isentropic enthalpy drop is the relevant drop and the spouting velocity is v_{1s}.

The *speed ratio* is the ratio of the blade speed to the spouting velocity:

$$\lambda = u/v_{1s}.$$

The optimum speed ratio with an impulse turbine is

$$\lambda_o = \frac{\phi_s \sin \alpha_1}{2} \approx 0.465.$$

With $\alpha_1 = 75°$ corresponds $\beta_1 \approx 62°$. Rotor loss can be expressed, with $w_2 = \phi_r w_1$, by

$$\frac{w_1^2}{2} - \frac{w_2^2}{2} = \left(\frac{1}{\phi_r^2} - 1\right)\frac{w_2^2}{2} = \xi_r\frac{w_2^2}{2}.$$

Thus:

$$\phi_r = \frac{1}{\sqrt{1 + \xi_r}}.$$

For $\Delta\beta = 124°$, Soderberg's formula gives $\xi = 0.1304$ and $\phi_r = 0.941$. The corresponding optimum efficiency with $\phi_r = 0.941$ is

$$(\eta_i)_o = \frac{1 + \phi_r}{2}\phi_s^2 \sin^2 \alpha_1 \approx 0.84.$$

The loss formula enables the analysis of the influence of the nozzle angle. For $\alpha_1 = 80°$ follows $\phi_s = 0.962$, $\lambda_o = 0.475$, $\beta_1 = 70.5°$, $\phi_r = 0.930$, $(\eta_i)_o = 0.866$. This example demonstrates that the largest possible α_1-angle is advantageous for efficiency. But the axial velocity decreases with a larger nozzle angle. This implies a lower flow rate through the machine, which may be a drawback for some applications. Even if the flow rate limitation is acceptable, the nozzle angle cannot be increased until 80°. The accompanying vane cannot be materialised. About 75° is the highest achievable nozzle angle.

Figure 6.4 represents the velocity triangles with optimal operation for $\alpha_1 = 75°$. The outflow velocity v_2 deviates somewhat from the axial direction in the running sense. Some rotor asymmetry, in the sense of $|\beta_2| > \beta_1$, increases efficiency. This is applied in practice. The outflow is then approximately axial and the efficiency amounts to about 0.85.

It is common to express the *work coefficient* of a stage, also called *stage loading coefficient*, by

$$\psi = \frac{\Delta W}{u^2}.$$

A similar coefficient is defined with the isentropic enthalpy drop supplied to the stage (here: total-to-static). The isentropic enthalpy drop is often called the *isentropic head* and even just the *head*. So, the term head is used, similarly as with a constant-density fluid, to express the work capacity of the fluid.

The *head coefficient* is

$$\psi_s = \frac{\Delta h_s}{u^2}.$$

With $\Delta W = \eta_i \Delta h_s$, it follows that $\psi = \frac{\eta_i}{2\lambda^2}$ and $\psi_s = \frac{1}{2\lambda^2}$.

For a Laval stage, the optimum value of the work coefficient is $\psi_o \approx 1.95$. The optimal value of the head coefficient is $(\psi_s)_o \approx 2.30$.

6.5   The Pressure-Compounded Impulse Turbine or Rateau Turbine

6.5.1   Principle

Pressure compounding is series arrangement of stages with a pressure drop in each of them. The objective is to distribute the overall enthalpy drop over a number of stages so that the enthalpy drop per stage becomes tractable. In the previous section we learned that the work coefficient of an impulse stage with loss of outlet kinetic energy is about 1.95. The blade speed acceptable for a turbine depends on the steam temperature. At a temperature of the 500 °C order, it is limited to about 200 m/s. This rather low value comes from the use of steel materials, which are much cheaper than the high-temperature materials used in gas turbines. Cheaper materials are imperative with steam turbines due to their very large dimensions. For $u = 200$ m/s and $\psi = 1.95$ is $\Delta W \approx 1.95 \times (200)^2$ J/kg ≈ 75 kJ/kg. At the end of the expansion, near to the vacuum, the steam temperature decreases to about 50 °C. The blade speed may then be much larger (see Sect. 6.8) and also the enthalpy drop may then be larger (see exercises). But, the example of the possible enthalpy drop of a first stage shows that multiple stages are mostly needed.

Figure 6.9 represents the series arrangement of two impulse stages. The first stage is fed by nozzles mounted on a distributor chamber (often called steam chest). Steam thus enters the turbine casing at a lower temperature and a lower pressure than the steam supplied. The distributor is a relatively small chamber, more advantageously subjected to pressure load than the much larger turbine casing. From the second stage onwards, nozzles are mounted in stator discs, called *diaphragms*, sealed to the shaft with labyrinth rings with very small clearance. In an impulse stage, the whole

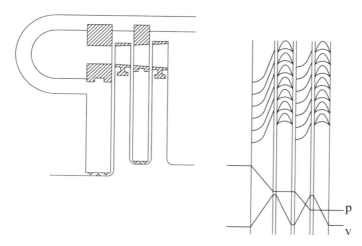

Fig. 6.9 Pressure compounding (series arrangement) of impulse stages

pressure drop is on the stator. The role of the diaphragms is to reduce the clearance surface around the shaft. Rotor blades rows need not be sealed to the casing, as there is no pressure drop through the rotor. With the nozzles on diaphragms, the rotor blades are on discs connected to the rotor hub. The rotor is then a *disc rotor*. The alternative is a *drum rotor* (see later).

6.5.2 Efficiency

Series arrangement allows use of outlet kinetic energy. The repeating stage efficiency is rendered by Eq. (6.10). The isentropic enthalpy drop through the stage (total-to-total or static-to-static) is $h_o - h_{1s}$ (Fig. 6.5). The nozzle velocity is

$$\frac{v_1^2}{2} = \phi_s^2 \left(\frac{v_s^2}{2} + \frac{v_2^2}{2} \right) \text{ with } \frac{v_s^2}{2} = h_0 - h_{1s},$$

where v_s is, again, called the *spouting velocity*.

According to Fig. 6.4 it follows, with symmetric rotor blades, that

$$v_{1u} = v_1 \sin \alpha_1, \quad v_{1a} = v_1 \cos \alpha_1, \quad w_{1u} = v_{1u} - u,$$

$$v_{2a} = \phi_r v_{1a}, \quad w_{2u} = -\phi_r w_{1u}, \quad v_{2u} = u + w_{2u}, \quad \eta_i = \frac{2u(v_{1u} - v_{2u})}{v_s^2}.$$

The motivation for using symmetric blades is the same as with a single-stage impulse turbine. Velocity components relative to v_s and the internal efficiency can be determined iteratively from the foregoing relations, for a given nozzle exit angle α_1, and given speed ratio $\lambda = u/v_s$. Iterations start with $\phi_s = \phi_r = 1$. Flow angles are determined after a first calculation, loss coefficients (Soderberg) follow and thus also rotor and stator velocity coefficients. Values of v_{1u}/v_s, v_{2u}/v_s and η_i as functions of λ, with $\alpha_1 = 75°$, are given in Table 6.1.

The optimum speed ratio is much higher than the optimum with loss of outlet kinetic energy. It rises from 0.47 to 0.72, but the meaning of the reference velocity v_s is not the same in both cases. With equal pressure levels, v_s is lower here. This partially explains the higher value of λ. The optimum speed ratio shifts to a higher value, mainly because rotor and stator flow turnings then decrease, and the loss coefficients with them. The outlet kinetic energy increases, but this is not a penalisation now. For the same reason, the nozzle exit angle does not have to take the maximum technically realisable value. We assume here the rather high value of $\alpha_1 = 75°$, because this leads to low axial velocity and hence large height of the blades. This is appropriate for flow with high steam density as in the high-pressure part of the turbine. But actually, the nozzle angle may be optimised too (see Chap. 15).

The optimum efficiency is almost 9 percentage points better than with loss of outlet kinetic energy: it increases from 84% to almost 93%. The dependence of the

Table 6.1 Internal efficiency as a function of speed ratio for an impulse turbine stage with useful exit kinetic energy (Soderberg eq. with AR = 4; $\alpha_1 = 75°$); values in bold are for maximum efficiency and for 1.5 points lower efficiency

$\lambda = u/v_s$	η_i	v_{1u}/v_s	v_{2u}/v_s
0.40	0.8557	0.952	−0.117
0.50	0.9027	0.965	0.062
0.54	**0.9134**	**0.975**	**0.129**
0.55	0.9154	0.978	0.146
0.60	0.9225	0.995	0.226
0.65	0.9260	1.015	0.303
0.68	0.9271	1.029	0.348
0.70	0.9275	1.039	0.377
0.72	**0.9276**	**1.050**	**0.406**
0.74	0.9275	1.061	0.434
0.77	0.9271	1.078	0.476
0.80	0.9264	1.096	0.517
0.85	0.9245	1.127	0.584
0.90	0.9219	1.161	0.649
1.00	0.9152	1.233	0.775

efficiency on the speed ratio is weak near the optimum, so that the speed ratio may be strongly reduced without significant decrease of the internal efficiency. The efficiency lowers with about one and a half percentage point with a speed ratio lowered until 0.54. The work coefficient is higher with a lower speed ratio. The turbine can then be built with fewer stages, which results in lower wheel friction loss and lower leakage loss. The consequence is that the speed ratio for optimum overall efficiency (leakage and wheel friction included) is much lower than that for optimum internal efficiency.

The foregoing reasoning does not allow the determination of the precise value of the optimum speed ratio. We may assume that it is around 0.54. The velocity triangles are drawn in Fig. 6.10 with $\lambda = 0.54$ (compare to Fig. 6.4). The work coefficient is $\psi = \eta_i/2\lambda^2 \approx 1.56$. Clearly, with some blade asymmetry, the work coefficient can be increased somewhat and the outflow brought nearer to the axial direction.

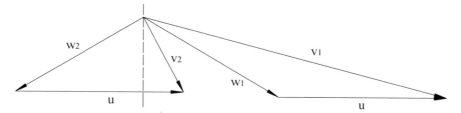

Fig. 6.10 Velocity triangles of an impulse turbine with useful outlet kinetic energy: $R_s = 0$, $\psi = 1.56$ (work coefficient for optimum total-to-total efficiency)

6.6 The Velocity-Compounded Impulse Turbine or Curtis Turbine

Velocity compounding means that steam from the nozzles is used in several rotor blade rows, without pressure drop in the components downstream of the nozzles. Figure 6.11 illustrates the principle. Having worked in a first blade row, steam is turned by stator vanes without any pressure drop, so that it can be supplied to a second blade row running in the same sense as the first one. Therefore, the two rotor blade rows can be mounted together on the same wheel. The kinetic energy from the nozzles is reduced in two partial stages. This explains the term velocity compounding. This turbine type was patented in 1895 by C. G. Curtis.

Figure 6.12 represents the velocity triangles. We analyse for lost kinetic energy at

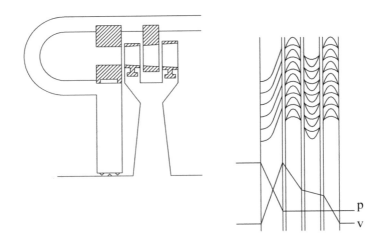

Fig. 6.11 Velocity compounding: Curtis turbine

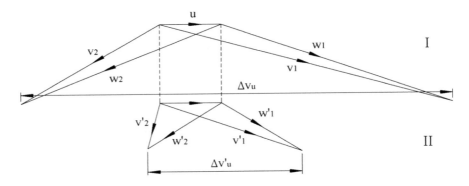

Fig. 6.12 Velocity triangles with a Curtis turbine

stage exit. The velocity component v_{1u} must be about 4 times larger than the blade speed u to make the flow kinematically possible. For the same reasons as with the Laval turbine, rotor blades and inversing stator vanes must be symmetric.

The tangential exit velocity of the first rotor blade row is

$$v_{2u} = w_{2u} + u = -\phi_{r1}(v_{1u} - u) + u.$$

The work in the first rotor blade row is

$$\Delta W_1 = u(v_{1u} - v_{2u}) = (1 + \phi_{r1})u(v_{1u} - u).$$

The tangential entry velocity of the second rotor blade row is

$$v'_{1u} = -\phi_{inv}v_{2u} = \phi_{inv}(\phi_{r1}(v_{1u} - u) - u).$$

The work in the second rotor blade row is

$$\Delta W_2 = (1 + \phi_{r2})u(v'_{1u} - u).$$

The total work is

$$\Delta W = (1 + \phi_{r1})u(v_{1u} - u) + (1 + \phi_{r2})u(\phi_{inv}\phi_{r1}v_{1u} - \phi_{inv}\phi_{r1}u - \phi_{inv}u - u).$$

With assumed equal velocity coefficients, this expression is reduced to

$$\Delta W = (1 + \phi)\, u\left[(1 + \phi^2)v_{1u} - (\phi^2 + \phi + 2)u\right].$$

With constant ϕ and constant v_{1u}, the work is, as with the Laval stage, a parabolic function of u. Maximum work is attained for

$$u = \frac{1 + \phi^2}{2(\phi^2 + \phi + 2)}v_{1u} = \frac{1 + \phi^2}{2(\phi^2 + \phi + 2)}\phi_s \sin \alpha_1 v_{1s}.$$

With $\alpha_1 = 75°$, $\phi_s = 0.965$ and $\phi = 0.94$ follows $\lambda_o \approx 0.230$.
And $(\eta_i)_o = \frac{\Delta W_o}{v_s^2/2} = 2(1 + \phi)\lambda_o\left[\frac{1+\phi^2}{2}\phi_s \sin \alpha_1\right] \approx 0.785$.
The corresponding work coefficient is $\psi_o = \frac{\Delta W}{u^2} = \frac{\eta_i}{2\lambda_o^2} \approx 7.40$.

The optimum speed ratio of the Curtis stage is thus about half of that of the Laval stage (0.230 against 0.466). The work coefficient is about 3.8 times larger (7.40 against 1.95). The foregoing derivations may be refined by calculating the velocity coefficients depending on the turnings in the blade rows. This affects details of the results obtained, but not the overall conclusion. The efficiency of a Curtis stage is lower than that of a Laval stage, basically because the conversion of the nozzle kinetic energy causes three losses, respectively in the first rotor blade row, the inversing stator vane row, and the second rotor blade row. The rotor work is about 4 times larger than with the Laval stage. The Curtis principle can be extended to 3 and more rotor blade rows. For $\alpha_1 = 90°$ and taking no losses into account, the following relations are optimal:

Laval: $v_{1u}/u = 2$, $\psi = 2$, $\lambda_o = 1/2$.

Curtis 2: $v_{1u}/u = 4$, $v'_{1u}/u = -v_{2u}/u = 2$, $\psi = 6 + 2 = 8$, $\lambda_o = 1/4$.

Curtis 3: $v_{1u}/u = 6$, $v'_{1u}/u = -v_{2u}/u = 4$, $v''_{1u}/u = -v''_{2u}/u = 2$, $\psi = 10 + 6 + 2 = 18$, $\lambda_o = 1/6$.

Curtis stages with three rotor blade rows have never been used in practice, because the efficiency is even much lower than with two rotor blade rows. We will discuss the application of a Curtis stage with two rotor blade rows in Sect. 6.8.4, but Curtis stages are only very exceptionally applied in modern machines.

6.7 The Reaction Turbine

6.7.1 Degree of Reaction

Figure 6.13 represents the h–s diagram of a turbine stage with a degree of reaction around 50%. The degree of reaction that we used until now is defined by

$$R = \frac{h_1 - h_2}{h_{01} - h_{02}}, \tag{6.15}$$

with $h_1 - h_2 = \frac{w_2^2}{2} - \frac{w_1^2}{2}$ and $h_{01} - h_{02} = u(v_{1u} - v_{2u}) = u(w_{1u} - w_{2u})$.

Fig. 6.13 h–s diagram of a reaction turbine

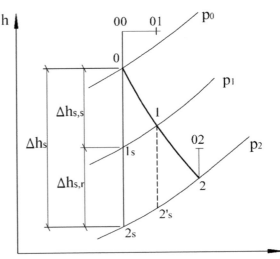

This degree of reaction can thus be expressed as a function of the velocity components and is therefore named the *kinematic degree of reaction.*

With an assumed constant axial velocity (approximately met), it follows that

$$R = \frac{w_{2u}^2 - w_{1u}^2}{2u(w_{1u} - w_{2u})} = -\frac{w_{2u} + w_{1u}}{2u} = -\frac{w_{mu}}{u}.$$

There is a second definition of the degree of reaction, called *thermodynamic degree of reaction,* expressed with static enthalpy drops on the isentropic process:

$$R_s = \frac{\Delta h_{s,r}}{\Delta h_s} = \frac{\Delta h_{s,r}}{\Delta h_{s,s} + \Delta h_{s,r}}. \tag{6.16}$$

The two definitions of degree of reaction are equal for a repeating stage with lossless flow. With losses and with degrees of reaction not near to 0 or 1, $R \approx R_s$. For degrees of reaction near 0 or 1 the difference is more significant. For instance, $R_s = 0$ but $R < 0$ for a stage with constant pressure in the rotor. We will also use the term *isentropic degree of reaction* to refer to R_s. The reason for using the definition (6.16) is the same as with fans (definition of the pressure degree of reaction): precise determination of the kinematic degree of reaction is not possible in practice due to typically rather large errors with temperature measurements.

6.7.2 Efficiency

Figure 6.14 represents velocity triangles for 50% degree of reaction.

We assume a repeating stage and denote the isentropic static-to-static enthalpy drop over the stage by $\Delta h_s = h_0 - h_{2s}$ (Fig. 6.13). Using isentropic efficiency, the stator exit velocity follows from

$$\frac{v_1^2}{2} = \eta_{ss}\left[\frac{v_2^2}{2} + (1 - R_s)\Delta h_s\right],$$

$$\frac{v_{1u}^2}{2} + \frac{v_a^2}{2} = \eta_{ss}\left[\frac{v_{2u}^2}{2} + \frac{v_a^2}{2} + (1 - R_s)\Delta h_s\right],$$

$$v_{1u} = \sqrt{\eta_{ss}\left[v_{2u}^2 + v_a^2 + (1 - R_s)v_s^2\right] - v_a^2}. \tag{6.17}$$

Fig. 6.14 Velocity triangles for a turbine with degree of reaction of 50% ($R = 0.50$, $\psi = 1.20$; ψ is larger than the value $\psi = 0.95$ for optimum efficiency)

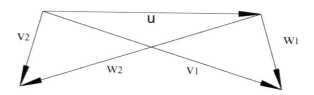

It is obvious that the positive root has to be chosen.

$$w_{1u} = v_{1u} - u.$$

Ignoring the enthalpy increase by the divergence of the isobars, the rotor exit velocity follows from

$$\frac{w_2^2}{2} = \eta_{sr}\left[\frac{w_1^2}{2} + R_s \Delta h_s\right],$$

$$w_{2u} = -\sqrt{\eta_{sr}(w_{1u}^2 + v_a^2 + R_s v_s^2) - v_a^2}. \tag{6.18}$$

The negative root has to be chosen in this case.

$$v_{2u} = w_{2u} + u, \quad \eta_i = \frac{2u(v_{1u} - v_{2u})}{v_s^2}.$$

Iterative determination of the velocity components relative to v_s, with chosen R_s and λ, is quite simple. We take $\alpha_1 = 72°$ as an example and start iterations with $v_{2u} = 0$, $\eta_{ss} = 1$, $\eta_{sr} = 1$. After a first calculation, we determine the flow angles and the loss coefficients with Soderberg's formula. Stator and rotor efficiencies follow from $\eta_{ss} = 1/(1 + \xi_s)$, $\eta_{sr} = 1/(1 + \xi_r)$. Table 6.2 lists the results for $R_s = 0$, 0.25, 0.50 and 0.75.

Values in bold in Table 6.2 are those with optimum internal efficiency and with a lower speed ratio corresponding to one and a half percentage points lower efficiency. Maximum efficiency is obtained with $R_s = 0.5$. The reason is that the velocity triangles then are symmetric to each other, as can be derived from Fig. 6.14. Losses are proportional to the square of the exit velocities of rotor and stator, which leads to minimum sum of losses with symmetric velocity triangles.

As from the earlier analysis for $R_s = 0$, we observe that the speed ratio can be decreased substantially compared to the optimum, without strong reduction of the internal efficiency. The number of stages is lower with a lower speed ratio. Turbines with degree of reaction significantly above zero are constructed with a drum rotor, because sealing is then also necessary on the rotor blades (see the later Figs. 6.18 and 6.19). But also with drum rotors, leakage losses decrease with fewer stages and end-wall friction loss is lower due to a shorter rotor. So, the optimum overall efficiency is reached at a lower speed ratio than for optimum internal efficiency. As an estimate for the optimum, we can adopt once more a decrease of internal efficiency with one and a half points.

The global results from Table 6.2 are summarised in Table 6.3. Also shown is the performance for $\lambda = 0.60$ and for axial inlet and outlet. Note that the results for $R_s = 0$ do not accord completely with those in Table 6.1. Stator angles are slightly different, the axial velocity is constant in the present analysis and blade symmetry is not imposed.

Table 6.2 Total-to-total internal efficiency as a function of speed ratio with varying degree of reaction for repeating stages; $a = v_{1u}/v_s$, $b = v_{2u}/v_s$ (Soderberg correlation AR $= 4$; $\alpha_1 = 72°$)

$R_s = 0.0$				$R_s = 0.25$			
λ	η	a	b	λ	η	a	B
0.40	0.8605	0.959	−0.117	0.50	0.8898	0.830	−0.060
0.45	0.8836	0.960	−0.022	0.55	0.9042	0.834	0.012
0.50	0.9006	0.968	0.068	0.60	0.9149	0.841	0.079
0.53	**0.9080**	0.975	0.119	**0.62**	**0.9182**	0.845	0.104
0.60	0.9184	0.998	0.233	0.65	0.9223	0.852	0.142
0.65	0.9218	1.019	0.310	0.70	0.9371	0.865	0.203
0.70	0.9232	1.044	0.385	0.75	0.9301	0.880	0.260
0.72	**0.9232**	1.055	0.414	0.80	0.9318	0.898	0.312
0.75	0.9230	1.072	0.456	**0.87**	**0.9325**	0.926	0.390
0.80	0.9217	1.102	0.526	0.90	0.9324	0.939	0.421
0.90	0.9166	1.169	0.659	1.00	0.9303	0.986	0.521

$R_s = 0.50$				$R_s = 0.75$			
λ	η	a	b	λ	η	a	B
0.60	0.9070	0.678	−0.078	0.60	0.9150	0.519	−0.244
0.65	0.9187	0.678	−0.028	**0.64**	**0.9220**	0.508	−0.212
0.70	**0.9280**	0.681	0.019	0.70	0.9284	0.495	−0.168
0.75	0.9348	0.687	0.063	0.75	0.9316	0.488	−0.134
0.80	0.9393	0.694	0.106	0.80	0.9338	0.483	−0.101
0.85	0.9418	0.702	0.148	0.85	0.9354	0.480	−0.070
0.90	0.9428	0.712	0.188	0.90	0.9364	0.479	−0.041
0.94	**0.9430**	0.721	0.219	0.95	0.9369	0.480	−0.013
1.00	0.9423	0.736	0.264	**0.97**	**0.9369**	0.481	−0.002
1.05	0.9411	0.749	0.301	1.05	0.9361	0.485	0.039
1.10	0.9396	0.764	0.336	1.10	0.9348	0.488	0.063

Table 6.3 Internal efficiency with varying degree of reaction

R_s	λ_o (−1.5%)	η_o	b_o	η ($\lambda = 0.60$)	λ ($b = 0$)	η ($b = 0$)
0	0.53	0.908	0.119	0.918	0.46	0.888
0.25	0.62	0.918	0.104	0.915	0.54	0.902
0.50	0.70	0.928	0.019	0.907	0.68	0.925
0.75	0.64	0.922	−0.212	0.915	0.97	0.937

Conclusions are:

- Optimum overall efficiency is reached with $R_s = 0.5$ and the corresponding speed ratio $\lambda \approx 0.70$. Inflow and outflow are very near to the axial direction.
- For a somewhat lower speed ratio, as $\lambda = 0.60$, internal efficiency is at its lowest with $R_s = 0.5$. This shows that the optimum in the efficiency variation is weak.
- With axial inflow and outflow, the optimum speed ratio increases with increasing degree of reaction.

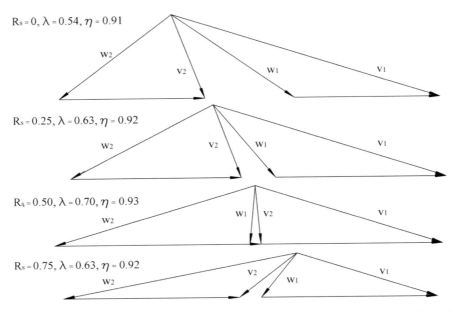

Fig. 6.15 Velocity triangles with maximum overall efficiency ($\alpha_1 = 72°$, constant Δh_s); internal efficiency values are specified

Velocity triangles for maximum overall efficiency (optimum internal efficiency -1.5 points) are shown in Fig. 6.15 (drawn for the same value of v_s, thus same Δh_s). The maximum efficiency depends only weakly on the degree of reaction, but the shape of the rotor blades depends strongly on the degree of reaction. The highest efficiency is reached with $R_s = 0.50$, however. The corresponding work coefficient $\psi = (\eta_i/2\lambda^2) \approx 0.95$ and the entry and exit velocities of the stage are near to axial. This is very convenient, as it enables a turbine without special first and last stages.

The internal efficiency in Fig. 6.15 is lower with a lower degree of reaction, but the corresponding work coefficient is larger. This results in fewer stages, causing lower wheel friction loss. Leakage is also smaller with a low degree of reaction when the machine is made with a disc rotor, as shown in Fig. 6.9, but wheel friction with a disc rotor is higher than with a drum rotor. Wheel friction and leakage make that the overall efficiency with low degree of reaction is not necessarily lower than with 50% degree of reaction. Especially with high steam density and short blades, leakage loss may be a significant loss component such that a low degree of reaction may become the most efficient.

Figure 6.16 presents the velocity triangles with axial inflow and outflow. For low degree of reaction, there is a small efficiency loss compared to the optimum with the triangles in Fig. 6.15.

The small variation of internal efficiency in Figs. 6.15 and 6.16 has as consequence that it is not possible to decide on an optimum degree of reaction with a simple analysis of internal efficiency, as done here. The overall optimum is case-dependent because

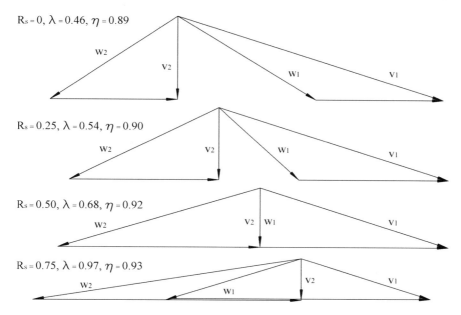

$R_s = 0$, $\lambda = 0.46$, $\eta = 0.89$

$R_s = 0.25$, $\lambda = 0.54$, $\eta = 0.90$

$R_s = 0.50$, $\lambda = 0.68$, $\eta = 0.92$

$R_s = 0.75$, $\lambda = 0.97$, $\eta = 0.93$

Fig. 6.16 Velocity triangles with axial inlet and outlet ($\alpha_1 = 72°$, constant Δh_s); internal efficiency values are specified

leakage and wheel friction intervene. It is also not possible to decide on the precise shape of the velocity triangles corresponding with a chosen degree of reaction. In practice, the precise shape follows from an optimisation study, which nowadays is largely made with CFD tools. The Figs. 6.15 and 6.16 thus only show tendencies.

The foregoing results have historically led to two radically different design options. With some manufacturers, the choice was for $R_s = 0.50$ and corresponding $\lambda = 0.7$ ($\psi = 0.95$). The motivation for this choice is maximum efficiency and axial inflow and outflow. With other manufacturers, the choice was a low degree of reaction, namely $R_s \approx 0.10$ to 0.15 and corresponding $\lambda \approx 0.57$ ($\psi \approx 1.40$). The internal efficiency is then somewhat lower (about one percentage point), but the turbine has fewer stages. The types are historically named *reaction turbine* and *impulse turbine,* with the meaning that a reaction turbine has a degree of reaction around 50% and that an impulse turbine has zero degree of reaction or a low degree of reaction $R_s = 0.10$ to 0.15.

In modern reaction type turbines, the degree of reaction is not exactly 50% on the mean radius. The degree of reaction is optimised and varies from stage to stage and varies also along the radius in a stage (see Sect. 6.9). With a reaction type turbine stage is meant that the degree of reaction is significantly above zero.

As discussed earlier, the blade shape is disadvantageous with $R_s = 0$ (Figs. 6.7 and 6.8), even if rounding of the leading edge is possible. With a degree of reaction somewhat above zero, the rotor blade becomes asymmetric and creation of a leading edge stagnation region increases the force capacity. Therefore, the degree of reaction is chosen low but not zero, even with partial admission. Although impulse type stages have a somewhat lower efficiency, they are meaningful for some applications. We discuss some examples in Sects. 6.8.3 and 6.8.4.

6.7.3 Axial Inflow and Outflow

From the foregoing analysis follows that the inlet and outlet flows of an optimised stage are near to the axial direction. A priori assumption of axial outflow enables a very simple analysis of the axial turbine. Such an analysis is often done for demonstration of some approximate dependencies.

Rotor work is then

$$\Delta W = u(v_{1u} - v_{2u}) = u v_{1u}.$$

The kinematic degree of reaction is

$$R = \frac{\frac{1}{2}w_2^2 - \frac{1}{2}w_1^2}{\Delta W} \text{ or } 1 - R = \frac{\frac{1}{2}v_1^2 - \frac{1}{2}v_2^2}{\Delta W}.$$

With axial outflow and constant axial velocity through the rotor is $v_2 = v_{1a}$, so that

$$1 - R = \frac{v_{1u}^2}{2u v_{1u}} \text{ or } R = 1 - \frac{1}{2}\frac{v_{1u}}{u}. \tag{6.19}$$

The work coefficient is

$$\psi = \frac{\Delta W}{u^2} = \frac{v_{1u}}{u}. \tag{6.20}$$

The expressions (6.19) and (6.20) are similar to the expressions obtained with the analysis of a radial fan in Chap. 3. The assumptions for the fan were: entry velocity in the meridional plane and constant meridional component of the velocity in the rotor. The assumptions are similar here. From Eqs. (6.19) and (6.20) follows

$$\psi = 2(1 - R). \tag{6.21}$$

So, a low degree of reaction corresponds to a high work coefficient and vice versa. For $R = 0$ is $\psi = 2$. For $R = 0.5$ is $\psi = 1$. For $R = 1$ becomes $\psi = 0$.

Further it follows that

$$\lambda^2 = \frac{u^2}{2\Delta h_s} = \frac{\eta_i}{2\psi},$$

or, with $\eta_i \approx 0.92$:

$$\lambda = \frac{\sqrt{\eta_i/2}}{\sqrt{1-R}} \approx \frac{0.48}{\sqrt{1-R}}. \tag{6.22}$$

This expression reproduces very well the results in Fig. 6.16:
$R = 0, \lambda = 0.48; R = 0.25, \lambda = 0.55; R = 0.50, \lambda = 0.68; R = 0.75, \lambda = 0.96.$

6.8 Steam Turbine Construction Forms

Steam turbines are constructed in a large variety of forms and many design decisions depend on requirements of the application and on economic factors. It is therefore not feasible to discuss extensively construction choices of steam turbines in a course text on fundamental aspects. We discuss some general features. We make a distinction between large steam turbines with power in the order of 1000 MW of above it, mid-size machines with power about 200–800 MW and small machines with power less than 100 MW.

6.8.1 Large Steam Turbines for Coal-Fired Power Stations

Figure 6.17 is a possible lay-out of a steam turbine in the 1000–1200 MW order for use in coal-fired power stations. The machine has 5 parts: a high-pressure part (HP), an intermediate-pressure part (IP) and 3 low-pressure parts (LP). These parts are physically separate machines with the shafts coupled to each other, driving a single generator. The turbine is thus composed of modules, often called cylinders.

The high-pressure part has a single flow. There is reheat after the HP part. The intermediate part and each low-pressure part have two flows. Multiplication of the number of flows as the expansion advances is necessary, because of the strong decrease

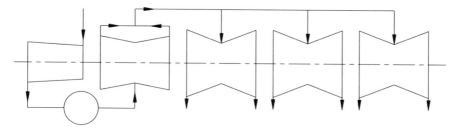

Fig. 6.17 Possible lay-out of a large steam turbine for a coal-fired power station: one single-flow HP part, one double-flow IP part and three double-flow LP parts (some machines have two LP parts); reheat between HP and IP parts

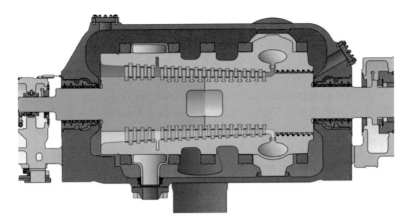

Fig. 6.18 HP cylinder of a large steam turbine (*Courtesy* GE Steam Power)

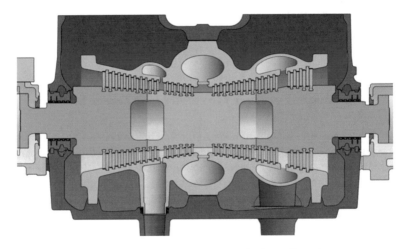

Fig. 6.19 IP cylinder of a large steam turbine (*Courtesy* GE Steam Power)

of steam density. As already said in Sect. 6.1, the most advanced supply conditions (2021) are 330 bar, 650 °C (ultra-supercritical). The reheat temperature is then 670 °C. The overall length of a 1200 MW machine is about 40 m. The rotational speed is 3000 rpm in a 50 Hz system and 3600 rpm in a 60 Hz system, called full speed (examples of half speed come later).

Figures 6.18 and 6.19 show longitudinal sections of HP and IP cylinders of reaction type made with drum rotors. Note the steam extractions (the outlet of the IP cylinder is not visible). The stages are said to be of reaction type, but in modern large machines, the degree of reaction is optimised for each stage and may vary quite strongly from entry to exit of a module. In particular, the optimum degree of reaction may be rather low in the first stages of an HP module because of the benefit of reducing leakage flows in high-density steam. Moreover, with a low degree of reaction in the first stage, the temperature at the entry to the rotor is reduced.

The rotors of Figs. 6.18 and 6.19 are pure drums, but rotors may also be made with somewhat reduced diameter at the positions of the stator rows such that they get the appearance of a disc rotor. This construction form requires more axial length but lowers the leakage flow in the stator parts.

The HP and IP modules have a double casing. The casings are divided in an upper and a lower half by a horizontal split. The outer casings of both modules and the inner casing of the IP module are joint by flanges and bolts. An opened IP module is then as shown in the later Fig. 6.26. The parts of the inner casing of the HP cylinder are joined by shrink rings.

Figure 6.20 is an inner casing with shrink rings. These rings are expanded by inductive heating for placing and removing them. Because bolt heads need space, the assemblage by shrink rings is done for limiting the radial dimensions of the casings, which is advantageous concerning strength. The alternative, shown in Fig. 6.21 is with the inner casing halves joined by countersunk bolts and the outer casing in the form of a barrel. The inner casing is slid in it and it is closed by a circular cover.

The construction of HP and IP cylinders with two casings is meant to bring the outlet pressure between the two casings. This does not reduce the pressure difference between the inside and the outside of the machine, but replacing a very thick shell by two thinner ones is advantageous concerning thermal stresses. These are the stresses due to unequal heating, especially delicate during starting and stopping of the turbine.

HP and IP rotor parts are traditionally forged from ingots with parts joined by welding, as in Figs. 6.18 and 6.19. An alternative is composing the rotor from more parts. The fabrication of these parts is then easier, but there are more welds. HP and IP rotors are made from chromium-based high-strength steel alloys with up to 12%

Fig. 6.20 Inner casing of an HP cylinder with shrink rings (*Courtesy* GE Steam Power)

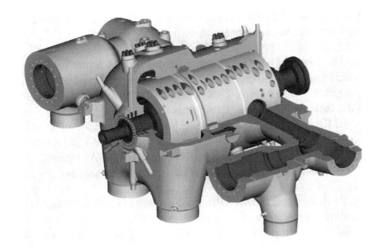

Fig. 6.21 Inner casing of an HP cylinder with countersunk bolts and outer casing of barrel type (*Courtesy* Siemens Energy)

Cr and 1–1.5% Mo. A similar alloy is used for the cast inner casings. The outer casings are cast from an alloy with much less content of Cr.

With reaction blading, there is an axial force on the rotor in the through-flow sense. In the double-flow modules of Fig. 6.17, the axial forces in the opposite flow senses balance each other. With the single-flow HP module, the axial force is balanced by an equilibrium piston. This is the cylindrical part at the right-hand side of the rotor of Fig. 6.18, sealed at its periphery by labyrinth combs. The steam pressure after the first stator row is at one side and the much lower exit pressure at the other side (the equilibrium piston is also visible in the later Fig. 6.28).

Stator vanes and rotor blades are mounted in tangential slots in HP and IP cylinders, as illustrated in Fig. 6.22. Rotor and stator blades are milled from forged pieces of chromium-steel, with up to 12% Cr for rotor blades. Rotor blades have a hammer root. A cavity is required for mounting the blade roots in the tangential slots. It is filled afterwards with a bolted metal piece. Mounting of stator vanes does not require roots, as in Fig. 6.22, but there are variants with roots on stator vanes.

Figure 6.22 also shows that blades and vanes may be shrouded with cover bands fitting to labyrinth combs. There are variants with combs on the cover bands and with smooth casing or smooth rotor parts. Mostly, cover bands are formed by blade heads, as shown in Fig. 6.23. These are shroud parts milled together with the blades.

The roots of the rotor blades represented in Fig. 6.23 (right) are fir-tree roots mounted in axial slots on a wheel (an impulse stage, see Sect. 6.8.4).

Figure 6.24 shows an LP rotor of a 3000 rpm turbine. The rotor rows have cover bands except for the last stage. With very high tip speeds, cover bands cannot be applied to the last stage for reasons of strength. Moreover, the last-stage blades are very long. The clearance between the rotor and the casing is thus, relatively seen,

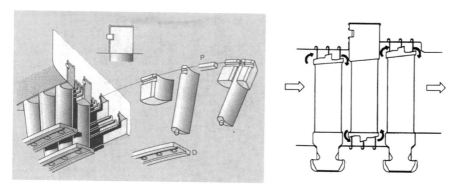

Fig. 6.22   Vanes and blades in tangential slots (left: *Courtesy* GE Steam Power; right: from [2], permission by SAGE Publishing)

Fig. 6.23   Left: IP stator vane (shown upside down) and rotor blade with integrated cover band and tangential slot mounting as in Fig. 6.22; right: rotor blades with integrated cover band and axial slot mounting on the wheel of an impulse type first stage (*Courtesy* GE Steam Power)

very small. Thus, clearance loss does not contribute much to the total loss. Rotor hubs of full-speed machines (3000 rpm in a 50 Hz system) are made by welding forged parts, as for HP and IP parts. Casings are made from welded steel plates.

Blades are mounted with tangential slots in the front stages and with axial slots in discs in the rear stages, curved for the last stage and straight for those before the last one. The longest last-stage LP blades in 3000 rpm machines are nowadays (2021) about 1.45 m long. But such blades are made from titanium-aluminium alloy, which is lighter than a steel alloy. With a hub-to-tip ratio of 0.40, the tip diameter is 4.85 m and the tip speed about 750 m/s.

Fig. 6.24 Rotor of LP part of a large steam turbine (*Courtesy* Siemens Energy)

Double-reheat turbines have a very-high pressure part (VHP), and HP, IP and LP parts, with reheat between the VHP and HP and between the HP and IP. The most advanced supply conditions (2021) are 330 bar, 620 °C, with reheats to 630 °C. The turbine efficiency (shaft output) is somewhat above 54%, while with single reheat it is somewhat below 54%. With the largest turbines, up to 1350 MW, the VHP and HP are separate modules. With the in-line coupling, as in Fig. 6.17, called tandem-compound, the overall length becomes then very large, about 70 m, generator included. It is then more convenient to split into two lines, one with the VHP (single flow) and HP (double flow) and one with 2 IP (double flow) and 3 LP (double flow) modules. This arrangement is called cross-compound. The power split is about 40 and 60%. The first line turns at full speed (3000 rpm in a 50 Hz grid) and is placed close to the hot part of the boiler. The second turns at half-speed (1500 rpm in a 50 Hz grid). The VHP and HP parts of the first line are similar to the HP and IP parts of a single-reheat turbine. The exit sections of the LP parts are then much larger, which allows larger mass flow rates, and thus larger power. The longest last-stage LP blades in 1500 rpm machines are nowadays (2021) about 1.85 m long (chromium-steel). With a hub-to-tip ratio of 0.45, the tip diameter is 6.75 m and the tip speed about 525 m/s.

6.8.2 *Large Steam Turbines for Nuclear Power Stations*

Figure 6.25 shows the rotor of a combined HP-IP part of a large steam turbine of impulse type of the 1000–1750 MW order for use in nuclear power stations. In the example, the HP side of the combined part has 9 stages and the IP part has 4 stages (flows in opposite senses). Depending on the power, the turbine may have two (1000 MW order) or three (1750 MW order) LP-modules. Steam conditions with a nuclear power station are significantly lower than with a coal-fired plant. E.g., the values of Chooz-B (France) are 71 bar, 287 °C at HP entry (saturated steam); 10.1 bar, 268 °C at IP entry (superheated) and 3.3 bar at LP entry. There is moisture separation and reheat between the HP and IP parts.

Due to the rather low supply pressure, the HP-IP module can be made with a single casing. Blades are quite long in the HP-IP part. This comes from the necessity for a large flow rate, because the enthalpy drop of the steam over the entire turbine is rather low. So, large through-flow areas are necessary. Therefore, turbines in nuclear power stations turn at half speed: 1500 rpm in a 50 Hz system. As said in the previous section, the longest last-stage LP blades in 1500 rpm machines are nowadays (2021) about 1.85 m long and the tip diameter is about 6.75 m.

Rotor rows in the HP-IP part have cover bands, composed by blade heads. In the HP-IP part, the degree of reaction is low at the hub, but it increases towards the periphery (see Sect. 6.9.2 for the degree of reaction along the radius).

So, saying that the HP and IP parts are of impulse type means that the degree of reaction is low at the hub, but it is not low on the average radius. In the LP parts, the degree of reaction is also low at the hub, but it is near to 90% at the casing in the last

Fig. 6.25 Rotor of the combined HP-IP part of a large impulse type steam turbine (*Courtesy* GE Steam Power)

stage (see Exercise 6.10.3). The HP-IP module is made with a disc rotor. The blades are mounted with feet in axial slots (visible in Fig. 6.25).

The general appearance of a half-speed LP rotor is similar to a full-speed one, as shown in Fig. 6.24, but the diameters are larger. The longer blades in the last stage (1850 mm instead of 1450 mm) require reinforcing elements, touching each other, called snubbers (see the later Fig. 6.31). These elements increase the stiffness of the rotor (higher eigen frequencies), but cause aerodynamic losses. Half-speed LP rotors are made by discs shrunk on a hub or by welding forged parts. LP-casings are made from welded steel plates. Blades are mounted with axial slots in disc rotors (curved for the last stage).

6.8.3 Mid-Size Steam Turbines

Mid-size turbines with power above about 200 MW are made with separate modules, similar to large turbines, but limited to two modules for the lowest power. The supply conditions are usually lower than with large turbines, even as low as 100 bar and 565 °C. The applications are combined gas and steam turbine power plants and combined heat and power plants. There are reheat and non-reheat types. The efficiency is lower than with large machines, typically around 47.5%. Such turbines are also offered for coal-fired mid-size power plants of the order of 400–800 MW,

Fig. 6.26 Impulse type VHP part of a mid-size steam turbine for an ultra-supercritical cycle in a combined heat and power station (older design: commissioned in 1997); nozzle diaphragm (*Courtesy* GE Steam Power)

but high efficiency then requires highly supercritical supply conditions, e.g. 300 bar and 600 °C. The possible layout is then by one HP (single flow), one IP (single flow) and one LP (double flow) modules, with efficiency 50–51%.

In the past, such machines have even been built with a VHP part and two reheats, similar to very large turbines. Some of these are still in use, but new machines of this type are not offered anymore because of the high supplementary cost compared to the limited efficiency gain. The VHP module of a double-reheat turbine of 400 MW functioning in a combined heat and power plant in Denmark with firing by wood chips and natural gas (Skaerbaek; commissioned in 1997) is shown in Fig. 6.26. Steam conditions are 285 bar and 580/580/580 °C. We take it as an example because the VHP module is of impulse type. The lower half of a nozzle diaphragm is also shown. The VHP section has an inner and an outer casing, both horizontally split with parts joined by flanges and bolts. The choice for impulse stages comes from the moderately large power combined with a high inlet pressure, which causes a rather low volume flow rate at the turbine entry. It is then advantageous to choose a low degree of reaction in the first module, because the optimal speed ratio is then lower. This choice results in a smaller diameter, which allows longer blades (better efficiency) and allows the construction of the casings by flanges and bolts. Obviously, the same argument holds for any mid-size turbine with a high supply pressure. So, the first module of a mid-size turbine is often of impulse type.

6.8.4   Industrial Steam Turbines

Industrial steam turbines are single-casing machines. Applications are mechanical drive of turbo-compressors (natural gas pipelines) and large pumps (feed-water pumps in power plants), small combined steam-and-gas turbine power stations (condenser turbines) and combined heat-and-power generation in industrial processes (condenser turbines and backpressure turbines). The largest machines reach about 200 MW, but the smallest turbines have a power less than 1 MW.

Figure 6.27 is a small industrial turbine of impulse type (remark the nozzle diaphragms). A remarkable feature is that the diameter of the first stage is much larger than that of the second stage. With these feature, one knows that the first stage has partial admission.

With small power and a rather high steam supply pressure, it is normally impossible to design the first stage with full admission. Blade height must be sufficient in order to reach acceptable efficiency of the flow through the rotor and the stator. With small blade height, the friction area is large compared to the through-flow area. Especially the effect of the end boundary layers (on hub and shroud) becomes then important.

The most efficient flow within a turbine is obtained for a large aspect ratio: height-to-chord ratio. Customarily, blade heights are not made less than about 25 mm. A larger blade height is then obtained by partial admission. This requires an impulse design.

Fig. 6.27 Industrial steam turbine of impulse type (*Courtesy* Siemens Energy)

Partial admission, however, impairs stage efficiency. A windage flow originates from the centrifugal force in rotor channels without through-flow. This circulating flow consumes power. Further, the flow is less efficient in flowed rotor channels, as there is a start-up and a run-out phenomenon. The windage flow also interferes with the exit flow, reducing the efficiency of the outlet kinetic energy in the next stage. Therefore, mostly, full admission is aimed at from the second stage on. This normally requires a diameter reduction to make the through-flow area sufficiently small, as in Fig. 6.27. In the second and further stages, the degree of reaction may then be kept low, as in Fig. 6.27, but it may also be around 50%. With all stages of impulse type, the axial force on the rotor is small. An equilibrium piston is then not necessary and a thrust bearing suffices.

With very small machines, full flow may stay impossible after a first impulse stage, even with a strong diameter reduction. Historically, a Curtis stage was then chosen as first stage. Since a Curtis stage produces nearly four times as much work as a Laval stage at the same blade speed, a much higher pressure drop is possible. It becomes then easier to achieve full admission in the second and further stages. Nowadays, this possibility is not used anymore, because of the lower efficiency of a Curtis stage compared to a Laval stage. Normally, it is more efficient to feed a steam turbine at a somewhat lower pressure. This reduces the cycle efficiency, but allows higher turbine efficiency. The optimum of the product must be aimed at.

Many single-casing turbines are contractions of large machines into one casing, in the sense that HP, IP and LP parts can be distinguished with increasing hub diameter of these parts. As an example, Fig. 6.28 shows the rotor of an industrial condenser turbine of reaction type, with steam extractions. Variants exist with backpressure. There is an equilibrium piston at the left-hand side.

Many machines have a similar build as shown in the previous figure, but with a first stage of impulse type, with not necessarily a larger diameter than the subsequent stages. The first stage is also often of impulse type in mid-size turbines. It has a role

Fig. 6.28 Rotor of an industrial steam turbine of reaction type; equilibrium piston on the left-hand side (*Courtesy* MAN Energy Solutions)

in the power control of the turbine and the stage is then usually called the control stage.

Large turbines function with sliding-pressure control which means that the power is varied by changing the pressure of the boiler. There is always a throttle valve at the turbine entry, but this functions only for fine tuning. Power control by varying boiler pressure is slow. Most industrial turbines and some mid-size turbines need fast power variation. This may be done by a throttle valve, but this is energetically inefficient. It is then mostly advantageous to provide the machine with a first impulse stage fed by several supply channels. There are usually four nozzle boxes providing steam to four rotor blade segments that each cover less than a quarter of the circumference. For power control from low to high, valves to the boxes are opened gradually, the one after the other. The power of the first stage increases then mainly by enlarging the arc of admission, which is more efficient than by pure throttling. Downstream of the first stage, there is a mixing zone and the subsequent stages function by throttling.

6.9   Blade Shaping

6.9.1   HP and IP Blades

Figure 6.29 sketches a cylindrical section of a stator vane row in the HP part of a steam turbine with subsonic flow, the pressure distribution on the vanes on the mean radius and the variation of the outflow angle along the height (angle to the tangential direction). The pressure distribution allows deducing that the outflow Mach number is about 0.5. The average outflow angle is 71°. The peaks and valleys in the angle variation follow from the presence of secondary flow.

Secondary flow is defined as the flow caused by interaction between the core flow within the vane row and the boundary layers on the hub and casing end sides (the

Fig. 6.29 Cylindrical
section of a stator cascade;
mid-height pressure
distribution; outflow angle
along the height (*Courtesy*
Siemens Energy)

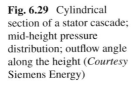

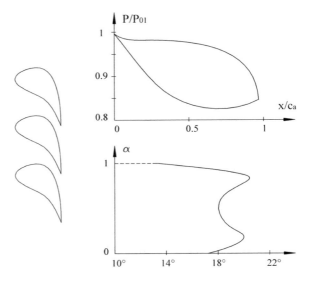

end walls). The core flow is called the primary flow (a precise definition is given in
Chap. 13). Figure 6.30 illustrates the development of the secondary flow, in the same
way as discussed in Chap. 2 (Sect. 2.3.2) for a curved duct. Because the velocity in
the core flow is larger than in the end wall boundary layers, the centrifugal force by
streamline curvature is stronger in the core flow for equal curvature. The difference
causes a double vortex flow, where the parts together are denominated the *passage
vortex*.

In the end wall boundary layers, the passage vortex pushes the through-flow more
to the tangential direction, which increases the flow turning. Conversely, it pushes

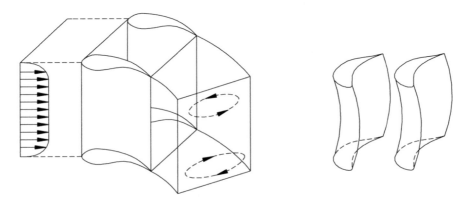

Fig. 6.30 Passage vortex generation; bowing of vanes (or blades) reducing profile loss in the end
wall zones and intensity of the passage vortex

the through-flow more to the axial direction just above the end wall boundary layers, which decreases the flow turning.

The corresponding variation of the outflow angle is visible in Fig. 6.29 (right, bottom). By the variation in flow curvature, there is approximately constant pressure in the direction perpendicular to an end wall, which is necessary for force equilibrium in this direction.

The interaction between the primary flow and the end-wall boundary layers generates additional vortex structures, not represented in the figure. These will be discussed in Chap. 13. The whole of these vortices is denominated *secondary flow* and the passage vortex is its main component. The secondary flow causes non-homogeneity of the outflow. The flow homogenises downstream of the vane row with a corresponding mixing loss. With low-height vanes, as may occur in the HP part of a steam turbine, losses in the boundary layers on the end walls and losses due to the secondary flow cause a serious loss increase for the vane row.

The losses due to the end-wall boundary layers are reduced by bending the vanes, as represented in Fig. 6.30. This bending is called *bowing*. In the end zones, the suction side is turned to the centre of the vane passage. The consequence is that a streamline in an end wall boundary layer, near the suction side of the vane, is pushed towards the centre of the passage at the vane row entrance. In principle, there is an inverse effect at the pressure side, but the displacement possibility of a streamline is limited there due to the acute angle of the vane. One expects that the streamline near the suction side returns towards the end wall at the vane row exit. In reality this does not fully happen because the passage vortex is formed in approaching the exit plane of the vane row. The passage vortex hinders this return (see the rotation sense in Fig. 6.30).

A first net effect of bowing is an increase of the acceleration at the suction side of the vane somewhat more to the centre of the blade passage and a reduction of it in the end zone. The consequence is decrease of the profile lift (tangential force), and so the primary loss (the loss due to the boundary layers on the vane), in the end wall zone. This is beneficial, as the profile loss in the end wall zone is high due to flow perturbation by the end wall boundary layer.

The profile lift increases somewhat away from the end wall, but the associated increase of the primary loss can be minimised by adapting the profile shape to the increased lift.

A second net effect of bowing is that some mass flow is displaced from the end wall boundary layer towards the centre of the vane channel. By this shift, the intensity of the passage vortex diminishes and it spreads over a larger spanwise distance. So, the non-homogeneity of the outflow is lower and mixing loss downstream of the vane row decreases.

From these effects one understands that bowing may be beneficial, but it should not be too strong, because the flowed surface (mostly called the wetted area) increases and by the spanwise lift variation vortices are formed in the wake of a vane (trailing vortices). It means that optimisation of bowing is quite delicate. Also, bowing on a vane row changes the inflow of the downstream blade row.

Concave bending at the suction side is advantageous with rotor blades as well, for the same reasons as for stator vanes. So, the optimisation of a complete stage or even a number of stages has to be considered.

6.9.2 LP Blades

The blades are long in the LP part of a large steam turbine. Secondary losses therefore do not play a crucial role. The main problem with LP blades is the large variation of the degree of reaction along the blade span. The flow is very tangential at the stator exit. This induces a strong centrifugal force in the space between stator and rotor, causing an important radial pressure gradient with high pressure at the casing and low pressure at the hub. The flow is almost axial at the stator entry and the rotor exit and there is no radial pressure gradient. The pressure gradient in between the stator and the rotor thus imposes a high degree of reaction at the casing and a low one at the hub.

The enthalpy drop over the stage should approximately be the same at all radii and thus the speed ratio increases from the hub to the casing. The consequence is a natural correspondence between the variations along the radius of the speed ratio and the degree of reaction, in the sense that the speed ratio does not differ much from the optimal value appropriate for the degree of reaction, as shown in Fig. 6.16. So, Fig. 6.16 represents the variation tendency of the velocity triangles from hub to casing, where typically $R_s \approx 0.15$ at the hub, and the degree of reaction increases with the radius. The longest blades reach nowadays (2021) $R_s \approx 0.85$ at the casing.

The triangles in Fig. 6.16 are drawn with constant axial velocity for a specified degree of reaction and this axial velocity decreases with increasing degree of reaction. In a real LP-stage, the axial velocity is nearly constant along the radius at entry and exit of the stage, with nearly the same value. But, at stator exit and rotor entry, the axial velocity decreases from hub to casing due to the centrifugal force associated to the tangential velocity component (for realistic triangles, see Fig. 6.36 in the exercise section).

Nowadays (2021), the maximum blade length is about 1.40 m with a 2.00 m hub diameter (turbines at 3000 rpm: coal-fired) and 1.80 m with a 3.00 m hub diameter (turbines at 1500 rpm: nuclear). The most extreme hub-to-tip radius ratio is thus about 0.40, as in Fig. 6.36.

Figure 6.16 (also Fig. 6.36) demonstrates that the rotor blade profile strongly changes from hub to tip, with a large turning at the root and a very small one at the tip. Figure 6.31 sketches the variation of the flow angles at stator exit and rotor entry and the Mach number at rotor entry for the final stage of a large steam turbine with straight stacking of vane and blade sections and rotor hub-to-tip radius ratio 0.45 (data provided by GE Steam Power). A rotor blade is also shown in Fig. 6.31.

The problem with very long blades is that the Mach number at rotor entry is large at the hub and at the casing, and may approach unity. This generates shock waves (see Fluid Mechanics), with associated shock loss. Therefore, a final stage in an LP part

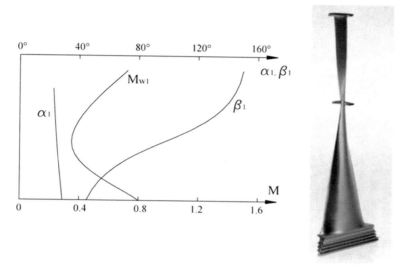

Fig. 6.31 Variation of flow angles and Mach number along the radius (vertical direction) for the final stage of an LP part of a large steam turbine with straight stacking of vane and blade sections (angles to tangential direction); right: rotor blade with snubber reinforcements (*Courtesy* GE Steam Power)

is adapted in order to minimise shock loss. The other stages in the LP part and the stages in the HP and IP parts also show variation of the degree of reaction along the height, with as consequence twisting of rotor blades (see Fig. 6.23). This variation does not cause high Mach number problems at the hub and the casing, however. The problem only arises in the final stage of an LP module.

Three adaptations reduce the rotor entry Mach number at the hub and the casing. The traditional one is setting the stator outflow more axial at the hub (opening the vanes) and setting it more tangential at the casing (closing the vanes). The changed velocity triangles are drawn with dashed lines in Fig. 6.32 (left). The Mach number decrease of the relative flow at rotor entry is obvious. The feasibility of this adaptation presumes that the pressure distribution adjusts to the altered velocity distribution. This actually happens, but we do not prove this here (the study requires an analysis of radial equilibrium, which is discussed in Chap. 13).

The vane angle adaptation decreases the variation of the degree of reaction from hub to casing. The flow rate increases near the hub and decreases near the casing. The stator outflow angle has a geometric limit, around 75° to the meridional plane. Because the limit is critical at the casing, the adaptation causes a smaller stator angle at the hub than practically attainable and thus smaller stage work than physically possible. Normally, this is no drawback, because the stator angle is typically not set to its maximum geometric value in order to allow a larger flow rate (see Fig. 6.31).

A change in the variation of the degree of reaction may also be attained by exerting a radial force against the centrifugal force. Figure 6.32 (top, right) demonstrates that

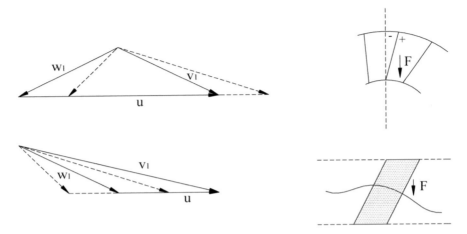

Fig. 6.32 Mach number reduction at hub and casing at rotor entry: (1) opening of the nozzles at the hub (bottom left) and closing them at the casing (top left); (2) lean within the orthogonal plane, generating pressure force towards the hub; (3) sweep within the meridional plane, generating pressure force towards the hub

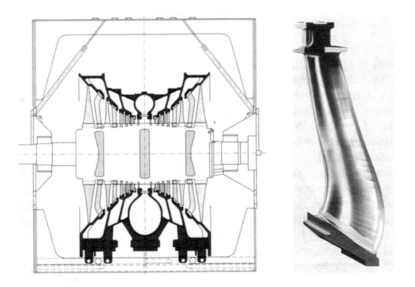

Fig. 6.33 3D-shape of the stator vane of the final stage in the LP part of a large steam turbine (*Courtesy* GE Steam Power)

inclining the stator vanes in the orthogonal plane, called *lean*, generates a radial force, directed inwards, due to the pressure difference between the pressure and suction sides of the vanes. This can also be achieved by inclining in the meridional plane, called *sweep* (Fig. 6.32, bottom, right). Sections more towards the casing get the expansion later, creating a higher pressure level on the same axial position. Combination of lean and sweep results in a rather complex 3D-shape of the stator vane, as shown in Fig. 6.33.

The adaptations discussed are applied to all stages, but not leading to such extreme shapes as in the last stage of an LP-part. With short blades, the main motivation is lowering the degree of reaction at the casing, which reduces leakage losses.

Thanks to aerodynamic optimisation, the efficiency in the LP part of a turbine yields about 92% for dry flow. In reality, condensation reduces the efficiency to about 90%. The efficiency of an entire large steam turbine comes to about 92.5% (HP 94%, IP 96%, LP 90%).

6.10 Exercises

A basic Mollier-diagram for water-steam can be found on: http://www.engineeringt oolbox.com/mollier-diagram-water-d_308.html. Steam properties may be calculated with: https://beta.spiraxsarco.com/Resources-and-design-tools.

6.10.1 Compare the energy loss coefficient (6.11) and the pressure loss coefficient (6.12) applied to the expansion of air ($\gamma = 1.4$) in a nozzle with infinitesimal efficiency 0.9 for varying exit Mach number $M_1 = 0.25, 0.50, 1$ and 2. Use the formulae of Chap. 4 and ignore the kinetic energy at the entry to the nozzle. Observe that the energy loss coefficient decreases weakly and the pressure loss coefficient increases quite strongly with increasing Mach number. Derive that the energy loss coefficient is linked to the isentropic efficiency by $\eta_s = 1/(1 + \xi)$, which explains the weak dependence.

The Mach number effect on the pressure loss coefficient may be neutralised by using the logarithm of pressure instead of pressure. Such a modified coefficient is then linked to the polytropic efficiency by $\eta_\infty = 1/(1 + \omega_{log})$. Observe that the isentropic efficiency and the polytropic efficiency can be determined by measurement of the total pressure at the entry to the nozzle and the total pressure and static pressure at the exit from the nozzle. Temperature measurement is not necessary.

6.10.2 Figure 6.25 shows the HP-IP rotor of the Arabelle impulse type steam turbine (GE Steam Power). The machine is built with two LP parts for electric power 900–1400 MW and 3 LP parts for 1300–1800 MW. The machines are not strictly unique and are adapted somewhat to the application. We consider here Flamanville 3 (France) with 1750 MW electric power. The steam conditions are: supply at 75 bar, 290 °C (saturated) with expansion to 11 bar in the HP part; moisture separation and reheat on 11 bar to 275 °C (superheated); condenser pressure is 46 mbar. The mass flow at turbine entry is 2500 kg/s (the flow rate is somewhat lower in the IP and LP parts) and the rotational speed is 1500 rpm. The tip diameter and hub diameter of the last stage are 6.00 and 2.50 m (blade length is 1750 mm).

The hub diameter of the first stage in the HP part is 1.75 m (the hub diameter varies slightly in the HP-IP part). The corresponding blade speeds are 471.24, 196.35 and 137.45 m/s. The blade speed in the HP part is very low, causing a low stage work. The reason is that the shafts of the HP-IP part and the 3 LP parts are coupled in-line to drive a single generator (tandem-compound machine). A more comfortable blade speed at the hub of the HP-IP part would be the double speed, being about 275 m/s, obtainable by splitting the turbine in an HP-IP part turning at 3000 rpm and LP-parts turning at 1500 rpm (cross-compound machine).

Determine the number of stages in the HP part, assuming that the degree of reaction is 10% at the hub and that the hub diameter is 1.75 m. Design the hub section of the first stage of the HP part with zero degree of reaction (Laval stage). The real machine has a small positive degree of reaction at the hub, but we take zero degree of reaction for the ease of computation. Take $\alpha_1 = 75°$.

A: $(\Delta h_s)_{stage} \approx 38$ kJ/kg, $(\Delta h_s)_{HP-part} \approx 340$ kJ/kg: 9 stages (see Fig. 6.25). The outlet kinetic energy of the stage is useful. So, calculation of the velocity triangles requires iteration. Assuming a repeating stage and symmetric rotor blades, the velocity triangles result in $\beta_1 = -\beta_2 = 60.83°$. For determination of the blade shape, according to Fig. 6.7, we take $\beta_1 = -\beta_2 = 60°$. We denote the inner and outer radii of the blade channel by r_1 and r_2 and the channel width by $b = r_2 - r_1$. From the mass flow rate follows that the height of the stator vanes at exit = height of the rotor blades at entry ≈ 183 mm. We choose the axial chord of the rotor blades equal to 100 mm (we may scale afterwards) and the minimum blade thickness to 2 mm.

The Zweifel tangential force coefficient set to unity is

$$C_{Fu} = \frac{2s\ w_{2a}\Delta w_u}{c_a w_2^2} = 1.$$

Ignoring losses: $\Delta w_u = -2w_{2u} = 2w_2|\sin\beta_2|$ (in reality somewhat larger) and ignoring blade thickness: $s = b/\cos\beta_2$ (in reality somewhat larger) and $c_a = 2r_2|\sin\beta_2|$. Inserting these expressions, the Zweifel coefficient becomes:

$$C_{Fu} = \frac{2s\ w_2\cos\beta_2 2w_2|\sin\beta_2|}{c_a w_2^2} = 4\frac{b\cos\beta_2|\sin\beta_2|}{\cos\beta_2 2r_2|\sin\beta_2|} = 2\frac{b}{r_2}.$$

Another way for deriving the maximum acceptable width of the rotor blade channel is by a local diffusion factor applied to the decelerating part of the suction side boundary layer in Fig. 6.8 (part DF):

$$D_{loc} = \frac{w_{max} - w_2}{w_{max}} < 0.5.$$

w_{max} is the maximum velocity at the suction side, being $r_2 w_2/r_1$ according to the free-vortex flow. The expression of the local diffusion factor becomes:

$$D_{loc} = \frac{r_2/r_1 - 1}{r_2/r_1} = \frac{r_2 - r_1}{r_2} = \frac{b}{r_2}.$$

Both $C_{Fu} = 1$ and $D_{loc} = 0.5$ result in $b/r_2 < 0.5$. Because the chord is reduced by rounding the leading edge, we take $b/r_2 = 0.4$.

The channel width becomes $b = 23.09$ mm, rounded to 23 mm. Taking the blade thickness at the trailing edge 2 mm, leads to a pitch of 50 mm. The number of blades becomes 110.

Figure 6.34 shows the rotor blade geometry. The leading edge is rounded because the inlet Mach number is well below unity. A straight part is added to the blades at the trailing edge. The axial chord is 90 mm (it may later be scaled, if necessary).

We begin dimensioning the nozzle vanes with an outlet width of 23 mm (= width of the rotor blade passages). With thickness 2 mm, the pitch is then 96.6 mm and the corresponding number of vanes is 57. The number of rotor blades 110 is too close to $2 \times 57 = 114$. So, we reduce the number of vanes to 48, for instance. The pitch is then 115.5 mm. The outlet width becomes 28 mm. With a Zweifel coefficient of unity, the corresponding axial chord is 58 mm. We take this value as a basic value (we may scale afterwards).

Figure 6.35 shows the construction of a preliminary nozzle vane shape. The blade thickness in the trailing edge zone is set to 2 mm. Point E is chosen on the pressure side. The camber line EF is a curve with angle of the tangent linearly varying as a function of axial distance (x) from 0° to 75°. This is a parabola with equation $y/x_E = (\tan \alpha_1/2)(x/x_E)^2$. A thickness distribution and a leading edge zone are added. The locations of points C and E are varied in order to obtain a smooth profile. The resulting axial chord is 70 mm, which brings the Zweifel coefficient to 0.825. This is an acceptable value.

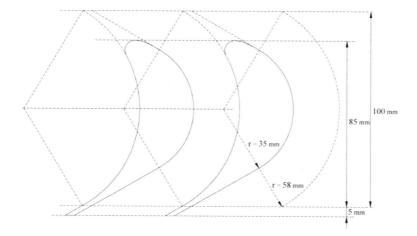

Fig. 6.34 Rotor blade profile; initial shape

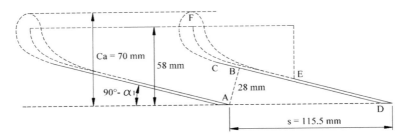

Fig. 6.35 Stator vane profile; subcritical flow; initial shape

The geometries of Figs. 6.34 and 6.35 may serve as initial geometries in a CFD package for optimisation of the vane and blade shapes. Comparison with the shape shown in Fig. 6.29 reveals that the geometrically determined shape of Fig. 6.35 is rather far away from an optimised shape. But the optimum shape strongly depends on exit angle and outlet Mach number.

6.10.3 Design the last stage of the LP part of the Arabelle turbine described in the previous exercise.

(a)   Determine the number of stages necessary in the expansion from 11 bar, 275 °C to 46 mbar. Take as hub diameter of the first IP stage 2.50 m (Fig. 6.25) and assume a degree of reaction of 10%. Consider also 10% degree of reaction for the hub part of the last stage in the LP part and take again as hub diameter 2.50 m. Take into account that in the real machine, the hub diameter increases somewhat in the successive stages of the IP part and then decreases in the successive stages of the LP part.

   A: The isentropic enthalpy drop of the IP part + LP part is 875 kJ/kg. For $R = 0.1$, the isentropic enthalpy drop in the first and last stages is about 77 kJ/kg. With this value, the necessary number of stages is 11.4. The number of stages may certainly be reduced to 10, taking into account the larger diameter in the middle stages. In reality, there are 9 stages (4 + 5). These look thus somewhat more loaded than optimal.

(b)   Determine the variation of the velocity as a function of the radius in the space between the stator and the rotor of the last stage of the LP part. Consider constant total enthalpy along the radius downstream (and upstream) of the stator vanes. Take as stator outflow angle 70°, constant along the radius (this is lower than the geometric maximum of about 75° in order to enlarge the mass flow rate). Estimate the stator loss coefficient with the formula of Soderberg, ignoring secondary loss. Take as simplification that streamsurfaces are cylindrical (radial velocity component equal to zero), such that there is equilibrium between the radial pressure gradient and the centrifugal force (called simple radial equilibrium). Derive that the tangential velocity varies according to $v_{1u} \sim r^{-a}$, with $a = (\phi_s \sin \alpha_1)^2$, where ϕ_s is the velocity coefficient of the nozzle vanes. Neglect the difference between infinitesimal and isentropic efficiency such that $\eta_\infty = \phi_s^2$.

A: Cylindrical streamsurfaces: force equilibrium:

$$\frac{1}{\rho}\frac{dp}{dr} = \frac{v_{1u}^2}{r}.$$

Constant enthalpy downstream of nozzle vanes ($v_{1r} = 0$):

$$h_0 = h_1 + \frac{1}{2}v_{1u}^2 + \frac{1}{2}v_{1a}^2.$$

Constant vane angle:

$$\tan \alpha_1 = v_{1u}/v_{1a}.$$

Combination of the above, with the derivative of the enthalpy equation in the radial direction:

$$\frac{dh_0}{dr} = 0 = \frac{dh_1}{dr} + \left(1 + \frac{1}{\tan^2 \alpha_1}\right)v_{1u}\frac{dv_{1u}}{dr}$$

Loss in the stator according to Soderberg:

$$\xi = 0.025\left[1 + \left(\frac{\delta^\circ}{90}\right)^2\right], \text{ with } \delta^\circ = \alpha_1 = 70^\circ, \ \phi_s = 1/\sqrt{1+\xi} = 0.9805.$$

Infinitesimal efficiency of the nozzle vanes:

$$\eta_\infty = (-dh)/\left(-\frac{1}{\rho}dp\right) \text{ or } \frac{dh}{dr} = \eta_\infty\frac{1}{\rho}dp.$$

Downstream of the nozzle vanes:

$$\frac{dh_1}{dr} = \eta_\infty\frac{1}{\rho}\frac{dp_1}{dr} = \eta_\infty\frac{v_{1u}^2}{r}$$

Combination with the enthalpy equation:

$$0 = \eta_\infty\frac{v_{1u}^2}{r} + \frac{1}{\sin^2 \alpha_1}v_{1u}\frac{dv_{1u}}{dr} \text{ or } \eta_\infty \sin^2 \alpha_1\frac{v_{1u}}{r} + \frac{dv_{1u}}{dr} = 0.$$

This equation is satisfied for $v_{1u} \sim r^{-a}$, with $a = \eta_\infty \sin^2 \alpha_1$. We take $\eta_\infty = \phi_s^2$: $a = 0.8490$.

(c) For equal work on all radii and axial outflow of the stage, the tangential component of the stator exit velocity should vary according to $v_{1u} \sim r^{-1}$. Thus with axial outflow at the hub and $v_{1u} \sim r^{-a}$, v_{1u} at the casing is larger than according to $v_{1u} \sim r^{-1}$ and v_{2u} is positive. In order to reach axial outflow in the mean sense, with the objective of minimising the tangential kinetic energy in the outflow, v_{2u} has to be set to a negative value at the hub. This can be obtained by enlarging somewhat the work coefficient. For $R = 0.10$, the speed ratio for axial outflow is:

$$\lambda = \frac{u}{v_s} = \frac{0.48}{\sqrt{1-R}} \approx 0.50, \text{ thus } \psi_s = \frac{\Delta h_s}{u^2} = \frac{1}{2\lambda^2} \approx 2.$$

Choose for $R_s = 0.10$: $\psi_s = 2.20$. This lowers somewhat the efficiency at the hub.

(d)   Determine the velocity triangles at the hub for $R_s = 0.10$ and $\psi_s = 2.20$. Take constant axial velocity at the hub ($v_{1a} = v_{2a}$). Consider a repeating stage. The isentropic degree of reaction is then $R_s = (h_{1s} - h_{2s})/(h_0 - h_{2s})$. With $u = 196.35$ m/s and $\psi_s = 2.20$: $\Delta h_s = 84.82$ kJ/kg. We round to 85 kJ/kg and distribute as $\Delta h_{ss} = h_0 - h_{1s} = 76.5$ kJ/kg and $\Delta h_{sr} = h_{1s} - h_{2s} = 8.5$ kJ/kg. Calculate the last stage without consideration of the recovery of the outlet kinetic energy in the diffuser downstream of the last stage. This means that it is assumed that the magnitude of the axial velocity downstream of the stage has been chosen already. In reality, the optimisation of the outflow velocity determines the value of the nozzle angle. So, it is assumed here that $\alpha_1 = 70°$ is the result.

A: The calculations require iteration. $\beta_1 = 53.2°$, $\beta_2 = -57.4°$; $R = 0.081$; $\Delta W = u(v_{1u} - v_{2u}) = 79.45$ kJ/kg; $\eta_{tt} = 0.935$.

(e)   Determine the velocity triangles as a function of the radius, taking into account: $v_{1u} \sim r^{-a}$, $v_{1a} = v_{1u}/\tan \alpha_1$, $v_{1s} = v_1/\phi_s$, $\Delta h_{ss} = \frac{1}{2}v_{1s}^2$, $\Delta h_s = 85$ kJ/kg and $\Delta h_{sr} = \Delta h_s - \Delta h_{ss}$. Take as inflow velocity of the stator the value of v_{2a} at the hub. Calculate loss coefficients with the formula of Soderberg, ignoring secondary loss. Take as approximation that v_{2a} is constant along the radius. With this last assumption, a mass flow balance is not necessary. This can be used afterwards to determine the height of the streamtubes at the entry and exit of the rotor. Calculate for the values of radius: $r = 1.25, 1.60, 2.15, 3.00$ m.

A: the results are shown in Table 6.4 and the velocity triangles in Fig. 6.36.

6.10.4 Design a small industrial steam turbine with 3000 rpm rotational speed. Steam at turbine inlet is at 100 bar and 500 °C. Take as flow rate $\dot{m} = 100$ kg/s

Table 6.4 Blade of the last stage of the LP part

R	u	v_{1u}	v_{1a}	w_{1u}	w_{2u}	v_{2u}	η_{tt}	R	ψ
m	m/s	m/s	m/s	m/s	m/s	m/s	–	–	–
1.25	196.4	382.9	139.4	186.5	−217.5	−21.1	0.933	0.08	2.06
1.60	251.3	310.5	113.0	59.2	−261.4	−10.0	0.948	0.44	1.28
2.15	337.7	241.6	87.9	−96.1	−338.7	−0.9	0.964	0.72	0.72
3.00	471.2	182.1	66.3	−289.2	−461.7	9.6	0.956	0.89	0.37

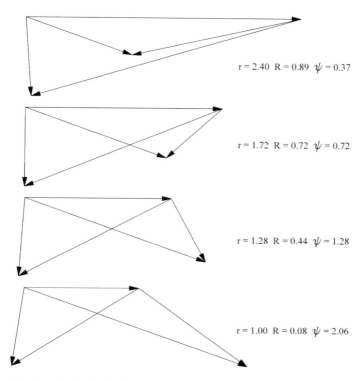

r = 2.40 R = 0.89 ψ = 0.37

r = 1.72 R = 0.72 ψ = 0.72

r = 1.28 R = 0.44 ψ = 1.28

r = 1.00 R = 0.08 ψ = 2.06

Fig. 6.36 Velocity triangles for the last stage of the LP part

(calculations can easily be adapted for another flow rate). The stator outflow angle is $\alpha_1 = 75°$. Design the turbine such that the expansion is realised until 25 bar backpressure. Achieve this in 3 different ways: (1) by Curtis stages; (2) by Laval stages (Rateau turbine); (3) as a reaction turbine with 50% isentropic degree of reaction. Make the calculations for axial inflow and outflow of the stage. Take into account that the blade height should be minimally about 25 mm for ideal functioning. Take also into account that, at 500 °C, the blade speed must not exceed about 200 m/s and that it may be larger at lower temperature.

(a) Determine the main features and dimensions: the number of stages required; the average diameter (possibly different per stage); the blade height (possibly different per stage); the velocity triangles of the first stage; the h–s diagram of the first stage with an indication of losses; the internal power of the first stage.

   A: Curtis: 1 stage, $h = 25$ mm, $\theta \sim 2 \times 90°$; Laval: 4 stages, $h_1 = 25$ mm, $\theta_1 \sim 2 \times 80°$, $h_4 = 25$ mm, $\theta_4 = 360°$; 50% reaction: 7 stages, $h_1 \sim 15$ mm (ideal minimum height cannot be obtained).

(b) Design the nozzles and the rotor blades. Take the Curtis stage as an example. Represent the superheated steam as an ideal gas and determine the heat capacity

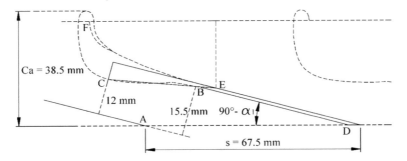

Fig. 6.37 Stator vane profile; supercritical flow; initial shape

ratio by fitting the isentropic density-pressure relation (**A:** $\gamma = 1.28$). Take the throat width of the nozzles approximately equal to half the nozzle height.

A: Curtis: $Z_s = 32$, $b_{throat} = 12$ mm, $b_1 = 15.5$ mm. $Z_r = 150$, $c_a \approx 35$ mm, $b_1 = 7.45$ mm. The rotor blades may be drawn as in Fig. 6.34, except that the leading edge should stay sharp as the inlet Mach number is above unity. The stator vanes may be drawn as in Fig. 6.35, but the nozzle channels are converging–diverging here: see Fig. 6.37. The positions of the points B, C and E should be chosen such that the resulting profile is smooth.

References

1. Dixon SL, Hall CA (2014) Fluid mechanics and thermodynamics of turbomachinery, 7th ed. Elsevier, ISBN 978-0-12-415954-9
2. Havakechian S, Greim R (1999) Aerodynamic design of 50 percent reaction steam turbines. J Mech Eng Sci IMechE 213:1–25
3. Korpela SA (2011) Principles of turbomachinery. Wiley, ISBN 978-0-470-53672-8
4. Lewis RI (1996) Turbomachinery performance analysis. Wiley, ISBN 0-470-23596-9

Chapter 7
Dynamic Similitude

Abstract Similitude means that two machines may be similar in the sense that there exist proportionality factors for geometry, velocities and forces. The performance characteristics of one of the machines can then be derived from the known characteristics of the other. The concept forms the basis of the initial design of a turbomachine, which is the determination of main parameters as the rotational speed, the size and some geometric ratios. In this chapter, we discuss the theory of dynamic similitude and some applications, including the basic design of a radial fan.

7.1 Principles of Dynamic Similitude

7.1.1 Definition of Dynamic Similitude

Two flows are dynamically similar when they are geometrically similar, when velocities at corresponding points have the same direction and have a constant ratio and when forces at corresponding points have the same direction and have a constant ratio. In other words, flows are dynamically similar when similitude is as well geometric, kinematic and dynamic.

With the simplest flows, i.e. steady flows of constant-density fluids in stationary geometries, dynamic similitude implies kinematic similitude, because the velocity field then unambiguously results from the integration of the force field. Further, kinematic similitude implies geometric similitude, because streamlines then unambiguously result from the integration of the velocity field. Therefore, the term dynamic similitude is used as a common denomination of the three elements within similitude. Henceforth we simply use the term similitude.

The notion of dynamic similitude is limited to the flow itself. Besides fluid transport, other transport phenomena may occur within a flow. Heat transport is an example. The notion of similitude may be extended to these phenomena. Additional transport phenomena are not directly relevant for the fundamental study of turbomachines. For turbomachines, rotor work is an additional phenomenon compared to simple flows. In the present chapter, the similitude study is limited to constant-density fluids. The extension to compressible fluids is made in Chap. 15.

© The Author(s), under exclusive license to Springer Nature Switzerland AG 2022   257
E. Dick, *Fundamentals of Turbomachines*, Fluid Mechanics and Its Applications 130,
https://doi.org/10.1007/978-3-030-93578-8_7

7.1.2 Dimensionless Parameter Groups

Corresponding lengths have a constant ratio within similar flows. Corresponding velocities have a constant ratio as well. As the dimension of velocity is length divided by time, the combination of velocity and length implies that corresponding time intervals have a constant ratio. Analogously, from the constant ratio of forces and from the dimension of force being mass times length divided by the square of time ($N = kg\ m/s^2$), it follows that similar masses have a constant ratio.

The fundamental quantities, i.e. length, mass and time, by which all mechanical quantities may be expressed, thus show a constant ratio at corresponding points within similar flows. The consequence is that a *dimensionless group of parameters* formed by combining quantities describing the flow must have the same value in corresponding points. For example, pressure (N/m^2) has the same dimension as the product of a density (kg/m^3) and the square of a velocity (m^2/s^2). A dimensionless pressure coefficient (7.1) thus has the same value in corresponding points in similar flows, with p_r being a reference value for pressure, e.g. inlet pressure:

$$C_p = (p - p_r)/(\rho v^2). \tag{7.1}$$

Example (7.1) demonstrates that dimensionless groups of parameters necessarily are formulated as

$$\pi = A^a B^b C^c \ldots, \tag{7.2}$$

with A, B, C ... being the parameters and a, b, c, ...exponents. The exponents are integers in the example, but this is not absolutely necessary, as broken powers of dimensionless parameter groups are dimensionless as well. The term π-group is often used for a dimensionless group.

7.1.3 Similitude Conditions

We take flow of a constant-density fluid within a stationary channel as an example to derive how similitude conditions may be identified. The quantities describing the flow are geometry, velocities and forces. We consider a second flow being similar to the first one, as sketched in Fig. 7.1.

Ratios of corresponding lengths have a constant value, and this is also the case for mass ratios and time ratios. By choosing a length unit, a mass unit and a time unit that are characteristic for the flow (intrinsic units) and by expressing the equations describing the flow with this unit system, the resulting number equations must thus be identical for both flows. This allows identification of similitude conditions.

As unit of length e_ℓ we take e.g. the inlet width L. As velocity unit we take e.g. the inflow velocity $e_v = V$. From a consistent system of units follows a time unit:

Fig. 7.1 Similar flows and intrinsic units

$$e_v = \frac{e_\ell}{e_t} \rightarrow e_t = \frac{L}{V}.$$

The mass unit may be determined by choosing the fluid density as the density unit: $e_\rho = \rho$. As $e_\rho = e_m/e_\ell^3$, a mass unit follows with a consistent unit system. The flow equations are

$$\nabla \cdot \vec{v} = 0, \quad \frac{\partial \vec{v}}{\partial t} + \vec{v} \cdot \nabla \vec{v} + \frac{1}{\rho} \nabla p = -g\, \vec{1}_z + \nu \nabla^2 \vec{v}.$$

The pressure unit is chosen, because of consistency, as $e_p = e_\rho e_v^2 = \rho\, e_v^2$. The dimensionless equations follow from

$$\vec{v} = [\vec{v}]\, e_v, \quad t = [t]\, e_t, \quad \nabla = [\nabla]\, \frac{1}{e_\ell},$$

with quantities between brackets being dimensionless. The momentum equation becomes:

$$\left[\frac{\partial \vec{v}}{\partial t}\right] \frac{e_v}{e_t} + [\vec{v} \cdot \nabla \vec{v}]\, \frac{e_v^2}{e_\ell} + \left[\frac{1}{\rho}\nabla p\right] \frac{e_p}{\rho\, e_\ell} = -g\, \vec{1}_z + \nu\, [\nabla^2 \vec{v}]\, \frac{e_v}{e_\ell^2}.$$

The dimensional factors on the left-hand side are all $e_v^2/e_\ell = V^2/L$. By multiplication with L/V^2 we obtain a dimensionless equation with right-hand term:

$$-\frac{g\, L}{V^2}\, \vec{1}_z + \frac{\nu}{V\, L}\, [\nabla^2 \vec{v}].$$

Two dimensionless groups are formed, that must have the same value for both flows considered. We note that the continuity equation does not cause similitude conditions because of its homogeneity. Similitude conditions are expressed by

$$\mathrm{Re} = \frac{V\, L}{\nu} = \text{constant (Reynolds number)}, \quad Fr = \frac{V}{\sqrt{g\, L}} = \text{constant (Froude number)}.$$

The similitude conditions may be interpreted as conditions that forces of different origins must have a fixed ratio.

We distinguish (acceleration is interpreted as inertia force):

$$\underbrace{\frac{\partial \vec{v}}{\partial t} + \vec{v}.\nabla \vec{v}}_{inertia\,force} + \underbrace{\frac{1}{\rho}\nabla p}_{pressure\,force} \quad = \quad \underbrace{-g\,\vec{1}_z}_{gravity\,force} + \underbrace{\nu\,\nabla^2\,\vec{v}}_{viscous\,force} \; .$$

So: $Re =$ inertia force/viscosity force, $Fr^2 =$ inertia force/gravity force.

If the pressure unit had not been chosen a priori as ρV^2 because of the consistency of the unit system, a third similitude condition would be

$$Eu = e_p/\rho V^2 \; \text{(Euler number)} = \text{pressure force/inertia force}.$$

For instance, we could have chosen the pressure difference over the channel as pressure unit. This would constitute a principal error. This means that the Euler number is not a similitude condition. But it is a dimensionless group. Consequently, for all flows that are similar (=family of similar flows), there must be an unambiguous relation of the form.

$$Eu = \frac{\Delta p}{\rho V^2} = f(\text{Re}, Fr).$$

All other possible dimensionless groups are also functions of Re and Fr.

The above example implies that similitude conditions result from the requirement that *forces with different origins must have a fixed ratio*. In the same way, *velocities with different origins must have a fixed ratio*. With turbomachines, there are always two velocities with different origins: the through-flow velocity and the blade speed. In a general flow problem, there are dynamic and kinematic similitude conditions.

7.1.4 Purpose of Similitude Analysis

Similitude analysis demonstrates that a dependent parameter cannot depend individually on the independent parameters of a problem. The relation must necessarily hold between a *dimensionless group containing the dependent parameter and dimensionless groups of independent parameters*. Similitude analysis thus reduces the number of degrees of freedom of the relations.

The result demonstrates that a physical relation cannot depend on the unit system chosen to measure the variables. Therefore, it is advantageous to determine the dimensionless quantities that group the independent parameters when performing a flow analysis, either experimentally or numerically. Formulating a problem as in Fig. 7.1 in a dimensionless form generates the solution for an arbitrary size (geometric), an arbitrary inflow velocity (kinematic) and an arbitrary density (dynamic) in one effort. So, ∞^3 problems are analysed in one effort.

7.1.5 Dimensional Analysis

The above reasoning demonstrates that considering similar flow problems means exactly the same as the dimensionless formulation of a flow problem. Dimensional analysis is herewith a tool to determine the independent dimensionless groups without much prior physical knowledge. The only prerequisite is the ability to list the independent parameters. In the above example these are:

fluid: ρ, μ (or $\nu = \mu/\rho$)
geometry: L
kinematic: V
dynamic: g

Apart from that, the intervening fundamental dimensions must be listed. These are found in the list of the international system of units, being length, mass and time for the flow problem considered. It is obvious now that 5 independent parameters with 3 fundamental dimensions only allow the construction of 2 independent dimensionless groups. A dimensionless group has the form $\Pi = \rho^a \nu^b L^c V^d g^e$. The dimension of the group is verified with Table 7.1.

Being dimensionless requires compliance with the following conditions:

$$-3a + 2b + c + d + e = 0$$
$$a = 0$$
$$-b - d - 2e = 0$$

The system of three homogeneous equations in five variables allows ∞^2 solutions. This corresponds to 2 independent groups. Determination of these groups is not unambiguous. It is customary to take only one 'clearly acting' factor per group. For example, we may opt for a group containing ν but not g and for a second group containing g but not ν.

The first group is found by $b = 1$ and $e = 0$, from which $d = -1, a = 0, c = -1$; and the second group by $b = 0$ and $e = 1$, from which $d = -2, a = 0, c = 1$.

The resulting dimensionless groups are:

Table 7.1 Dimension table of a parameter group formed by the independent parameters of the flow problem of Fig. 7.1

	ρ	ν	L	V	g
L	-3	2	1	1	1
M	1				
t		-1		-1	-2
	a	b	c	d	e

$$\Pi_1 = \frac{\nu}{LV} \quad \text{or} \quad \frac{LV}{\nu} \quad \text{and} \quad \Pi_2 = \frac{Lg}{V^2} \quad \text{or} \quad \frac{V}{\sqrt{gL}}.$$

The general result is that there are $m - n$ dimensionless groups with m independent parameters and n fundamental dimensions. This result constitutes the Vaschy-Buckingham theorem, often called the π-theorem.

7.1.6 Independent and Dependent Parameter Groups

The listing of problem-describing parameters requires a certain understanding of the problem. For instance, we know that the performance of a turbomachine with a constant-density fluid is determined by the change of the mechanical energy E_m $= (p/\rho) + (v^2/2) + gz$. So ΔE_m must be specified as a single quantity. If the three components of ΔE_m were specified separately as parameters, an Euler number and a Froude number would appear incorrectly in the similitude conditions.

7.1.7 Dimensionless Parameter Groups for Turbomachines with a Constant-Density Fluid

For a pump (or fan), a possible set of independent parameters is.

fluid: ρ, ν
geometry: D (relevant rotor diameter)
operating point: $\Omega, \dot{m}$

The relevant diameter D is normally the outer diameter of the rotor. To determine the operating point, we assume that the rotational speed is imposed by the driving motor and that the flow rate is adjusted by valves. The head is then a dependent parameter. We remark that rotational speed and head may be considered as independent and flow rate as dependent. It is important to understand that two parameters determine the operating point.

There are 3 fundamental dimensions: L, M, t. Consequently, there are $5 - 3$ $= 2$ independent dimensionless groups. These may be determined by dimensional analysis. Acting factors must be chosen for that. They may be ν and Ω. On the base of the insight that we meanwhile have acquired, these groups can be formed directly, as the problem encompasses a dynamic similitude condition (associated to viscosity) and a kinematic similitude condition (associated to rotational speed).

A measure for the through-flow velocity is $V = \dot{m}/(\rho D^2)$. A Reynolds number may thus be formed by

$$\text{Re} = \frac{VD}{\nu} = \frac{\dot{m}}{\rho D \nu} = \frac{\dot{m}}{D \mu} = \frac{Q}{D \nu}. \tag{7.3}$$

The peripheral speed of the rotor follows from $u = \Omega D/2$. The kinematic similitude condition thus results in the group

$$\phi = \frac{V}{u} = \frac{2\dot{m}}{\rho D^3 \Omega}.$$

It is customary to omit numerical factors and to take

$$\Phi = \frac{\dot{m}}{\rho \Omega D^3} = \frac{Q}{\Omega D^3} \quad \text{(flow factor).} \tag{7.4}$$

A dependent group, based on the mechanical energy change is

$$\psi = \frac{\Delta E_m}{u^2} \quad \text{or} \quad \Psi = \frac{\Delta E_m}{\Omega^2 D^2} \quad \text{(head factor).} \tag{7.5}$$

It would be hazardous to express the mechanical energy rise as a manometric head H_m, with the dimension of a height. By lack of attention, a group H_m/D could be formed, which has no physical meaning.

Henceforth, dimensionless groups formed by externally measurable quantities as Q, ΔE_m, Ω and D will be denoted by capitals. Such groups are officially called factors. Groups that are based on internal quantities will be denoted by lower-case letters. For instance, the ratio of an axial velocity component (with axial machines) or a radial component (with radial machines) to the blade speed by

$$\phi = \frac{v_a}{u} \quad \text{or} \quad \phi = \frac{v_{r2}}{u_2}.$$

Such a quantity is officially called flow coefficient (not flow factor). For a dimensionless representation of work, a work coefficient is defined by

$$\psi = \frac{\Delta W}{u^2}.$$

The dimensionless groups ϕ and ψ, based on internal quantities, have already been applied spontaneously in earlier chapters. The terminology distinction between internal and external dimensionless groups, as introduced here, is commonly not made in practice. The terms flow factor and flow coefficient are mostly used indistinctly, as the terms work factor and work coefficient. Head factor and head coefficient are used as well. Inclusion of numerical factors in dimensionless groups sometimes occurs. A flow factor may be defined as $\Phi' = Q/\left(\frac{\pi D^2}{4} \cdot \frac{\Omega D}{2}\right)$ and a head factor as $\Psi' = 2\Delta E_m/(\Omega D/2)^2$. This requires some attention.

For the Reynolds number according to (7.3), the through-flow velocity has been chosen as a base quantity. The peripheral rotor speed may be chosen as well, leading to $Re = \Omega D^2/\nu$. The last expression is even commonly favoured.

7.1.8   Strong and Weak Similitude Conditions

Not all mathematically determined similitude conditions are equally relevant. This is demonstrated by the friction factor for circular ducts, shown in Fig. 2.17 of Chap. 2. With a given relative roughness, the friction factor becomes independent of the Reynolds number for a sufficiently high value of the Reynolds number. The physical reason is that, at a high Reynolds number, the (laminar) sublayer at the wall becomes so thin that the roughness peaks of the wall directly affect the core flow. Friction is then only determined by roughness and not by viscosity.

The example demonstrates that, with a high Reynolds number, this number is not a strong condition for similitude. In turbomachines with low-viscosity fluids such as water ($v \approx 10^{-6}$ m²/s) or air ($v \approx 15 \times 10^{-6}$ m²/s), the Reynolds number is mostly so high (order of magnitude of some hundred thousands) that is does not constitute a condition for similitude in a first approach. A limited correction for the effect of the Reynolds number is necessary (see below).

Analogous behaviour occurs with the Mach number for flows of compressible fluids. The Mach number is the ratio of the flow velocity to the speed of sound. The finite propagation velocity of sound waves results from the compressibility of the fluid. In an incompressible fluid, this velocity is infinitely high (theoretically). This demonstrates that, with a sufficiently low Mach number, this number does not constitute a strong similitude condition. We refer to Chap. 4, where it has been demonstrated that, with a Mach number until about 0.3, the Bernoulli and Saint–Venant equations produce the same results with a good approximation.

7.2   Characteristic Numbers of Turbomachines

7.2.1   Definition of a Characteristic Number

A characteristic turbomachine number is a dimensionless combination of parameters taken from *flows with optimum efficiency*. Such a number is unique for the machine and for geometrically similar machines. It characterises the shape of the machine. We should first demonstrate the relevance of such a definition. Therefore we consider the Q-H curve of a driven turbomachine with a constant-density fluid (pump or fan), as sketched in Fig. 7.2.

From the foregoing we know that, with neglect of the influence of the Reynolds number, the only similitude condition, with fixed machine shape, is the constancy of the flow factor $\Phi = Q/\Omega D^3$. The head factor $\Psi = \Delta E_m/\Omega^2 D^2$ is then a dependent parameter. With a change of rotational speed, all flows with constant Φ, and thus constant Ψ, are similar. The shape of the velocity triangles stays then the same. The flow rate in these flows changes proportionally to the rotational speed. The energy rise is proportional to the square of the rotational speed. So, all flows on a parabola through the origin in Fig. 7.2 are similar.

Fig. 7.2 Similar flows at varying rotational speed for a pump or a fan

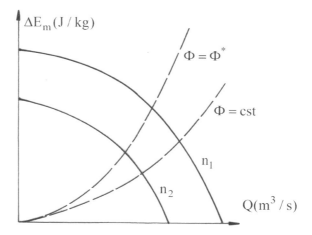

These similar flows have the same (internal) efficiency, as the latter is a dependent dimensionless group (see also the reasonings in Sect. 3.8 of Chap. 3 and Sect. 5.7.4 of Chap. 5). One of the parabolas ($\Phi = \Phi^*$) groups the flows with optimum efficiency. As a consequence, the corresponding values of the flow factor Φ and the head factor Ψ are unique for the machine (and geometrically similar ones). These numbers are thus characteristic numbers for the machine shape.

The above reasoning leading to the similarity parabolas implies that the Q-H curves of a driven turbomachine with a constant-density fluid all coincide for various rotational speeds, if rendered dimensionless in the form Ψ as a function of Φ. The latter is only correct if the effect of the Reynolds number is negligible (see Sect. 7.4.1 on effect of the Reynolds number).

7.2.2 Specific Speed and Specific Diameter

Two other characteristic numbers are customarily constructed from the numbers Φ ($=\Phi^*$) and Ψ ($= \Psi^*$), containing respectively Ω but not D, and D but not Ω. These numbers are denominated *specific speed* (Ω_s) and *specific diameter* (D_s). The diameter is eliminated from Φ and Ψ by the combination

$$\frac{\Phi^2}{\Psi^3} = \frac{Q^2}{\Omega^2 D^6} \cdot \frac{\Omega^6 D^6}{\Delta E_m^3} = \frac{\Omega^4 Q^2}{\Delta E_m^3}, \quad \text{thus:} \quad \Omega_s = \frac{\Omega \sqrt{Q}}{(\Delta E_m)^{3/4}}. \tag{7.6}$$

The rotational speed is eliminated by the combination

$$\frac{\Phi^2}{\Psi} = \frac{Q^2}{\Omega^2 D^6} \cdot \frac{\Omega^2 D^2}{\Delta E_m} = \frac{Q^2}{D^4 \Delta E_m}, \quad \text{thus:} \quad D_s = \frac{D(\Delta E_m)^{1/4}}{\sqrt{Q}}. \tag{7.7}$$

Each family of similar flows within a turbomachine is characterised by a certain value of Ω_s and D_s. The specific speed and the specific diameter of the machine are the values for the flows with optimum efficiency.

The numbers Ω_s and D_s are more convenient numbers than Φ^* and Ψ^*, as a machine is designed for target values of Q and ΔE_m. So the specific speed determines the corresponding rotational speed. The specific diameter determines the corresponding diameter, thus the machine dimensions.

According to similitude theory, only one dimensionless group is independent. Generally, the specific speed is preferred, because the three quantities in this group, flow rate, energy change and rotational speed can be measured externally. A rotor diameter can only approximately be determined externally, which implies that an exact determination of the numbers Φ^*, Ψ^* and D_s is not possible without information about the actual rotor diameter.

An interpretation of Ω_s and D_s follows from considering a reference machine, yielding the required flow rate and energy change with $\Phi = 1$ and $\Psi = 1$. The rotational speed and the diameter of the reference machine meet $\Omega_r D_r^3 = Q$ and $\Omega_r^2 D_r^2 = \Delta E_m$, resulting in

$$\Omega_r = \frac{(\Delta E_m)^{3/4}}{\sqrt{Q}} \quad \text{and} \quad D_r = \frac{\sqrt{Q}}{(\Delta E_m)^{1/4}}, \quad \text{thus:} \quad \Omega_s = \frac{\Omega}{\Omega_r} \quad \text{and} \quad D_s = \frac{D}{D_r}.$$

The adjective 'specific' follows from this interpretation. More general terms are *speed number* and *diameter number*. These terms are sometimes used. Numerical factors are sometimes included into the definitions of Ω_s and D_s, namely when this is done with Φ and Ψ. The former definitions of Φ' and Ψ' imply for the rotational speed and the diameter of the reference machine:

$$\Omega_r' = \sqrt{\pi} \frac{(2\Delta E_m)^{3/4}}{\sqrt{Q}} \quad \text{and} \quad D_r' = \frac{2\sqrt{Q}}{\sqrt{\pi}(2\Delta E_m)^{1/4}}.$$

Speed number (σ) and diameter number (δ) are then defined by

$$\sigma = \frac{\Omega}{\Omega_r'} = \frac{\Omega\sqrt{Q}}{\sqrt{\pi}(2\Delta E_m)^{3/4}} \approx 0.335\,\Omega_s, \delta = \frac{D}{D_r'} = \frac{\sqrt{\pi}D(2\Delta E_m)^{1/4}}{2\sqrt{Q}} \approx 1.054\,D_s.$$

Generally, the notions specific speed, specific diameter, speed number and diameter number are used as defined here.

We further note that $\sigma\delta = \frac{\Omega D/2}{\sqrt{(2\Delta E_m)}} = \frac{u}{V_e} = \lambda$, with V_e being an energy reference velocity. The ratio λ is called *speed ratio*. This notion has already been used in the previous chapter (with the energy reference velocity derived from isentropic enthalpy changes).

The foregoing expression is also

$$\Omega_s D_s = \frac{\Omega D}{\sqrt{\Delta E_m}} = \lambda \, 2\sqrt{2} = \frac{1}{\sqrt{\Psi}}. \tag{7.8}$$

This expression demonstrates again that, if desirable, speed ratio or head factor may be used as a shape parameter, instead of specific speed. Speed ratio (see Chap. 6) or work coefficient (see Chap. 3) are sometimes more visual.

Older technical literature on pumps and hydraulic turbines often uses a dimensional form of specific speed, mostly denoted by n_s or n_q, defined by

$$n_q = n\sqrt{Q}/H_m^{3/4},$$

where the units of n, Q and H_m are rpm, m^3/s and m. This number is not dimensionless and proportional to Ω_s according to $[n_q] \approx 53 \, [\Omega_s]$.

7.2.3 Relation Between Characteristic Numbers and Machine Shape

In the previous sections we reasoned in an abstract way that the numbers $\Phi*$, $\Psi*$, Ω_s, D_s and λ characterise the shape of a machine. We reason now concretely by means of the specific speed (7.6) of a centrifugal pump, with the meridional section drawn by the full lines in Fig. 7.3. The mechanical energy change ΔE_m is, through the internal efficiency, linked to the rotor work $\Delta W = u_2 v_{2u} - u_1 v_{1u}$. The specific speed may be increased in several ways.

A first way is keeping the rotational speed constant, keeping the velocity triangles constant and increasing the flow rate. The mean inner and outer diameters of the rotor stay then the same. A higher flow rate with the same velocity components requires a larger rotor width. This may result in the dashed shape in Fig. 7.3.

Fig. 7.3 Specific speed increase of a centrifugal pump in two ways

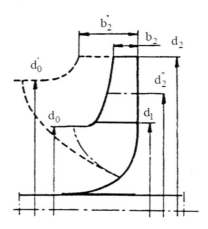

A second way is keeping the flow rate constant, keeping the velocity triangles constant and increasing the rotational speed Ω. The mean inner and outer diameters of the rotor must then decrease. The corresponding outer diameter in Fig. 7.3 may thus be the diameter indicated by d_2'' and the resulting meridional section the dash-dotted form. Since the flow rate and the velocities are unchanged, the diameter of the suction eye d_0 remains unchanged.

Increasing the specific speed to the same new value in the two ways described makes the dashed shape and the dash-dotted shape geometrically similar.

If the specific speed is increased strongly in the second way (constant d_0), but combined with diminishing the rotor work, a diagonal shape and finally an axial shape are obtained, as shown in Fig. 7.4. This illustrates that *specific speed is a*

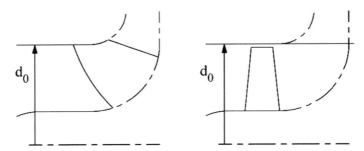

Fig. 7.4 Diagonal shape (mixed flow) and axial shape attained by specific speed increase

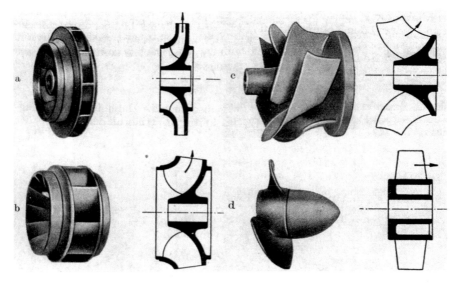

Fig. 7.5 Pump rotor shapes. **a** $\Omega_s = 0.2$–0.7; **b** $\Omega_s = 0.7$–1.5; **c** $\Omega_s = 1.5$–3; **d** $\Omega_s = 2.5$–6 (from Fuchslocher-Schulz [1]; permission by Springer)

shape parameter with low values corresponding with radial forms and high values with axial forms. The other numbers are shape parameters as well.

Figure 7.5 shows four pump rotor shapes with corresponding specific speed values. The evolution of the specific speed may be understood by noting that, for a given rotational speed and rotor peripheral diameter, a radial form has a lower flow rate (smaller entry section) and a higher energy rise (higher ratio of external to internal rotor diameter).

7.2.4 Design Diagrams

Flows with optimum efficiency may be determined for a given turbomachine. The shape parameters (characteristic numbers) take a certain value with these flows. Ω_s and D_s may be determined that way. The same Ω_s and D_s values may also correspond to the optimum flows in a machine with a slightly different shape (a strong deviation is not possible). The optimum efficiencies of both machines are not the same then. So, there is a shape corresponding to the maximum efficiency for each combination of Ω_s and D_s.

The maximum possible efficiency for given values of Ω_s and D_s may be drawn as sketched for single-stage pumps and compressors in Fig. 7.6. For the compressors, the volume flow rate is determined with the inlet density and the head with the isentropic enthalpy rise.

In principle, such a diagram may be composed by studying all machines on the market and by including each time the machine with the best efficiency for given Ω_s and D_s for the optimal flows. But this is unfeasible in practice. Diagrams of Fig. 7.6 kind in the literature are always calculated. The calculation method is verified for a

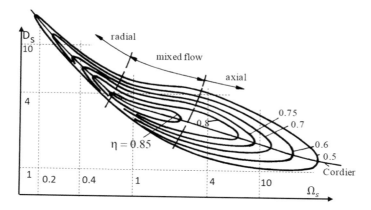

Fig. 7.6 Maximum attainable efficiency for single-stage pumps and compressors at given Ω_s and D_s; adapted from Balje [2]

limited number of test machines, making the diagrams only partially reliable (see also Sect. 7.4.1: effect of Reynolds number and Sect. 7.4.2: effect of relative roughness).

One machine shape yielding maximum efficiency corresponds to a chosen specific speed. In Fig. 7.6 diagram, a line may be drawn through the corresponding values of the specific diameter. This line is called the *Cordier line*.

We note that optimisation for a given specific diameter generates a slightly different line. But, as discussed hereafter, optimisation is normally done for a given specific speed.

As a function of specific speed, there is always an absolute maximum efficiency around $\Omega_s = 1$ (mixed-flow form). Low specific speed machines have a lower efficiency, because their channels are long and narrow, causing increased friction losses. High specific speed machines have a lower efficiency, because rotor work then gets low, which causes increased relative weight of losses.

Pump design starts from the specified flow rate and energy rise. The choice for the rotational speed that optimises the specific speed may be attempted as a first step. From $\Omega = \Omega_s \, \Delta E_m^{3/4}/\sqrt{Q}$ emerges that, with a high head and a low flow rate, the corresponding rotational speed may be large and may become unrealisable. For a pump driven by an electric motor, the maximum rotational speed at 50 Hz power supply is about 2900 rpm (asynchronous motor). In such a case, a suboptimal specific speed might be opted for. The opposite occurs with a low head and a high flow rate. The optimum rotational speed might be too low to be realisable in practice. Synchronous rotational speeds are fractions of 3000 rpm and currently can be: 3000, 1500, 1000 and 750, asynchronous rotational speeds being about 3% lower. A gearbox between the motor and the machine may be opted for.

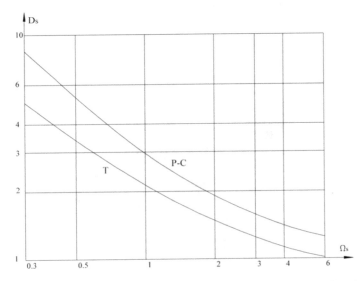

Fig. 7.7 Universal Cordier diagram: Turbines (T), Pumps and Compressors ($P–C$); adapted from Dubbel [3]

The reasoning demonstrates that the choice of rotational speed is a priority. This also applies to other turbomachines. For this reason, optimisations are always done for a chosen specific speed. When no solution with an acceptable efficiency can be found within a design diagram, series operation or parallel operation has to be considered (see Sect. 7.5).

When collecting the Cordier lines of all types of turbomachines, two lines are obtained in a $\Omega_s - D_s$ diagram, one for all driven machines (pumps, compressors) and one for all driving machines (turbines) (Cordier 1953). Figure 7.7 sketches this universal diagram. The diagram concerns machines with full through-flow. An additional degree of freedom comes from partial through-flow. The diagram has always to be applied when designing a turbomachine, but it should be mentioned that turbomachines are not always designed for optimum efficiency.

7.2.5 Shape of Characteristic Curves

Turbomachine characteristics have a shape depending on the value of the specific speed. We illustrate this for pumps. In Chap. 1 we learned that rotor work may be written as the sum of three kinetic energy difference terms:

$$\Delta W = \frac{u_2^2 - u_1^2}{2} + \frac{v_2^2 - v_1^2}{2} + \frac{w_1^2 - w_2^2}{2}.$$

The lower the specific speed, the more radial-type is the meridional shape and the larger is the relative contribution of the centrifugal term (first term) to the rotor work. This term does not vary with the flow rate. The consequence is that the lower the specific speed, the flatter the head is as a function of the flow rate and the flatter the efficiency is as a function of the flow rate. This is illustrated in Fig. 7.8, based on pump characteristics published in the books by Dietzel [4] and Pfleiderer [5].

At a high specific speed, the energy rise as a function of the flow rate typically has a maximum and a minimum. This is due to recirculating flows with operation at very low net flow rate, as sketched in Fig. 7.9. With mixed-flow and axial machines, the consequence is an increase of the centrifugal term in the rotor work at very low flow rate. Recirculating flows at inlet and outlet occur with purely radial rotors as well, but do not influence much the energy rise in the through-flow with backward inclined or backward curved blades (there is large influence with forward curved blades: see Fig. 3.18 in Chap. 3).

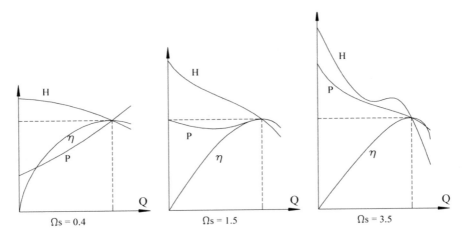

Fig. 7.8 Change of shape of performance characteristics of pumps with change of specific speed; head H, power P, efficiency η

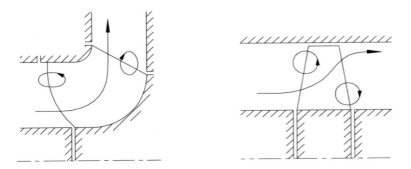

Fig. 7.9 Recirculating flows with operation at very low net flow rate in mixed-flow and axial pumps; adapted from Pfleiderer [5]

7.2.6 Power Specific Speed

Rotor power is the product of mass flow rate and rotor work. Thus, specific speed may also be written as

$$\frac{\Omega\sqrt{P_{rot}}}{(\Delta E_m)^{5/4}\sqrt{\rho}}.$$

For hydraulic turbines, a *power specific speed* is defined by

$$\Omega_{SP} = \frac{\Omega\sqrt{P_{shaft}}}{(g\,H_m)^{5/4}\sqrt{\rho}},$$

with P_{shaft} the power measured at the shaft and H_m the manometric head.

This form of specific speed is commonly used, as flow rate determination with a hydraulic turbine is not easy. Rotational speed, head and shaft power are the externally measurable quantities. The power specific speed is, strictly speaking, not a dimensionless group of flow quantities. The rotor power should be used for that, as mechanical losses do not obey the similitude laws of flows. Since rotor power is not directly measurable and the mechanical and volumetric efficiencies are very close to unity with hydraulic turbines, shaft power is used.

7.3 Application Example of Similitude: Variable Rotational Speed with a Pump

We consider a pump application as sketched in Fig. 7.10.

The energy rise between the discharge side (pressure side: p) and the suction side (s) of the pump is

$$\Delta E_m = \frac{p_p - p_s}{\rho} + \frac{v_p^2 - v_s^2}{2} + gh_m,$$

where h_m (machine) is the height difference between the discharge and suction flanges of the pump. The required energy increase follows from the connected pipes and reservoirs. The work equations for the suction and the discharge pipes are

$$\frac{p_a}{\rho} + 0 + 0 = \frac{p_s}{\rho} + \frac{v_s^2}{2} + gh_s + q_{irr,s},$$

$$\frac{p_p}{\rho} + \frac{v_p^2}{2} + 0 = \frac{p_a}{\rho} + 0 + gh_p + q_{irr,p}.$$

The sum of these expressions is

Fig. 7.10 Pump between suction and discharge reservoirs

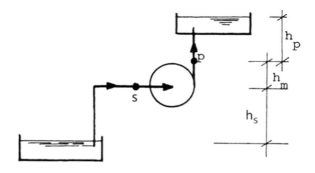

Fig. 7.11 Operating point;
pump characteristic P, load
characteristic L

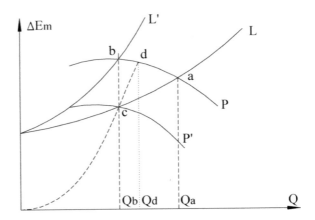

$$\frac{p_p}{\rho} - \frac{p_s}{\rho} + \frac{v_p^2}{2} - \frac{v_s^2}{2} = g(h_s + h_p) + q_{irr,sp}.$$

Thus:

$$\Delta E_m = g h_{geo} + q_{irr,sp},$$

where h_{geo} is the geometric height difference between the levels in the suction and discharge reservoirs.

The required energy rise as a function of the flow rate is denominated the *load characteristic,* but the more general term *system characteristic* is often used. Pump and load characteristics are sketched in Fig. 7.11. The load characteristic has a static part (gh_{geo}), which is the value for zero flow rate and a dynamic part which rises approximately with the square of the flow rate. The dynamic part is typically called the *head loss,* meaning the dissipation of mechanical energy.

The intersection of both characteristics determines the operating point of the pump. When designing the installation, a pump with an optimum operating point close to the target operating point should be chosen. Also, the pipe dimensions should be optimised in the sense that a balance should be made between the energy cost by dissipation within the pipes and the investment cost of the pipes, taking their expected life time into account.

An example of an installation according to Fig. 7.10 is water distribution. The application may be of large scale with a water distribution company, of small scale with the water supply to a horticulture farm and of very small scale with home distribution of rain or well water. The discharge reservoir is not necessarily built as suggested in the figure, namely an open reservoir at a height. Often, it is a pressurised reservoir. This makes no energetic difference, however. The load characteristic keeps the shape as in Fig. 7.11.

Assume that the installation is designed for flow rate Q_a delivered by the pump to the discharge reservoir. Water distribution from the discharge reservoir mostly shows a very strong flow rate variation, typically from zero to a maximum value. The pump in Fig. 7.10 is not intended to follow these flow rate changes. The pump functions around the operating point (*a*), as indicated in Fig. 7.11, with a discharge reservoir level fluctuating between a minimum and a maximum (or, equivalently, the pressure in the closed reservoir fluctuating between a minimum and a maximum). Thus, the load characteristic (*L*) of the pump in Fig. 7.11 is not fixed, but moves up and down.

The pump is switched on when the reservoir level reaches a minimum and is switched off when the level reaches a maximum. Determination of the design operating point thus requires analysis of the frequency distribution of the flow rates extracted from the discharge reservoir. This study determines the design flow rate of the pump and the capacity of the discharge reservoir.

With some pump applications, a large-capacity discharge reservoir as in Fig. 7.10 is not wanted, or even no reservoirs are wanted. Pumps in chemical plants or pumps for hot water distribution in central heating systems are examples. Such pumps must allow flow rate variation. The traditional solution is constricting the discharge pipe through a control valve, changing the load characteristic in Fig. 7.11 from L to L'. The flow rate is then reduced from Q_a to Q_b. There are possibly control valves on various branches in the pipe system, as with thermostatic valves on individual radiators of a central heating system. Constricting is disadvantageous from an energy point of view. With larger systems it is therefore advisable to make the rotational speed of the pump variable and to reduce the pump characteristic to P' by reducing the rotational speed. The required rotational speed may be calculated by considering a similitude parabola through the origin and through point *c*. This intersects the pump characteristic P in point *d*.

The required rotational speed follows from $\dfrac{n_c}{n_a} = \dfrac{Q_b}{Q_d}$.

Variable pump speed requires rotational speed control on the driving motor. A frequency controlled asynchronous motor is the most current system. Optimising the system and the control strategy can sometimes become quite complicated, as often a combination of buffer capacity, constricting valves and rotational speed control is required. Variable rotational speed is very beneficial from an energetic point of view, when the static head (gh_{geo}) in the load characteristic is low. The operating points are then close to a similitude parabola. An example is water circulation within a central heating system. Central heating circulators for somewhat larger applications are nearly always provided with variable rotational speed motors nowadays. Currently there is an evolution towards permanent magnet motors with a wet rotor (see Sect. 8.7.2 on canned pumps in Chap. 8). Fan applications have a very small static head as well. Fans are therefore almost always equipped with variable rotational speed motors, typically frequency controlled asynchronous motors.

Even when energy considerations are of lower relevance, as with home distribution of rain and well water, pumps with variable rotational speed are applied. The control strategy is quite simple then, namely keeping the discharge pressure constant and switching off at zero flow. Variable rotational speed pumps have the advantage that

Fig. 7.12 Left: domestic pump unit with pressure reservoir and fixed motor speed; Right: without reservoir and electronic motor speed control (*Courtesy* Wilo)

only a very small reservoir is needed, which is just meant for detection of pressure. At present, both pumps with variable rotational speed and very small reservoir and pumps with fixed rotational speed and a larger reservoir are marketed.

Figure 7.12 shows the two types of domestic pump units. The buffer capacity of the larger reservoir comes from water stored in a rubber bellows, enclosed in an air-filled tank. The variable rotational speed system is most often chosen, because the larger reservoir needs some care, as air leaks from the reservoir and bellows rupture are possible flaws.

7.4   Imperfect Similitude

7.4.1   *Effect of Reynolds Number with the Same Fluid*

With operation at low flow rate, the Reynolds number may become rather small. By the mere application of the kinematic similitude condition, similitude is then not entirely correct. Especially, efficiency diminishes when the Reynolds number is reduced to a small value. For example, in the map of characteristics of a centrifugal pump for various rotational speeds, as sketched in Fig. 7.13, the efficiency contours only follow the similitude parabolas in the centre of the diagram.

At low flow rates, curves close due to the Reynolds number effect. Curves close as well at high flow rates. Another phenomenon occurs there. Velocities increase at

Fig. 7.13 Efficiency contours of a centrifugal pump; adapted from Dietzel [4]

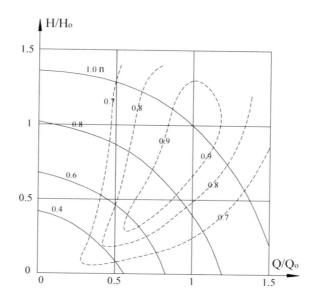

high flow rates. As a consequence, the pressure at the suction side of the pump drops. Vapour bubbles are generated when the pressure gets lower than the vapour pressure. This phenomenon is called *cavitation*. The vapour bubbles occupy space, causing the fluid to accelerate with higher friction losses. This diminishes the efficiency compared to the value expected from kinematic similitude. The Reynolds number influence occurs with fans, but not the cavitation effect. So, efficiency contours stay open for high flow rate.

7.4.2 Effect of Relative Roughness

With higher Reynolds number ($Re > 10^5$) and higher relative roughness ($k/D > 10^{-3}$), the friction factor in the Moody diagram (Fig. 2.17 in Chap. 2) is mainly determined by the relative roughness.

In some turbomachines, relative roughness is low and roughness only has a very limited effect on the performance parameters. Centrifugal fans are examples. These machines are made from smooth rolled steel sheets and mostly have rather large dimensions (industrial fans: rotor diameter 0.5–2 m). Hydraulic turbines are other examples. The rotors of these machines are cast, but they are sufficiently large (rotor diameter 1–10 m) to allow finishing by milling and grinding.

Pump parts are cast and pump dimensions are mostly rather limited (rotor diameter 0.1–0.5 m), precluding final polishing. Absolute roughness with pumps is approximately constant. Pump dimensions thus affect efficiency, with lower efficiency for smaller pumps. So, a diagram such as Fig. 7.6 cannot have general validity.

Fig. 7.14 Effect of machine
size on efficiency with
pumps (Worthington
diagram); adapted from
Pfleiderer and Peterman [6]

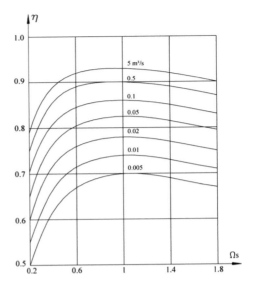

The diagram is calculated for an average machine size. To exemplify this aspect,
Fig. 7.14 renders the internal efficiency of pumps as a function of specific speed at
various machine sizes, expressed by the design flow rate. This diagram, known as the
Worthington diagram, is found in nearly every book on pumps. We note that machine
size has no effect on the specific speed at which optimum efficiency is attained. In
other words, the universal Cordier diagram (Fig. 7.7) keeps its validity.

7.4.3 Effect of Viscosity

Pump characteristics are mostly available for water. Use of a more viscous fluid
impairs the performance because of larger friction losses (head and efficiency
decrease) and larger flow displacement by boundary layers (flow rate decreases).

Almost all books on pumps contain a nomogram by the Hydraulic Institute in New
York [7], allowing corrections for increased viscosity. It can also be found on many
internet sites (look for Hydraulic Institute viscosity correction chart). The entries
of the chart are the flow rate and the head for the viscous fluid, together with the
fluid viscosity. The outputs are correction factors for flow rate (C_Q), head (C_H) and
efficiency (C_E). The relations are $Q_{vis} = C_Q Q_w$, $H_{vis} = C_H H_w$, $\eta_{vis} = C_E \eta_w$. The
relations allow to calculate the flow rate and the head with water (subscript w), from
which the pump can be chosen. With the known efficiency with water, the efficiency
with the more viscous fluid can be estimated.

Quite often, in chemical industry, an identical pump or compressor is used with
different fluids. Performance is normally determined by the manufacturer with water
and air as fluids. Calculating the expected performance with another fluid is very

important for the manufacturer because of the guarantee to be given to a user. Therefore, a calculation method for radial compressors was developed by a group of manufacturers (Casey [8], Strub et al. [9]). The procedure is quite rational and may be applied to pumps too, although it has not been made for pumps.

The proposed correction factor for internal efficiency is

$$\frac{1 - \eta_i}{1 - \eta_{it}} = \frac{0.3 + 0.7 \, (f/f_\infty)}{0.3 + 0.7 \, (f_t/f_\infty)}. \tag{7.9}$$

In this formula, f is the friction factor from the Moody diagram (Fig. 2.17) and f_∞ is the friction factor for very high Reynolds number for the same relative roughness. The formula presumes that 30% of the losses do not depend on the Reynolds number. The subscript t refers to the test result with air.

To determine the Reynolds number, an average velocity of $0.5 \, u_2$ within the rotor and the stator is assumed, as well as an average hydraulic diameter of rotor and stator channels being $2 \, b_2$. So $Re = (u_2 b_2)/\nu$.

The technical roughness of a surface, called arithmetic roughness, is the average value of the deviation to a mean surface:

$$R_a = \frac{1}{\ell} \int\limits_o^\ell |y| \, dx.$$

Roughness values must be measured within the rotor on a blade, on the hub disc and the shroud, near the rotor outlet, and within the stator on the sidewalls and in the centre of a vane (in the case of a vaned diffuser), near the stator inlet. An average value is determined. The roughness in the Moody diagram is the equivalent sandgrain roughness (k), by which is meant the diameter of closely packed sand grains resulting in the same skin friction. The ratio to the arithmetic roughness is around 10 for values of the arithmetic roughness lower than about 20 μm (Adams et al. [10]). So, we may assume

$$\varepsilon = \frac{k}{D_h} = \frac{10 \, R_a}{2 \, b_2} = 5 \frac{R_a}{b_2}.$$

It is assumed that half of the efficiency change affects the energy rise, by

$$\frac{\Delta E_m}{\Delta E_{mt}} = 0.50 + 0.50 \frac{\eta_i}{\eta_{it}}, \tag{7.10}$$

and the other half the work by

$$\frac{\Delta W}{\Delta W_t} = 0.50 + 0.50 \frac{\eta_{it}}{\eta_i}. \tag{7.11}$$

It is further found that the flow rate and the energy rise are related as with a rotational speed change, according to

$$\frac{Q}{Q_t} = \sqrt{\frac{\Delta E_m}{\Delta E_{mt}}}. \tag{7.12}$$

Corrections applied to flow rate and energy rise imply a correction of the kinematic similitude condition.

This means that, with different fluids, operation points with optimum efficiency do not exactly correspond according to kinematic similitude.

From the quoted study emerges that the calculation method is accurate on condition that the loss change $(1 - \eta_i)$ is limited to about 20%.

Possible values for radial compressors are $u_2 = 300$ m/s; $d_2 = 400$ mm; $b_2 = 20$ mm; $R_a = 2$ μm (milled surfaces); $v = 15 \times 10^{-6}$ m²/s (air). This results in $Re = (u_2 b_2)/v = 4 \times 10^5$, $\varepsilon = 5R_a/b_2 = 5 \times 10^{-4}$. A point in the transition area (both effect of Re and ε) corresponds in the Moody diagram. Possible values with radial pumps are $u_2 = 30$ m/s; $d_2 = 200$ mm; $b_2 = 10$ mm; $R_a = 20$ μm (smooth sand cast surfaces); $v = 10^{-6}$ m²/s (water), resulting in $Re = (u_2 b_2)/v = 3 \times 10^5$, $\varepsilon = 5R_a/b_2 = 0.01$. The point in the Moody chart is in the rough surfaces area (no effect of Re). A point within the transition area is obtained with a fluid that is ten times as viscous (light oil).

7.4.4 Rotor Diameter Reduction: Impeller Trimming

The rotor diameter may be reduced by turning on a lathe in order to lower the head and flow rate of a pump (impeller trimming). When designing a pump installation, it would be a mere coincidence to obtain an exactly appropriate pump from a manufacturer. In general, a pump with a slightly too high head and too high flow rate must be chosen. The head and flow rate may then be diminished by reduction of the rotor diameter.

This way, a manufacturer can cover a certain head and flow rate range with the same pump. Figure 7.15 is an application field of a pump type. The meaning of the pump numbers is the diameter of the entry of the suction part of the pump, the diameter of the exit of the discharge part of the pump and the rotor diameter, all in mm. When trimming, normally only the blade tips are cut. The hub disc and the shroud, if any, often must be kept, as they contribute to the rotor sealing and the guiding of the fluid to the post-connected stator.

Assuming that trimming does not change the rotor blade exit angle and rotor exit width, which is normally met with a good approximation, flow rate and energy rise change according to $Q \sim u_2 d_2 \sim d_2^2$, $\Delta E_m \sim u_2^2 \sim d_2^2$, on the basis of kinematic similitude on the velocity triangle at the rotor exit. The similitude is not perfect, however. The decrease of the blade length diminishes the work done on the fluid. In

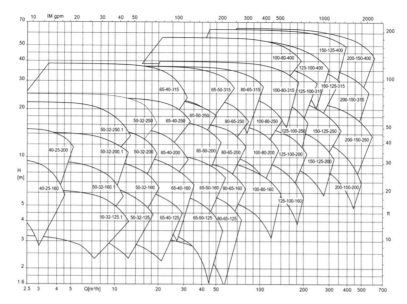

Fig. 7.15 Application field of a single-stage centrifugal pump type for rotational speed 1450 rpm (*Courtesy* KSB pumps)

other words, the slip between flow and geometry increases. This causes the energy rise to be lower than predicted by the simple reasoning. Furthermore, flow rate reduction at a constant rotational speed generates incidence at the entry of the rotor and the entry of the post-connected stator. So, the efficiency diminishes somewhat. Due to the combined effects of slip increase and incidence, the rotor must be trimmed down less than predicted by the simple similitude reasoning. A reduction factor has to be applied to the theoretical diameter decrease. This factor depends somewhat on the specific speed. An average value is about 0.75 (Pfleiderer [5]).

7.4.5   Reduced Scale Models

When developing machines of large size (hydraulic turbines, large pumps), tests are sometimes performed on a reduced-scale model. When using the same fluid, the Reynolds number diminishes. In addition, perfect geometric similitude is impossible due to no constant relative roughness and no constant relative clearance values. As most machines are manufactured with the same techniques (e.g. casting) or the same machine tools, wall roughness is typically of the same order of magnitude. So, absolute roughness is approximately constant and relative roughness is not. Clearances are typically kept as small as possible, requiring a certain minimum value to be workable. Further, there is never an exact geometric similitude as some parts (e.g. blades) have a minimum thickness imposed by the production technique.

In order to take into account the Reynolds number decrease and the increase of relative roughness, the relative clearances and the relative thicknesses, correction formulae known as *model laws* may be applied. Model laws are inspired by the observation that $1 - \eta$ is proportional to the friction losses and that these, according to the Moody diagram, vary within the transition area with the Reynolds number to a power between -0.25 and -0.1. Clearly, the exponent is a rough estimate. Thus, model laws have a weak foundation and should not be overvalued.

Moody's model law for hydraulic turbines is

$$1 - \eta = (1 - \eta_0) \left(\frac{D}{D_0}\right)^{-0.05} \left(\frac{\mathrm{Re}}{\mathrm{Re}_0}\right)^{-0.2}. \tag{7.13}$$

Pfleiderer's model law for centrifugal pumps is

$$1 - \eta = \left(1 - \left[\frac{1 - 70/d^{1.5}}{1 - 70/d_0^{1.5}}\right] \eta_0\right) \left(\frac{\mathrm{Re}}{\mathrm{Re}_0}\right)^{-0.1}. \tag{7.14}$$

In the first formula, D is the rotor diameter. In the last formula, d is the diameter of the suction eye, expressed in mm. The subscript 0 refers to the model. Both formulae contain a correction to the effect of the Reynolds number in order to take the imperfect geometric similitude into account. For pumps, Re is determined with the peripheral rotor speed ($\mathrm{Re} = u_2 d_2 / v$), for turbines with the spouting velocity ($\mathrm{Re} = \sqrt{2g H_m} d_2 / v$).

7.5 Series and Parallel Operation

If no single-rotor machine with an acceptable rotational speed can be found for a particular combination of energy change and flow rate, series or parallel operation may be considered. Series or parallel arrangements may be made within the same machine. Series arrangement increases the overall energy exchange. Parallel arrangement increases the flow rate. These internal arrangements are common with pumps (see next chapter). Internal series arrangement is very common with compressors, steam and gas turbines. Series or parallel operation of separate machines may be chosen to enable a flexible use of the machine group.

7.5.1 Parallel Operation of Fans

Figure 7.16 (left) shows the flow rate addition of two identical fans in parallel operation. The flow rate is doubled with the same energy rise. With the load characteristic $L1$ (typically no static part with fans), the operating point is B for an individual fan. When one of the fans stops functioning, with reversed flow impeded, the other fan operates in point C. Impeding reversed flow with fans is often difficult because of ducts of large size. Assuming that $L2$ is the resistance characteristic of a freely turning fan with reversed flow, the load of the working fan then consists of the parallel arrangement of the duct network with characteristic $L1$ and the fan with resistance characteristic $L2$. Characteristic $L2$ must, as concerns the working fan, be plotted

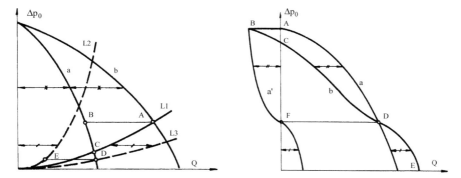

Fig. 7.16 Parallel operation of two identical fans (left) and two fans of largely different size (right)

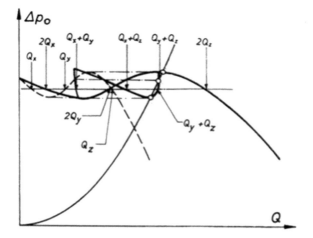

Fig. 7.17 Parallel operation of identical fans with a minimum and a maximum in the characteristic

toward the side with positive flow rates. The working fan thus has a load characteristic $L3$, found by adding the abscissas of $L1$ and $L2$. The operating point of the
working fan is now D, while a flow rate corresponding to E flows back through the
freely turning fan.

When the fans have different characteristics (strongly different characteristics a
and a' in Fig. 7.16, right), the characteristic of the parallel arrangement is again
found by adding flow rates. The part BF represents the resistance characteristic of
the working small fan, when fluid is forced to flow back. If backflow through the
small fan is not impeded, the resulting characteristic is BCDE. If no reversed flow is
possible (non-return valve), ADE is the resulting characteristic.

The parallel operation becomes more complicated with fans having a minimum
and a maximum in their characteristic. Parallel operation of two identical fans with
forward curved blades (Fig. 7.17) results in several possible branches for the characteristic of the parallel arrangement. Certain load characteristics intersect the fan
group characteristic in several points. The fans in parallel operation may then shuttle
between several operating points (exceptional behaviour).

7.5.2 Parallel Operation of Pumps

With parallel arrangement of pumps, typically head losses (H_{li}) and geometric heads
(H_i) are different in the different parallel paths. Figure 7.18 shows an example with
two pumps. The parallel operation may be studied by determining the difference of

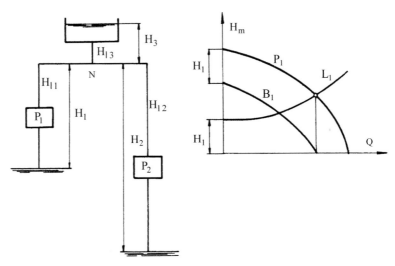

Fig. 7.18 Parallel operation of pumps; right: net characteristic of a pump and the load by a branch
in the pipe network

the head produced by a pump (P1) and the head consumed by the connected pipe (L1) between the suction reservoir and the node in the network (N). The thus obtained net characteristic of the left-hand branch is denoted by B1 in Fig. 7.18. The characteristic of the parallel operation of the two branches may then be determined as before, i.e. by summing the flow rates of the net characteristics B1 and B2 for equal head. The load on the resulting characteristic is the branch from the node N to the discharge reservoir.

7.5.3   Series Operation of Fans

When two fans are operating in series (different characteristics a and a' in Fig. 7.19), the characteristic of the series arrangement is found by adding the pressure rises for given flow rate. The EG part in the characteristic of the small fan represents the negative pressure difference over the fan required to enforce a flow rate through this working fan larger than what it can deliver without load. The BG part of the composed characteristic is not useful. The series operation has only very marginal meaning for the load characteristic $L1$ (the small fan only contributes a little). It has no advantage for the load characteristic $L2$ (flow is sucked through the small fan).

Fig. 7.19 Series operation of two fans with largely different characteristics

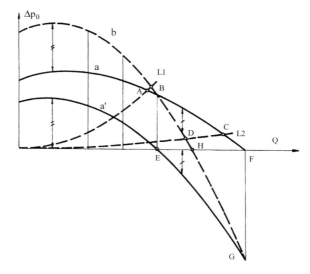

7.6 Turbomachine Design Example: Centrifugal Fan

This book aims at fundamental understanding of turbomachines. The basic dimensioning of most machines is possible with the topics analysed. Detailed design requires a more advanced study, but the full design of a rather simple machine, as a radial fan, is achievable with the topics treated. As an example, we discuss the design of the radial fan of the laboratory test in Chap. 5.

Before starting, it has to be stressed that design is not purely the result of a sequence of calculation phases. It also includes taking decisions that are not unique and that may be made for ease of construction or ease of application of the machine.

The targeted performance data of the fan are $Q = 0.8$ m^3/s and $\Delta p_0 = 3500$ Pa. The fan is meant to draw ambient air and feed a duct system for ventilation of an industrial workshop. We immediately mention that attaining a total pressure rise of 3500 Pa is quite challenging. From the fan chapter we learned that a head coefficient of a radial fan with backward curved blades can attain about $\psi = 0.5$. With $\rho \approx$ 1.20 kg/m^3 air density it follows that

$$\psi = \frac{\Delta p_0/\rho}{u_2^2} = 0.5 \rightarrow u_2 \approx 75 \text{ m/s}.$$

So, the required peripheral speed is rather high.

The design starts from the Cordier diagram as given in Fig. 7.7. In specialised literature, such a diagram is complemented with geometric data as diameter ratio (d_1/d_2) and width to diameter ratio (b_2/d_2) as functions of specific speed (Eck [11], Pfleiderer [5]). These supplementary data are not strictly necessary, however.

The couple on the Cordier line, leading to a rotational speed of about 2900 rpm is $\Omega_s = 0.70$, $D_s = 3.80$. This results in $d_2 = 0.46$ m, $n = 2966$ rpm, which we may reduce somewhat to $d_2 = 0.45$ m, $n = 2900$ rpm, so $u_2 = 68.33$ m/s. The required ψ then becomes 0.625. If it later turns out that the pressure rise or the flow rate are somewhat too small, we may then enlarge somewhat the diameter or the rotational speed.

With Eq. (3.41), for maximum deceleration in the rotor eye ($b_1/d_1 = 0.40$):

$$(d_1)_o = \sqrt{2}\left(\frac{Q}{\Omega}\right)^{1/3} = 0.195 \text{ m}.$$

We take $d_1 = 0.20$ m. Then, from Eq. (3.42): $v_1^b = \frac{1}{2}Q^{1/3}\Omega^{2/3} = 20.97$ m/s.

With $d_0 = 0.18$ m: $v_0 = \frac{4Q}{\pi d_0^2} = 31.44$ m/s. The value of v_0 is acceptable, because a duct velocity may be up to 20 m/s, but with direct aspiration from a room, the inflow velocity may be as high as 30 m/s. The through-flow velocity of the rotor cannot be much lower than 20.97 m/s (velocity after rotor entry), because then $v_1^b/v_0 = 0.667$ (the limit value is about 0.6).

We take as estimates of the volumetric efficiency $\eta_v = 0.90$ and the blade obstruction factor at rotor entry $\tau_1 = 0.85$.

$$Q_{rotor} = \pi d_1 b_1 \tau_1 v_1^b \approx Q/0.9.$$

The result is $b_1 = 79.4$ mm. But with $b_1/d_1 = 0.40$, the value is $b_1 = 80$ mm. We take $b_1 = 80$ mm. This lowers somewhat v_1 to $v_1^b = 20.80$ m/s.

With $u_1 = 30.37$ m/s: $\tan \beta_1^b = -u_1/v_1^b : \beta_1^b \approx -55.6°$. This value differs only a little from the theoretically optimum value (−55°). $w_1^b = \sqrt{u_1^2 + (v_1^b)^2} = 36.17$ m/s.

$$\Delta p_0/\rho = 2917 \text{ J/kg}.$$

With $\eta_i \approx 0.85$: $\Delta W = 3432$ J/kg; then $\psi = \frac{\Delta W}{u_2^2} = 0.735$ and $\Delta W = u_2 v_{2u}$, thus: $v_{2u} = 50.22$ m/s. With an estimated work reduction factor of $\varepsilon = 0.85$, we obtain $v_{2u}^b = 59.08$ m/s.

Suppose that we can realise the deceleration ratio $(w_2/w_1)_b = 0.70$ in the rotor, then $w_2^b = 25.32$ m/s. From $v_{2u}^b = u_2 + w_2^b \sin \beta_2^b$ follows $\beta_2^b = -21.4°$. We may thus opt for straight blades with $\beta_1^b = -57°$ and $\beta_2^b = -22°$ (see Fig. 7.20).

With an estimated blade obstruction factor at rotor exit of $\tau_2 = 0.95$, we obtain then $v_{2r}^b = 23.48$ m/s, $v_{2r} = 22.30$ m/s, $w_2 = 28.86$ m/s. Width at rotor exit: $Q_{rotor} = \pi d_2 b_2 v_{2r} : b_2 = 28.2$ mm.

$C_M = \frac{w_{2r} v_{2u}}{\sigma_M w_2^2} \approx 1.0$ (Eq. 3.32), gives $\sigma_M \approx 1.36$, with $\sigma_M = \frac{Z M_{st}}{2\pi r_2 b_2 r_2}$.

$M_{st} = \int b r dr \approx \frac{1}{2}(b_1 r_1 + b_2 r_2)(r_2 - r_1)$: $Z = 13.6$. We take $Z = 16$, which then reduces the slip at the rotor exit and thus increases the rotor work. Choice of plate thickness: $t = 3$ mm.

We can now correct the velocity triangles, using improved values of the obstruction factors. At this stage, we still take $\eta_v = 0.90$ and $\eta_i = 0.85$.

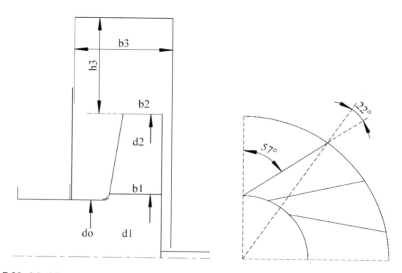

Fig. 7.20 Meridional and orthogonal sections of the rotor

Inlet triangle:

$$\beta_1^b = -57°, \tau_1 = \frac{\pi d_1 - Zt/\cos\beta_1^b}{\pi d_1} = 0.860,$$

$$v_1^b = \frac{Q}{\pi d_1 b_1 \tau_1 \eta_v} = 20.57\,\text{m/s};\ u_1 = 30.37\,\text{m/s};\ w_1^b = 36.68\,\text{m/s}.$$

Outlet triangle:

$$\beta_2^b = -22°, \tau_2 = \frac{\pi d_2 - Zt/\cos\beta_2^b}{\pi d_2} = 0.963,$$

$$w_2^b = 0.70\,w_1^b = 25.68\,\text{m/s};\ v_{2r}^b = w_2^b\cos(-22°) = 23.81\,\text{m/s};$$

$$v_{2r}^b = \frac{Q}{\pi d_2 b_2 \tau_2 \eta_v} : b_2 = 27.4\,\text{mm};$$

$$u_2 = 68.33\,\text{m/s};\ v_{2u}^b = u_2 + w_2^b\sin\beta_2^b = 58.71\,\text{m/s}.$$

Slip:

$$\xi = \lambda(2.5 + \beta_2^o/60) = 0.75 \times (2.5 - 0.367) = 1.60;$$

$$Pf = \frac{\xi}{Z}\frac{2b_2 r_2^2}{(b_1 r_1 + b_2 r_2)(r_2 - r_1)} = 0.157,$$

$$\varepsilon = \frac{1}{1 + Pf} = 0.865,\ v_{2u} = 50.76\,\text{m/s};$$

$\Delta W = 3468.2$ J/kg, $\Delta p_0 = \eta_i \rho \Delta W = 3537.6$ Pa. This value is still as wanted.

The inlet area of the fan is $\pi d_0^2/4$, which is about 160 mm × 160 mm or 150 mm × 170 mm. Thus, for reaching comparable exit and entry velocities of the fan, we may opt for a scroll with an exit area $b_3 \times h_3 = 150$ mm × 150 mm, followed by a diverging channel until a section of 150 mm × 170 mm. A diffuser can then be added for connection to the ductwork for ventilation. With $h_3 = 150$ mm, $r_3 = r_2 + 150$ mm = 375 mm. The flow angle imposed by the volute at the exit of the rotor follows from: $\ln(\frac{r_3}{r_2}) = \frac{2\pi}{\tan\alpha_2}$. The result is $\alpha_2 = 85.35°$. The opening angle of the scroll is then 4.65°. The scroll is very compact (the opening angle is lower than the recommended range of 5° to 6°). The theoretical flow rate at the scroll exit, with $v_u r$ = constant, is $\int_{r_2}^{r_3} \frac{v_{2u} r_2}{r} b_3\,dr = 0.881$ m³/s. The actual flow rate is about 10% smaller due to boundary layer obstruction, thus about 0.8 m³/s. So, the scroll is very well matched to the rotor.

The width leap from the rotor exit to the scroll entry becomes very large: 27.4–150 mm. It is then better to widen the rotor exit, for instance to 60 mm. The strongest possible deceleration is then realised in the rotor, but with separated flow at the rotor exit. It is more efficient to allow separated flow in the rotor because the friction loss within the rotor then decreases, without increase of the dump diffusion loss at the entry to the volute. We may estimate the maximum velocity reduction to 0.7

Fig. 7.21 Orthogonal section of the volute

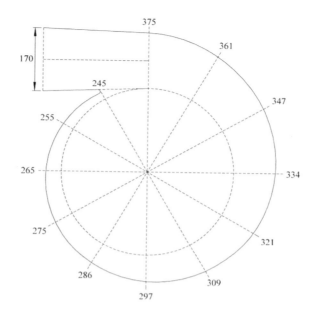

between the inlet flow just after rotor entrance and the outlet flow just before rotor exit, so before slip. Enlargement of the rotor exit width is commonly used with fans with slightly backward leaning blades or forward curved blades. For forward curved blades, typically the outlet width is taken equal to the inlet width, because the diameter ratio is near to unity: see Fig. 3.18.

We take a tongue clearance of about 10% of the radius: about 22.5 mm. The opening angle of the scroll is taken somewhat larger than the theoretical value for the first sector of 30°, because the flow expansion by the width leap cannot be sudden. This way, the tongue clearance becomes 20 mm. For the rest of the scroll, the theoretical opening angle is used.

Figures 7.20 and 7.21 show the geometry. A final calculation of the performance of the fan can now be made (see Exercise 7.7.5). In particular, the rotor slip, the leakage flow rate and the losses have to be estimated. There is no information in the literature on how rotor slip has to be estimated for a rotor with a strongly separated flow. Therefore, we will assume that the slip may be calculated with the geometric data of the rotor and applied to the core flow.

7.7 Exercises

7.7.1 Measured data of a pump are $Q = 0.20$ m³/s, $H_m = 30$ m, $\eta_{global} = 0.75$ for $n = 1450$ rpm. The fluid is water with $\nu = 10^{-6}$ m²/s. Determine flow rate, head and efficiency of the pump when doubling the rotational speed for a similar flow.

Assume a volumetric efficiency and a mechanical efficiency of respectively 1 and 0.9 and assume that these values do not change.

Correct the internal efficiency for imperfect similarity by Pfleiderer's formula (Eq. 7.14). Correct head and flow rate according to the formulae of Casey and co-authors (Eqs. 7.10–7.12). Consider the method of Casey et al. (Eq. 7.9) as an alternative for the efficiency correction. Assume $R_a = 20 \, \mu m$ for that. Determine an approximation for the rotor diameter and the rotor width at outlet assuming a work coefficient 0.4 and a flow angle $-70°$ at the rotor exit.

Observe that the correction on the efficiency for the changed rotational speed is very small with Eq. (7.14) and even smaller with Eq. (7.9).

A: $Q = 0.401 \, m^3/s$, $H_m = 120.80 \, m$, $d_2 \approx 400 \, mm$, $b_2 \approx 26.5 \, mm$.

7.7.2 Take 400 and 25 mm as approximations of the rotor diameter and the width at rotor exit for the pump of the previous exercise. Determine for $n = 1450$ rpm the head and the flow rate for operation with a light oil with $v = 200 \times 10^{-6} \, m^2/s$ in the operating point corresponding to $Q = 0.20 \, m^3/s$ and $H_m = 30 \, m$ with water ($v = 10^{-6} \, m^2/s$). Correct for imperfect similitude with the formulae of Casey and co-authors (Eqs. 7.10–7.12). Use the formula of Haaland to describe the Moody diagram (Sect. 2.3.1 in Chap. 2).

Observe the large efficiency correction with a 1/200 viscosity ratio.

A: $H_m = 28.83 \, m$, $Q = 0.196 \, m^3/s$.

7.7.3 A 1/10 scale model has to be built of a hydraulic turbine with 40 MW shaft power, 50 m manometric head, 100 rpm rotational speed and overall efficiency 0.90. The available head in the laboratory is 6 m. Take thus 6/50 as scale ratio for head.

Determine the rotational speed of the model, the flow rate and the shaft power, with similar flow. Correct the efficiency for imperfect similitude by Moody's formula (Eq. 7.13). Apply the correction to the overall efficiency, in other words assume volumetric and mechanical efficiencies both equal to unity.

Determine the flow rate and the rotational speed, taking into account that the head and flow rate obtained by pure kinematic similitude are not correct due to imperfect similitude. Estimate the effects of imperfect similitude by the method of Casey et al. (Eqs. 7.10–7.12). Note that head and work have to be interchanged in these formulae for use with turbines.

Observe the very large efficiency correction with a 1/10 scale ratio.

A: $H_m = 6.0 \, m$, $Q = 0.292 \, m^3/s$, $n = 333.8$ rpm, $\eta = 0.780$, $P = 13.42 \, kW$.

7.7.4 A throttle valve in the discharge pipe of a pump is a traditional way of flow rate adjustment. Another traditional way is fitting a bypass line to the pump connecting the discharge side to the suction side with a valve in the bypass line. The net flow rate of the pump is then reduced by circulating part of the flow rate. Determine the net characteristic of the pump combined with the valve in both cases. Observe that with the valve in series the net head is mainly reduced for high flow rate (the net characteristic is steeper) while with the valve in parallel the net head is mainly reduced for low flow rate (the net characteristic is flatter).

7.7.5 Calculate flow rate and total pressure rise in the design point of the fan of Figs. 7.20 and 7.21 with the methods of Chap. 3. Compare with the target values $Q = 0.80 \, m^3/s$ and $\Delta p_0 = 3500$ Pa at 2900 rpm. Take density equal to 1.20 kg/m^3.

Consider that, at the design operating point, the inlet flow of the rotor is aligned with the blade direction after the flow has entered. Determine the design rotor flow rate under this condition.

Take into account that the deceleration in a rotor channel cannot be stronger than 0.70. Interpret that this applies to the flow just after rotor entry (w_1^b) and the flow just before rotor exit, so before slip (w_2^b). Observe that the deceleration limit is reached. Take as model for the rotor exit flow a core flow with uniform flow velocity (w_2^b) next to a zero flow velocity zone (jet and stagnant wake model). Assume that Pfleiderer's work reduction factor formulae, with geometrical data as input, apply to the core flow (Eqs. 3.24 and 3.25). Determine the rotor work with this assumption.

Consider the dump deceleration at the entry to the volute as between the core flow in the rotor channels and a uniform flow immediately downstream of the rotor exit, filling the full width of the volute. Due to the partial filling of the rotor channels, the dump loss is considerably larger than with full through-flow.

Calculate the leakage flow between the volute and the rotor inlet assuming that the total pressure at the volute side of the gap is the static pressure at the volute entrance after dump deceleration (station 2') and that the pressure at the rotor side of the gap is the inlet pressure of the rotor (station 1). Ignore friction in the rotor and contraction in the leakage flow. Assume that the gap width is 2 mm.

Calculate the incidence loss at the entrance to the volute, assuming that angular momentum is conserved between the volute entry, after tangential deflection due to incidence (station 2''), and the volute exit (station 3). The velocity distribution at the volute exit follows from the net flow rate.

Determine the total pressure rise of the fan taking into account the dump loss and the incidence loss at the volute entrance. Neglect the other losses. This leads to some overestimation of the total pressure rise. Take as density $\rho = 1.20$ kg/m³.

A: $Q_{rotor} = 0.853$ m³/s, $\varepsilon_{Pfleiderer} = 0.816$, $\Delta W = 3290.1$ J/kg, $q_{irr}^{dump} = 180.27$ J/kg, $(\Delta p / \rho)_{gap} = 2086.5$ J/kg, $Q_{leak} = 0.0763$ m³/s, $\eta_v = 0.911$, $Q_{net} = 0.777$ m³/s, $q_{irr}^{incid} = 4.75$ J/kg (very low), $\Delta p_0 = 3726$ Pa, $\eta_i = 0.944$.

The calculated flow rate is 3% lower than the target value. The calculated pressure rise is 6.5% larger than the target value, but the inlet loss in the rotor eye and friction losses in the rotor, the scroll and the diffuser are not taken into account. The internal efficiency should decrease until 0.887 for obtaining the target value of the total pressure rise. So, presumably, with ignored losses subtracted, the calculated pressure rise will still be somewhat larger than the target value. The calculations thus suggest that the couple of target performance data, $Q = 0.8$ m³/s and $\Delta p_0 = 3500$ Pa, are approximately reached in reality for a flow rate somewhat larger than the one for incidence-free rotor entry (see next exercise).

7.7.6 Calculate the flow rate and total pressure rise in the design point of the fan of Figs. 7.20 and 7.21 from the measured values at 2750 rpm (Chap. 5). Scale the laboratory test results from 2750 to 2900 rpm with kinematic similitude, assuming that efficiency is unchanged.

A: The scaled measured values of flow rate and total pressure rise at the best efficiency operating point are 0.725 m³/s and 3590 Pa. The best efficiency flow rate is always somewhat lower than the design flow rate. With the flow rate at the design operating point estimated 10% higher than at optimum efficiency, the estimated design flow rate is 0.80 m^3/s. The corresponding interpolated value of the total pressure rise is 3470 Pa. The obtained values of flow rate and total pressure rise are very near to the target values $Q = 0.8$ m^3/s and $\Delta p_0 = 3500$ Pa.

7.7.7 Determine the nondimensional groups that are relevant for the performance analysis of a hydraulic turbine. In the present chapter, all reasonings are on power receiving machines. With such machines, the driving motor imposes the rotational speed. The input to the turbomachine is thus the blade speed. The results by the turbomachine are the flow rate and the head. The natural coefficients for description of the performance are thus the flow coefficient, which is the ratio of a through-flow velocity to the blade speed and the work coefficient, which is the ratio of the rotor work to the blade speed squared.

A: The input to a hydraulic turbine is the head and the results are the rotational speed and the flow rate (or power). From the head follows the spouting velocity $v_0 = \sqrt{2g H_m}$, which is the jet velocity that would be obtained by the head imposed to a nozzle without losses.

The natural performance coefficients are thus the ratio of the blade speed to the spouting velocity (called speed ratio) and the ratio of a through-flow velocity to the spouting velocity. A performance map of a hydraulic turbine is thus often represented by lines of constant vane angle or blade angle and efficiency contours in a plane with coordinates $(\Omega D)/v_0$ and $Q/(D^2 v_0)$.

References

1. Fuchslocher/Schulz (1967) Die Pumpen, 12th edn. Springer. no ISBN
2. Balje OE (1981) Turbomachines: a guide to design, selection and theory. Wiley, ISBN 0-471-06036-4
3. Dubbel H (2001) Taschenbuch für den Maschinenbau, 20th edn. Springer. ISBN 3-540-67777-1
4. Dietzel F (1980) Turbinen, pumpen und verdichter. Vogel Verlag. ISBN 3-8023-0130-7
5. Pfleiderer C (1961) Die Kreiselpumpen für Flüssigkeiten und Gase, 5th edn. Springer. no ISBN
6. Pfleiderer C, Peterman H (1991) Strömungsmaschinen, 6th edn. Springer. ISBN 3-540-53037-1
7. Hydraulic Institute standards for centrifugal, rotary and reciprocating pumps (1983) Hydraulic Institute, New York
8. Casey MV (1985) The effects of Reynolds number on the efficiency of centrifugal compressor stages. J Eng Gas Turbines Power 107:541–548
9. Strub LA, Casey MV et al (1987) Influence of the Reynolds number on the performance of centrifugal compressors. J Turbomach 109:541–544
10. Adams T, Grant C, Watson H (2012) A simple algorithm to relate measured surface roughness to equivalent sand-grain roughness. Int. J Mech Eng Mechatron 1:66–71
11. Eck B (1972) Ventilatoren, 5th edn. Springer. ISBN 3-540-05600-9

Chapter 8
Pumps

Abstract The fundamentals of pump operation were treated in the foregoing chapters. Three particular aspects are discussed in the present chapter. The first is evaporation of the fluid when the pressure inside the pump becomes lower than the vapour pressure. Cavities with vapour then emerge and the phenomenon is generally described as cavitation. The second concerns starting up. Priming is necessary. This is filling the pump and the suction pipe with the liquid to be pumped. Some pumps are self-priming. The third topic is the intersection of the pump characteristic with the load characteristic, which mostly has a large static part. The intersection does not necessarily result in a stable operating point. Cavitation, priming and stability are discussed in the present chapter. Further topics are shaping of the components, constructional aspects and pump examples. The focus is on centrifugal pumps, because this pump type is the most common one.

8.1 Cavitation

8.1.1 Cavitation Phenomenon and Cavitation Consequences

Cavitation occurs when a liquid evaporates locally with generation of a vapour cavity due to static pressure decrease below vapour pressure. The vapour pressure of a liquid depends strongly on the temperature and it may be very low at atmospheric temperature. At 15 °C, the vapour pressure of water is about 2 kPa. It is nevertheless possible that such a low pressure is attained in the entrance zone of a pump. The intervening phenomena are decrease of static pressure due to suction height, pressure drop by losses and pressure decrease by flow acceleration.

Cavitation-generated vapour bubbles are transported by the flow and condense at places where the static pressure raises above the vapour pressure. Above some minimum bubble size, this occurs in a fast collapse, termed *implosion*. The colliding surfaces of the bubble generate strong liquid jets impacting on material walls. The material surface crumbles locally and pits are formed that gradually grow into perforations. The phenomenon is termed *pitting* or *cavitation erosion*. Fatigue-resisting materials resist cavitation erosion the best. When cavitation is unavoidable, it is advisable to manufacture the pump from a stainless steel type, e.g. 18/8 chromium-nickel steel. But cavitation leading to erosion is mostly avoided.

Ship propellers are also exposed to cavitation, but vapour bubbles can be kept away from material surfaces. So cavitation does not necessarily cause erosion.

Vapour bubbles hardly affect the flow with incipient cavitation (beginning of cavitation), so that the performance characteristics of a pump do not change. Developed cavitation, i.e. with a large vapour zone, affects the flow. The through-flow velocity is locally enlarged by the obstruction by the vapour zone and the flow direction is changed. Both effects increase the losses. Pump efficiency and head decrease with developed cavitation.

8.1.2 Types of Cavitation

With flow over free-standing hydrofoils or over convex walls, incipient cavitation mostly occurs at some distance from the walls. Several types may be distinguished: bubble cavitation, sheet cavitation, cloud cavitation and vortex cavitation. Bubble cavitation refers to individual bubbles within the fluid. These bubbles originate around nuclei of air mixed with vapour. Bubbles joining together to a larger zone are termed sheet or cloud cavitation, depending on whether the zone adopts the form of a thin surface or rather that of a full space. Vortex cavitation arises within the low-pressure core of tip vortices. Bubbles join then to lines following the vortex core. Cavitation in the tip vortex of a ship propeller blade is an example.

With cavitation by bubbles, whether or not joining in lines, sheets or clouds, the presence of cores of gas and vapour is crucial. Gases, as air, only dissolve very little in water. Cores, called nuclei, form due to surface tension. Inside a core, gas is mixed with vapour from the fluid. The pressure within a core exceeds the pressure in the surrounding fluid because of surface tension. The smaller the radius of a (spherical) core, the larger the pressure difference is. Cavitation starts when the cores expand due to evaporation of the fluid at their border surfaces. Therefore, due to surface tension, the pressure around a core must be below the vapour pressure, as vapour pressure, by definition, is the pressure at which the liquid evaporates in equilibrium with the vapour, i.e. with a very large contact surface, so without the effect of surface tension. The required pressure drop below vapour pressure, at 20 °C in water, is approximately 1 bar with core radius 1 μm, 0.1 bar with core radius 10 μm, 0.01 bar with core radius 100 μm [1].

In the presence of only small cores, tension must thus be exerted on the liquid before cavitation bubbles can form. Cores are rather large in industrial water, with the radius of the largest cores typically in the 30 μm order. The required pressure decrease below the vapour pressure is then not very large, about the magnitude of the vapour pressure itself at 20 °C. One could say that cavitation starts when the liquid pressure drops to zero pressure. The exact pressure within the liquid at which cavitation occurs is termed the critical pressure (critical regarding cavitation). This value applies under static conditions. Inertia and viscous effects intervene. These generate delay between the location where the critical pressure is reached and the location of bubble formation.

The equation describing the radius of a bubble, as a function of surface tension and viscous forces, is known as the Rayleigh-Plesset equation [1, 2]. It allows the calculation, with numerical methods, of bubble generation, if the critical pressure is known. The liquid state determining the critical pressure is termed *water quality*. It mainly concerns the size of the largest cores and the thermodynamic state, first of all the temperature. The critical pressure is difficult to predict, which makes the calculation of bubble cavitation (and derived forms) a delicate matter.

The study of incipient cavitation (bubble cavitation) is not crucial with pumps, as incipient cavitation neither causes erosion nor performance deterioration. For these phenomena to occur, a significantly large cavitation zone is required. Inside a pump, a large vapour zone is always attached to the suction side of a blade and, therefore, the term *attached cavitation* is used. With an attached cavitation zone, it may be assumed that the pressure within and around the zone is vapour pressure with no interfering effect of surface tension. The cavitation thus becomes independent of the water quality.

Henceforth we assume that cavitation occurs as soon as the local pressure drops to the vapour pressure. This is justified for developed cavitation. Further, we assume that the vapour pressure in the cavitation zone is the same as at the pump entry. This is a justified assumption for a cold fluid. For a fluid near evaporation, the vapour pressure in the cavitation zone may be lower than at the pump entrance because the latent heat necessary for vaporisation lowers the local temperature.

8.1.3 Blade Pressure Distribution in Presence of Cavitation

Figure 8.1 sketches pressure distributions in absence and in presence of cavitation in the front zone of the suction side of a rotor blade of a centrifugal pump for flow with zero incidence.

The leading edge is often an ellipse with a 2/1 radius ratio (it is circular in Fig. 8.1), followed by a constant-thickness zone. In absence of cavitation, the minimum value of the pressure coefficient is then approximately $(C_p)_{min} = -1$ for zero incidence. With a better hydrodynamic profile, which is possible with large centrifugal pumps, and always possible with axial pumps, the acceleration at the leading edge, and so the pressure drop, may be lower, thus lowering the value of $-(C_p)_{min}$ for zero incidence to about 0.5.

Positive incidence causes additional acceleration on the suction side, increasing $-(C_p)_{min}$ by 0.2 to 0.3 per degree of incidence until a limit value is attained when the flow separates, which occurs at 5° to 6° incidence. With negative incidence, a suction zone may occur on the pressure side and $-(C_p)_{min}$ may increase rather strongly. The value of $-(C_p)_{min}$ depends thus strongly on the angle of incidence.

Fig. 8.1 Suction side
pressure distribution on the
leading edge zone of a blade
profile of a centrifugal pump;
i: in absence of cavitation (or
with incipient cavitation); e:
with a cavitation zone that
large that erosion occurs; 3:
with a cavitation zone
causing 3% head drop;
sketch with constant pressure
in the oncoming flow and
three levels of vapour
pressure

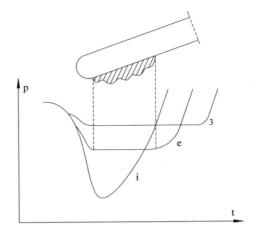

Figure 8.1 also sketches pressure distributions in the presence of a cavitation zone.
There is vapour pressure within the zone, levelling out the minimum in the pressure
distribution. A vapour zone that is sufficiently large for causing erosion typically
has a $-(C_p)_{min}$ plateau value around 0.3. When the cavitation zone further expands
until it obstructs the through-flow, the head and efficiency decrease. The $-(C_p)_{min}$
for performance deterioration is lower than the value for start of erosion, but this
value may largely vary [3].

The cavitation zone in Fig. 8.1 is highly unsteady in reality. Fluid parts continuously enter the zone and evaporate and fluid parts also continuously leave the zone
and implode.

8.1.4   Cavitation Assessment: Required Net Positive Suction Height

Two concepts are used for assessing cavitation risk and cavitation degree. The *cavitation number* is a pressure coefficient formed with the pressure and the velocity of the
oncoming flow, like a pressure coefficient for description of the pressure distribution
on a blade:

$$\sigma = \frac{p_1 - p_v}{\frac{1}{2}\rho w_1^2}. \tag{8.1}$$

The vapour pressure is denoted with p_v. Pressure and velocity just upstream of the rotor entry are p_1 and w_1. The cavitation number is an internal quantity, which, strictly, cannot be determined by external measurements. But, p_1 and w_1 may be estimated. The pressure at the pump entry may be used as an approximation of p_1, if there is only a small pressure change from pump entry to rotor entry due to inlet losses and flow acceleration. For an inlet flow without pre-whirl, the velocity w_1 may be determined by

$$w_1^2 = u_1^2 + v_1^2, \tag{8.2}$$

where u_1 is the blade speed at rotor entry and v_1 follows from the flow rate and the through-flow area.

The pressure p_1 for which there is incipient cavitation on the blade determines a critical value of the cavitation number:

$$\sigma_i = \frac{(p_1 - p_v)_i}{\frac{1}{2}\rho w_1^2}. \tag{8.3}$$

The subscript i indicates incipient cavitation. The value of σ_i may be determined from the pressure distribution on the rotor blade. The pressure coefficient is

$$C_p = \frac{p - p_1}{\frac{1}{2}\rho w_1^2} \quad \text{thus} \quad \sigma_i = -(C_p)_{\min}. \tag{8.4}$$

As argued in the previous section, incipient cavitation is not crucial. We study erosion cavitation. The corresponding value of the cavitation number is $\sigma_e \approx 0.3$.

The concept of NPSH (*net positive suction head*) is used as an externally determinable quantity. NPSH is the difference between the total pressure at the suction flange and the vapour pressure, expressed in height. When the difference is expressed in energy, the term NPSE (net positive suction energy) is used. When determining the pressure at the suction flange, it is corrected to the centre of the rotor entry. NPSH becomes then approximately independent of the position of the pressure reading point.

$$NPSE = gNPSH = \frac{p_s}{\rho} + \frac{v_s^2}{2} - \frac{p_v}{\rho}. \tag{8.5}$$

The subscript s refers to the suction flange. When defining NPSH, the total pressure is used, because the dynamic pressure is also available as a buffer against cavitation.

In order to avoid cavitation, NPSH has to take a minimum positive value, denominated NPSH-required: $NPSH_r$. The subscript r stands for *required*. This minimum value follows from the sum of the pressure changes between the suction flange and the location with minimum pressure on the rotor blades. The total pressure at the rotor entry is lower than the total pressure at the pump entry, due to losses within the entry zone:

$$\frac{p_s}{\rho} + \frac{v_s^2}{2} = \frac{p_1}{\rho} + \frac{v_1^2}{2} + \xi \frac{v_1^2}{2}. \tag{8.6}$$

The loss coefficient ξ is low, typically about 0.1.

The pressure drop between the rotor entry and a location on the rotor blade is determined by the pressure coefficient:

$$-C_p = \frac{p_1 - p}{\frac{1}{2}\rho w_1^2}. \tag{8.7}$$

When cavitation occurs, the pressure equals the vapour pressure in the cavitation zone. We may then replace $-C_p$ by σ according to (8.1).

The NPSH value required to avoid cavitation has to satisfy

$$\text{NPSH} > \text{NPSH}_r \quad \text{with} \quad gNPSH_r = (1 + \xi)\frac{v_1^2}{2} + \sigma \frac{w_1^2}{2}. \tag{8.8}$$

The required NPSH is always positive and this feature is expressed by the terminology. The highest, and thus most critical value of NPSH$_r$, is attained for the most unfavourable combination of v_1 and w_1.

From now on, we reason with a pump with the entry in line with the shaft, as in Fig. 1.3 for an axial pump and Fig. 1.13 for a centrifugal pump. The reasoning is the same for a mixed-flow pump (see the further Fig. 8.13). The most unfavourable combination of v_1 and w_1 is then at the rotor shroud or the pump casing.

We note the highest value of NPSH$_r$ by

$$gNPSH_r = \lambda \frac{v_1^2}{2} + \sigma \frac{w_1^2}{2}, \tag{8.9}$$

with $\lambda \approx 1.1$. The velocity v_1 just upstream of the entry of the bladed rotor part may be estimated from the flow rate by $Q = k(\pi d_1^2/4)\, v_1$. The diameter d_1 is the tip diameter of the bladed rotor part at the rotor entry. The factor k takes into account that the delivered flow rate is somewhat smaller than the rotor flow rate and that the rotor flow rate calculated by the area $\pi d_1^2/4$ and the velocity v_1 on this entire area is a slight overestimation. Figures like 1.3, 1.13, 7.5, 8.13 and many later figures in this chapter show that the factor k does not vary much and is in the order of 0.80 to 0.90, except for the very wide radial rotors used for pumping slurry or solid materials (Fig. 8.28), where it is larger than unity. This means that there is generally a slight acceleration between the entry of the rotor eye and the entry of the rotor blade channels (see Sect. 8.1.5). The value of w_1 follows from the velocity triangle at the entry of the bladed part of the rotor. In particular, in absence of pre-whirl:

$$w_1^2 = u_1^2 + v_1^2, \quad \text{with} \quad u_1 = \frac{\Omega d_1}{2} \quad \text{and} \quad v_1 = \frac{4}{k\pi}\frac{Q}{d_1^2}. \tag{8.10}$$

Fig. 8.2 Variation of NPSH$_i$, NPSH$_e$ and NPSH$_3$

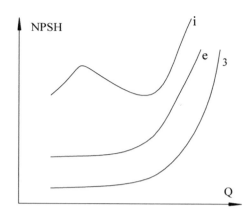

For the study of incipient cavitation, the cavitation number σ_i must be introduced in Eq. (8.9). This number varies strongly with incidence. The NPSH-value corresponding to start of erosion is denoted by NPSH$_e$. In practice, the effect on performance is mostly determined with a criterion of 3% head drop due to cavitation. We denote the corresponding value by NPSH$_3$.

Figure 8.2 sketches NPSH$_i$, NPSH$_e$ and NPSH$_3$ as functions of the flow rate for fixed rotational speed [4]. The deviating variation of NPSH$_i$ at very low flow rate is due to the recirculating flow at the pump entry (Fig. 7.9).

Manufacturers publish curves of NPSH$_3$, as this quantity can be measured. The value of NPSH$_i$ can only be determined optically, with the pump made observable through transparent parts, especially manufactured for this purpose. Determination of NPSH$_e$ requires long duration operation of a pump and damage observation.

A pump user should know that the NPSH$_e$ value is larger than the NPSH$_3$ value. Estimation of NPSH$_e$ thus requires a safety margin applied to the value of NPSH$_3$. We discuss this margin in a later section.

8.1.5 Optimisation of a Rotor Entry

With operation in the design point, the coefficients λ and σ in the expression (8.9) for NPSH$_r$ have a fixed value. The NPSH$_r$ may then easily get minimised.

First, we remark that the NPSH$_r$ cannot be improved by pre-whirl. With unaltered values of λ and σ, the NPSH$_r$ would lower somewhat with a moderate pre-whirl by lowering the magnitude of the relative velocity, but the value of λ increases by the presence of inlet guide vanes, which compensates the possible gain.

In absence of pre-whirl, Formula (8.9) takes the form

$$2gNPSH_r = ax^{-2} + bx,$$

with

$$x = d_1^2, \quad a = (\lambda + \sigma)\left(\frac{4}{k\pi}\right)^2 Q^2, \quad b = \frac{\sigma}{4}\Omega^2.$$

The optimum is for $x_o^3 = 2a/b$. Thus:

$$(d_1)_o = \left(8\frac{\lambda + \sigma}{\sigma}\right)^{1/6}\left(\frac{4}{k\pi}\right)^{1/3}\left(\frac{Q}{\Omega}\right)^{1/3}. \tag{8.11}$$

The corresponding NPSH$_r$ value is

$$(gNPSH_r)_o = \frac{3}{4}(2a)^{1/3}b^{2/3} = \frac{3}{4}[2(\lambda + \sigma)]^{1/3}\left(\frac{4}{k\pi}\right)^{2/3}\left(\frac{\sigma}{4}\right)^{2/3}Q^{2/3}\Omega^{4/3}. \tag{8.12}$$

With optimisation to erosion with $\lambda \approx 1.1$, $\sigma \approx 0.3$, and $k \approx 0.85$:

$$(d_1)_o \approx 2.10\left(\frac{Q}{\Omega}\right)^{1/3} \quad \text{and} \quad (gNPSH_r)_o \approx 0.25\, Q^{2/3}\Omega^{4/3}. \tag{8.13}$$

This result allows the definition of a specific speed, termed *suction specific speed*:

$$\Omega_{ss} = \frac{\Omega\sqrt{Q}}{(gNPSH_r)^{3/4}}. \tag{8.14}$$

With (8.13), the optimum suction specific speed is about 3 for $\lambda \approx 1.1$, $\sigma \approx 0.3$, and $k \approx 0.85$. The shape of the velocity triangle at the rotor entry is fixed for minimum NPSH$_r$. From (8.13) it follows that the corresponding u_1 and v_1 are

$$u_1 = \frac{\Omega d_1}{2} \approx 1.05\, Q^{1/3}\Omega^{2/3} \quad \text{and} \quad v_1 = \frac{4}{k\pi}\frac{Q}{d_1^2} \approx 0.34\, Q^{1/3}\Omega^{2/3}, \tag{8.15}$$

and

$$(\tan\beta_1)_o = -(u_1/v_1)_o = -\sqrt{2\frac{\lambda + \sigma}{\sigma}} \approx -3.05, \quad \text{from which} \quad \beta_1 \approx -72°.$$

The obtained blade angle is at the casing or the shroud. On the average radius, the magnitude of the angle may be lower (if there is radius variation). We observe that the optimum blade angle does not depend on the factor k.

The pump entry may also be optimised with respect to losses. A similar reasoning as with NPSH applies, but with λ and σ representing loss coefficients. For $\lambda = 0.1$ and $\sigma = 0.1$ (losses at the inlet of the rotor), the blade angle is $\beta_1 \approx -63°$. The corresponding values of diameter and velocity components are:

$$(d_1)_o \approx 1.80 \left(\frac{Q}{\Omega}\right)^{1/3}, \quad u_1 \approx 0.90 \; Q^{1/3}\Omega^{2/3}, \quad v_1 \approx 0.46 \; Q^{1/3}\Omega^{2/3}. \quad (8.16)$$

It is sometimes argued that optimum internal efficiency of a radial pump is attained by minimum relative velocity w_1 at the mean entry diameter [5, 6], as with radial fans. The corresponding values are then $\lambda = 0$ and $\sigma = 1$ and the blade angle becomes $\beta_1 \approx -55°$. The angle at the blade tip may then be close to $\beta_1 \approx -60°$.

But optimisation of a pump is more complex than that of a fan, and none of the obtained values of β_1 is a universal result. But, clearly, optimisation for efficiency leads to a higher through-flow velocity. The rotor entry is thus narrower than with optimisation for cavitation. Optimisation for efficiency is possible with uncritical cavitation. Examples are the second and following stages in a multistage pump. So, often, the entry of the first stage of a multistage pump is somewhat wider than that of the other stages.

The results allow to estimate possible through-flow velocities in a centrifugal pump rotor. We take as an example $u_2 = 30$ m/s (e.g. obtained with $d_2 = 200$ mm and $n = 2900$ rpm; see Exercise 1.9.8). With moderately backward curved blades, v_{2u} may be about 18 m/s. With a diameter ratio 0.4, $u_1 = 12$ m/s. To $\beta_1 = -72°$ corresponds $v_1 \approx 3.9$ m/s. This flow velocity is higher than what is typically allowed in a suction pipe, being 2 to 2.5 m/s. So, either the magnitude of β_1 has to be chosen larger than 72° (but it cannot be much larger for constructional realisation) or a converging duct part upstream of the pump is required. The through-flow velocity (meridional velocity component) may then be increased after the first stage in a multistage pump. To the blade angle $-63°$ corresponds the through-flow velocity of 6.1 m/s. A through-flow velocity of 6 m/s is somewhat too large, taking into account that the acceptable flow velocity in a discharge pipe is at maximum 3 to 3.5 m/s. The observation demonstrates that, mostly, a blade angle magnitude lower than 63° at the blade tip cannot be chosen. Thus, Eqs. (8.13) and (8.16) give good estimates of the possible values of the inlet diameter of a centrifugal pump rotor.

The example also shows that deceleration between the inlet of the eye of the rotor (velocity v_0) and the inlet of the blades of the rotor (velocity v_1) is not a realistic option. One might deduce from Eq. (8.9) that deceleration would be beneficial for reduction of the NPSH$_r$. This is not a correct conclusion because deceleration causes increase of the entrance loss, so a higher value of λ. In most pump rotors there is a slight acceleration in the entry zone (see Figs. 1.3, 1.13, 8.13). So, there is a big difference with rotors of centrifugal fans.

A high through-flow velocity is possible in case of direct suction as with the axial pump in Fig. 1.3. But, for limiting the trough-flow velocity of an axial pump with high peripheral rotor speed, it may be necessary to choose the diameter somewhat larger than by Eq. 8.13 and the axial velocity somewhat smaller than by Eq. 8.15 (see Exercises 8.9.9 and 8.9.10).

8.1.6   Net Positive Suction Head of the Installation

The total pressure at the suction flange results from the pressure within the suction reservoir decreased by the pressure drop by the suction height and the head loss within the suction pipe. The work equation between the suction reservoir and the suction flange is

$$\frac{p_r}{\rho} + 0 + 0 = \frac{p_s}{\rho} + \frac{v_s^2}{2} + g H_s + q_{irr,s}.$$

The pressure in the reservoir, above the liquid surface, is represented by p_r, which mostly is atmospheric pressure. H_s is the suction height, positively calculated when the centre of the rotor inlet is above the liquid level of the suction reservoir.

The NPSH supplied to the pump by the installation is called the available value and is denoted by $NPSH_a$.

Its value is

$$g N P S H_a = \frac{p_s}{\rho} + \frac{v_s^2}{2} - \frac{p_v}{\rho} = \frac{p_r}{\rho} - \frac{p_v}{\rho} - g H_s - q_{irr,s}. \qquad (8.17)$$

With atmospheric pressure in the reservoir ($p_r = 101.3$ kPa) and water at 15 °C ($p_v = 1.7$ kPa), $(p_r - p_v)/\rho g$ represents about a 10 m height. The geometric suction height thus cannot be larger than 10 m.

8.1.7   Avoidance of Cavitation

In order to avoid cavitation, the available NPSH has to be larger than the required NPSH. Generally, the available NPSH becomes critical for high flow rate and high rotational speed. An estimate of critical flow rates follows from Eq. 8.13, which produces a required NPSH of 2.60 m for $Q = 40$ m³/h combined with 2900 rpm and for $Q = 160$ m³/h combined with 1450 rpm. Taking into account suction pipe losses and some margin, the allowed suction height can be about 6 m with the obtained required NPSH. So, there is normally no problem of available NPSH.

When NPSH becomes critical, one has to rely on the data of $NPSH_3$ by the manufacturer. But manufacturers do not provide precise information on the difference between the NPSH required for avoiding erosion and the $NPSH_3$. They commonly advise to keep a safety margin of 0.5 m on the value of $NPSH_3$, but the European association of manufactures [4] warns that this margin is often not sufficient.

For estimation of the necessary margin, one may use standards. E.g., the latest version of the ANSI standard on the topic [7] lists safety factors depending on the pump type, the application field and the fluid. For cold water flow in a pressurised duct work, the recommend safety factor is 1.1, with a minimum margin of 1 m for pumps constructed according to standards (ANSI, API, EN, see Sect. 8.7) and a minimum margin of 1.5 m for non-standardised pumps. The recommended safety factor for boiler feed pumps (hot water) is 1.30. But, one always has to be careful with the safety factor. Many literature sources on pumps, e.g. [3, 8], recommend larger safety factors (about 1.25 for cold water). It is therefore advisable to take into account an estimate of the required NPSH by Eq. 8.14 or by analysis of the flow velocity at the pump entrance (see Exercises 8.9.1 and 8.9.12).

8.1.8 Increasing the Acceptable Suction Height

In case of difficulties with suction height, some obvious remedies are increasing the diameter of the suction pipe or elimination of bends in order to diminish the head loss (increasing $NPSH_a$) or choosing a pump with a lower specific speed (decreasing $NPSH_r$: Eq. 8.13: lower rotational speed for the same head). These changes raise costs: larger ducts and a larger pump.

Some pumps reach a somewhat higher suction specific speed than the usual $\Omega_{SS} \approx$ 3. The improvement is by local reduction of the relative velocity at the rotor entrance at the shroud side, obtained by a larger width at the impeller entry combined by a modified blade shape. But, the possible increase of Ω_{SS} stays limited.

In case a strong decrease of the required NPSH is necessary, a possible solution is mounting a helical rotor upstream of the radial rotor, as shown in Fig. 8.3. This helical rotor is termed an *inducer*. The blade surface of the inducer is large, allowing a significant force with a small pressure difference between pressure and suction

Fig. 8.3 Inducer upstream of an impeller (*Courtesy* Sundyne)

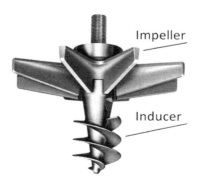

Impeller

Inducer

sides. The pressure rise by the inducer should be sufficient to avoid cavitation within the downstream centrifugal rotor.

With an inducer, Ω_{SS} may be raised to about 7. The inducer shown in Fig. 8.3 has one blade, but types exist with several blades. There are also types with the inducer integrated with the impeller, which then means that the inducer blades continue smoothly into some of the blades of the post-connected centrifugal rotor.

8.2   Priming of Pumps: Self-priming Types

The rotor work per mass unit, given by Euler's formula, does not depend on the fluid density. The corresponding pressure increase is proportional to the fluid density. The density of water is about 1000 kg/m³ and that of air about 1.20 kg/m³ under atmospheric conditions. The ratio is about 800. If a pump features a 40 m water column head, it yields 40/800 m = 50 mm water column in air. This pump can, with a rotor running in air, raise water within the suction pipe by 50 mm. Normally, this does not allow the water to reach the pump, which implies that pumping water cannot start.

In order to start operation, the pump and the suction pipe have to be filled with liquid. This preparation is called *priming*, which literally means making ready for operation. This is normally done at the first operation when there is a foot valve at the bottom end of the suction pipe. Even with a foot valve, air may enter the suction pipe and the pump at standstill due to loss of tightness of the foot valve. So, it is often necessary to remove air. Jet and water-ring vacuum pumps are sometimes applied with large pumps. Some pumps realise the air removal themselves. These are called *self-priming*. Four types are described below.

8.2.1   Side Channel Pump

This is the most classic self-priming pump (Fig. 8.4). It may be single-stage or multistage. An open rotor with many radial blades turns between two stator walls.

One wall has an inlet port (A), the other one an outlet port (B), both near the shaft. At a larger distance from the shaft, each wall features an incomplete ring-shaped channel (C or D), starting and deepening gradually at the place of the inlet port and becoming shallower and disappearing at the place of the outlet port (the pump also functions with one side channel and ports on the same side of the rotor).

With a first operation, the pump is filled with liquid. By rotor motion, a liquid ring is formed at the rotor periphery, becoming less thick in the radial direction as the side-channel is wider. In the zone where the side channel widens, the volume enclosed by the hub, the blades and the liquid ring increases, which generates suction of air through the inlet port. In the zone where the side channel becomes narrower, the enclosed volume decreases, causing air discharge through the outlet port. During

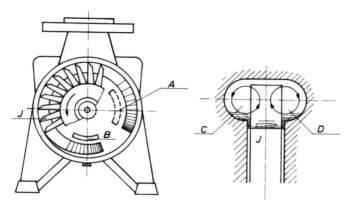

Fig. 8.4 Side channel pump

air evacuation from the suction pipe, the machine thus functions as a volumetric compressor.

When the rotor is completely filled with liquid, turbomachine operation starts. There is then a continuous connection between the inlet port and the outlet port through the side channel(s). Figure 8.4 demonstrates that a helical flow is generated within the side channels as a consequence of the centrifugal force in the spaces between the blades. The inlet (1) and outlet (2) velocity triangles during motion in and out the rotor are sketched in Fig. 8.5 for the side channel D. The triangles take into account that the outflow of the rotor is in the axial direction (blades aligned with the shaft) and that kinetic energy is transformed into pressure energy in the side channel. The consequence is that the inflow of the rotor is approximately in the axial direction.

The rotor work is

$$\Delta W = u_2 v_{2u} - u_1 v_{1u} \approx u_2^2 - u_1^2. \tag{8.18}$$

When the flow rate through the pump is smaller, the pitch of the helical motion within the side channels becomes smaller. The work (8.18) is done each time the liquid passes through the rotor. So, the head increases strongly when the flow rate diminishes. Efficiency is low, due to the short turns of the flow, leakage losses and friction losses within the water ring. The head of a side channel pump is 4 to 10 times

Fig. 8.5 Velocity triangles at the inlet and the outlet in the through-flow of the rotor of a side channel pump ($u_1 < u_2$)

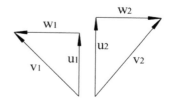

higher than that of a common centrifugal pump with the same peripheral speed, as
the water flows repeatedly through the rotor. Side channel pumps are designed for
the $0.03 < \Omega_s < 0.6$ range. Their efficiency can maximally attain about 50%. A rather
low efficiency is no drawback with small power applications, as a small centrifugal
pump does not attain a high efficiency either.

8.2.2 Peripheral Pump (Regenerative Pump)

Figure 8.6 sketches an orthogonal and a meridional section. The wet operation is
identical to that of the side channel pump and the Q-H characteristic is similarly
steep.

The capability to remove air from the suction pipe is due to splattering of liquid
within the space around the blades. This creates an air–water emulsion with a density
sufficient to be pumped. At the outlet side of the pump, air and liquid are separated.
In the pump shown, this is done by a cyclone pipe. The centrifugal force drives the
liquid to the outside. The air is led to the discharge pipe by the cyclone core. The
liquid flows back to the rotor. The cyclone is not essential. Air–liquid separation is
also realised within a sufficiently large chamber where the pumped emulsion comes
to low velocity. In contrast to the side channel pump, the peripheral pump does not
function as a compressor during air removal. The discharge pipe should thus allow
spontaneous evacuation of the air. If not, fitting a bypass line for discharging air to
the outside during priming may be necessary. The peripheral pump reaches a better
efficiency than the side channel pump.

Fig. 8.6 Peripheral pump

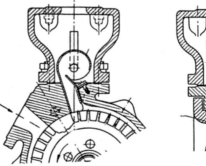
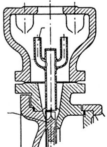

8.2.3 Self-priming Centrifugal Pump

The splattering principle with generation of an emulsion may be applied to a centrifugal pump with a classical rotor shape as well. Figure 8.7 shows two examples. A self-priming pump requires permanent presence of liquid inside the pump casing. Liquid, present at standstill, should not flow back to the suction pipe. Therefore, the suction pipe is positioned above the rotor and fitted with a non-return valve. Also, this liquid should not, at the pump start, be drained away through the discharge pipe. An air–liquid separator (A) is incorporated in the pump house for that purpose. The separator is a large space in which the pumped emulsion comes to a low velocity, allowing the air to escape from the liquid. The air leaves the pump through the discharge pipe and the liquid drops back to the bottom of the pump house and is led to the rotor. As for the peripheral pump, the discharge pipe should allow spontaneous evacuation of the air.

With the pump of Fig. 8.7 (left), liquid enters the rotor through an orifice B in the volute. Wet operation is that of a common centrifugal pump, with the exception of a small flow rate through the orifice (may be in one sense or the other). The type shown in Fig. 8.7 (right) is self-priming as well, without an orifice. This is due to the double volute. The rotor-mixed emulsion is drained through the internal volute and the separated liquid flows back through the external volute. Backflow through an orifice in the volute (Fig. 8.7, left) or the outer branch of the volute (Fig. 8.7, right) is delicate, however. The weight of the back flowing liquid should be sufficient to attain flow against the head realised by the emulsion.

Self-priming pumps are frequently used in agricultural applications, with open rotors and few blades (Fig. 8.7, right), but with a backflow orifice that directly carries the liquid upon the rotor such that the splattering is efficient. The drawback is then some circulating flow in wet operation. The pumps are suitable for contaminated liquids, but the particles should, of course, be smaller than the backflow orifice and the volute throat.

Fig. 8.7 Self-priming centrifugal pump; left: with backflow through an orifice; right: with backflow through the external part of a double volute; d indicates the discharge side

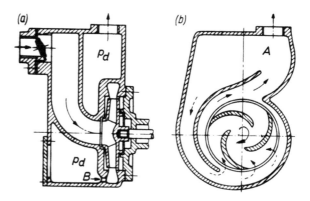

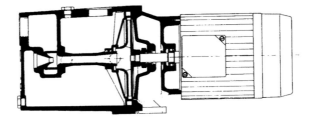

Fig. 8.8 Jet pump (former manufacturer Stork Belgium)

8.2.4 Jet Pump

This pump is by far the most used in household applications for rain or dwell water supply. A water jet pump (see Exercise 2.5.5) is internally mounted upstream of a centrifugal rotor. The water leaving the rotor runs into a pump house that is connected to the nozzle (bottom of the house in Fig. 8.8). The water jet entrains the air from the suction pipe (top left of the house in Fig. 8.8). The pumped water–air emulsion splits in the pump house. The air is forced into the discharge pipe. Wet operation is similar. The jet pump and the centrifugal rotor operate in series, which allows attaining a higher head. Cavitation cannot occur in the centrifugal rotor but may occur in the suction chamber or the mixing chamber of the jet pump.

8.3 Unstable Operation

Unstable operation may occur for flow rates lower than the flow rate at the maximum of the Q-H characteristic. When a pump delivers through a pipe without head loss to a reservoir from which a fixed flow rate is extracted, all operating points with positive flow rate at the left of the maximum are unstable (Fig. 8.9). The load characteristic is a horizontal line in that case (only static head).

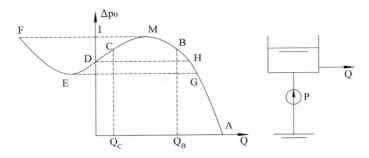

Fig. 8.9 Stable and unstable operation of a pump

When filling the reservoir, the operating point passes the right curve branch from A towards the maximum M. Assume that the flow rate delivered by the pump to the reservoir equals the flow rate extracted from it in point B. Point B is then a *stable operating point*. This may be verified by exerting a perturbation, e.g. a decrease of the flow rate extracted from the reservoir during a short period of time. The level within the reservoir rises then, which causes the flow rate delivered by the pump to decrease. This causes the level within the reservoir to decrease and the operating point to return to its original position.

Point C, at the left of the maximum, is an *unstable operating point*. Assume that the flow rate delivered by the pump to the reservoir equals the flow rate extracted from it in point C. The reaction of the pump to a perturbation is now the inverse of that in point B. When the flow rate extracted from the reservoir decreases during a small period of time, causing a rise of the reservoir level, the flow rate delivered by the pump increases and the operating point moves away from its original position. In this case, the operating point moves to the maximum M. With a short duration increase of the flow rate extracted from the reservoir, the operating point moves towards point D. As a consequence, the operating point C cannot be maintained.

With Q_C being the flow rate extracted from the reservoir, the operating point, during filling, passes the right curve branch from A to M. At the maximum of the characteristic, the flow rate delivered by the pump still exceeds the flow rate extracted from the reservoir. So the reservoir level keeps rising. One moment later, there is no longer an intersection with the characteristic for $Q > 0$. But there is an intersection with the continued characteristic for $Q < 0$. This branch represents the head by the pump when flow is forced through the pump against the normal flow sense. The operating point thus jumps to point F. This intersection is stable (may be verified by perturbation analysis). At the operating point F, the reservoir discharges through its delivery pipe and through the pump. So the reservoir level sinks. The operating point moves from F to E. The reservoir still discharges at the minimum E. The next moment there is no longer an intersection for $Q < 0$. The operating point leaps to point G. The reservoir level then rises again, which implies that the GMFE cycle keeps running and point C is never attained. If a non-return valve is provided in the suction pipe, the flow rate cannot turn negative. The cycle is then HMID.

When taking losses into account, the head imposed on the pump by the reservoir and the pipes rises with increasing flow rate. Further, the flow rate extracted from the reservoir normally increases when the level in the reservoir increases. Taking these dependencies into account, one sees that the separation point between stable and unstable operating points is not the maximum in the characteristic, but a point with a somewhat lower flow rate.

To be entirely sure that no unstable operating point can occur with a pump, the Q-H characteristic must not feature a maximum. This assumes sufficient backward curving of the blades. The slip intervenes as well. With fewer rotor blades, slip is larger and the characteristic curve is steeper. With the load characteristic commonly found in fans, i.e. a parabola through the origin, no unstable operating points occur.

A final remark concerns the shape of the cycle around an unstable operating point. In Fig. 8.9 it is assumed that the operating point jumps suddenly between branches while conserving head. In reality, the jump takes some time and the reservoir level sinks in the transition from M to F and rises in the transition from E to G. This causes the flow rate range in the cycle to be smaller than sketched in Fig. 8.9. The range decreases with decreasing capacity of the reservoir.

8.4   Component Shaping

8.4.1   Simply and Doubly-Curved Blades in Radial Rotors

The simplest rotor shape has simply curved blades (Fig. 8.10), meaning that all orthogonal blade sections are equal. Inlet and outlet angles follow from the velocity triangles. The blade shape at intermediate points should be determined so that the through-flow area of a rotor channel increases gradually and that the channel is not unnecessarily curved. The blade camber line is often a circular arc, meeting the flow directions at the inlet and the outlet, as in Fig. 8.10. Another possible curve is like a logarithmic spiral, but with $(\tan \beta)$ linearly varying as a function of the radius, according to Eq. 8.19. The equation can be integrated analytically.

$$\frac{r d\theta}{dr} = \tan \beta = \frac{\tan \beta_2 (r - r_1) + \tan \beta_1 (r_2 - r)}{r_2 - r_1}. \tag{8.19}$$

Figure 8.11 shows a rotor shape with non-orthogonal circumferential streamsurfaces due to change of width, but constant flow angles (β) on a given radius. The β-angle projected onto an orthogonal plane is β', given by

$$\tan \beta' = \tan \beta / \cos \delta, \tag{8.20}$$

where δ is the projection angle. With the inlet edge of the rotor blades on a constant radius, constant inlet velocity and constant flow angle β_1 along the edge, β_1' varies.

Fig. 8.10 Simply curved blade

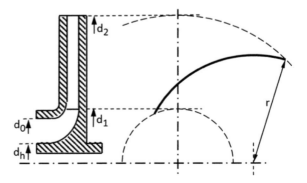

Fig. 8.11 Doubly-curved
blades due to varying width

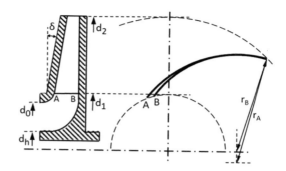

For the flow to enter without incidence, blades have to be doubly curved, which means curved in orthogonal sections and curved in cylindrical sections. Figure 8.11 represents this shape by projection of two camber lines onto an orthogonal plane.

Constant blade width, as on Fig. 8.10, is only efficient with very small specific speed, lower than about 0.25. The shape of Fig. 8.11 can be used up to specific speed values of about 0.50, but for larger specific speed, a meridional section of a rotor with curved circumferential streamsurfaces, as on Fig. 8.12, becomes necessary. The β_1-angle then also varies along the inlet edge and the blade shape is unavoidably doubly-curved. Doubly-curved blades are also required when the blades are extended into the suction eye starting from a shape as on Fig. 8.10, which means that the radius of the inlet edge is then not constant. The extension into the suction eye improves the efficiency and lowers the cavitation sensitivity, mainly because the flow velocity at the rotor entry then lowers.

Figure 8.12 shows how a skeleton surface of a rotor blade can be constructed. The meridional section is subdivided into a number of streamtubes with equal flow rate and equal static moment. The figure only shows three streamtubes, but in reality many more are used. The condition of equal static moments determines the projection of

Fig. 8.12 Meridional
section of a rotor with curved
circumferential
streamsurfaces (left) and
construction of a blade
camberline in a
streamsurface (right)

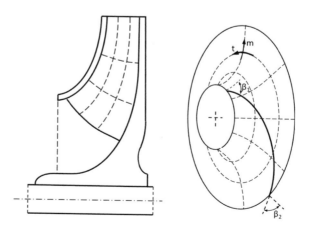

the blade leading edge on the meridional plane. On each streamsurface, a curvilinear coordinate system is defined with the m-direction in meridional planes and the t-direction in orthogonal planes (tangential direction). A camberline can then be drawn on a streamsurface with the inlet- and outlet angles β_1 and β_2, just as in a Cartesian coordinate frame. The skeleton surface results by stacking the camberlines and the blade is formed by adding a thickness. The stacking is often done with the blade trailing edge on a line parallel to the shaft, but for larger specific speed, this leads to strong leading edge curvature in the direction orthogonal to the streamsurfaces, caused by a larger meridional length at the hub. This transversal curvature can be reduced by a trailing edge on a line that forms an angle (often called the rake angle) with the shaft direction, as visible on the later Fig. 8.20.

The stacking is then done with camberlines near the shroud advanced in the rotation sense. The blade skeleton then features lean. The lean influences the flow distribution in the direction orthogonal to the circumferential streamsurfaces, similarly as with lean of blades in axial machines, discussed in Sect. 6.9.2. And, there is, of course, also the possibility of blade sweep. By combined lean and sweep, 3D optimisation of the blade shape may be done. We do not discuss this optimisation here, because the 3D effects on the flow are even more complex than with axial machines. We will briefly come back to the topic in Sect. 14.4.3 in Chap. 14 on radial compressors, but a detailed discussion of 3D shape optimisation is far beyond the possibilities of the present course text on fundamental aspects.

8.4.2 Blade Shapes of Mixed-Flow and Axial Pumps

Blade extension into the suction eye is required for larger values of the specific speed. Non-extended blades then become broader and shorter, impairing the guidance of the flow. The extension lengthens the blades. It also reduces the cavitation sensitivity, as the relative velocity at rotor entry decreases. So blades are extended into the suction eye and become doubly-curved above $\Omega_s \approx 0.5$ (blade extension may also be done for smaller specific speed). In order to obtain streamlines of comparable length, the outlet edge of a blade cannot be kept parallel to the shaft from $\Omega_s \approx 1.20$ onward (change in radius and deviation from the meridional plane). This way, a mixed-flow pump is obtained (Fig. 8.13).

Fig. 8.13 Mixed-flow pump

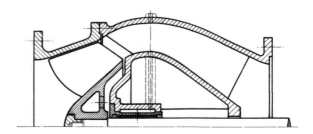

With further increase of the specific speed, it is advantageous to omit the rotor shroud. The reason is that the relative velocity becomes larger than the absolute one. So, there is less friction without shroud. Mixed-flow open-rotor pumps are applied with a specific speed of about 1.5–3. A radial diffuser followed by a volute or an axial diffuser (Fig. 8.13) may be mounted downstream of the rotor.

With further specific speed increase, the outlet diameter becomes equal to the inlet diameter, resulting in an axial rotor, followed by an axial diffuser. Axial pumps are applied for specific speeds above about 2.5. So, the application field of axial pumps partially overlaps the field of mixed-flow pumps.

8.4.3 Pump Inlet

The inlet conveys the liquid to the suction eye of the (first) pump rotor. A straight duct in line with the shaft is the most efficient, but is only possible when no bearing is required at the pump suction side or only a small bearing in a hub positioned by guide vanes. This is possible with most single-stage pumps: see Fig. 1.3 for an axial pump and Fig. 1.13 for a radial pump. Figure 8.14 is an example of a large single-stage axial pump for low head. The entry to the rotor is in line with the shaft and a bend is used as outlet. The pump has no stator vanes.

When a bearing is necessary at the entrance of a centrifugal pump, a volute-shaped suction chamber is often applied. The further Figs. 8.20 and 8.21 are examples. Figure 8.20 shows the suction chambers of a double-suction pump. A double-suction rotor is a form of internal parallel arrangement. Figure 8.21 shows the suction chamber of the first stage of multistage pump. Figure 8.15 sketches an orthogonal and a meridional section of the suction chambers of a double-suction pump.

The volute shape of the suction chambers is not complete. Part of the flow is led directly to the suction eye of the rotors. Pre-whirl by rotor entrainment may be prevented by vanes just upstream of the rotor entry.

Fig. 8.14 Axial pump with in-line entry (*Courtesy* Sulzer pumps)

Fig. 8.15 Volute-shaped
suction chambers of a
double-suction pump

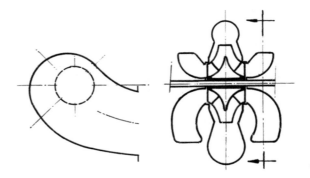

8.4.4 Pump Outlet

Fluid is conveyed from the rotor exit to the discharge pipe or to the next stage
and kinetic energy is converted into pressure energy (pressure recovery). With a
centrifugal pump, the first function is mostly realised by a volute followed by a
linear diffuser (see the further Fig. 8.17). The kinetic energy conversion may also be
done upstream of the volute with a vaned or a vaneless diffuser ring (radial diffuser).
These components are discussed hereafter.

8.4.5 Vaneless Diffuser Rings

A vaneless diffuser ring of a centrifugal pump is a space between two walls that
are orthogonal or approximately orthogonal to the shaft. The axial width (b) may
be constant or vary slightly as a function of the radius r. Assuming that no external
forces affect the fluid, the flow has constant angular momentum:

$$v_u r = cst. \tag{8.21}$$

The mass flow imposes

$$2\pi \, r \, b \, v_r = cst. \tag{8.22}$$

For $b = $ constant, it follows that

$$\tan \alpha = \frac{v_u}{v_r} = cst. \tag{8.23}$$

Within a diffuser with constant width, streamlines are logarithmic spirals. The
velocity magnitude decreases inversely with the radius. When much kinetic energy
has to be converted into pressure energy, the external diameter of the vaneless diffuser

becomes large. Due to the long streamlines, friction losses then become large. They even increase at a flow rate below the design value, because the flow becomes more tangential and streamlines lengthen. When the width increases with the radius, streamlines are even longer than with parallel walls. So, vaneless diffusers are only suitable for limited pressure recovery from kinetic energy.

8.4.6 Vaned Diffuser Rings

In a vaneless diffuser: $v_u r = cst$. A faster deceleration is achieved if v_u decreases faster. This requires tangential forces on the fluid by vanes (Fig. 8.16). The vanes are mounted more radially than the streamlines with vaneless diffusers. Two nearby vanes only overlap over a limited length.

Deceleration mainly occurs within the widening section BC of the channels formed by two adjacent vanes. The wall AB, constituting the supply to the overlap section, principally is a logarithmic spiral. Similarly, the wall CD is of logarithmic spiral form. The number of vanes should be sufficiently large such that the diverging section (BC) can be made with a limited radial extension of the vane ring.

The realisable flow deceleration is quite strong. E.g., depending somewhat on the depth and the depth increase in the flow sense, the widening channel on Fig. 8.16 can have an area ratio of about 1.85 and a length to entry diameter ratio of about 3. The velocity reduction is then about 0.55 and the corresponding pressure recovery about 0.65, according to Fig. 2.29 (the ideal value is about 0.70).

When operating in off-design conditions, the flow angle at the entry of the diffuser differs from the vane angle, with flow separation as a possible consequence. With a high flow rate the fluid must accelerate in order to pass through the throat area. So, a pressure drop occurs upstream of the throat position B. The pressure drop may be stronger than the pressure rise downstream of B. So, it may happen, from a certain flow rate onward, much higher than the design flow rate, that the vaned diffuser generates a pressure drop instead of a pressure rise.

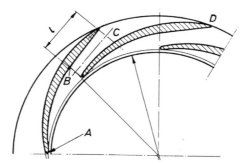

Fig. 8.16 Vaned diffuser ring

8.4.7 Volute (Spiral Case)

A volute collects the liquid leaving the rotor and realises velocity reduction.
Figure 8.17 sketches volutes fitting immediately to the rotor. There is no diffuser ring
between the rotor exit and the volute entry. This type is frequently used with single-
stage pumps. If an internal diffuser ring is mounted (vaneless or vaned), the following
discussion of the volute stays valid, however. Further deceleration is obtained in a
post-connected linear diffuser. This may be straight and mounted tangentially, as on
Fig. 8.17. But, it may also be curved (Fig. 8.18) such that the centre line of the pump
exit comes in the same plane as the pump shaft. With a pump entry in the direction
of the shaft, the suction and discharge pipes can then be in the same plane.

With the design flow rate, the volute imposes a uniform backpressure to the rotor.
With a lower flow rate, the pressure at the volute entrance decreases in the flow
sense and it increases with a flow rate above the design value. The backpressure is
then non-uniform, impairing the rotor efficiency. A second drawback may be a high

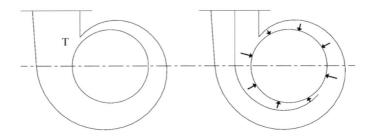

Fig. 8.17 Single volute and double volute for attenuation of radial force in off-design conditions;
tangential diffusers

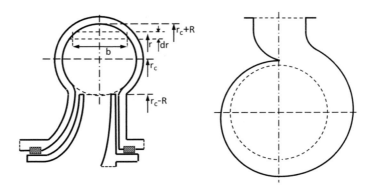

Fig. 8.18 Circular cross section of a volute (left) and curved exit diffuser for centre-line pump
outflow (right)

bearing load due to the radial force on the shaft. Both drawbacks are attenuated by two partial volutes (Fig. 8.17, right), instead of one volute.

The cross section is often circular, as sketched in Fig. 8.18. But deviating shapes are also used, and also asymmetric ones. Seen in an meridional section, the flow leaving the rotor enters as a jet and two vortex motions are formed in a symmetric volute section (this may be one in an asymmetric section). The kinetic energy associated to the radial velocity component of the rotor exit flow is almost completely dissipated, but the loss is not large because the rotor exit flow is rather tangential.

In absence of friction, the flow obeys constant angular momentum, $v_u r = cst$. A volute is preliminary calculated this way. A position along the rotor periphery is indicated by $\varphi = \theta/2\pi$, with θ the angle from the start of the volute, where the cross section area of the volute is (theoretically) equal to zero; thus $0 \leq \varphi \leq 1$. The flow rate through the volute section at the position φ is then $Q_\varphi = \varphi\,Q$, with Q the flow rate delivered by the pump. With Fig. 8.18, this flow rate is

$$Q_\varphi \approx \int_{r_c-R}^{r_c+R} v_u\, b\, dr = v_{2u} r_2 \int_{r_c-R}^{r_c+R} 2\sqrt{R^2 - (r - r_c)^2}\,\frac{dr}{r},$$

where r_c is the distance to the rotor axis of the centre of the circular volute section, R is the radius of this section, b is the local width and where $v_u r = v_{2u} r_2$ is used with r_2 the radius of the rotor. According to Pfleiderer [9], the result is:

$$Q_\varphi = \varphi\,Q = v_{2u} r_2 2\pi \left(r_c - \sqrt{r_c^2 - R^2}\right). \tag{8.24}$$

In Fig. 8.18, $r_c - R = r_2$, but this is not imperative. In particular, there may be a vaneless space between the rotor and the volute. We thus set $r_c - R = r_{2*}$ and define:

$$C = \frac{Q}{v_{2u} r_2 2\pi r_{2*}} = \frac{\eta_V v_{2m} b_2 2\pi r_2}{v_{2u} r_2 2\pi r_{2*}} = \eta_V \frac{v_{2m}}{v_{2u}} \frac{b_2}{r_{2*}}. \tag{8.25}$$

Equation 8.24 may then be written as:

$$\varphi\,C = \frac{r_c}{r_{2*}} - \sqrt{\left(\frac{r_c}{r_{2*}}\right)^2 - \left(\frac{R}{r_{2*}}\right)^2} = 1 + \frac{R}{r_{2*}} - \sqrt{1 + 2\frac{R}{r_{2*}}},$$

which is:

$$R/r_{2*} = \varphi\,C + \sqrt{2\,\varphi C}. \tag{8.26}$$

With and C according to (8.25), the basic shape of the volute is defined. To make it practical, a part of the theoretical volute has to be cut away by an entry radius larger than the rotor exit radius such that a tongue is made (T in Fig. 8.17). The clearance limits vibrations and noise by the outlet flow from the rotor channels, periodically hitting the opposite wall. An exaggerated clearance, however, causes a circulating liquid ring, lowering the efficiency.

The tongue clearance is typically chosen at 5% of the rotor radius. But the minimum width of the volute is the outlet width of the rotor, which results in a larger clearance with large specific speed. The clearance may then be lowered by changing the initial volute section shape to a rounded rectangle (see the further Fig. 8.24).

The cross section areas have to be enlarged by about 15% to compensate for boundary layers and secondary flows. But, this enlargements are not always done. The volute is then somewhat too narrow. The consequence is a shift of the operating point with maximum efficiency to a lower flow rate. The maximum efficiency is then somewhat lower but the range of good efficiency is larger.

The ratio of the average radius at the volute exit to the rotor radius (r_c/r_2) increases with increasing specific speed (larger value of C by Eq. 8.25). Figure 8.17 demonstrates that a radius ratio near to 1.50 is possible (but it is a very wide volute, as in Fig. 8.20). The velocity reduction is the inverse, thus about 0.67. Further deceleration is done in the post-connected linear diffuser. Figure 8.17 illustrates that a diffuser area ratio near to 2 is possible, combined with a length to entry radius ratio above 3. The total possible deceleration of the volute with added diffuser may thus be stronger than 0.40 (see Exercise 8.9.4), which is a very strong deceleration.

The curved diffuser in Fig. 8.18 is mainly chosen for the practical reason of a pump exit in the radial direction, but its efficiency is not lower than that of a straight diffuser. A beneficial effect is that the high-speed side of the exit flow of the volute comes in the concave part of the diffuser.

8.4.8 Return Channels

In multistage pumps, the liquid must be led from the diffuser ring outlet diameter to the smaller diameter of the suction eye of the next stage. With radial rotors, the liquid turns 180° in the meridional view, between the diffuser and the return channel. This is realised by a ring-shaped, vaned or vaneless chamber with a constant cross section (vaneless in Fig. 8.19). If this space features vanes, the liquid follows continuous channels from the diffuser (with vanes) until the suction eye of the next stage. This is hydrodynamically optimal, but it requires more complicated castings.

Remark that continuous diffuser and return channels are standard practice with mixed-flow stages in series arrangement (Fig. 8.13).

The flow in the return channels is often decelerating and they are mostly made with increasing axial width in the flow sense. The return vanes end radially or are somewhat more curved ($\varepsilon = 5°$) in order to compensate for the flow deviation.

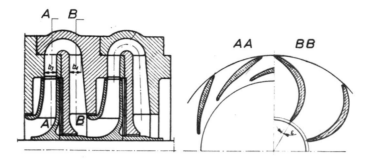

Fig. 8.19 Return channels in multistage centrifugal pumps

8.5 Internal Parallel or Series Arrangements

8.5.1 Reasons for Internal Parallel or Series Arrangements

Flow rate and head of a pump are imposed by the application. The rotational speed follows from the choice of the specific speed. In a first design stage, a specific speed is chosen for good efficiency (Fig. 7.6). Deviation from this choice is mostly required to attain a rotational speed that is suitable for a driving motor. If the specific speed becomes too high or too low, parallel or series arrangement is necessary. Internal forms of these arrangements are discussed hereafter.

8.5.2 Internal Parallel Arrangement

Internal parallel operation is normally done with 2 rotors forming a double-suction rotor, with volute-shaped suction chambers (sections are shown in Fig. 8.15) and a common spiral case outlet. Figure 8.20 shows an example (remark the rake angle of the rotor blades). The inlet to the two suction chambers is at the back of the casing and the outlet of the discharge volute is at the front. Double-suction rotors have the additional advantage that no axial force is exerted on the shaft (see Sect. 8.6.3). Double-suction pump casings are made with an axial split (plane containing the shaft). In practice, this is mostly a horizontal split, since typically these pumps are made with a horizontal shaft. Suction and discharge pipes are connected to the lower part of the casing in order to allow easy opening of the pump for maintenance.

Fig. 8.20 Axially split
casing double-suction pump
(*Courtesy* Andritz pumps)

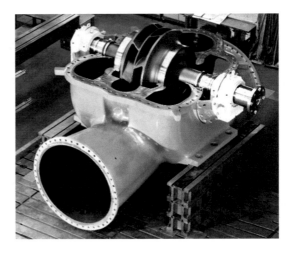

8.5.3 Internal Series Arrangement: Multistage Pumps

Internal series arrangement results in a multistage pump. Each stage encompasses a
rotor, a diffuser and return channels. Rotors should be mounted close to each other
in view of shaft rigidity. Return channels thus feature short bends. This impairs the
efficiency. In the traditional design, called a ring-section pump (Fig. 8.21), the stator
is composed of discs, axially tightened by bolts. The vaned diffuser and the return
channels are cast in a disc (Fig. 8.19). During the assembly, stator and rotor elements
are slid alternately on the shaft. With this construction, all rotors face in the same
sense so that there is a resulting axial force on the shaft. The axial force is balanced
by the force on a disc or a drum (see Sect. 8.6.4).

Ring-section pumps are made in many variants concerning the positions of the
inlet and outlet ports. E.g., on Fig. 8.21, one sees that an inlet or an outlet easily can

Fig. 8.21 Multistage
ring-section pump; inlet at
the right hand; outlet at the
left hand; balance drum at
the left of the rotor (*Courtesy*
Sulzer pumps)

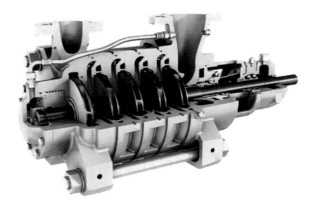

be turned by $90°$. The inlet may also be in the axial direction, with the bearing at the inlet side in a hub positioned by stationary vanes. Ring-section pumps are also made with a vertical shaft.

An alternative is a construction with an axial split, similar to the pump shown in Fig. 8.20. For balancing the axial force, rotors may then be mounted in two groups with opposing sense. For instance, a six-stage pump may be divided into two parts with three rotors. After a first group of rotors, the liquid is then collected by a volute and led by a conduit to the suction eye of the first rotor of the second group. Mostly, the connecting conduit is integrated in a casing part (realised by a passage in the cast piece). Suction and discharge pipes are connected to the lower part of the casing in order to allow opening of the pump without the need to disconnect the pipework.

8.6 Constructional Aspects

The practical realisation of a pump requires the study of a large number of aspects, as the choice of the materials, the way of construction of the components, the assembly of the components, the choice of the driving motor, the installation, and many more. Moreover, pumps are manufactured in many variants. In this section, we treat a very limited number of constructional aspects. For further study, we refer to books on practical aspects [8, 10] and to a book by Grundfos pumps that can downloaded free of costs: search for Grundfos pump handbook.

8.6.1 Shaft Sealing

Sealing may be by a compression packing or a mechanical seal. With a compression packing, rings with square cross section are pressed around the shaft with a tubular piece called the gland in an annular stator space called the stuffing box. A packing ring is a braided rope, mostly made from teflon and graphite. It is appropriate for pressures under 15 bar. A small leakage flow should be allowed for lubrication and for removing friction heat. This type of sealing is also called a gland packing or a stuffing box packing.

If the pumped liquid contains solid particles, it may be necessary to rinse the packing rings with pure liquid from an external source. This rinsing prevents solid particles from penetrating the packing. The rinse liquid is pressed into the seal through a ring with radial perforations, placed halfway the package rings. Liquid injection may be useful as well if the pump is operated with a suction pressure below atmospheric pressure, to prevent air intake. The injected liquid may then be tapped from the pressure side of the pump.

A mechanical seal consists of two rings with polished surfaces sliding over each other, one of them fixed to the stator, the other one to the rotor, pressed together by a spring. The rings are manufactured from a hard material (e.g. silicon carbide or tungsten carbide). A liquid film should be formed between the rings for lubrication. This liquid mostly evaporates partly. A mechanical seal is not completely tight.

8.6.2 Bearings

Rotational speed, required life and required rigidity determine the choice of the type of bearing. Ball bearings and roller bearings of various types are applied, always with one of the bearings for balancing the axial force (e.g. an angular contact ball bearing or tapered roller bearing). Oil-lubricated sliding bearings are used in case of high rotational speeds, large dimensions and large loads.

8.6.3 Axial Force Balancing with Single-Stage Pumps

Rotors with one-sided suction exert an axial force on the shaft, oriented to the suction side. If the force is not very significant, it may be balanced by a ball or a roller bearing, in other cases by a thrust bearing. Another solution consists in generating a hydrostatic balancing force.

Figure 8.22 shows that the rotor is subjected to rotor outlet pressure (p_p) on the back of the rotor disc and on the front of the shroud. In the rotor eye, the pressure is suction pressure (p_s). The resulting axial force is approximately

$$F_{ax} = (A_e - A_s)(p_p - p_s),\tag{8.27}$$

with A_e the eye area and A_s the cross section area of the shaft, if the shaft passes through the suction eye. In case of an overhung rotor, Eq. (8.27) becomes:

$$F_{ax} = (A_e - A_s)p_p + A_s p_{atmos} - A_e p_s.\tag{8.28}$$

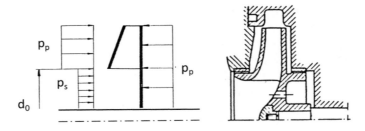

Fig. 8.22 Generation of axial force (left) and rotor balance chamber (right)

Both expressions are approximate. The pressure on the back of the rotor is not constant. Dragged by the rotor disc, the liquid outside the rotor is subjected to a centrifugal force, which generates a pressure gradient. By the turning of the flow within the suction eye, suction pressure is not constant, either.

When two rotors are mounted on the same shaft, with their suction eyes into opposite senses, the axial force is completely neutralised when both rotors operate in parallel. With rotors operating in series, a significant reduction of the resultant axial force is obtained, also outside design conditions.

For an individual rotor, the axial force may be largely balanced by a chamber made as shown in Fig. 8.22 (right). The chamber is separated from the rotor outlet area by a collar on the rotor disc, fitting with a clearance seal (small gap) to a wear ring on the casing (there is a very small leakage flow). The balance chamber is put on suction pressure by holes in the rotor disc (most common) or by an external pipe connecting the chamber to the suction part of the pump. In the first case, the leakage flow perturbs the main flow somewhat.

Remark that also a clearance seal with wear ring is used to separate the outlet area of the pump rotor from the suction eye (there is also a small leakage flow).

8.6.4 Axial Force Balancing with Multistage Pumps

Changing the suction sense of some rotors is a solution, but it complicates the construction of a multistage pump. A balance chamber may be added on each one-sided rotor. But with multistage pumps, a single balance disc (Fig. 8.23) or a balance drum (Fig. 8.21) are typical.

The balance disc (D) is mounted on the shaft in between two chambers (2 and 3). Chamber 3 is connected to the suction part of the pump by an external pipe (see Fig. 8.21). The outlet pressure of the last rotor in chamber 1 is reduced to an intermediate pressure in chamber 2 by flow through the clearance e_{12}. A pressure

Fig. 8.23 Balance disc for a multistage pump

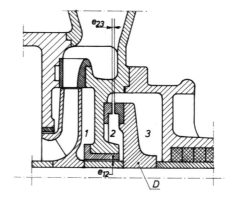

difference is maintained between chambers 2 and 3 by flow through the clearance e_{23}. The balancing disc exerts, with the pressure difference $p_2 - p_3$, an axial force in opposite sense to the axial force on the rotors.

Axial force compensation requires appropriate disc diameter and clearance dimensions. The balancing disc relieves the high-pressure sealing around the shaft, but generates leakage. The shaft is movable in the axial direction (sliding bearings or roller bearings), which makes the balance disc in Fig. 8.23 self-adjusting. The clearance width e_{23} is variable. As clearance e_{12} is fixed, p_2 decreases when the shaft moves to the right, i.e. when the axial force exerted on the rotors decreases. Wear rings on both sides of the gaps have to be replaced when these wear out. In principle, the disc can only be used with pure liquids.

The balance drum in Fig. 8.21 is not self-adjusting. The possible remaining axial force is balanced by the double-row ball bearing at the right-hand side. Remark the conduit from the suction side of the pump to the back of the balance drum.

8.6.5   Wear Rings

Clearance seals are used in pumps for separating the outlet area of the rotor from the suction eye and for forming chambers for balancing the axial force. The clearances are very small, typically of the order of 1.5–2‰ of the diameter in order to keep the leakage flow small. Although there is theoretically no metal contact, the surfaces of the clearances wear due to the high shear stress in the leakage flow. Therefore, often on both sides of the gap, rings in hard metal are placed. These rings are called wear rings. When the leakage flow becomes too large (two to three times the design value), these rings are replaced.

In many pumps, there are no wear rings on the impeller side (Figs. 8.22 and 8.23), which makes the pump more reliable. The casing wear rings may then be replaced together with the impeller. This is often practical since the time between necessary replacements may be very large, of the order of several years, and the impeller blades also wear over such a long time. It is also possible to replace a casing wear ring by a slightly thicker version and reduce the diameter of the corresponding collar on the impeller by turning on a lathe.

8.7   Pump Examples

8.7.1   Norm Pumps or Standard Pumps

Some pump types are manufactured according to norms concerning main dimensions and materials (ANSI, API, EN). An example is the norm EN733 for single-stage end-section centrifugal pumps for pressure up to 10 bar. The norm specifies diameters of

Fig. 8.24 Standard single-stage end-suction pump (*Courtesy* Sulzer pumps)

the suction and pressure side ports of the pump and the nominal rotor diameter (the diameter without rotor trimming, but the diameter does not have to be exactly the nominal diameter), from which has to be chosen as a function of the nominal flow rate and head.

Such pumps are usually called norm pumps or standard pumps. The advantage with standardised dimensions of suction and discharge flanges and base plate builds is that a pump may be replaced with a comparable type of another manufacturer without the need to change the pipework and the pump foundation.

Figure 8.24 is an example of a standardised single-stage end-suction water pump according to the norm EN733. Figure 7.15 in Chap. 7 is an example of an application field of such a type of pump. These pumps are mostly indicated with a three-number code of the form 80-65-160, where the numbers are the diameters of the suction and the pressure ports and the nominal rotor diameter, all in mm.

The pump in Fig. 8.24 has a short internal suction pipe with constant diameter, but this is not a universal feature. Due to standard values of suction port diameters, an exact match with the diameter of the rotor eye is not always possible. The internal suction channel may thus also be slightly converging or diverging.

8.7.2 Sealless Pumps: Circulation Pumps, Chemical Pumps

Circulation pumps are used with steam generators with forced circulation and domestic heating boilers. The head is generally low, but pressure and temperature may be very high. Pumps with the shaft passing to the exterior and conventional shaft sealings may have large leakage. With some applications as circulation pumps for domestic use and chemical pumps, leakage is just not allowed.

Sealless pumps are common, with liquid penetrating around the rotor of the driving electric motor (Fig. 8.25). The traditional motor is a squirrel cage induc-

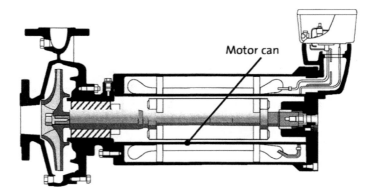

Fig. 8.25 Canned motor chemical pump (*Courtesy* Grundfos)

tion motor with stator windings isolated from the rotor with a non-magnetisable can.
The pumped liquid lubricates the sliding bearings.

Due to the can, the gap between the stator winding and the rotor is rather large,
which lowers the motor efficiency. A modern evolution is application of permanent-
magnet synchronous motors. These have higher efficiency as the gap size is then not
very critical.

Another kind of sealless pump is with a magnetic coupling formed by permanent
magnets on both ends of a split shaft (Fig. 8.26). The inner and outer magnets are
separated with a non-magnetisable can. Canned motor pumps and magnetic drive
pumps are often used for pumping chemicals, as complete avoidance of leakage is
then necessary.

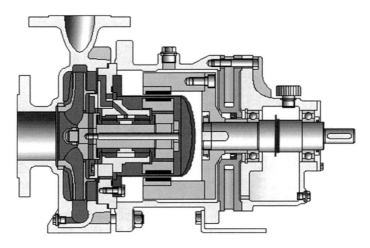

Fig. 8.26 Magnetic drive chemical pump (*Courtesy* Klaus Union)

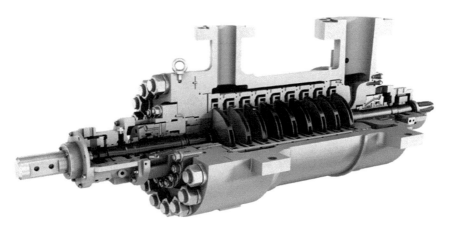

Fig. 8.27 Barrel-type high-pressure pump (*Courtesy* Sulzer pumps)

8.7.3 High-Pressure Pumps

High-pressure pumps are used, for instance, as boiler feed pumps (pressure up to 550 bar) or in chemical plants. Avoiding leakage of ring-section pumps, composed of rings tightened with external bolts may then be difficult, especially when the fluid is hot. For such applications, the ring sections are mounted in a cylindrical tube casing. Inlet and outlet parts are connected to the envelope of this casing and the rings are pressed by a cover plate. These pumps may be opened without disconnecting the pipes. They are mostly called barrel-type pumps.

Figure 8.27 is an example. The assembled rotors and stator rings are slid in the outer cylindrical casing (the barrel) and pressed by the cover at the left side. On the left side of the rotor, there is a an axial thrust balancing drum, and extremely on the left of the shaft, there is a thrust bearing.

8.7.4 Borehole Pumps

The main requirement is a small external diameter in order to lower the pump into a borehole pipe. The pump is thus always of the multistage type, even with relatively low head. Mixed-flow stages according to Fig. 8.13 are set in series. The pump may be driven by a motor mounted on the top of the borehole pipe, through a long shaft, borne in rubber bearings, lubricated and cooled by the pumped liquid. When this liquid contains solid particles, the bearings may be mounted in a separate tube, filled with pressurized water or oil. Mostly, the motor is submersible, connected directly to the pump, usually a wet rotor squirrel cage motor (see canned pump). The smallest possible diameter is about 60 mm.

8.7.5 Vertical Submerged Pumps

On ships, limitation of floor space is more stringent than limitation of height. There-fore, vertical pumps are typically used. These pumps are often submersible and placed in a high cylindrical container with a rather small diameter. So, these pumps resemble borehole pumps, but have a much larger diameter. Multistage pumps with mixed-flow rotors and with radial rotors exist. Also single-stage pumps with radial, mixed-flow and even axial rotors are employed.

The suction pipe is connected to the upper part of the container, but the suction part of the pump is near the bottom, so that the inlet pressure is higher than the pressure in the suction pipe. This may increase the available net positive suction height with several meters. A similar arrangement is used with condensate extraction pumps in thermal power stations, as the fluid is near to vapour pressure. Similar arrangements may also be used for the same reason in chemical plants.

8.7.6 Slurry Pumps

Clogging of the rotor channels must be avoided by wide passages. Impellers with one or two blades are common (Fig. 8.28). Vaned or vaneless diffuser rings are not used because of the clogging risk. A ring-shaped stator channel with constant cross section around the rotor is typical. If a volute is used, the tongue must leave sufficient clearance to allow the passage of the largest solid particles. Compression packings are rinsed with pressurised clean water in order to prevent sand entrance. The presence of solid elements in the pumped liquid decreases the pump performance. This is also the case after strong wear. Wear-resistant materials are used and several stator parts are protected by replaceable wear plates.

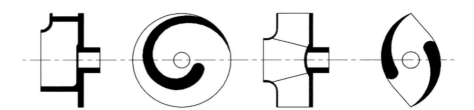

Fig. 8.28 Rotor shapes with slurry pumps

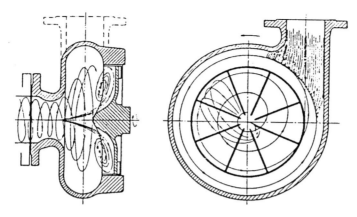

Fig. 8.29 Vortex pump

8.7.7 Pumping of Solid Materials

Solid materials, mixed with a liquid, can be pumped with special pumps. For pumping sand, gravel, clay and chalk, with dredge pumps, pumps are quite conventional ones, except that passages are wide and wear-resistant materials and replaceable wear plates are used. Efficiency and head are lower than with pure water. These pumps are combined with a cutter wheel that stirs up the solid material.

Pumping fruit, potatoes and even fish is possible with a pump with sufficiently wide impeller passages of the types of Fig. 8.28. A pump for the same purpose is a vortex pump as sketched in Fig. 8.29.

The pump has a recessed open impeller and a large distance between the blades and the stator walls. The traditional rotor has pure radial blades which create a vortex motion in a constant cross section stator.

The working principle is the same as with the side channel or the peripheral pump (Fig. 8.5). The efficiency may amount to about 60%. Modern types often have backward curved blades. This improves the efficiency as part of the rotor work comes then from functioning as a conventional centrifugal pump.

8.7.8 Partial Emission Pumps

Such a pump is meant for high head and small flow rate. The traditional type has a rotor with pure radial blades turning at high rotational speed in a constant cross section stator chamber (Fig. 8.30). The discharge is through a diffuser with small cross section area, placed tangentially to the stator chamber. The entrance of the diffuser only allows outlet flow of the rotor over a fraction of its periphery.

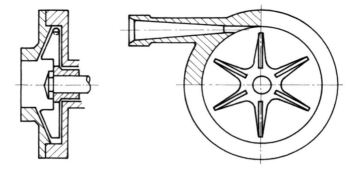

Fig. 8.30 Partial emission pump

The partial emission realises the low flow rate. Some pumps of this type are made with an inducer as shown in Fig. 8.3. A small-scale example is the pump for water spraying on the windshield of a car. Some pumps look similar but have backward curved blades as with conventional centrifugal pumps. This improves the efficiency. Their peripheral speed is much lower. These are pumps for very small flow rate, but they do not realise a very high head.

8.7.9 Pumps for Viscous Fluids

Pumps for viscous fluids do not differ strongly from pumps for low viscosity fluids. Blade passages are mostly wider and the efficiency is lower (see Chap. 7). Centrifugal pumps maintain an acceptable efficiency up to viscosities in the order of 150 mm^2/s (150 times the viscosity of water). Operation stays possible up to 500 mm^2/s. Channel impeller pumps, in the form as used for slurry pumps (Fig. 8.28, right) may even be used up to 1000 mm^2/s.

8.7.10 Vertical Propeller Pumps

Figure 1.3 of Chap. 1 is a vertical submersible axial pump for mounting in a wide discharge tube. The pump rests by its own weight on a flange at the bottom of the tube and is submerged in a pit. Such pumps are called propeller pumps. They are not purely axial because the hub diameter varies so that there is also centrifugal functioning at the hub. This increases considerably the rotor work. The rotor blades are adjustable, which allows functioning over a range of flow rate. Figure 8.31 shows a second example, where the variation of the rotor hub is larger than in Fig. 1.3. We illustrate the functioning of such pumps in the Exercises 8.9.9 to 8.9.12.

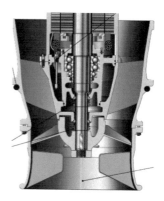

8.8 Determination of Main Dimensions and Performance Prediction

8.8.1 Main Dimensions

The specialised literature on pumps contains diagrams with characteristic geometric ratios, e.g. the ratio of the entry diameter to the exit diameter of the rotor or the ratio of the exit width to the exit diameter, as functions of specific speed. There are also diagrams of speed ratios, as the ratios of the meridional component of the rotor entry and exit velocities to the energy reference velocity $\sqrt{2g H_m}$. But, not all diagrams are strictly necessary for an estimation of characteristic dimensions and velocity triangles of a pump.

The crucial relations are the expressions (8.13) and (8.16) and the Cordier diagrams of Figs. 7.6 or 7.7. First, the rotational speed has to be chosen in order to obtain the best possible efficiency. In principle, this means a specific speed close to unity, but practical reasons often necessitate strong deviation from this value.

With the flow rate and the rotational speed, the inlet diameter of the rotor follows from the expressions (8.13) and (8.16). The outlet diameter follows from the Cordier diagram. The velocity triangles at rotor entry and exit can then be constructed, taking into account the targeted rotor work.

The Cordier diagrams of Figs. 7.6 or 7.7 are not accurate. Therefore, it is best to use a Cordier line from a specialised book. In Sect. 7.6 of Chap. 7, we demonstrated that an accurate Cordier line is not necessary for the design of a radial fan. But choosing dimensions of a pump so that it becomes realisable is much more delicate and quite high precision of the Cordier line is necessary.

As an example, Table 8.1 represents a Cordier line in the form of a head coefficient $\Psi = 2g H_m / u_2^2$ as a function of the specific speed, obtained by reading a diagram in the book of Gülich [3]. The tip value of the blade speed at the rotor exit has to be used. We refer to the Exercises 8.9.7 to 8.9.10 for illustrations.

Table 8.1 Head coefficient as a function of specific speed; from Gülich [3]

Ω_s	$2g\,H_m/u_2^2$	Ω_s	$2g\,H_m/u_2^2$
0.2	$0.97 - 1.15$	2.0	$0.46 - 0.59$
0.4	$0.93 - 1.10$	2.5	$0.34 - 0.42$
0.6	$0.87 - 1.03$	3.0	$0.27 - 0.32$
0.8	$0.81 - 0.97$	3.5	$0.22 - 0.26$
1.0	$0.75 - 0.90$	4.0	$0.19 - 0.22$
1.3	$0.66 - 0.80$	4.5	$0.15 - 0.175$
1.65	$0.56 - 0.69$	5.0	$0.12 - 0.14$

8.8.2 *Performance Evaluation*

The rotor work follows from the velocity triangles. Determination of the velocity triangle at the rotor exit of a centrifugal pump requires an estimation of the slip factor. The Wiesner formula (Eqs. 3.20 and 3.21) is the best suited. For axial pumps, estimation of the deviation between the flow and the blade orientation at the trailing edge is necessary (the equivalent of the slip).

Mixed-flow pumps can be similar to radial pumps (Fig. 8.13), but propeller pumps with hub radius variation (Fig. 8.31) also have mixed-flow character.

For analysis of the first category, the blade loading may be estimated by local versions of the slip by Eq. 3.24 and the moment solidity by Eq. 3.31 (Chap. 3), according to a derivation by Pfleiderer [9]. The reasoning is on a streamtube in a meridional section with an infinitesimal varying height Δn under the assumption of constant product $\Delta n\,\Delta p$. Pfleiderer argues that this approximation is better for a pump than that of constant blade pressure difference Δp, used for fans. The result is an estimation of the local Δp at the rotor exit by

$$\overline{\Delta p}/\rho = w_2(w_s - w_p)_2 = \frac{v_{2m} 2\pi\, r_2}{Z\,M_S}\,\frac{\Delta W}{\Omega}. \qquad (8.29)$$

$M_S = \int_{1\to 2} r\,dm$ is the static moment of the streamline in the meridional plane, with r the radius and dm an infinitesimal length along the streamline. The local Pfleiderer factor (Eq. 3.24) and the local moment solidity (Eq. 3.31) thus become

$$Pf = \xi\,\frac{r_2^2}{Z\,M_S} \quad \text{and} \quad \sigma_M = \frac{Z\,M_S}{2\pi\,r_2^2}. \qquad (8.30)$$

Hub sections of propeller pumps with large radius variation may be analysed in a similar way (see Exercise 8.9.12). The analysis of tip sections is with relations derived from axial cascades.

Determination of the head, the flow rate and the input power requires correlations for estimation of the mechanical losses, the leakage flows and the internal losses in the flow. These correlations are complex and delicate. Therefore, we refer to specialised literature [3, 5, 6]. In the exercises, we use approximate mean values of the mechanical, volumetric and internal efficiencies.

8.9 Exercises

8.9.1 Some pump manufacturers publish characteristic curves, including the required NPSH as a function of the flow rate. Take as an example the vertical DPV type of the manufacturer of stainless steel pumps for drinking water in water distribution and food industry: www.dp-pumps.com (Druijvelaar pumps). This pump type is manufactured as single stage and multistage pump with in-line suction and discharge sides. The smaller versions are made with centrifugal rotors and the larger versions with mixed-flow rotors.

Study the largest version, which is 125B. This pump is only made for rotational speed 2900 rpm and may have 1 to 4 stages. The nominal flow rate is 125 m^3/h with overall efficiency 81% and $NPSH_3 = 4.8$ m. Verify Ω_{SS} for the nominal flow rate by Eq. 8.14 (A: $\Omega_{SS} = 3.15$). Observe that the value with the $NPSH_3$ is only a little higher than $\Omega_{SS} \approx 3.0$, meant for the erosion value of NPSH. This observation gives confidence in the published NPSH values. The manufacturer recommends to add 1 m to the $NPSH_3$ as an estimate for the erosion value. For 140 m^3/h, the $NPSH_3$ becomes 6 m and the efficiency drops a little to 80%. For 90 m^3/h, the efficiency drops to 75% and the corresponding $NPSH_3$ is 3 m.

We conclude that the pump can function with good efficiency in the flow rate range 90–140 m^3/h, provided that the available NPSH can be about 7 m. Taking losses into account, the pump can thus extract from an open reservoir with a suction height up to about 2.5 m. This seems perfectly practicable.

8.9.2 Return to Exercise 1.9.8 of Chapter 1 and perform a mean-line analysis, taking obstruction by blade thickness and slip at the rotor exit into account.

The figure in Chap. 1 shows a meridional and an orthogonal section. At the mean entry of the bladed rotor part, the diameter is $d_1 = 92$ mm and the rotor width is $b_1 = 22$ mm. At the rotor exit, the diameter is $d_2 = 196$ mm and the width is $b_2 = 14$ mm. On the mean streamsurface, the blade angle, with respect to a meridional plane, is $\beta_1 = -68°$ at the rotor entry and $\beta_2 = -60°$ at rotor exit. The number of blades is 6 and the blade thickness is 3 mm. The rotor inflow is in the meridional direction in the absolute frame (no pre-whirl). The pump runs at 2900 rpm and pumps water ($\rho = 1000$ kg/m^3).

Determine the velocity triangles at rotor entry and exit. Distinguish between triangles just inside and just outside the rotor. Assume alignment of flow and blades at the rotor entry, just inside the rotor (design flow rate). Estimate the slip with the formula by Wiesner (Eqs. 3.20 and 3.21). Determine the rotor flow rate and the rotor work ΔW. Split the rotor work into the part by the Coriolis force and the part by the lift force. Determine the pressure energy rise and the kinetic energy rise in the flow through the rotor. Calculate the degree of reaction R.

Estimate the delivered flow rate, the realised head and the shaft power of the pump with estimates of internal efficiency 85%, volumetric efficiency 96%, mechanical efficiency 92%.

A: $u_1 = 13.97$ m/s, $v_1{}^b = 5.64$ m/s, $v_1 = 4.71$ m/s, $w_1 = 14.74$ m/s, $u_2 = 29.76$ m/s, $v_{2m}{}^b = 3.69$ m/s, $v_{2m} = 3.47$ m/s, $\sigma_{slip} = 0.80$, $v_{2u} = 17.75$ m/s, $w_2 = 12.51$ m/s, $v_2 = 18.08$ m/s, $\Delta W = 528.14$ J/kg, $\Delta W_C = 130.8\%$, $\Delta W_L = -30.8\%$, $\Delta E_p = 375.7$ J/kg, $\Delta E_k = 152.4$ J/kg, $R = 0.71$, $Q = 28.73$ l/s, $H_m = 45.76$ m, $P_{shaft} = 17.2$ kW.

8.9.3 Calculate the specific speed of the pump of the previous exercise for the design values of rotational speed, flow rate and head. Verify the head coefficient $2gH_m/(u_2)^2$ against Table 8.1. On the figure in Chap. 1, one sees that the tip diameter at the entry of the bladed part of the rotor is $(d_1)_{tip} \approx 103$ mm. Verify the entry diameter with the nondimensional expression $(d_1)_{tip}/(Q/\Omega)^{1/3}$, according to the formula 8.13. Calculate the moment solidity (Eq. 3.31) and verify the number of rotor blades with the Pfleiderer moment coefficient (Eq. 3.32).

A: $\Omega_S = 0.53$, $2gH_m/(u_2)^2 \approx 1.01$, Table 8.1: 0.89 − 1.05 (thus OK); $(d_1)_{tip}/(Q/\Omega)^{1/3} \approx 2.26$, thus somewhat larger than the value 2.10 by Eq. 8.13; $\sigma_M \approx 0.44$, $C_M \approx 0.90$, with $(C_M)_{lim} \approx 1.40$, the number of blades is appropriate.

8.9.4 Determine a volute and a tangential diffuser adapted to the rotor of Exercise 8.9.3. Use the formulae 8.25–8.26 Take the rotor diameter as diameter of the base circle (196 mm). Take into account that for matching with the rotor flow, the volute sections have to be enlarged by approximately 15%. Observe that the length available for the straight diffuser is about 125–150 mm. Observe that an area ratio of about 1.85 is possible according to Fig. 2.29 (Chap. 2). Calculate the velocity at the exit of the diffuser and the ratio of this velocity to the exit velocity of the rotor. Observe the very strong realised deceleration.

A: $R/r_2 \approx 0.26$, $d_3 = 55$ mm, $L/R \approx 5$, $d_4 = 75$ mm, $v_4 = 6.50$ m/s, $v_4/v_2 = 0.36$.

8.9.5 Estimate the internal efficiency of the pump of Exercise 8.9.2 when the flow rate is half that of the design flow rate. The design flow rate internal efficiency is 85%. With halved flow rate, losses by friction and by dump are reduced by a factor 4. But there is incidence at the rotor entry and the volute entry. Estimate a loss by incidence as the kinetic energy of the tangential deflection velocity. Take into account that the tangential velocity imposed by the volute to the rotor, $(v_{2u})_{stator}$, is proportional to the flow rate. Assume that the slip factor is a constant.

Calculate the realised head. Assume that the leakage flow rate is proportional to the square root of the head. The volumetric efficiency at the design flow rate is 96%. Estimate the volumetric efficiency for halved flow rate. Estimate the global efficiency for halved flow rate with $\eta_{mech} = 0.92$ ($\eta_{glob,design} = 0.751$).

A: Internal loss at design $= 79.22$ J/kg, $(v_{def})_{rotor} = 6.98$ m/s, $(v_{2u})_{rotor} = 20.75$ m/s, $(v_{2u})_{stator} = 8.87$ m/s, $(v_{def})_{stator} = 11.88$ m/s, sum losses $= 114.75$ J/kg, $\Delta W = 617.59$ J/kg, $H_m = 51.26$ m, $\eta_i = 0.814$, $Q_{leak} = 1.27$ l/s, $\eta_V = 0.915$, $\eta_{glob} = 0.686$.

8.9.6 Estimate the mean diameter at the entry of the bladed rotor part and the corresponding inlet width on the figure of the pump of Exercise 8.9.2 for doubling the flow rate. Take into account that according to formula 8.13, the tip value of the inlet diameter should vary proportional to the flow rate to the power 1/3. Make a similar estimate for a halved flow rate. Observe that for the doubled flow rate, it becomes necessary to vary the exit rotor diameter, with a larger value at the shroud.

A: $(d_1)_m = 102$ mm with $b_1 = 40$ mm, $(d_1)_m = 78$ mm with $b_1 = 13$ mm.

8.9.7 Figure 7.15 in Chap. 7 is an application field of a single-stage end-suction centrifugal pump type constructed according to the norm EN733 (KSB pumps: Etanorm pump), similar to the pump in Fig. 8.24.

Determine approximate dimensions of the rotor of the pump 150-125-400 with the simple methods of this course book. The chosen pump is for large head and large flow rate. Some geometric data are on the web site of KSB. Do not use these, but compare afterwards. Take as estimates: internal efficiency 85%, volumetric efficiency 96%, mechanical efficiency 92%. The rotor inflow is without pre-whirl.

From Fig. 7.15, one reads that the design data of the pump are approximately 330 m^3/h for flow rate and 54 m for head. The rotational speed is 1450 rpm.

Choose the tip diameter of the rotor entry by Eqs. 8.13 and 8.16, but limit the difference with the pump entry diameter to 10%. Set then the mean diameter of the rotor entry to about 90% of the tip diameter (**A:** $d_{1tip} = 165$ mm, $d_{1m} = 150$ mm).

The specific speed is about 0.42. Use Table 8.1 to choose the rotor diameter, but limit the difference with the nominal diameter to 10% (**A:** $d_2 = 420$ mm).

The inlet and outlet flow velocities of the pump are $v_s = 5.19$ m/s and $v_d = 7.47$ m/s. These are large velocities, but further analysis reveals that high entry and exit flow velocities are necessary for kinematic realisation of the studied pump.

Determine an appropriate value of the meridional velocity component just before rotor entry, v_1, using Eq. 8.15. Take the meridional velocity component just after rotor entry, $(v_1)^b$, equal to v_1. Compensate thus the blade thickness obstruction by adapting the inlet width and do not allow acceleration at the entry of the rotor, because it is advantageous for NPSH. But do not allow a difference of more than 20% with the inlet velocity of the pump v_s. Determine the entry width. Take 0.85 as blade obstruction factor at the rotor entry.

Determine the velocity triangle just inside the rotor entry. Determine the velocity triangle just outside the rotor exit and the exit width. Take $(v_{2m})^b/(v_1)^b = 0.75$ as ratio of the meridional velocity components inside the rotor and take 0.95 as blade obstruction factor at the rotor exit.

Remark the strong deceleration to be realised in the volute and the diffuser downstream of the rotor. Compare now with values from the website of KSB ($d_{1tip} = 162.4$ mm, $d_2 = 419$ mm to 330 mm, $b_2 = 25.9$ mm, $\eta_{glob} = 0.82$, $P_{shaft} = 60$ kW). The obtained estimates are not identical, because the real values are the result of further steps. The overall efficiency of the pump is much better than assumed, because the pump is large and we use average values of efficiency components.

A: $u_1 = 11.93$ m/s, $v_1{}^b = 4.36$ m/s, $v_1{}^b/v_s = 0.84$, $\beta_1{}^b = -69.0°$, $b_1 = 54.7$ mm, $u_2 = 31.89$ m/s, $v_{2m}{}^b = 3.27$ m/s, $v_{2m} = 3.11$ m/s, $u_2 v_{2u} = 623.22$ J/kg, $v_{2u} = 19.55$ m/s, $v_2 = 19.79$ m/s, $v_d/v_2 = 0.38$, $\beta_2 = -75.9°$, $\alpha_2 = 81.0°$, $b_2 = 23.3$ mm.

8.9.8 Repeat the task of the previous exercise for the pump 50-32-125.1 in the application field of Fig. 7.15. Use the same procedures and the same assumptions.

The design data of the pump are approximately 10 m^3/h for flow rate and 5 m for head. The rotational speed is 1450 rpm. The specific speed is about 0.43. The considered pump is for small flow rate and small head in the application field. Observe that the chosen pump is for smaller flow rate than the pump 50-32-125. The coding 125.1 thus means that the rotor is narrower.

The flow velocities at the pump entry and exit are $v_s = 1.42$ and $v_d = 3.45$ m/s. These are small values, but the analysis reveals that low velocities are necessary for kinematic realisation of the studied pump. Compare with values from the website of KSB ($d_{1tip} = 52.4$ mm, $d_2 = 139$ mm to 104 mm, $b_2 = 6.8$ mm, $\eta_{glob} = 0.61$, $P_{shaft} = 220$ W). The overall efficiency of the pump is much lower than assumed, because it concerns a small pump and we use average values of efficiency components.

A: $(d_1)_m = 50$ mm, $u_1 = 3.80$ m/s, $v_1^b = 1.36$ m/s, $v_1^b/v_s = 0.96$, $\beta_1^b = -70.3°$, $b_1 = 15.9$ mm, $d_2 = 130$ mm, $u_2 = 9.87$ m/s, $v_{2m} = 0.97$ m/s, $u_2 v_{2u} = 57.71$ J/kg, $v_{2u} = 5.85$ m/s, $v_2 = 5.93$ m/s, $v_d/v_2 = 0.58$, $\beta_2 = -76.5°$, $\alpha_2 = 80.6°$, $b_2 = 7.3$ mm.

8.9.9 On the website of Xylem Water Solutions (www.xylem.com), on can find details on the Flygt submersible propeller pump type, as shown in Fig. 1.3 in Chap. 1 (submersible column pump). We choose the P7076, which is a version for low head. The rotor with 3 blades has a diameter of 560 mm. The rotational speed is 585 rpm. The pump fits in a discharge tube of 1000 mm diameter.

The performance characteristics are published on the website. The maximum efficiency for the blade angle 18° (largest possible flow rate) is 74.8% for flow rate 3800 m³/h and head 3.1 m. From a drawing of a cross section, on can deduce that the ratio of the hub diameter to the rotor diameter is 0.27 at the rotor inlet and 0.38 at the rotor outlet. The hub of the pump is approximately conic.

Verify that the head coefficient is somewhat larger than in Table 8.1, when calculated with the blade speed at the tip of the rotor (**A:** $\Omega_S \approx 4.85$, $\Psi \approx 0.21$).

With pumps without shroud (axial, mixed-flow and centrifugal), it is not meaningful to define a volumetric efficiency. There is leakage flow through the gap between the rotor blade tips and the casing, but the leakage flow mixes with the main flow. The associated loss has thus to be included in the internal efficiency. We may thus calculate with $\eta_V = 1$. Take for mechanical efficiency $\eta_m = 0.96$.

Estimate the through-flow velocities at the rotor inlet and rotor outlet from orthogonal sections for the specified flow rate. Take the obtained values as approximations of the meridional flow components in the inlet and outlet sections. Verify the rotor diameter with the nondimensional expression $(d_1)_{tip}/(Q/\Omega)^{1/3}$, according to Eq. 8.13 and the inflow velocity with the nondimensional expression $(v_1)/(\Omega^{2/3}Q^{1/3})$, according to Eq. 8.15. Observe that the nondimensional diameter is somewhat larger than by Eq. 8.13 and the nondimensional inflow velocity is somewhat smaller than by Eq. 8.15 (**A:** $v_{1m} = 4.62$ m/s, $v_{2m} = 5.01$ m/s, $(d)_{non} = 2.17$, $(v)_{non} = 0.29$).

Construct the velocity triangles at the hub at the rotor inlet and outlet and at the shroud at the rotor inlet and outlet (just before entry and just after exit). Assume inflow without pre-whirl and the same work in all streamtubes.

Determine the fractions of the work by the Coriolis force and the lift force at the hub and determine the degree of reaction. Observe that the contribution by the Coriolis force is more than 50%. Determine the degree of reaction at the rotor tip.

Observe that the rotor blade loading at the hub is high, but not extremely high, when judged with criteria for axial cascades, which are: maximum turning about 40°

and strongest deceleration about 0.70. The necessary turning in the subsequent stator seems too high with a magnitude of α_2 of about 50°.

But, downstream of the rotor is a vaneless space in which the hub diameter becomes about 0.55 times the rotor diameter. The through-flow section stays approximately constant. The tangential velocity component at the hub thus reduces considerably (from 5.99 m/s to about 4.14 m/s), while the meridional component stays approximately the same (5.01 m/s). This reduces the flow angle to about 40°, which makes the turning in the vaned diffuser possible.

The observations are that the most critical zone in an axial pump is the hub zone and that the work capacity increases by radius variation in the hub zone.

A: Tip: $u_1 = 17.15$ m/s, $w_1 = 17.76$ m/s, $\beta_1 = -74.9°$, $u_2 = 17.15$ m/s, $v_{2u} = 2.28$ m/s, $v_2 = 5.50$ m/s, $w_2 = 15.70$ m/s, $\beta_2 = -71.4°$, $\Delta\beta = 3.5°$, $w_2/w_1 = 0.90$.

Hub: $u_1 = 4.63$ m/s, $w_1 = 6.54$ m/s, $\beta_1 = -45.1°$, $u_2 = 6.52$ m/s, $v_{2u} = 5.99$ m/s, $v_2 = 7.81$ m/s, $w_2 = 5.04$ m/s, $\beta_2 = -6.1°$, $\Delta\beta = 39.0°$, $w_2/w_1 = 0.77$, $\alpha_2 = -50.1°$.

Hub: Coriolis part $= 53.9\%$, lift part $= 47.1\%$, $R = 49.3\%$; Tip: $R = 88.6\%$.

8.9.10 On the website of KSB pumps (www.ksb.com), on can find details on the submersible propeller pump Amacan P, shown in Fig. 8.31. We choose the PB4 900–540, which is a version with a comparable flow rate as the pump in the previous exercise, but a large head. The rotor with 4 blades has a diameter of 540 mm. The rotational speed is 960 rpm. The pump fits in a discharge tube of 900 mm diameter.

The performance characteristics are published on the website. The maximum efficiency is 85.3%, for the blade angle 16°, with corresponding flow rate 4200 m³/h and head 8.8 m. From Fig. 8.31, which shows a cross section, on can deduce that the ratio of the hub diameter to the rotor diameter is 0.24 at the rotor inlet and 0.48 at the rotor outlet. The hub of the pump is approximately spherical.

Verify that the head coefficient accords with Table 8.1, when calculated with the blade speed at the tip of the rotor (A: $\Omega_s \approx 3.85$, $\Psi \approx 0.23$).

Estimate the through-flow velocities at the rotor inlet and rotor outlet from orthogonal sections for the specified flow rate. Take these values for meridional flow components in the inlet and outlet sections. Take $\eta_V = 1$ and $\eta_m = 0.96$.

Verify the rotor diameter with the nondimensional expression $(d_1)_{\text{tip}}/(Q/\Omega)^{1/3}$, according to Eq. 8.13 and the inflow velocity with the nondimensional expression $(v_1)/(\Omega^{2/3}Q^{1/3})$, according to Eq. 8.15. Observe that the nondimensional diameter is larger than by Eq. 8.13 and the nondimensional inflow velocity is smaller than by Eq. 8.15 (A: $v_{1m} = 5.41$ m/s, $v_{2m} = 6.62$ m/s, $(d)_{\text{non}} = 2.39$, $(v)_{\text{non}} = 0.24$).

Construct the velocity triangles just before rotor entry and just after rotor exit, at the hub and the shroud. Assume inflow without pre-whirl and the same work in all streamtubes. Determine the work fractions by the Coriolis force and the lift force at the hub. Determine the degree of reaction at the rotor hub and tip.

A: Tip: $u_1 = 27.14$ m/s, $w_1 = 27.68$ m/s, $\beta_1 = -78.7°$, $u_2 = 27.14$ m/s, $v_{2u} = 3.58$ m/s, $v_2 = 7.53$ m/s, $w_2 = 24.47$ m/s, $\beta_2 = -74.3°$, $\Delta\beta = 4.4°$, $w_2/w_1 = 0.88$, $R = 0.86$.

Hub: $u_1 = 6.51$ m/s, $w_1 = 8.47$ m/s, $\beta_1 = -50.3°$, $u_2 = 13.03$ m/s, $v_{2u} = 7.46$ m/s, $v_2 = 9.97$ m/s, $w_2 = 8.65$ m/s, $\beta_2 = -40.1°$, $\Delta\beta = 10.2°$, $w_2/w_1 \approx 1.0$, $\alpha_2 = -48.4°$,

Hub: Coriolis part $= 131\%$, lift part $= -31\%$, $R = 0.64$.

8.9.11 Compare the design strategies of the hub zone of the propeller pump for low head of Exercise 8.9.9 and that for high head of Exercise 8.9.10. With low head, the functioning is half of axial and half of centrifugal type. The local work coefficient $\Delta W/(u_2)^2 \approx 0.92$ is much higher than possible with a purely axial or a purely centrifugal pump, but there is no simple method for analysis of the mixed functioning. With high head, the Coriolis part of the work at the hub is about 130% and the lift part is negative. This is similar to a centrifugal pump and is due to the strong radius variation (the design strategy is the same by Flygt for high head). The blade loading may thus be judged with methods of centrifugal rotors (Eqs. 8.29–8.30). The local work coefficient $\Delta W/(u_2)^2 \approx 0.57$ is moderately high for a centrifugal pump (see Table 8.1). Estimate the static moment of a meridional streamline near the hub by remarking that the shape is a segment of a circle. Estimate the moment solidity and the Pfleiderer moment coefficient. Observe the large moment coefficient.

A: $M_S \approx 0.87\,(r_2)^2$, $\sigma_M \approx 0.55$, $C_M \approx 1.20$.

8.9.12 On the website of KSB, one finds the required NPSH of the PB4 900-540 propeller pump of Exercise 8.9.10. For blade angle 16° and 960 rpm, the NPSH$_3$ is 9.5 m for flow rate 4200 m^3/h. With a margin of 1.5 m, the estimated erosion value of the NPSH is thus 11 m. The corresponding Ω_{SS} is 3.24.

Further insight comes by calculation of the corresponding σ-value in Eq. 8.9 for NPSH, with $\lambda = 1$ (no inlet loss). With the values of v_1 and w_1 at the tip of the rotor, obtained in Exercise 8.9.10, we find $\sigma = 0.244$. The values of Ω_{SS} and σ are somewhat better than typical with a centrifugal pump, but are possible with a propeller pump due to better hydrodynamic blade profiling. Nevertheless, it seems wise to take a somewhat larger safety margin than 1.5 m.

For flow rate 4500 m^3/h, the NPSH$_3$ raises to 10.6 m. To be fully safe, we set the required NPSH to 12.5 m. For flow rate 3800 m^3/h, the efficiency decreases to about 80%. It seems thus appropriate to use the pump in the flow rate range 3800–4500 m^3/h and to provide a submersion depth of 2.5 m for cold water in a pit exposed to atmospheric pressure. The submersion depth advised by the manufacturer for avoiding air ingestion by an aspiration vortex is about 2 m. A submersion depth of 2.5 is thus also safe for this aspect.

References

1. Franc JP, Michel JM (2004) Fundamentals of cavitation. Kluwer. ISBN 1-4020-2232-8
2. Brennen CE (2013) Hydrodynamics of pumps. Cambridge University Press. ISBN 978-1-107-40149-5
3. Gülich JF (2020) Centrifugal pumps, 4th edn. Springer. ISBN 978-3-030-14787-7
4. Europump (1999) Net positive suction head for rotodynamic pumps, a reference guide. Elsevier. ISBN 978-1-85617-356-8

5. Japikse D, Marscher W, Furst RB (1997) Centrifugal pump design and performance. Concepts ETI. ISBN 0-933283-09-1
6. Tuzson J (2000) Centrifugal pump design. Wiley. ISBN 0-471-36100-3
7. ANSI/HI (American National Standard Institute/Hydraulic Institute) (2017) Rotodynamic pumps guideline for NPSH margin. ISBN 978-1-935762-57-7
8. Volk M (2014) Pump characteristics and applications, 3th edn. CRC Press. ISBN 978-1-4665-6309-4
9. Pfleiderer C (1961) Die Kreiselpumpen für Flüssigkeiten und Gase, 5th edn. Springer, no ISBN
10. Girdhar P, Moniz O (2005) Practical centrifugal pumps. Elsevier. ISBN 0-7506-6273-5

Chapter 9
Hydraulic Turbines

Abstract In this chapter, we discuss the different types of hydraulic turbines for electric power plants. We analyse their main characteristics in order to understand in which range of head and flow rate they can be used efficiently. We also discuss bulb turbines for tidal energy plants and reversible pump-turbines for pumped-storage plants.

9.1 Hydraulic Energy

Hydraulic turbines convert the gravitational potential energy, available by water flow between two places with a difference in altitude, in mechanical energy. Except for some old small-scale plants, the application is production of electricity.

Natural altitude differences are very seldom just serviceable. In most cases, a lake is created by a dam in a river. In mountain areas, a supply pipe may connect the lake with a power station at a much lower altitude. The available head is then much larger than the dam height. With some rivers, however, the power station can only be built within the dam or very near to it and the available head is comparable to the dam height. With rivers in plain areas, normally, the dam also enables ship traffic by lock chambers. Available heads may thus strongly differ, from the order of 1000 m to some few meters. It is obvious that a smaller head causes higher investment costs per unit of power. The minimal economically usable head is about 4–5 m.

River water originates from rain or snow at higher altitude. The mechanism is evaporation of water at lower altitude, rising of the vapour and condensation of it in the colder atmosphere at higher altitude. The energy source with this process is the sun. Hydraulic energy is thus part of a thermal cycle driven by solar energy. Throughout the world there is about 1150 GW installed hydraulic power (2020), not included pumped-storage plants (about 160 GW) (see website of IHA, International Hydropower Association). About 16% of all electricity is generated by hydraulic energy and about 30% of the practically exploitable hydraulic energy potential is already in use.

Hydraulic turbines vary strongly in size. The typical power is 100–400 MW, but there also exist 4–5 MW turbines and even much smaller ones. Hydraulic power

E. Dick, *Fundamentals of Turbomachines*, Fluid Mechanics and Its Applications 130,
https://doi.org/10.1007/978-3-030-93578-8_9

stations with power under 10 MW are classified as small ones and those with power not exceeding 1 MW are called micro power stations.

At present, the largest power station is the Three Gorges plant on the Yangtze river in China with 22,500 MW installed power, 32 turbines with maximum power of 700 MW and 2 smaller turbines of 50 MW, nominal head 81 m, lake surface area of about 1000 km². The second largest power station is the Itaipu plant at the Brazil-Paraguay border with 14,000 MW installed power, 20 turbines with maximum power of 700 MW, nominal head 118 m, lake surface area of about 1350 km².

9.2 Hydraulic Turbine Types

9.2.1 Large Turbines (> 10 MW)

Three types are used for large-scale applications (see websites of manufacturers: e.g., Andritz hydro, Voith hydro).

A *Pelton turbine*, as drawn in Fig. 9.1, is an impulse turbine (zero loss-free degree of reaction) with an injector (nozzle) generating a water jet. This jet propels a runner turning in the atmosphere. The blades have the shape of a double spoon. Figure 9.1 represents a machine with a horizontal shaft and one injector. Up to two injectors are possible with a horizontal shaft (one propelling the bottom side, as shown, and one propelling vertically downwards). Two wheels may be mounted on the same shaft (one on both sides of the generator). A Pelton turbine may also be built with a vertical shaft. There may then be up to 6 injectors.

A *Francis turbine*, represented in Fig. 9.2, is a radial machine with a medium to high degree of reaction (0.55–0.75). With the turbine shown, water is supplied by a volute (also called spiral casing). This is a typical design (a water chamber may be

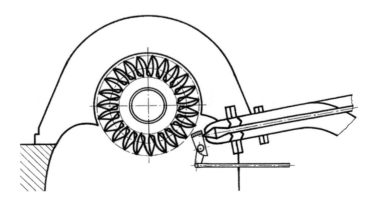

Fig. 9.1 Pelton turbine with horizontal shaft and one injector

Fig. 9.2 Francis turbine
(high specific speed type);
adjustable stator vanes

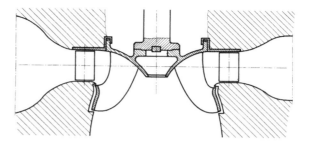

used with small turbines). The water flows through guide vanes, adjustable for flow
rate control (the lever mechanism is not drawn).

The vanes are actually variable vanes, because adjustment is possible during
operation. But, with hydraulic turbines, the term adjustable is mostly used. The
vanes can be closed to zero flow rate. The rotor usually has 12–16 blades. The inflow
of the rotor is radial (no axial velocity component), but the outflow is approximately
axial (small radial velocity component). Such a rotor is called radial. Blades of radial
rotors cannot be adjusted. Strongly different rotor forms are used (see further below).
Downstream of the rotor is a draught tube. The represented draught tube is modelled
in the concrete structure. It can also be made of steel.

A *Kaplan turbine*, represented in Fig. 9.3, is an axial machine with a high degree of
reaction (0.75 and higher). The turbine shown has a volute supply and a radial stator
with adjustable vanes (actually: variable). Upstream of the adjustable stator vanes,
there are fixed stator vanes that play a role in supporting the structure. These are called
stay vanes and are often also employed with Francis turbines. The through-flow of
the rotor is axial and the rotor has adjustable blades (actually: variable). Downstream
of the rotor is a draught tube.

Flow rate control is mainly by positioning the rotor blades by a lever mechanism
inside the rotating hub, activated by a rod in the hollow shaft, driven by a hydraulic
actuator. The shaft with bearings is schematically represented by the densely hatched
area in Fig. 9.3 and the rod inside the rotating shaft is drawn. The stator vanes play

Fig. 9.3 Kaplan turbine;
adjustable rotor blades and
stator vanes

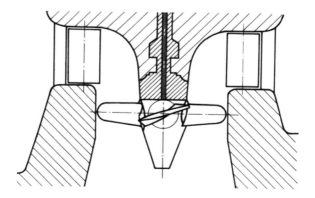

only a minor role (see below). Stator vanes can, as with Francis turbines, be closed to zero flow rate (the lever mechanism is not represented in Fig. 9.3).

Large turbines are mounted vertically and have a volute, as represented in Fig. 9.3. Small turbines (see Chap. 1) may be made with an inclined shaft and with an axial stator and axial water supply. Small machines may have a horizontal shaft as well (see later: tube turbines). The rotor usually has 4–6 blades. Kaplan turbines are sometimes described as propeller turbines due to the resemblance of the rotor to a ship propeller.

There are also medium-size mixed-flow rotor forms with adjustable rotor blades. Mixed-flow machines have a lower degree of reaction than axial machines and have properties in between those of Francis and Kaplan turbines.

Hydraulic turbines are of single-stage type, due to their very low specific energy. For example, the specific energy corresponding to a 1000 m head is $gH \approx 10,000$ J/kg $= 10$ kJ/kg, whereas the order of magnitude with gas turbines and steam turbines is 1000 kJ/kg. With complete energy use in a nozzle, as with a Pelton turbine, 10,000 J/kg results in a jet velocity of about 140 m/s. The optimal blade speed is then about 70 m/s (degree of reaction $= 0$, speed ratio as with an impulse type steam turbine). With a 2 m diameter, the corresponding rotational speed is about 670 rpm. This is a rather low value. Typical rotational speeds with Pelton turbines thus are 500, 428, 375 rpm, with lower rotational speeds as heads get lower. With Pelton turbines, rotational speeds become impracticable at low head, but, of course, the smaller the turbine, the lower the head may be.

The example demonstrates that hydraulic turbines are single-stage machines and that lower heads require higher rotational speeds than with impulse-type machines. As we will study later, the relation between the degree of reaction and the optimal speed ratio with hydraulic turbines is the same as with steam turbines. Lower heads thus require turbines with a medium–high degree of reaction (Francis turbine). Still lower heads require a high degree of reaction (Kaplan turbine).

Figures 9.1, 9.2, and 9.3 further reveal that, at equal sizes, Kaplan turbines handle the highest flow rates and Pelton turbines the lowest ones. The head-flow combinations mean that a Pelton turbine has a low specific speed: $\Omega_s = \Omega \sqrt{Q}/(gH)^{3/4}$. Values range from about 0.05 to 0.165 with a single-injector turbine. Specific speed ranges from about 0.30 to 2.10 with Francis turbines, whereas Kaplan turbines cover a 1.65–6.00 range (see below). The specific speed range from 0.165 to 0.30 is covered by multi-injector Pelton turbines. Figure 9.4 represents the range of application for the various types ($n_q \approx 53 \, \Omega_s$). The term bulb/tube turbine refers to axial machines with an axial stator and axial water supply.

9.2.2   Small Turbines (< 10 MW)

The three types of large machines are used with small-scale applications as well. The term small scale may refer to a turbine generating a few MW, but also to one yielding only some tens of kW. The diagram in Fig. 9.4 does not always apply anymore then. It is obvious that a Pelton turbine may be used with a 10 m head if dimensions are

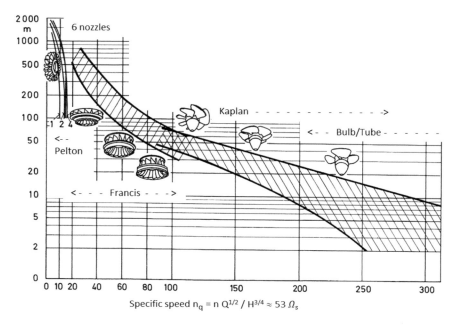

Fig. 9.4 Application range of various types of hydraulic turbines: head as a function of specific speed (adapted from Dietzel [2]; permission by Vogel Fachbuch)

sufficiently small. In the calculation example, the same rotational speed is obtained with a 10 m head and a 0.2 m diameter. Small Pelton turbines for high head exist as well. Pelton turbines are rather easily made with small dimensions. Small Francis and Kaplan turbines are only possible for low head. As already mentioned, Francis turbines may be simplified by using a water chamber for lower head (applicable up to 4–5 m head).

For low power, an axial turbine is mostly integrated in a pipe, called a tube turbine. Small axial turbines mostly have a fixed stator, as the flow rate is mainly determined by the position of the rotor blades (see below). Machines with fixed rotor blades and adjustable stator vanes exist as well. There are also small axial machines with fixed rotor blades and fixed stator vanes. No flow rate variation is then possible (at a constant head and a constant rotational speed).

Apart from the three types discussed, there is a cross-flow type, mostly called a Banki-turbine (after its inventor). Figure 9.5 represents an orthogonal section. The functioning is similar to that of a cross-flow fan (Chap. 3), but turbine-wise.

Blades have a radial direction at the inside of the rotor. Relative velocities at the rotor entry and exit are identical. As a result, the degree of reaction is zero. Cross-flow turbines are a further development of undershot water wheels (see Exercise 1.9.10). The machine has a draught tube enabling utilisation of the downward head. The rotor partially runs in air. A sniffer valve supplies a small air flow, as air is entrained in the

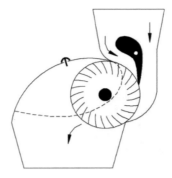

Fig. 9.5 Cross-flow turbine

draught tube. Cross-flow turbines can be constructed very easily, but their efficiency is significantly lower than that of the other hydraulic turbine types, namely about 80% compared to about 90%.

9.3   Pelton Turbines: Impulse Turbines

9.3.1   Performance Characteristics

Figure 9.6 is a sketch of the flow over the rotor buckets of a Pelton turbine (the term bucket is typically used).

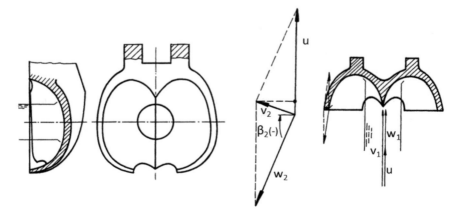

Fig. 9.6   Flow over a Pelton turbine bucket

The jet velocity is $v_1 = \phi_s \sqrt{2g\,H_m}$. H_m is the manometric head, determined with a manometer at the injector entry, with the stagnant atmosphere at the ejector exit as reference. ϕ_s is a velocity coefficient expressing the friction in the injector. By writing v_1 as above, the downward head between the injector exit and the tailwater is not included in the manometric head, which is a typical practice. Since there is no possibility for confusion, we represent the manometric head by H from now on.

The jet velocity v_1 has the same direction as the blade speed u:

$$w_1 = v_1 - u.$$

The top angle of the central ridge of the bucket does not equal zero, which causes a small incidence. The incidence loss is integrated in the rotor loss coefficient.

At the rotor exit, the relative flow speed w_2 forms an angle β_2 with the shaft direction. The flow angle is approximately equal to the outlet blade angle. From the outlet blade angle follows ($\beta_2 < 0$, $\beta_2 \approx -80°$):

$$v_{2u} = u + w_{2u} = u + w_2 \sin \beta_2, \quad \Delta W = u(v_{1u} - v_{2u}) = u(w_1 - w_2 \sin \beta_2).$$

Let $\phi_r = w_2/w_1$ be a velocity coefficient, taking rotor blade losses into account. Then:

$$\Delta W = u(v_1 - u)(1 - \phi_r \sin \beta_2).$$

The rotor work varies parabolically as a function of the blade speed u with a maximum for $u = \frac{1}{2}v_1 = \frac{1}{2}\phi_s\sqrt{2g\,H}$. The optimal speed ratio is ($\phi_s \approx 0.97$):

$$\lambda = \frac{u}{\sqrt{2g\,H}} = \frac{1}{2}\phi_s \approx 0.48. \tag{9.1}$$

Nozzle efficiency may be defined by

$$\eta_s = \frac{\frac{1}{2}v_1^2}{g\,H} = \phi_s^2 \approx 0.94,$$

and rotor efficiency by

$$\eta_r = \frac{\Delta W}{\frac{1}{2}v_1^2} = 2(1 - \phi_r \sin \beta_2)\frac{u}{v_1}\left(1 - \frac{u}{v_1}\right).$$

The optimum rotor efficiency is ($\phi_r \approx 0.96$, $\beta_2 \approx -80°$):

$$\eta_{r,o} = 2(1 - \phi_r \sin \beta_2)\left(\frac{1}{4}\right) \approx 0.97.$$

Fig. 9.7 Power, torque and
efficiency as functions of
speed ratio with a Pelton
turbine ($v_0 = \sqrt{2gH}$ is the
spouting velocity)

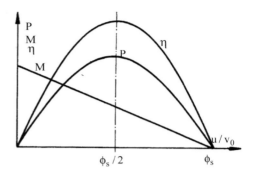

Overall (global) efficiency is $\eta_g = \eta_s \eta_r \eta_m$.

From the foregoing values and a mechanical efficiency of about 0.99, it follows
that the overall efficiency at optimum speed ratio is about 0.90. It even may be better
in practice, namely up to 0.92.

Rotor power is $P_{rot} = \dot{m}\Delta W$,

and shaft power P_{shaft} equals rotor power multiplied by mechanical efficiency.

Figure 9.7 represents the variation of power and efficiency, as functions of the
blade speed u and the variation of the torque exerted on the rotor by the fluid. The
torque is the rotor power divided by the angular velocity:

$$M = \dot{m}r(1 - \phi_r \sin \beta_2)(v_1 - u).$$

The torque is a linear function of the blade speed. It decreases from the maximum
value for $u = 0$ to zero for $u = v_1$. The maximum speed that may be reached by the
turbine, as far as the turbine is not driven, thus equals double the design speed. If the
turbine were designed to resist a centrifugal force four times as high as the design
centrifugal force, no damage could occur if the turbine runs idling. Pelton turbines
are not built so strong for reasons of cost. A device for limiting over-speed is thus
needed (see Sect. 9.3.4).

The foregoing conclusions rest on an assumed constant mechanical efficiency.
Mechanical losses are mainly by rotor windage. These vary with the cube of the
rotational speed and thus increase with increasing speed. As a consequence, the
optimum speed ratio is somewhat lower than derived here, namely about 0.45–0.47,
with lower values for turbines with a lower power (one injector, smaller injector
diameter). The maximum speed with an idling turbine (runaway speed) is about 1.8
times the design speed.

9.3.2 Specific Speed

We take the notation d for the exit diameter of the nozzle and D for the diameter of the rotor at the jet centre line. The flow rate is obtained by

$$Q = v_1 \frac{\pi d^2}{4} = \phi_s \sqrt{2gH} \frac{\pi d^2}{4}.$$

The rotational speed is derived from the blade speed by $\Omega = 2u/D$. Thus:

$$\Omega_s = \frac{\Omega \sqrt{Q}}{(gH)^{3/4}} = \frac{u}{\sqrt{2gH}} \sqrt{\phi_s \pi} \, 2^{3/4} \frac{d}{D}.$$

With the optimum speed ratio $\lambda = u/\sqrt{2gH} \approx 0.48$ and $\phi_s \approx 0.97$:

$$\Omega_s \approx \sqrt{2} \frac{d}{D}. \tag{9.2}$$

For good flow conditions, the diameter ratio d/D has to be between about 1/24 and 1/8. Small d/D values generate long jets with a large contact surface with the air compared to the cross section. This causes relatively high friction and drop formation. Large d/D values generate jets with bad flow guidance by the buckets.

There is a constructional limit to the specific speed due to the stress by the centrifugal force on the buckets. With increasing Ω_s at constant Q and H values, Ω must increase. The centrifugal force on a bucket is $m \, \Omega^2 r$. The mass of the bucket (m) stays constant since the mass only depends on d, so on Q and v_1, which both are constant. Further, $u = \Omega r$ must be constant, as v_1 is constant and u/v_1 as well. The centrifugal force is thus proportional to Ω. If Ω_s increases, D must decrease, because of $\Omega_s \, d/D$. The machine diameter D cannot decrease below a minimum value determined by the circumferential length required for mounting the buckets since their size and number are determined by flow requirements. The required mounting length increases with the head, as forces on the buckets increase proportionally to the head.

Table 9.1 lists the maximum values of d/D and Ω_s as functions of the head, according to Vivier [4]. These values apply to larger units (10–200 MW). With smaller turbines, the limitations caused by the centrifugal force may already occur

Table 9.1 Maximum values of d/D and Ω_s depending on head for Pelton turbines (one rotor and one injector)

H (m)	400	600	1000	1500	2000
$(d/D)_{max}$	1/8	1/10	1/14	1/19	1/24
$(\Omega_s)_{max}$	0.165	0.130	0.095	0.070	0.055

with lower head. The foregoing discussion applies to turbines with one rotor and one injector. With z_1 rotors with z_2 injectors each, the flow rate is multiplied by $z_1 z_2$ and thus the specific speed with $\sqrt{z_1 z_2}$.

9.3.3   Determination of the Main Dimensions

Flow rate and head are imposed by the application. Specific speed is chosen with regard to the strength limitations by the centrifugal force. Specific speed is mostly chosen as high as possible in order to attain relatively high and electrically serviceable rotational speeds (375–500 rpm). With a large flow rate, several rotors and several injectors may be needed. With the jet velocity $v_1 = \phi_s \sqrt{2gH}$, where $\phi_s = 0.96$ to 0.98, the diameter of the injector throat follows from:

$$ Q = \frac{\pi d^2}{4} v_1, $$

with Q the flow rate per injector. All bucket dimensions are subsequently derived from the jet diameter. The condition for maximum efficiency, $u/v_1 \approx 0.48$, determines the blade speed and thus the rotor diameter. The number of blades depends on the condition that at least one blade must receive the jet. When a blade leaves the jet, the following one should enter with some overlap.

Figure 9.8 shows a rotor. Rotors are mostly manufactured of stainless steel because of good resistance of this material to erosion. Other erosion-exposed parts are made of stainless steel as well. The cut-out at the tip of the buckets is intended to enable a smooth transition of the water jet from one bucket to the next one.

Fig. 9.8 Pelton turbine rotor
(*Courtesy* Andritz Hydro)

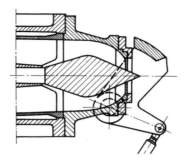

Fig. 9.9 Adjusting needle and jet deflector with a Pelton turbine

9.3.4 Flow Rate Control and Over-Speed Protection

Control of the flow rate, and thereby power at constant head, is achieved with a needle that adjusts the through-flow area of the injector (Fig. 9.9). Care must be taken for too abrupt changes of the position of the needle.

The closing time should be sufficiently large (typically 15–30 s) in order to avoid water hammer in the supply duct. If flow rate reduction is required with a fast turbine load decrease, a jet deflector is used. The deflector is turned between the injector and the buckets. This takes at most a few seconds. The deflected jet causes strong erosion of the surrounding parts. This device should thus only be applied in anticipation of flow rate adjustment by a slow movement of the needle.

9.4 Francis and Kaplan Turbines: Reaction Turbines

9.4.1 Shape of the Velocity Triangles: Kinematic Parameters

Figure 9.10 is a sketch of possible velocity triangles at rotor entry and exit for an average streamline with radius ratio $u_2/u_1 = 0.5$.

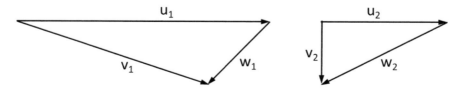

Fig. 9.10 Possible velocity triangles at rotor entry and exit with Francis turbines (not fully optimal) for $u_2/u_1 = 0.5$

Five parameters are required to determine the shape of the velocity triangles. For the inlet triangle these may be: tangential speed coefficient $\zeta = v_{1u}/u_1$ and flow coefficient $\phi = v_{1m}/u_1$. One of these parameters may be replaced by the stator vane angle α_1. For the outlet triangle they may be: radius ratio u_2/u_1, ratio of the meridional velocity components v_{2m}/v_{1m} and outflow angle α_2.

Parameters determining the shape of the velocity triangles are called kinematic parameters. In Fig. 9.10, these are $v_{1u}/u_1 = 0.75$ and $v_{1m}/u_1 = 0.25$ for the inlet triangle and $u_2/u_1 = 0.5$, $v_{2m}/v_{1m} = 1$, and $\alpha_2 = 0$ for the outlet velocity triangle. That this choice is reasonable, although not fully optimal, may be demonstrated by considering the losses and deriving some guidelines for optimisation.

9.4.2   Optimisation of the Velocity Triangles

The volute and the adjustable guide vanes constitute the distributor. The flow is accelerating, causing only moderate friction loss, of the 5% order of the kinetic energy at the stator exit $(v_1^2/2)$. We represent the distributor losses by $\xi_s \frac{1}{2} v_1^2$, with $\xi_s \approx 0.05$. The rotor loss is mainly by friction on the blades and may be represented by $\xi_r \frac{1}{2} w_2^2$. The rotor loss coefficient ξ_r is typically also around 5% (see Exercise 9.9.4), but may be much larger for blade channels of small height and large blade surface area, which occur with small specific speed (see Fig. 9.4). The kinetic energy at the rotor exit $(v_2^2/2)$ is not completely lost, as the draught tube also functions as diffuser. But, the recovery is limited to about 75%. Diffuser loss is represented by $\xi_d \frac{1}{2} v_2^2$, with $\xi_d \approx 0.25$.

The work may thus be represented by

$$\Delta W = gH - \xi_s \frac{v_1^2}{2} - \xi_r \frac{w_2^2}{2} - \xi_d \frac{v_2^2}{2}. \tag{9.3}$$

The kinematic parameters result by minimising the sum of the losses. With the relation (9.3), it is impossible to perform an exact optimisation, when roughly estimated loss coefficients are assumed. But some guidelines may be derived from it.

Due to the limited recovery in the diffuser, the kinetic energy at the diffuser exit is a significant loss. Therefore, the optimum set of parameters cannot be far from the one that minimises the kinetic energy at the rotor exit. As drawn in Fig. 9.10, the rotor exit velocity v_2 should thus be in the axial direction (at least near to the axial direction). Further, the magnitude of v_2 is minimised by a low flow coefficient v_{1m}/u_1. Therefore, the stator exit angle α_1 must be large. But, this angle cannot be exaggerated, as a larger α_1 means a more narrow volute or larger turning in the stator vanes and thus larger distributor losses. It also implies larger through-flow areas, which increases friction surfaces in both the rotor and the stator. An exact value for the stator angle α_1 cannot be determined with a simple reasoning. In real Francis turbines, α_1 varies between about 60° and 75° (see the later Table 9.2). The angle is about 71.5° in Fig. 9.10.

It is advantageous to choose the ratio of the meridional velocity components v_{2m}/v_{1m} lower than 1. This way, v_2 is reduced, but the required outlet area then increases. We therefore provisionally assume the ratio to be 1, but we keep in mind that a reduction is advantageous, if it is realisable. But, in practice, a value much lower than unity is not possible (see the later Fig. 9.13).

The work by the Coriolis force is $u_1^2 - u_2^2$. The work by the lift force is $u_1 w_{1u} - u_2 w_{2u}$. With the triangles in Fig. 9.10, the lift force work is exactly zero. It may be made positive by larger flow acceleration in the rotor, which is, e.g., obtainable by increased u_2/u_1. This enables reduction of the number of blades, resulting in a smaller friction surface area. But, a larger flow turning increases the rotor loss coefficient and the exit kinetic energy. Thus, acceleration and turning should not be too large. In Fig. 9.10, the velocity ratio w_2/w_1 is about 1.58, which is already quite large. Some positive lift force work is advantageous, but the lift force work for optimum efficiency stays small (see the later Fig. 9.13).

9.4.3 Degree of Reaction and Speed Ratio

We assume axial rotor exit velocity ($v_{2u} = 0$). We also assume constant meridional velocity components ($v_{2m} = v_{1m}$). It then follows that

$$\Delta W = u_1 v_{1u} - u_2 v_{2u} = u_1 v_{1u},$$

$$R = \frac{\frac{1}{2}(u_1^2 - u_2^2) + \frac{1}{2}(w_2^2 - w_1^2)}{u_1 v_{1u}} \quad \text{or} \quad 1 - R = \frac{\frac{1}{2}(v_1^2 - v_2^2)}{u_1 v_{1u}} = \frac{1}{2}\frac{v_{1u}}{u_1}.$$

Thus:

$$R = 1 - \frac{1}{2}\frac{v_{1u}}{u_1}. \tag{9.4}$$

The work coefficient is

$$\psi = \frac{\Delta W}{u_1^2} = \frac{u_1 v_{1u}}{u_1^2} = \frac{v_{1u}}{u_1}, \quad \text{or} \quad \psi = 2(1 - R). \tag{9.5}$$

The speed ratio is $\lambda = \frac{u_1}{\sqrt{2gH}} = \frac{u_1}{\sqrt{2\Delta W}}\sqrt{\eta_i} = \frac{\sqrt{\eta_i}}{\sqrt{2\sqrt{\psi}}} = \frac{\sqrt{\eta_i}}{2\sqrt{1-R}}$

With $\eta_i \approx 0.92$ (optimal efficiency), thus follows

$$\lambda \approx 0.48/\sqrt{1 - R}. \tag{9.6}$$

The relations (9.5) and (9.6) are very universal. They are compatible with the Pelton turbine result (9.1) and with the results for steam turbines in Chap. 6. The conditions for achieving the relation (9.6) are assumption of an axial exit flow and a constant meridional component of the velocity in the rotor.

In real machines these conditions are very well met. With the relation (9.4), the degree of reaction may be read from the velocity triangles. For example, the velocity triangles in Fig. 9.10 are drawn with $v_{1u}/u_1 = 0.75$. The corresponding degree of reaction is $R = 0.625$.

9.4.4 Examples

Figure 9.11 shows the rotor of a Francis turbine and Fig. 9.12 that of a Kaplan turbine. Figure 2.30 of Chap. 2 is an example of a draught tube of a Francis turbine. The draught tube of a Kaplan turbine is similar.

Fig. 9.11 Francis turbine rotor $\Omega_s \approx 1$ (*Courtesy* Andritz Hydro)

Fig. 9.12 Kaplan turbine rotor $\Omega_s \approx 3$ (*Courtesy* Andritz Hydro)

9.4.5 *Velocity Triangles with Varying Degree of Reaction*

In Fig. 9.13, velocity triangles are drawn for degrees of reaction varying from $R = 0.55$ to $R = 0.85$, with realistic proportions, derived from parameter variations in the books of Dietzel [2] and Vivier [4] (see the blade shapes in Figs. 9.11 and 9.12). The theoretical expression of the degree of reaction (9.4) is used. A variable meridional velocity component causes some deviation of the real degree of reaction. The triangles are drawn with $\alpha_2 = 0$. The used ratios for $R = 0.55$ to $R = 0.85$ are $v_{1m}/u_1 = 0.24, 0.28, 0.32, 0.36$; $v_{2m}/u_1 = 0.22, 0.27, 0.32, 0.36$; $u_2/u_1 = 0.45, 0.60,$ $0.75, 1$. These ratios illustrate the tendencies, but are not universally valid. The lift force work is positive with Francis turbines, but lower than 15% of the total work (see Exercise 9.9.6). The triangles for $R = 0.85$ are for a Kaplan turbine ($u_2 = u_1$, 100% lift force work).

The flow turning and the acceleration in the rotor are always moderate. A higher degree of reaction causes less freedom to decrease v_{2m} with respect to v_{1m}. Constant mass flow rate then requires an increase of the outlet area. This is more difficult with a diameter ratio near to unity (see Fig. 9.14).

Realisation of a Francis turbine with a degree of reaction lower than $R = 0.55$ is possible. The rotor turning must be high in that case in order to obtain a sufficiently large exit diameter. This reduces the efficiency (see Exercise 9.9.5). Therefore, turbines with a degree of reaction lower than 0.55 are not used in practice.

The flow coefficient v_{1m}/u_1 increases somewhat with increasing degree of reaction. It varies approximately from 0.24 to 0.32 with Francis turbines and changes more strongly with Kaplan turbines, namely $v_{1m}/u_1 \approx 0.30$–0.40, with a degree of reaction varying from 0.75 to 0.90. A simple argumentation cannot demonstrate that this is optimal. The approximate stator angle values are given in Table 9.2.

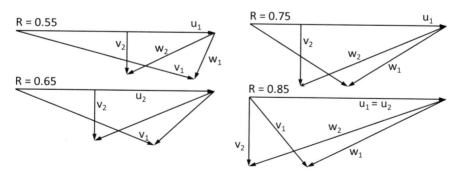

Fig. 9.13 Velocity triangles for varying degree of reaction with Francis and Kaplan turbines (approximate)

Table 9.2 Stator outflow angle as a function of degree of reaction with Francis and Kaplan turbines

$R = 0.55$	$R = 0.65$	$R = 0.75$	$R = 0.85$
$\alpha_1 = 75°$	$\alpha_1 = 68°$	$\alpha_1 = 57°$	$\alpha_1 = 40°$

9.4.6 Specific Speed and Meridional Shape of Francis Turbines

The specific speed is $\Omega_s = (\Omega\sqrt{Q})/(gH)^{3/4}$. We express Ω and Q as functions of geometric parameters. With b_1 being the stator vane height at the stator exit and d_1 the rotor entry diameter:

$$\Omega = \frac{2u_1}{d_1} = \frac{2 \times 0.48}{d_1\sqrt{1-R}}\sqrt{2gH},$$

and

$$Q = \pi d_1 b_1 v_{1m} = \pi d_1 b_1 \, 0.25 \, \frac{0.48}{\sqrt{1-R}}\sqrt{2gH},$$

where $v_{1m}/u_1 = 0.25$ is used. After substitution, the result is

$$\Omega_s \approx \frac{1}{(1-R)^{3/4}}\sqrt{\frac{b_1}{d_1}}.$$

This expression demonstrates that the ratio b_1/d_1 and the degree of reaction must increase in order to increase the specific speed. As demonstrated in the previous section, the diameter ratio increases with increasing degree of reaction. These influences explain the meridional shapes shown in Fig. 9.14.

As with pumps with doubly-curved blades, blade shapes may be derived from blade profiles, calculated in several streamtubes, drawn in a meridional section. The design requires analysis of the equilibrium between the streamtubes and the flow

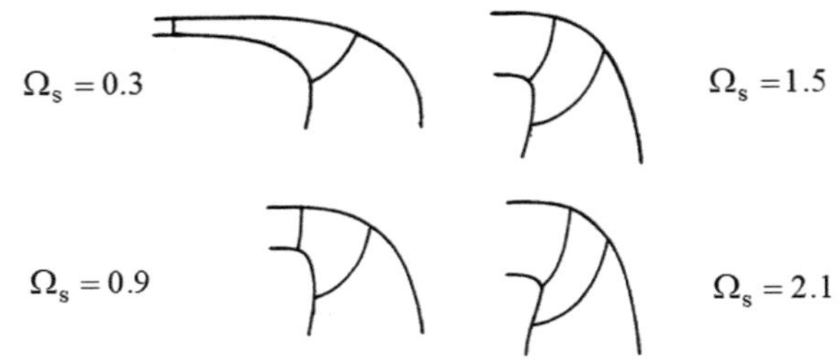

Fig. 9.14 Meridional section shape depending on specific speed with Francis turbines (adapted from Vivier [4])

$\Omega_s = 0.3$

$\Omega_s = 1.5$

$\Omega_s = 0.9$

$\Omega_s = 2.1$

turning in each streamtube. The study results in the chord, the inlet angle and the outlet angle of the blade segment in each streamtube. On the base thereof the blade may be drawn (with a method by Kaplan [2, 4]).

9.4.7 Efficiency Related to Specific Speed

In order acquire insight in the efficiency variation with Ω_s, Table 9.3 lists some characteristic data based on the books of Dietzel [2] and Vivier [4], and the kinetic energy at the rotor exit, derived from the velocity triangles in Fig. 9.13.

With increasing specific speed, the kinetic energy to be recovered in the draught tube is a larger fraction of the head. This is a factor which lowers the efficiency for large specific speed. The larger relative velocity in the rotor with increasing specific speed and associated larger degree of reaction (Fig. 9.13) has the same effect.

On the other hand, Fig. 9.14 demonstrates that the blade surface area in the rotor is large with a low specific speed, which causes large friction losses. Efficiency thus has a maximum as a function of specific speed (see Exercise 9.9.4).

The maximum overall efficiency with Francis turbines is about 93.8% (Itaipu). The corresponding internal efficiency is 95.1%, with 98.6% generator efficiency. The optimum specific speed is somewhat above unity (about 1.20). The internal efficiency drops to about 92% at specific speed 0.3 and to about 90% at specific speed 4 (Kaplan turbine).

The specific speed is limited by cavitation risk, because $v_2^2/2gH$ increases with Ω_s. For given head, the pressure at the rotor exit thus decreases with Ω_s, which explains the maximum allowable specific speed as a function of the head in Fig. 9.4.

Table 9.3 Degree of reaction and rotor exit kinetic energy with Francis and Kaplan turbines

FRANCIS							
Ω_s	0.3	0.6	0.9	1.2	1.5	1.8	2.1
R	0.53	0.56	0.59	0.63	0.67	0.71	0.75
$v_2^2/2gH$	0.022	0.027	0.033	0.043	0.057	0.073	0.095
KAPLAN							
Ω_s	2	3	4	5			
R	0.80	0.85	0.88	0.90			
$v_2^2/2gH$	0.13	0.20	0.28	0.36			

9.4.8 Flow Rate Control with Reaction Turbines

Figure 9.15 sketches the velocity triangles at rotor entry for degrees of reaction 0.55, 0.75 and 0.90 and changes to these triangles aiming at halving the flow rate.

A change of the stator angle causes almost no incidence on the rotor at $R = 0.55$. The angle change is drawn with constant magnitude of v_1. In reality, the magnitude increases somewhat such that the incidence is even smaller than drawn (see Exercise 9.9.7). The flow rate may thus be controlled by adjusting the stator vanes only.

The incidence with reduced flow rate stays small with larger degree of reaction, due to increased magnitude of v_1 with reduced flow rate (see Exercise 9.9.7). The main effect of reduced flow rate with a fixed position of the rotor blades is positive post-whirl at the rotor exit, as can be concluded from Fig. 9.13. The post-whirl reduces the efficiency and the efficiency penalisation at reduced flow rate increases with increasing degree of reaction.

The triangles for degree of reaction $R = 0.75$ in Fig. 9.15 show that post-whirl can be avoided by adjustment of both stator and rotor angles.

The rotor angle cannot be adjusted with Francis turbines (but there are diagonal machines with adjustable rotor blades). The angles of both stator vanes and rotor blades are adjustable with Kaplan turbines.

With a Kaplan turbine, the flow rate may be controlled by only adjusting the rotor angle, as drawn in Fig. 9.15 for $R = 0.90$. By adjusting the rotor blades, there is no incidence at rotor entry. But, with fixed stator vanes, there is negative post-whirl at the rotor exit, as can be concluded from Fig. 9.13. The post-whirl penalises the efficiency, but not very strongly (see Exercise 9.9.8). The post-whirl can be avoided by adjustable rotor blades and stator vanes.

Figure 9.16 renders efficiency versus flow rate for the various turbine types. The Pelton turbine has a very flat efficiency curve. The dependency is stronger with the Francis turbine, increasing with specific speed from 0.5 to 2. The Kaplan turbine with adjustable stator vanes and rotor blades also has a rather flat efficiency curve. The dependency on flow rate increases if only the rotor blades are adjustable.

The variation with the Kaplan turbine with $\Omega_s = 3$ and only adjustable rotor blades is comparable to that of a Francis turbine with $\Omega_s = 1$ (not shown). An axial

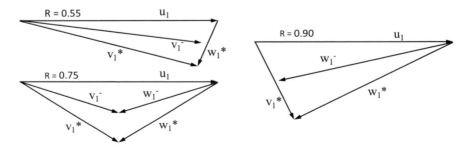

Fig. 9.15 Sensitivity to rotor or stator angle changes for reaction turbines

Fig. 9.16 Efficiency versus flow rate of the various turbine types. (adapted from Vivier [4])

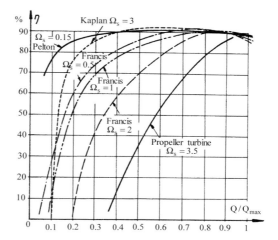

turbine with adjustable stator vanes, but fixed rotor blades (called propeller turbine in the figure), has a very sharp efficiency variation.

9.5 Bulb and Tube Turbines

Kaplan turbines for low head are made with a vertical shaft, with the generator above the tailwater level (Fig. 9.17). The generator is indicated by A (alternator) and the turbine by T. With low-head applications, the supply channel to the turbine is wide. Only part of the flow is guided to the rotor by a volute. The other part is admitted directly. Uniform flow distribution requires an appropriate stay vane ring. The quite strong changes of flow direction, especially in the draught tube, cause relatively high losses when the head is low.

With low head, a horizontal shaft of the turbine-generator set, aligned with the axis of the channel that connects both dam sides is more advantageous (Fig. 9.18). The generator (A) is positioned under the water surface then, which requires mounting in a watertight envelope. The bulb (B) that is formed that way is fixed to the channel walls with profiled struts (C and E). The bulb also houses the adjusting mechanism for the stator vanes (D) and rotor blades (R). The bulb is accessible through the hollow strut E, in which power cables run. The axial rotor with adjustable blades (R) is at the downstream side of the bulb.

Bulb turbines are also applied to exploit tidal energy. A river estuary is closed with a dam containing a series of bulb turbines. At low tide, turbines run with the through-flow in the estuary-sea sense. This is the normal flow sense. At high tide, turbines run in the sea-estuary sense. Rotor blades and stator vanes are turned by about 180° for that, and the sense of rotation is reversed. This impairs the efficiency somewhat.

Fig. 9.17 Kaplan turbine
with low head

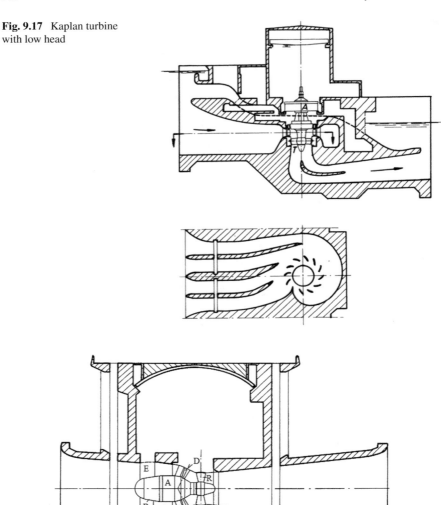

Fig. 9.18 Bulb turbine with low head

Parts a and b in Fig. 9.19 represent the vane and blade positions corresponding with turbine operation in both senses.

When the height difference becomes small towards the end of a turbine phase, it may be profitable to start pumping in the flow sense of this turbine phase. This operation allows an increase of the available head in the following turbine phase. Both turbine and pump functions can be performed by the same bulb machine.

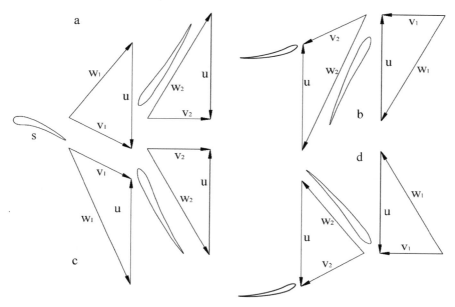

Fig. 9.19 Velocity triangles for turbine operation in normal sense (**a**) and reverse sense (**b**) and for pump operation in normal sense (**c**) and reverse sense (**d**)

Pump operation results by leaving the stator vanes in their position, turning the rotor blades into a position that is approximately mirrored with respect to the meridional plane and reversing the sense of rotation. The positions are shown on parts c and d of Fig. 9.19.

A remark is that pump operation is obtained as well with the blade positions in part a in Fig. 9.19, with reversed sense of rotation. Pumping then occurs in the sea-estuary sense. But leading edges and trailing edges are then not adapted to the flow, which lowers the efficiency. Pump operation as sketched in parts c and d of the figure is somewhat less efficient than the original turbine operation. The torsion of the blades is fixed. Blades may be positioned correctly for an average flow, but positions are not fully correct for all streamtubes.

Pump operation also allows converting electric energy into hydraulic energy by pumping water during periods of low electricity consumption. Functioning as a pumped-storage plant is then obtained. The Rance tidal power/pumped-storage plant in France (near Mont Saint-Michel) operates with these two modes (24 groups of 10 MW).

Small-size plants (10–500 kW) use bulb turbines with an angled gearbox in the bulb and the generator outside the bulb, above the tailwater level. Small-scale applications also allow housing the turbine in a bended tube, with the shaft passing to the outside and the generator in a watertight cellar (Fig. 9.20). Tube type turbines are completely axial ($u_1 = u_2$). The degree of reaction is high. Flow rate control is normally by adjustable rotor blades only.

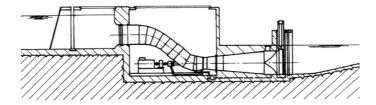

Fig. 9.20 Tube turbine with horizontal shaft

9.6 Reversible Pump-Turbines

Pumped-storage plants first accumulate hydraulic energy by pumping water during periods of low electricity consumption. Water is then supplied to turbines during peak periods. Pumped-storage plants differ in some aspects from the mere combination of a hydraulic power plant and a pump station. This is mainly the case when the four basic components (pump, motor, turbine, generator) are reduced to three (pump, turbine and motor-generator) or even two (reversible pump-turbine and motor-generator).

The disadvantage of the three-machine combination is that the pump is idling during turbine operation, and vice-versa. This disadvantage may be reduced by disconnecting either the turbine or the pump. In practice, only the pump shaft is provided with a coupling, as the pump has much more friction loss than the turbine (the pump is typically multistage and the turbine is single-stage). The common shaft of the three machines may be horizontal or vertical. With a vertical shaft, the electric machine is at the top, the turbine in the middle and the pump at the bottom. This is in view of the priming of the pump and its larger sensitivity to cavitation.

With a combined motor-generator and a combined pump-turbine, two machines are sufficient for the pumped-storage plant. The main advantage is then saving of investment cost. The design of an efficient rotor for both turbine and pump operation is not obvious, however. The discussion of bulb turbines already revealed that an axial hydraulic machine may operate both as turbine and as pump. Both operations are also possible with a radial machine, but with more efficiency penalisation.

Figure 9.21 (left) represents optimal shapes for turbine and pump rotors. The differences are mainly by the limitation to the deceleration in the pump rotor. A rotor serving both purposes must be a compromise between both shapes and be rather close to the optimal pump rotor shape. Figure 9.21 (right) shows the combined rotor. When functioning as a pump, the suction eye resembles the exit side of a turbine rotor. When functioning as a turbine, the rotor entry resembles the exit side of a pump rotor. There are fewer blades than typically with a Francis turbine.

The disadvantages of a two-machine plant are that pump and turbine efficiencies are lower than with a three-machine plant and that changing the operation type requires inversing the sense of rotation, and thus causes interruption time. Due to the slip at rotor exit during pump operation, with no equivalent flow deviation for turbine rotor entry (see Exercise 9.9.9), optimum pump operation requires a higher rotational speed than optimum turbine operation. This is traditionally achieved by pole change

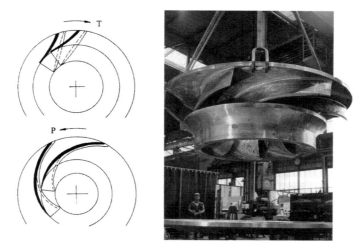

Fig. 9.21 Pump-turbine; left: optimal individual pump and turbine shapes; right: combined pump-turbine rotor (*Courtesy* Andritz Hydro)

in the motor-generator. In modern plants, the motor-generator is a doubly-fed induction machine with variable speed. It has a wound rotor connected to the grid through a frequency converter (see next chapter on wind turbines). Variable rotational speed also improves the efficiency with varying level difference between the reservoirs. This increases the pump-turbine cycle efficiency to about 80%, where without variable speed it is around 75%.

9.7 Pumps Functioning as Turbine

As discussed in the previous sections, pumps may operate as turbine by reversing the sense of rotation and the sense of through-flow, albeit that turbine operation is not perfect. With axial pumps and mixed-flow pumps, the turbine efficiency may be improved by replacing the inlet mouth of the pump by a diffuser. With centrifugal pumps, the role of the diffuser is much less critical due to the much larger head processed, but a diffuser may be added.

Pumps functioning as turbine reach a slightly better maximum efficiency as they do as pump, but this efficiency is much lower than that of a genuine turbine, namely about 80% internal efficiency instead of higher than 90% for a genuine turbine.

The head and flow rate for maximum efficiency as turbine are higher than for maximum efficiency as pump. For a single-stage centrifugal pump of medium specific speed with an axial inlet, the flow rate ratio is about 1.30 and the head ratio is about 1.50 (see Exercise 9.9.9). The ratios depend somewhat on pump type and on specific speed [3].

The turbine characteristic is quite different from that as pump. The pump characteristic has the shape as shown in Fig. 1.18. The shape of the turbine characteristic is as shown in Fig. 1.20, also for a radial machine, thus approximately a parabolic dependence of head on flow rate. Consequently, the turbine characteristic does not change much by changing the rotational speed, because flow rate then varies linearly with the rotational speed while head varies quadratically.

The application is mainly that of throttling device with energy recovery in water handling systems or industrial flow processes. These are applications with rather large processed head. The suitable pump types are thus centrifugal ones. Some manufacturers offer some pump types for such applications, mostly a single-stage end-suction type, a single-stage double-flow type and a multistage high-pressure type (see websites of manufacturers, e.g. Andritz pumps, KSB). Other manufacturers offer genuine turbines for these applications (e.g. Sulzer). Genuine turbines reach higher efficiency but are more expensive.

9.8   Cavitation in Hydraulic Turbines

Cavitation phenomena in hydraulic turbines are much more versatile than in pumps, due to the much larger dimensions which makes that cavitation bubbles away from surfaces have a significant effect, while only cavitation by large attached zones is relevant with pumps. A particular difficulty is that cavitation by non-attached bubbles depends on water quality. With Francis turbines, the relevant phenomena are inlet edge cavitation at the suction or pressure side due to incidence for head larger or smaller than the design value, cavitation at the hub continuing in a whirl cavity in the draught tube for flow rate different from the design value, cavitation by vortices inside the blade passages caused by recirculation zones at low flow rate and travelling bubble cavitation on the blade suction side for design values of head and flow rate [1]. With Kaplan and bulb turbines, the most significant are the inlet edge cavitation types, hub cavitation for flow rate larger than the design value, and an additional type by tip clearance vortices. In design conditions, the travelling bubble cavitation with Francis turbines and the hub cavitation with Kaplan turbines are the most critical. All these cavitation types are very sensitive to water quality. Thus, normally, model tests are necessary and these are delicate [1].

9.9   Exercises

9.9.1 The Pelton turbine rotor, shown in Fig. 9.8, is from the San Carlos hydraulic plant in Colombia. The data are: head 587 m, power per turbine 175 MW, rotational speed 300 rpm, (outer) rotor diameter 4100 mm, 6 injectors per rotor. Assess the suitability of these features, given the overall efficiency of 91%.

A: The diameter for optimal speed ratio ($\lambda = 0.48$) is 3.28 m, being 80% of the outer diameter. The ratio is realistic. The flow rate calculated from the power is 5.57 m^3/s per injector. The specific speed per injector is 0.112. The injector exit diameter calculated from the flow rate is 261 mm ($\phi_s = 0.97$). The diameter ratio is $d/D \approx 0.080$. The values in Table 9.1 show that the specific speed has not been set to the maximum possible value.

9.9.2 The Francis turbine rotor, represented in Fig. 9.11, is from the Karakaya power plant in Turkey. The data are: $H = 156.7$ m, $n = 150$ rpm, $P_e = 340$ MW, $d_1 = 5400$ mm, $b_1 = 1200$ mm. Assess the suitability of the rotor shape. The overall efficiency is 93.5%.

A: The calculated specific speed is about 1. The ratio of width to diameter accords with the meridional shapes shown in Fig. 9.14.

9.9.3 The Kaplan turbine rotor, represented in Fig. 9.12, is from the Jebba power plant in Nigeria. Features are $H = 29.3$ m, $n = 93.75$ rpm, $P_e = 102.7$ MW, $d_{outer} = 7100$ mm. Asses the suitability of the rotor shape. The overall efficiency is 92%.

A: The calculated specific speed is 2.75. This is a possible value.

9.9.4 Estimate the internal efficiencies of the Francis and Kaplan turbines of which the velocity triangles are sketched in Fig. 9.13. Take as distributor loss coefficient $\xi_s = 0.05$ and as diffuser loss coefficient $\xi_d = 0.25$. Estimate the rotor loss coefficient ξ_r by Soderberg's formula: $\xi = 0.025(1 + 3.2/AR)(1 + (\delta°/90)^2)$, with $\delta°$ the flow turning in degrees and AR the aspect ratio of the blade channels: height divided by axial chord. With Table 9.3 and Fig. 9.14, we may estimate the aspect ratios for $R = 0.55, 0.65, 0.75$: AR $= 0.4, 2, 3$. Take $AR = 4$ for the Kaplan turbine with $R = 0.85$. Study the distribution of the losses.

A: The calculated internal efficiencies for $R = 0.55, 0.65, 0.75, 0.85$ are: 0.936, 0.949, 0.928, 0.864. These are quite realistic values, despite the rough estimates of the loss coefficients. The distributor loss (as fraction of the head) decreases with increasing degree of reaction and the diffuser loss increases with increasing degree of reaction. The largest loss is always the rotor loss. It varies through a minimum as a function of the degree of reaction. It is large for small degree of reaction due to the small aspect ratio of the blades. It is large for large degree of reaction, due to large velocity level in the rotor.

9.9.5 Verify by extrapolating the kinematic features rendered in Fig. 9.13, that a Francis turbine with $R = 0.50$ is possible. Draw the velocity triangles. Observe that the ratio of the outlet diameter to the inlet diameter is rather small, which makes the construction difficult. Moreover, the flow turning in the rotor is large.

The results of the optimisation in Fig. 9.13 are $v_{1m}/u_1 \approx 0.25$, $v_{2m}/v_{1m} \approx 1$, $v_{2u} \approx 0$. With these nearly constant values, there are only two kinematic parameters that can vary widely: v_{1u}/u_1, thus degree of reaction (Eq. 9.4), and diameter ratio u_2/u_1. The obtained Francis rotor can be made more practical by enlarging the diameter ratio. The rotor flow turning and exit velocity then increase, which increases the rotor loss. The construction of the rotor thus becomes easier at the expense of some efficiency loss. Therefore, such a rotor is normally not used in practice, which sets the practical limit of the degree of reaction to 0.55.

Verify that it is possible to construct an efficient Francis turbine with a very low degree of reaction. Take as an example $R = 0.15$ and draw the velocity triangles for $v_{1m}/u_1 = 0.25$, $v_{2m}/v_{1m} = 1$, $v_{2u} = 0$. Deceleration in the rotor is just avoided by $w_2 = w_1$. The diameter ratio is then 0.70. Enlarge the diameter ratio to 0.75 in order to obtain a small flow acceleration. The flow turning in the rotor is very large, comparable to that in an impulse type steam turbine. Estimate the internal efficiency with the loss coefficients of the previous exercise. Take $AR = 3$, as in the previous exercise for $R = 0.75$ and diameter ratio 0.75. Observe that the efficiency for $R = 0.15$ is comparable to that for $R = 0.75$. The good efficiency is mainly due to the very large work coefficient of 1.70 for $R = 0.15$. With a low degree of reaction, the rotational speed of the turbine becomes low which makes it impractical for electricity production. But, in the past, low-speed Francis turbines have been built for milling and sawing applications.

A: $u_2/u_1 = 0.375$, $u_2/u_1 = 0.75$, $\eta_i = 0.925$.

9.9.6 Verify that the lift force part in the rotor work is 12.5% for the Francis turbine with $R = 0.75$, of which the velocity triangles are represented in Fig. 9.13. Determine the diameter ratio (u_2/u_1) with a 50% lift force part. Note the much larger diameter ratio with as consequence the larger magnitude of w_2 and the larger rotor flow turning, thus reduced efficiency. It is thus not beneficial to aim at a large contribution of the lift force to the work. The work fraction of the lift force is comparably small for $R = 0.55$ and $R = 0.75$.

A: The required diameter ratio is 0.87 (it is 0.75 in Fig. 9.13).

9.9.7 Argue that for the Francis turbine with $R = 0.65$, of which the velocity triangles are represented in Fig. 9.13, there is only small incidence at rotor entry when the flow rate is halved by closing the stator vanes (rotor blades are fixed). Assume first that the internal efficiency is unchanged, so that the rotor work remains the same. Remark the strong positive post-whirl, which causes an efficiency drop. Estimate then the reduction of the rotor work as equal to the kinetic energy associated to the post-swirl. Express the incidence by the difference between the tangential component of the stator exit velocity (v_{1u}) and the value for incidence-free rotor entry. Repeat the reasoning for $R = 0.55$ and $R = 0.75$.

A: The tangential incidence velocities for $R = 0.65$ are $+3\%$ and -1.5% of u_1 and are thus very small. The values for $R = 0.55$ are $+5.1\%$ and $+2.6\%$. For $R = 0.75$, these are $+3.13\%$ and -3.91%. There is thus never significant incidence at the rotor entry. The efficiency penalisation is caused by the post-whirl, which increases with increasing degree of reaction.

9.9.8 Estimate the efficiency reduction of the Kaplan turbine with $R = 0.85$, of which the velocity triangles are represented in Fig. 9.13, when the flow rate is halved by closing the rotor blades, with fixed stator blades. Assume first that the internal efficiency is unchanged, so that the rotor work remains the same. Observe the negative post-whirl, which causes an efficiency drop. Estimate then the reduction of the rotor work as equal to the kinetic energy associated to the post-swirl. Observe that the efficiency penalisation stays limited.

A: The efficiency is reduced with a factor of about 0.96.

9.9.9 Estimate the ratios of flow rate and head in the best efficiency points of a centrifugal pump functioning as turbine and as pump. The pump has end-suction entry, no diffuser vanes, and a volute. The features are: $u_1/u_2 = 0.4$, $\beta_1^b = -60°$, $\beta_2^b = -50°$, $Z = 6$. Assume a constant meridional velocity component in the rotor. Determine the slip when functioning as pump with Wiesner's formula (Formula 3.21 in Chap. 3). Ignore obstruction by blade thickness. Assume matched rotor and volute for design flow conditions as pump, by which is meant rotor entry and volute entry without incidence. Assume that the relative vortex does not affect the rotor entry flow when functioning as turbine. Assume that best efficiency is obtained with flow rate equal to 85% of the flow rate with rotor entry flow aligned with the blades, both for functioning as pump and as turbine. By the reduced flow rate, the head as pump increases while the head as turbine decreases. Estimate the internal efficiencies as pump and turbine by 0.82 and 0.85.

A: $Q_T/Q_P = 1.30$, $H_T/H_P = 1.50$. The flow angle α_2 at the rotor periphery for matched rotor and volute is $65.0°$.

9.9.10 The figure is a sketch of a Girard turbine. It is an axial hydraulic turbine of impulse type developed around 1850. The rotor turns in the atmosphere. Therefore, pressure is atmospheric pressure at the rotor entry and exit. The zero pressure difference through the rotor makes the analogy with a Laval impulse type steam turbine. The Girard turbine can be analysed similarly. The only essential difference comes from the gravitational potential energy in the rotor, which creates a slight acceleration such that the kinematic degree of reaction becomes slightly positive, where it is slightly negative with a Laval turbine.

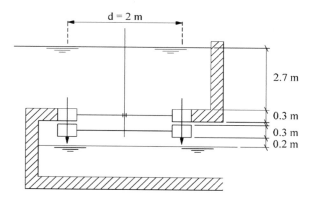

Analyse the flow through the Girard turbine with a mean line representation of the flow (one-dimensional representation). Assume admission over $360°$ to the stator. Take as outflow angle of the nozzle vanes $\alpha_1 = 60°$ and consider the operating point with axial absolute velocity at the rotor exit: $\alpha_2 = 0°$. Determine losses in stator and rotor cascades with Soderberg's formula. Take the aspect ratio of vanes and blades (height to axial chord) equal to unity. Assume constant axial component of the velocity in the rotor.

Determine axial and tangential components of the absolute velocity at the stator exit. Derive the exit velocity from the work equation in the stator. The height difference between the water surface in the water chamber and the exit of the stator vanes is 3.00 m.

Determine the blade speed such that the absolute velocity at rotor exit is in the axial direction. Derive this speed from the work equation in the rotor. Pay attention to the small acceleration in the rotor. The consequence is that rotor blades cannot be exactly symmetric for constant axial velocity. The computation requires iteration for the loss coefficient of the rotor. A starting value may be determined from a computation ignoring rotor loss. The height difference between entry and exit of the rotor is 0.30 m. Ignore the small gap between stator and rotor.

Determine the rotor work and the internal efficiency. The height difference between the rotor exit and the downward water level is 0.20 m. The head supplied to the machine is 3.00 m + 0.30 m + 0.20 m = 3.50 m.

Split the rotor work into the action and reaction parts. Determine the degree of reaction. Observe that it is a low positive number.

Determine the speed ratio. Calculate the spouting velocity from the supplied head of 3.50 m.

Make the balance of the rotor work, the losses in stator and rotor, the kinetic energy at the turbine exit and the downward head. The sum of the energy terms should be equal to the supplied mechanical energy.

Determine the fraction of the exit kinetic energy in the balance. Observe that the exit kinetic energy is the largest part of the losses.

Determine the number of stator vanes and rotor blades from a Zweifel coefficient value of 0.8.

A: $v_{1a} = 3.575$ m/s, $v_{1u} = 6.191$ m/s, $u = 3.213$ m/s, $\beta_1 = 39.8°$, $\beta_2 = -41.9°$, $\Delta W = 19.89$ J/Kg, $\eta_i = 0.579$, $R = 0.0365$, $\lambda = 0.388$, the exit kinetic energy is 18.61% of the supplied mechanical energy, $Z_s = 23$, $Z_r = 50$.

9.9.11 The internal efficiency of the Girard turbine studied in the previous exercise is only about 58%. The main reason is the quite large kinetic energy at the rotor exit. The efficiency may be improved by mounting the rotor in a cylindrical tube and adding a bent diffuser in the style of Fig. 9.17. Due to suction by the downward head and the pressure recovery in the diffuser, a pressure drop is then created through the rotor. Proper functioning thus requires mounting a sealing between the shroud of the rotor and the cylindrical tube.

Analyse the transformed machine with an unaltered stator. This means that the pressure at the stator exit (or rotor entry) remains atmospheric pressure and that the exit velocity of the stator does not change. Assume a pressure recovery coefficient of the diffuser equal to 0.60. Consider again the operating point with exact axial outflow of the rotor.

Determine the expression for the pressure at the rotor exit (= diffuser entry). Due to the suction by the downward head and the pressure recovery in the diffuser, this pressure is lower than atmospheric pressure. Calculate the pressure difference expressed by $(p_a - p_2)/\rho$.

Determine the blade speed such that the absolute velocity at the rotor exit is in the axial direction. This computation requires iteration for the loss coefficient of the rotor. Determine the loss coefficient with Soderberg's formula (Eq. 6.14 in Chap. 6) with aspect ratio equal to 1. A possible start is with a computation ignoring rotor loss. Take the height difference of 0.30 m between rotor entry and exit into account in the calculation of the pressure difference through the rotor.

Determine the rotor work and the internal efficiency. The head supplied to the machine is still 3.50 m. Observe the much improved efficiency.

Split the rotor work in the action and reaction parts. Determine the degree of reaction. Observe that it is now much higher than with the Girard turbine.

Determine the speed ratio. Calculate the spouting velocity from the supplied head of 3.50 m.

Make the balance of the rotor work, the losses in stator, rotor and diffuser. The sum of the energy terms should be equal to the supplied mechanical energy.

Observe, with respect to the Girard turbine, that the rotor loss is larger, but that the diffuser loss is much smaller than the sum of the losses due to exit kinetic energy and downward head without diffuser.

A: $v_{1a} = 3.575$ m/s, $v_{1u} = 6.191$ m/s, $u = 4.065$ m/s, $\Delta W = 25.168$ J/kg, $\eta_i = 0.733$, $R = 0.239$, $\lambda = 0.491$.

9.9.12 Results of the previous exercise are that the efficiency improves much by adding the diffuser and that the speed ratio increases due to the increased degree of reaction. The speed ratio may be further increased by opening the stator vanes. Analyse the effect of opening the vanes from $\alpha_1 = 60°$ to $\alpha_1 = 45°$ and to $\alpha_1 = 30°$. Keep the axial velocity. The flow rate is thus unchanged.

Determine axial and tangential components of the absolute velocity at the exit of the stator vanes.

Determine the expression for the pressure at the exit of the stator. This pressure is higher than atmospheric pressure. Calculate the pressure difference expressed in energy measure by $(p_1 - p_a)/\rho$.

Repeat the steps of the previous exercise. Observe that the efficiency increases by opening the stator vanes to $\alpha_1 = 45°$, but decreases by opening to $\alpha_1 = 30°$. This shows that there is an optimum stator vane angle.

A: $v_{1a} = 3.575$ m/s, $v_{1u} = 3.575$ m/s, $u = 7.405$ m/s, $\Delta W = 26.430$ J/kg, $\eta_i = 0.770$, $R = 0.758$, $\lambda = 0.892$.

$v_{1a} = 3.575$ m/s, $v_{1u} = 2.064$ m/s, $u = 11.322$ m/s, $\Delta W = 23.369$ J/kg, $\eta_i = 0.681$, $R = 0.909$, $\lambda = 1.366$.

References

1. Avellan F (2004) Introduction to cavitation in hydraulic machinery. In: Proceedings 6th international conference hydraulic machinery and hydrodynamics, p 11–22
2. Dietzel F (1980) Turbinen, Pumpen und Verdichter. Vogel Verlag, ISBN 3-8023-0130-7

3. Kramer M, Terheiden K, Wieprecht S (2018) Pumps as turbines for efficient energy recovery in water supply networks. Renew Energy 122:17–25
4. Vivier L (1966) Turbines Hydrauliques et Leur Régulation. Albin Michel, Paris, no ISBN (This rather old book is still relevant for theory, application and design of hydraulic turbines)

Chapter 10
Wind Turbines

Abstract In this chapter, we discuss the different types of wind turbines and the basic technical aspects of large wind turbines for electricity generation. We analyse the performance of wind turbines and discuss their adaptation to a wind regime.

10.1 Wind Energy

Wind is air circulation in the atmosphere caused by irregular warming by the sun. Wind energy systems use the kinetic energy of the wind. We generally speak of a Wind Energy Conversion System (WECS). Mostly, a system for electricity generation encompasses a rotor, a gearbox, a generator and a tower. The term Wind Turbine (WT) is commonly used to name the whole.

A particular feature of wind energy is its very low power density. The yearly average wind speed at 100 m height along the West European coast between Brittany and Denmark is about 8.0–8.5 m/s. The optimum energy yield is typically obtained by designing the system such that maximum power of the generator (called rated power) is reached at a wind speed (called rated wind speed) about 20% higher than the yearly average speed (see Sect. 10.4: wind regime). The energy flux of the undisturbed wind through a plane surface (area A) perpendicular to the wind direction is (mass flow rate × kinetic energy):

$$P_0 = \rho A v \left(\frac{1}{2}v^2\right) = \frac{1}{2}\rho v^3 A. \qquad (10.1)$$

To $\rho = 1.225$ kg/m^3 and $v = 10$ m/s corresponds 600 W/m^2. A wind energy system converts at maximum about 45% of the energy flux (see Sect. 10.3: performance). This results in a net power density of about 270 W/m^2. A rated power of 5 MW thus requires a through-flow area of about 18,500 m^2, corresponding to a diameter of about 153 m. This demonstrates that large power wind turbines have very large dimensions.

E. Dick, *Fundamentals of Turbomachines*, Fluid Mechanics and Its Applications 130, https://doi.org/10.1007/978-3-030-93578-8_10

The total installed wind power in the world for electricity generation was about 750 GW at the end of 2020 with about 90 GW added in 2020 (see website WWEA: World Wind Energy Association). The yearly growth rate of the installed power is thus very large.

10.2 Types of Wind Energy Conversion Systems

10.2.1 Drag Machines

There is a difference between machines driven by lift (force perpendicular to the relative velocity) or by drag (force in the direction of the relative velocity). A simple example of a drag machine is the cup anemometer (Fig. 10.1). The principle is that a cup whose concave side is facing the wind experiences a larger force than a cup with the convex side to the wind. There are many forms of such machines. Rather common is the Savonius rotor (Fig. 10.1) (Savonius 1924).

Drag machines have three major disadvantages. First, the average blade speed cannot exceed the wind speed. The consequence is a low rotational speed, except for very small machines. This is very disadvantageous for many applications. Further, the efficiency is low. The efficiency is expressed by a power coefficient:

$$C_P = \frac{P}{\frac{1}{2}\rho v_0^3 A}, \tag{10.2}$$

where P is the power produced by the machine, v_0 the undisturbed wind speed and A is the through-flow area perpendicular to the wind direction. The through-flow area is the area covered by the rotating parts of the machine. As already mentioned, the power coefficient of a modern lift-based wind energy conversion system is about

Fig. 10.1 Drag machines;
top left: cup anemometer;
right: Savonius rotor

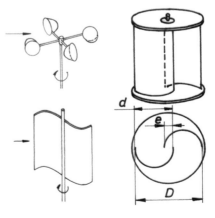

0.45. For a drag machine it is much lower, typically around 0.20, due to large forces in opposite sense to the rotation and thus dissipating energy. The third disadvantage with drag machines is their large solidity. With solidity is meant the ratio of the blade area to the through-flow area.

The blade area is usually counted as projected to a plane, thus the product of twice the width d and the height in Fig. 10.1. The solidity of drag machines typically exceeds unity. Drag machines are only rarely applied and only for low-power applications. An example is the Savonius rotor on the roof of vehicles to drive a fan in the interior.

10.2.2   High-Speed Horizontal-Axis Wind Turbines

The most common wind turbine type is an axial turbine with force generation by lift. Figure 10.2 (left) shows a three-blade horizontal-axis turbine. The shaft is approximately horizontal and aligned with the wind. Figure 10.2 (right) is a sketch of the blade in the lowest position with the velocity triangle and the force components in a section at approximately a quarter of the span from the hub. The wind velocity at the place of a rotor blade (v) is reduced compared to the free wind velocity (v_0) and deviates somewhat from the original direction. From the velocity triangle follows the origin of the lift (L) with a component in the sense of the rotation.

The tangential component of the lift (L_u) drives the blade section. The inverse of this component acts on the through-flow and causes the tangential flow deviation. The

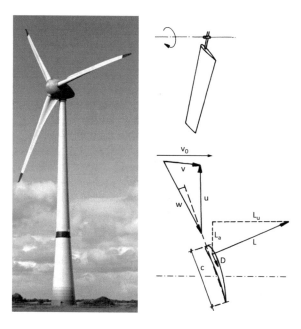

Fig. 10.2 Horizontal-axis wind turbine (HAWT): left: E126 (*Courtesy* Enercon); right: velocity triangle and force components in a section of the lowest blade

axial component of the lift (L_a) is much larger and its inverse causes the axial flow retardation. The drag force (D) is much smaller than the lift force and its tangential component opposes the motion. Nearer to the blade tip, the blade velocity (u) is larger, while the through-flow velocity (v) stays approximately the same. The blade section is thus more tangential and the blade speed at the rotor periphery is much higher than the wind speed. The tip speed ratio is defined by

$$\lambda_T = u_T/v_0, \tag{10.3}$$

where u_T is the blade speed at the rotor tip. The range of values for optimum operation is 6–8 (see Sect. 10.3.4).

10.2.3 Large Horizontal-Axis Wind Turbines for Electricity Generation

Figure 10.2 (left) is a front view of a three-blade horizontal-axis wind turbine for electricity generation (manufacturer Enercon). The rotor diameter is 127 m. The rated power is 3.5 MW (3 and 4 MW are also possible) for locations with a medium–high average wind speed (8.5 m/s at hub height; coast location).

At present (2021), turbines for electricity generation by utilities are mostly of the 2–5 MW order. But turbines of the 6–12 MW order are already available or are under development (see websites of manufacturers: Enercon, Vestas, Nordex, Siemens-Gamesa Renewable Energy, General Electric Renewable Energy). There is a strong tendency to the largest possible power.

Larger machines are cheaper per unit of power. Further, at larger hub height, the average wind speed is higher. For off-shore installation, the hub height is normally somewhat smaller than the rotor diameter (higher wind speed near the surface than for an inland location), for a coastal area about equal, and for an inland location, typically somewhat larger than the rotor diameter. But a smaller or larger hub height may be chosen in order to adapt to the local wind regime. Mostly, machines are installed in a group, called a wind farm, or in a row. Prevention of too strong mutual interference requires a spacing of at least 5 diameters.

Large turbines have three rotor blades. In the past, some large turbines had two blades, because two blades with a larger average chord cost less than three blades with a smaller average chord. At present, all large turbines have 3 blades (there are small turbines with 2 blades). A three-blade machine turns more steadily than a two-blade one. Fluctuation of power as a result of a larger wind force on an upper blade and a lower wind force on a lower blade is much smaller with a three-blade rotor.

Rotor blades are manufactured of glass fibre reinforced epoxy or polyester resin (epoxy is much lighter than polyester) or carbon fibre reinforced epoxy resin (carbon fibre is much stiffer and much lighter than glass fibre). With large rotor blades, typically, carbon fibre is used for the spar (the central element that takes most of the forces) and glass fibre for the surface parts, all with epoxy resin.

Blade profiles vary strongly along the span. At the tip, only a low lift coefficient is required because of the high relative flow velocity, but a very low drag coefficient is necessary. The profile required is similar to that of an aircraft wing, i.e. a laminar profile, but sensitivity for contaminants should be low. Towards the hub, a very high lift coefficient is necessary and a low drag coefficient is less important, i.e. a turbulent profile. The profiles near the hub are no standard NACA turbulent profiles, but profiles with an increased lift coefficient by aft-loading (significant pressure difference in the rear part; see Fig. 2.8 in Chap. 2). Along the whole span, profiles with a large relative thickness are needed, typically around 18% near the tip and 36% near the hub, because of strength.

The blades are rigidly fixed to the hub, but have a high bending flexibility. The rotor has a small cone angle in order to reduce the bending moment on the hub. The centrifugal force moment thus compensates the blade force moment. Sufficient clearance from the tower requires a small tilt angle of the shaft (small slope compared to the horizontal direction). With two-blade turbines, articulated rotors are typically applied. The rotor hinges then around an axis perpendicular to the turbine shaft. This enables absorption of the force difference due to the higher wind velocity on the upper blade and the lower wind velocity on the lower one.

The rotor shaft is held in a nacelle housing the gearbox (if present) and the generator. The rotational speed depends on the turbine size. The optimum tip speed ratio is, as already said, 6–8, with the lowest value for smaller turbines (somewhat larger solidity; see Sect. 10.3) and the highest value for larger turbines. E.g., with a diameter of 150 m for a turbine adapted to a medium–high average wind speed at hub height, about 10 m/s rated wind speed, and tip speed ratio 8, the tip speed is about 80 m/s and the corresponding rotational speed about 10 rpm. For a diameter of 45 m and a tip speed ratio of 6, the rotational speed is about 25 rpm.

The rotor is always set upwind of the tower. Downwind types were used in the past. A downwind rotor follows spontaneously changes in wind direction, but a big disadvantage is that the rotor blades turn through the wake of the tower. This causes efficiency loss and fatigue load. Upwind rotors require an active yaw mechanism. The nacelle is on the top of the tower, on a slewing ring with an internal gear, driven by yaw motors and controlled by the wind direction detected with a vane on the nacelle. Computer control is applied, as sudden changes of wind direction must not be reacted to immediately. Brakes are applied to spare the yaw mechanism. The tower is usually a steel tube tower. Large towers may have a concrete lower part. Small systems sometimes use a lattice tower.

Figure 10.3 represents the longitudinal section of a nacelle of a turbine with blade pitch control. It is a mid-size two-blade turbine of 500 kW that is no longer manufactured, but the components are similar in any wind turbine. Some parts are: hub cover, pitch control mechanism (the pitch angle is the angle between the blade chord and the axial direction on a reference radius), gearbox, generator, mechanical brake, yaw bearing, yaw motor.

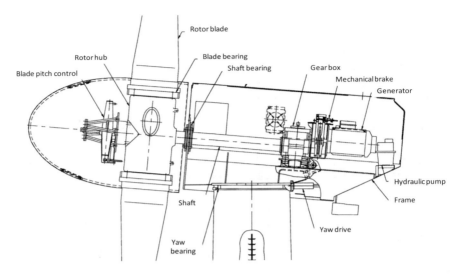

Fig. 10.3 Parts of a HAWT nacelle (former manufacturer NedWind)

At wind speed above the rated value, the possible rotor power exceeds the maximum generator power. The rotor power must then be limited. There are two common systems to achieve this: *pitch control* and *stall control*. Stall control is often used on mid-size wind turbines. We discuss it in the next section. With large wind turbines, only pitch control is used. With *pitch control*, the blades are turned towards the axial direction at higher wind velocities, causing a decrease of the angle of attack over the whole span. Pitch control requires rotor blade mounting to the hub by a bearing (slewing ring). A lever mechanism on the turning hub is activated by a rod in the hollow shaft, driven by a hydraulic actuator (as in Fig. 10.3; but level arms are not detailed) or the individual blades are provided with electrical or hydraulic pitch actuators (most common).

By positioning the blades aligned with the wind direction, the rotor is made powerless. This is called feathering the blades. At very low wind speed, the rotor is feathered and immobilised by a mechanical brake. The turbine is started by releasing the brake and turning the blades away from the feathered position. Feathering is also applied during storms. By feathering the rotor and immobilising it by the brake, rotor blade stresses are minimal. It must also be possible to stop the turbine in case of interruption of the grid connection. This is done with the brake, followed by feathering the rotor. In all cases, the rotor may be turned 90° out of the wind, as an ultimate measure, by means of the yaw mechanism.

Two variable-speed generator types are used. Both produce also reactive power. A first type is an asynchronous generator with a wound rotor connected to the grid through a frequency converter (called a doubly-fed induction generator). Due to the frequency control, the rotor magnetic field rotates with adjustable speed. The rotational speed is governed by the frequency difference between the stator and the rotor. A speed range of about ±30% compared to the synchronous speed (i.e. the speed corresponding to the grid frequency) is typical. At a sub-synchronous rotational speed, grid power flows to the rotor. At a super-synchronous rotational speed, rotor power flows to the grid. A ±30% speed change requires a frequency converter power of ±30% of the stator power. The second type is a synchronous generator connected to the grid through a frequency converter, converting the entire power. A typical ±30% speed change is also possible with this type. Both wound rotor generators with electric excitation and permanent magnet generators are applied. Synchronous generators can have a large number of pole pairs, and thus can turn at low rotational speed so that a gearbox may be unnecessary (direct drive).

Large wind turbines may be equipped with a direct-drive synchronous generator with a very large number of poles, with a one-stage planetary gearbox and a synchronous generator with a lower number of poles, or a three-stage gearbox with an asynchronous generator with rotor frequency control. The direct drive system needs a wide nacelle (Fig. 10.2). The system with the three-stage gearbox requires a long nacelle (Fig. 10.3). The nacelle of the mixed concept with a one-stage gearbox and a multi-pole generator is the most compact one.

10.2.4 Mid-Size and Small Horizontal-Axis Wind Turbines for Electricity Generation

The power of mid-size wind turbines is in the range of 200 kW–1 MW. These turbines are used on remote locations, typically combined with a diesel generator group. The turbines are simpler than the previously discussed large types with the objective to reduce the cost per unit of power.

The generator is normally an asynchronous machine with a squirrel-cage rotor with fixed rotational speed, except for slip variation. Pole changing is sometimes applied, allowing the generator to turn at lower speed with low wind velocities.

With a fixed rotational speed, the optimum power coefficient is only reached at one single wind speed. The energy yield is thus lower than with pitch control where the rotational speed may be kept optimal through a certain wind speed range. The main electrical disadvantage is the inability to produce reactive power. Capacitors are added to compensate reactive power consumption.

The power control is mostly by *stall control*. The blade speed and the relative velocity vary strongly along the span (see velocity triangle in Fig. 10.2). Equal energy capture in all streamtubes requires a constant product of the blade speed and the tangential component of the lift. The relative velocity is large near the tip, so that generation of the required lift needs only a small chord and a small angle of attack. A large chord and a large angle of attack are required near the hub. The blade chord is close to the tangential direction at the tip, but near the hub it is much nearer to the axial direction. Due to the variation of the angle of attack along the radius, the torsion of the blade is smaller than the angle variation of the relative velocity.

As the angle of attack is already very high at the hub in design conditions, a small increase of the wind velocity causes flow separation with an unaltered blade chord angle and blade speed. As wind velocity increases, the separation zone grows towards the tip. There is thus a spontaneous power limitation with increasing wind speed for constant rotational speed. A blade may be designed appropriately, so that the power is approximately limited to the rated power with wind speeds exceeding the rated value.

With stall control, the rotor is stopped during storms by a mechanical brake. The mechanical stress on the rotor blades is then higher than on a feathered rotor, but no strong problem arises with wind turbines up to the 1 MW size, in a moderate climate. The largest turbines in the mid-size range may be provided with pitch control or with *active stall control*. Stall control as described up to now is called *passive stall control*. There is also an active stall control system. Power limitation at higher wind speeds is achieved by stall, but the blade angle is adjusted such that the power gets exactly limited to the rated power.

Turbines with passive stall control have a mechanical brake and an aerodynamic brake. The aerodynamic brake functions by setting the rotor blade tip perpendicularly to the blade speed, reducing the power to a low value.

A turbine with stall control is started by releasing the brake. The rotor starts spontaneously if the design value of the tip speed ratio is not too large, which means values up to $\lambda = 6$ at optimum rotor operation. Rotors with higher design tip speed are normally not self-starting because the blades are mounted more tangentially.

Much simpler turbines are used for small-scale applications (200 W–10 kW). Very small turbines have high rotational speed, enabling direct coupling to the generator. These are mostly generators with permanent magnets. Small turbines are not computer-controlled. They are normally power-limited by turning out of the wind (see next section).

10.2.5 Low-Speed Horizontal-Axis Wind Turbines

For pumping, often a type as sketched in Fig. 10.4 is used. The working principle is identical to that of the previously discussed types, but the solidity is much larger. The solidity of a horizontal-axis wind turbine for power generation is 6–8%. The solidity of the machine represented in Fig. 10.4 is about 50%. With a larger solidity, the maximum power extraction from the wind is reached at a lower blade speed and the power coefficient is lower (see Sect. 10.3: performance). A high torque is obtained, as required for pump driving.

Turbines of this type are applied on locations where electric power supply is difficult, as on remote meadows and crop fields. Dimensions are limited. The rotor is kept in the wind by a vane. It is turned partially out of the wind at high wind speed . This may be realised by mounting the rotor on a frame with a hinge and a spring.

Fig. 10.4 Low-speed
horizontal-axis wind turbine
for pump drive

The rotor is turned either upwards or sideways by the axial force on the rotor or by the force on a plate perpendicular to the wind (not represented in Fig. 10.4).

10.2.6   Vertical-Axis Wind Turbines

Figure 10.5 represents a type with straight prismatic blades and a vertical shaft. A horizontal section is shown, with one blade in four positions. In the upwind and downwind positions, the velocity triangles cause a lift force with a tangential component in the sense of the blade speed. No lift is generated in the other two positions. The force on a rotor blade changes cyclically with this type. At least three rotor blades are required to achieve a constant torque.

Advantages of vertical-axis wind turbines are that they do not have to be turned to the wind direction and that the load is coupled at ground level. A disadvantage is that the turbine is not self-starting when the blade position is fixed. Therefore, sometimes, a vertical-axis wind turbine is combined with a Savonius rotor.

Self-starting may be achieved with prismatic blades by mounting these on pivots and by controlling their angle through rods. A simple mechanism is joining the rods in a point downstream of the shaft (Fig. 10.6). This point is set by a wind vane.

The centrifugal force causes large bending moments in vertical-axis wind turbines with straight blades. Turbine dimensions must be limited, therefore. The maximum blade length is about 15 m. A larger size is possible by bowing the blades into a troposkin shape (Greek for rotating rope). With this shape, only traction force occurs as a result of centrifugal force and gravity. This kind of turbine is commonly named a Darrrieus turbine (Darrieus 1931). A two-blade example is sketched in Fig. 10.6. Darrieus turbines are not self-starting. The largest turbine ever built had a height of 96 m, a diameter of 64 m and yielded 4 MW.

The fluctuating lift with vertical-axis wind turbines impairs the efficiency compared to that of horizontal-axis wind turbines, as there is drag, even with low lift.

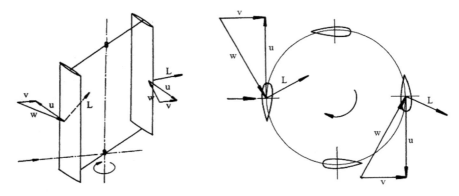

Fig. 10.5 Vertical-axis wind turbine with straight blades (VAWT)

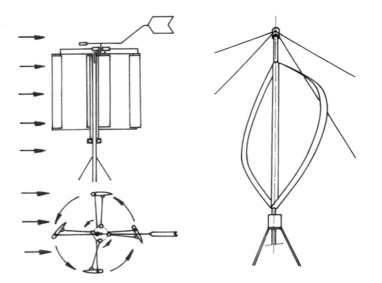

Fig. 10.6 Vertical-axis wind turbines; left: control mechanism with straight blades; right: troposkin shape

Further, straight-blade turbines always need drag generating struts. Upper and lower areas do not perform well with a troposkin shape. Due to the low blade speed, these blade parts cyclically enter in stall. The velocity triangles in Fig. 10.5 demonstrate that, at maximum lift ($\alpha \approx 12.5°$), the ratio of the blade speed on the equator to the local wind speed is about 4.5. With the deceleration of the wind at the turbine, the speed ratio based on the free wind speed is about 3. VAWTs thus typically run much slower than HAWTs.

Power limitation at wind speeds exceeding the rated value is problematic with a VAWT. There is a spontaneous mechanism of progressive stall with the troposkin shape, but it is impossible to limit the power to an approximate constant value as with a stall-controlled HAWT. A straight-blade VAWT cannot be limited in power by simple means. Machines with adjustable blade slope have been proposed, enabling decrease of the through-flow area, but these are mechanically vulnerable. Due to the lower power coefficient and the intrinsic difficulty of power limitation, VAWT types are only applied for small power.

10.3 Wind Turbine Performance Analysis

A simplified representation of the rotor of a horizontal-axis wind turbine is by a disc of infinitesimal thickness on which the energy extraction from the flow occurs abruptly and uniformly in the circumferential direction. Such a turbine rotor representation is called an actuator disc. It replaces the actual rotor by one with an infinite number of blades with infinitesimal chord. Consequently, information on the rotor solidity, i.e. the ratio of blade area to through-flow area, is lost.

A further simplified representation of the flow through the turbine rotor is by a one-dimensional loss-free flow. This means flow direction perpendicular to the rotor disc, uniform distribution of flow parameters perpendicular to the flow direction, neglect of tangential velocity components downstream of the turbine rotor, uniform energy extraction across the rotor disc and neglect of losses. For energy extraction without whirl downstream of the rotor, considering the Euler work equation, an infinitely high rotational speed of the rotor has to be assumed. Information concerning rotor speed is thus lost.

Despite the radical simplifications, basic results on the energy extraction by the turbine rotor can be obtained from momentum relations on a one-dimensional loss-free flow. We will derive then in two further steps how the effects of rotor speed, rotor solidity and blade drag can be taken into account.

10.3.1 Single-Streamtube Momentum Analysis

Figure 10.7 sketches the one-dimensional flow through the turbine rotor. Far upstream of the turbine, the wind velocity is v_0 and the pressure is p_0. The flow is retarded by the rotor disc, so that the velocity v_1 just upstream of the rotor is lower than v_0, and, consequently, the pressure p_1 is higher than p_0.

Just downstream of the turbine rotor, the velocity v_2 equals v_1 (continuity), but the pressure p_2 is lower than p_1. The pressure difference $p_1 - p_2$ thus expresses the energy extraction from the flow. Downstream of the turbine, the pressure recovers to p_0, which decreases the velocity to v_3. The velocity and pressure evolutions are represented in Fig. 10.8.

Fig. 10.7 One-dimensional streamtube through a horizontal-axis wind turbine

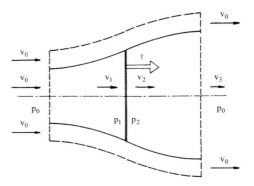

Fig. 10.8 Velocity and pressure evolution with one-dimensional flow

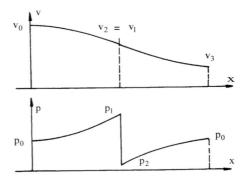

The relation between velocity and pressure follows from the work equations upstream and downstream of the disc (no work):

$$\frac{p_0}{\rho} + \frac{v_0^2}{2} = \frac{p_1}{\rho} + \frac{v_1^2}{2}, \quad \frac{p_2}{\rho} + \frac{v_2^2}{2} = \frac{p_0}{\rho} + \frac{v_3^2}{2},$$

from which by addition ($v_1 = v_2$):

$$\frac{p_2}{\rho} + \frac{v_0^2}{2} = \frac{p_1}{\rho} + \frac{v_3^2}{2}, \quad \text{or} \quad \frac{p_1 - p_2}{\rho} = \frac{1}{2}\left(v_0^2 - v_3^2\right). \tag{10.4}$$

The force on the rotor disc is

$$T = A(p_1 - p_2). \tag{10.5}$$

The momentum balance on the streamtube, with the force $-T$ on the flow, is

$$\dot{m}(v_3 - v_0) = -T, \tag{10.6}$$

where $\dot{m} = \rho v_1 A$ is the mass flow rate through the rotor disc.

When writing (10.6), it is assumed that the force by the pressure on the envelope of the streamtube equals zero. That supposes a symmetric pressure distribution in Fig. 10.8. This assumption may be avoided by applying a second, sufficiently wide, streamtube around the streamtube passing through the rotor, allowing the assumption of atmospheric pressure p_0 on the envelope of the outer tube (Fig. 10.7). There is no work in the second streamtube. As a consequence, the outlet velocity is v_0.

A momentum balance on both streamtubes together results in (10.6), without any further assumptions.

Combination of (10.4–10.6) results in

$$T = \rho v_1 A(v_0 - v_3) = A(p_1 - p_2) = \frac{1}{2} A \rho \left(v_0^2 - v_3^2 \right),$$

so that

$$v_1 = \frac{1}{2}(v_0 + v_3).$$

This means that the flow deceleration is symmetric, i.e. the same upstream and downstream of the rotor.

The velocity decrease is expressed by an interference factor a, so that

$$v_1 = v_0(1 - a), \quad v_3 = v_0(1 - 2a).$$

The power gained from the flow is

$$P = \dot{m} \left(\frac{1}{2} v_0^2 - \frac{1}{2} v_3^2 \right) = \rho A v_1 \frac{1}{2} \left(v_0^2 - v_3^2 \right).$$

The power coefficient is

$$C_P = \frac{v_1}{v_0} \left[1 - \left(\frac{v_3}{v_0} \right)^2 \right] = (1 - a)(4a - 4a^2) = 4a(1 - a)^2. \qquad (10.7)$$

The power coefficient reaches a maximum for

$$a = \frac{1}{3},$$

with maximum value

$$C_{P,\max} = \frac{16}{27} \approx 0.593. \qquad (10.8)$$

The obtained maximum value is called the Betz-limit. It is the approximate upper bound of the power coefficient of a wind turbine.

The derivation is well applicable to a horizontal-axis wind turbine, but the result applies, by extension, to any wind energy system, as only momentum and energy relations are used.

The force on the actuator disc follows from (10.6) in a dimensionless form:

$$C_T = \frac{T}{\frac{1}{2}\rho v_0^2 A} = 2\frac{v_1}{v_0}\left(1 - \frac{v_3}{v_0}\right) = 2(1-a)2a.$$

With $a = \frac{1}{3}$ follows:

$$C_T = \frac{8}{9} \approx 0.889.$$

A completely tight, plane plate, placed perpendicularly to the wind has a drag coefficient of about 1.28. The axial force on a working wind turbine is thus about 70% of the force exerted on a tight, plane plate. This puts a very large axial force on the rotor and the tower.

From Eq. 10.7, it follows that only a fraction of the kinetic energy flux in an undisturbed flow through the rotor disc can be captured. As the flow retards by the wind turbine rotor, the mass flow passing through the rotor decreases. There is thus an optimal retardation, found to be $a = \frac{1}{3}$ according to the preceding derivations.

10.3.2 Multiple-Streamtube Momentum Analysis

The analysis in the preceding section may be extended by considering a series of concentric annular streamtubes with an infinitesimal thickness and by writing the momentum and work relations for each streamtube. With this analysis, the radial distribution of flow parameters and the flow whirl downstream of the rotor can be taken into account. By the extension, the effect of the rotational speed is captured. The extension requires two assumptions which are plausible, but which need a posteriori verification. For this last aspect, we refer to books, e.g. [1].

Figure 10.9 represents an infinitesimal streamtube and the velocity triangles immediately upstream and downstream of the rotor disc. At the position of the rotor disc, the axial velocity component is represented by $v_{1a} = v_{2a} = (1-a)v_0$. There is no tangential velocity component immediately upstream of the rotor. The tangential velocity component immediately downstream of the rotor is represented by $-2b\,u$.

A first essential assumption is a symmetric pressure evolution in the streamtube. The momentum balance in axial direction is then the same as with the preceding analysis:

$$dT = (dA)(p_1 - p_2) = \rho v_{1a}(dA)(v_0 - v_{3a}), \quad \text{with} \quad dA = 2\pi r dr.$$

Fig. 10.9 Streamtube with infinitesimal thickness through a horizontal-axis wind turbine rotor

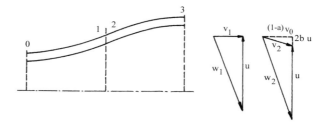

The pressure drop relation is:

$$\frac{p_1 - p_2}{\rho} = v_{1a}(v_0 - v_{3a}).$$ (10.9)

Without losses, the work equations upstream and downstream of the rotor (no work) are:

$$\frac{p_a}{\rho} + \frac{v_0^2}{2} = \frac{p_1}{\rho} + \frac{v_1^2}{2}, \quad \frac{p_2}{\rho} + \frac{v_2^2}{2} = \frac{p_a}{\rho} + \frac{v_3^2}{2}.$$ (10.10)

Thus:

$$\frac{p_1 - p_2}{\rho} = \frac{v_0^2 - v_1^2}{2} + \frac{v_2^2 - v_3^2}{2} = \frac{v_0^2 - v_{3a}^2}{2} + \frac{v_{2u}^2 - v_{3u}^2}{2}.$$ (10.11)

Combination of (10.9) and (10.11) results in

$$v_{1a}(v_0 - v_{3a}) = \frac{v_0 + v_{3a}}{2}(v_0 - v_{3a}) + \frac{v_{2u}^2 - v_{3u}^2}{2}.$$ (10.12)

Due to the radius increase within the streamtube downstream of the rotor, v_u decreases (constant rv_u): v_{3u} is thus smaller than v_{2u}. The simplest for the velocity expression (10.12) is to set the difference between v_{3u} and v_{2u} equal to zero. This may be achieved by assuming (second essential assumption) that the decrease of the kinetic energy associated to the tangential velocity component does not contribute to the pressure recovery downstream of the rotor. The decrease of this kinetic energy is then defined as a loss. This means replacement of the work equations (10.10) by:

$$\frac{p_a}{\rho} + \frac{v_0^2}{2} = \frac{p_1}{\rho} + \frac{v_1^2}{2}, \quad \frac{p_2}{\rho} + \frac{v_{2a}^2}{2} = \frac{p_a}{\rho} + \frac{v_{3a}^2}{2}.$$ (10.13)

With this simplification it follows that

$$v_{1a} = v_{2a} = \frac{1}{2}(v_0 + v_{3a}).$$ (10.14)

The flow deceleration is then, as with a one-dimensional streamtube analysis, the same upstream and downstream of the rotor.

We set

$$v_{1a} = v_{2a} = v_0(1 - a), \quad v_{3a} = v_0(1 - 2a), \quad v_{2u} = -u(2b).$$

The pressure drop relation (10.9) then becomes:

$$\frac{p_1 - p_2}{\rho} = v_0^2(1 - a)2a \tag{10.15}$$

The work equation in the relative frame between the positions just upstream and just downstream of the rotor (no work), in absence of losses, is

$$\frac{p_1}{\rho} + \frac{v_{1a}^2}{2} + \frac{u^2}{2} = \frac{p_2}{\rho} + \frac{v_{2a}^2}{2} + (1 + 2b)^2 \frac{u^2}{2}, \tag{10.16}$$

or

$$\frac{p_1 - p_2}{\rho} = 2b(1 + b)u^2. \tag{10.17}$$

Combination of (10.15) and (10.17) produces a relation between the interference factors:

$$a(1 - a) = b(1 + b)\lambda_r^2, \tag{10.18}$$

with λ_r the local speed ratio: $\lambda_r = u/v_0$.

The energy balance (10.16) may also be written as:

$$\frac{p_1}{\rho} + \frac{v_1^2}{2} = \frac{p_2}{\rho} + \frac{v_{2a}^2}{2} + 2bu^2 + \frac{(2bu)^2}{2} = \frac{p_2}{\rho} + \frac{v_{2a}^2}{2} + 2bu^2 + \frac{v_{2u}^2}{2}.$$

We recognise then the term $2bu^2$ as the Euler work: $\Delta W = u(2bu) = u(v_{1u} - v_{2u})$.

Work (power) gained from the streamtube is

$$dP = \rho(dA)v_0(1 - a)(2bu^2). \tag{10.19}$$

With (10.18), this is also

$$dP = \rho(dA)v_0^3 \frac{2a(1 - a)^2}{1 + b}. \tag{10.20}$$

Expression (10.20) replaces (10.7) in an infinitesimal form. Equation (10.7) is recovered for $b = 0$ (no post-whirl) and $a = $ constant along the radius.

Table 10.1 Interference factors in multiple-streamtube analysis

λ_r	a	b	$(C_P)_{loc}$
∞	1/3	0	0.5926
10	0.3331	0.00222	0.5913
6	0.3327	0.00613	0.5890
3	0.3307	0.02402	0.5787
2	0.3279	0.05235	0.5630
1	0.3170	0.18301	0.5000

The power (10.19) may be optimised for each streamtube by differentiation:

$$-b(da) + (1 - a)(db) = 0. \tag{10.21}$$

Expressions (10.18) and (10.21) together determine the optimal variation of the interference factors as functions of λ_r. The solution is (10.18), together with

$$a = \frac{1 + b}{3 + 4b}. \tag{10.22}$$

Table 10.1 shows some results, iteratively calculated by considering Eq. (10.18) as a quadratic equation in the factor b, from which the positive root is taken. The optimum interference factor a is now smaller than 1/3 and decreases as λ_r decreases. The corresponding interference factor b increases as λ_r decreases.

The local power coefficient in Table 10.1 is $(C_P)_{loc} = 4a(1 - a)^2/(1 + b)$, by Eq. 10.20. The power coefficient of the turbine is obtained by integration of this local factor. The result depends on the tip speed ratio $\lambda_T = u_T/v_0$, shown in Table 10.2.

The resulting value of C_P is smaller when λ_T is smaller. This result shows the effect of the post-whirl. With lower tip speed, the whirl behind the rotor is stronger due to the larger torque necessary for energy capture. This lowers the power coefficient by dissipation of part of the kinetic energy related to the tangential velocity component downstream of the rotor.

Drag on the rotor blades is not yet taken into account. Drag is added in the analysis in the next section. Without taking drag into account, the best turbine is obtained for the highest speed ratio (Table 10.2). This changes by the blade drag.

Table 10.2 Power as a function of tip speed ratio from multiple-streamtube analysis; results by Hunt [2]

λ_T	0.5	1	2	5	10
C_P	0.288	0.416	0.512	0.570	0.593

10.3.3 Blade-Element Momentum Analysis

The momentum analysis of the preceding section can be further extended by expressing the force on the blade elements in an infinitesimal streamtube and relating the interference factors to the force components. The effect of blade drag is then taken into account.

Figure 10.10 represents the velocity triangle at the rotor disc, where the axial and tangential interference factors are denoted by a and b. The axial and tangential components of the relative velocity are $(1 - a)v_0$ and $(1 + b)u$. The figure shows the lift and drag forces.

The axial force component dT and the moment around the shaft dM can be expressed as functions of lift coefficient, drag coefficient and flow angle ϕ. The following relations apply:

$$w = \frac{(1 - a)v_0}{\sin \phi} \quad \text{or} \quad w = \frac{(1 + b)u}{\cos \phi}, \quad \tan \phi = \frac{(1 - a)v_0}{(1 + b)u} = \frac{1 - a}{(1 + b)\lambda_r}, \quad (10.23)$$

$$dL = \frac{1}{2}\rho w^2 C_L(\alpha, \text{Re})c\,dr, \quad dD = \frac{1}{2}\rho w^2 C_D(\alpha, \text{Re})c\,dr, \quad (10.24)$$

$$dT = Z(dL \cos \phi + dD \sin \phi), \quad dM = Z(dL \sin \phi - dD \cos \phi)r. \quad (10.25)$$

C_L and C_D are the lift and drag coefficients that depend on the angle of attack $\alpha = \phi - \theta$ and the local Reynolds number. θ is the position angle of the blade element (pitch angle), Z is the number of blades and c is the local chord. The local speed ratio is $\lambda_r = u/v_0$. Hereafter, we use the local blade solidity, expressed by $\sigma_r = Zc(r)/2\pi r$.

The axial force component and the moment around the shaft are related to momentum changes in the axial and tangential directions. These were derived in

Fig. 10.10 Velocity triangle
at the rotor disc

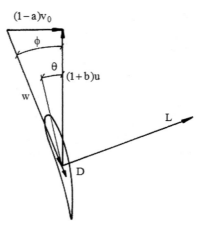

the previous section as

$$dT = \rho v_1 (dA)(v_0 - v_{3a}) = \rho v_0^2 (2\pi r dr)(1 - a)2a, \qquad (10.26)$$

$$dM = \rho v_1 (dA)(v_{1u} - v_{2u})r = \rho v_0 u(2\pi r dr)(1 - a)2br. \qquad (10.27)$$

By equating the terms of dT and dM in Eqs. (10.25–10.27), using the relations (10.23, 10.24), a set of two equations for the interference factors (a and b) is obtained. The factors can thus be expressed as functions of the local speed ratio λ_r, the local blade solidity σ_r, and the local lift- and drag coefficients $C_L(\alpha, \mathrm{Re})$ and $C_D(\alpha, \mathrm{Re})$. By choosing the most suitable expression of w in (10.23), a possible way of writing the relations is:

$$\frac{a}{1-a} = \frac{\sigma_L \cos\phi + \sigma_D \sin\phi}{\sin^2\phi} \quad \text{and} \quad \frac{b}{1+b} = \frac{\sigma_L \sin\phi - \sigma_D \cos\phi}{\sin\phi \cos\phi}, \qquad (10.28)$$

with

$$\sigma_L = \sigma_r C_L/4 \quad \text{and} \quad \sigma_D = \sigma_r C_D/4. \qquad (10.29)$$

Due to the relation between the local flow angle $\phi(r)$ and the interference factors by Eq. (10.23), the expressions are implicit and solutions for the interference factors can only be obtained numerically. A possible procedure is by iterative determination of the right hand sides of (10.23) and (10.28), starting from $a = 1/3$ and $b = 0$. For illustrations, we refer to the Exercises 10.5.5 and 10.5.6.

We remark that with $C_D = 0$, elimination of C_L between the relations (10.28), using the expression of $\tan\phi$ in (10.23), results in (10.18). The analysis with blade element forces is thus consistent with the analysis of the previous section.

The power is obtained by integration of the moment expression (10.27). This expression requires circumferentially averaged interference factors, which follow from the rotor representation by an infinite number of infinitesimal blades. But the actual force on a blade segment depends on local interference factors at the place of a blade in a rotor with a finite number of blades. The local factors vary in the circumferential direction. A relation between averaged factors and local factors is thus necessary. The difference between averaged and local factors is particularly important at blade tips. Calculation of the difference is, therefore, commonly called correction for tip effect or correction for tip loss.

There are several interpretations of the correction for tip loss. A simple method consists in taking the expressions (10.23) and (10.28) for the local interference factors, and replacing the interference factors by averaged ones in the expressions (10.26) and (10.27) when evaluating the resulting axial force (T) on the rotor disc and the resulting moment (M) on the shaft. A method by Prandtl is representation of the averaged factors by multiplication of the local factors by a function of the rotor radius ratio, the number of blades, and the flow angle. This function is derived from the flow induced by the vortex representation of the rotor blades. It stays very close to unity up to a radius ratio of about 0.75 and evolves towards zero at the rotor periphery such that the extracted power becomes zero at the blade tips. We refer to books on wind turbine aerodynamics [1, 3] for the analysis of the induction effects of the lift force of a blade and details about the possible correction calculations for tip loss (several proposals; see also reference [4]).

10.3.4   Optimisation of a Rotor Blade

A rotor blade for optimum performance can be designed with the equations derived in the previous section for known relations of C_L and C_D as functions of the angle of attack. The optimisation requires numerical techniques for which we refer to books [1–3]. Hereafter, we use results from the book by Hunt [2].

Figure 10.11 (left) represents the optimisation result of a HAWT for the design tip speed ratio $\lambda_T = 6$. The resulting maximum C_P is 0.45 (shaft power). A more complete C_P-curve is shown in the further Fig. 10.13.

Figure 10.11 also shows the performance of a Darrieus rotor and a Savonius rotor. There are similar analysis techniques, based on multiple streamtubes, for vertical-axis wind turbines (Hunt [2]).

The optimum C_P of a HAWT is nearly the same for a design tip speed ratio in the range $\lambda_T = 6$–8. With optimisation for a lower tip speed ratio, the C_P is lower due to stronger post-whirl. With optimisation for a very high tip speed ratio, e.g. exceeding 10, there is performance loss due to the larger effect of the rotor drag.

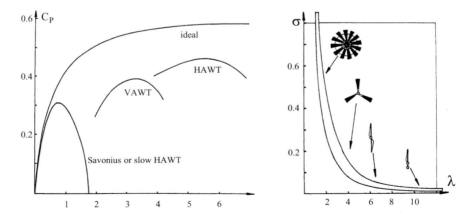

Fig. 10.11 Power coefficient with various wind turbine types (left); relation between solidity and design tip speed ratio for horizontal-axis wind turbines (right); adapted from Hunt [2]

Figure 10.11 (right) represents the solidity for optimum power as a function of the design tip speed ratio. The lower the design tip speed ratio, the higher is the torque on the wind turbine rotor and the larger is the corresponding blade surface.

The figure is adapted from the book by Hunt, published in 1981. The band in the shown dependence comes from variation of profile data. The shown relation is still valid, but the figure suggests that the number of blades has to be 2 for tip speed ratio 6–8. This was common at the time of publication of the book, but is no longer required because three-blade turbines can now (2021) be made for the corresponding (low) solidity due to much stronger blade materials. The figure suggests a one-blade turbine for tip speed ratios above 10. Such a type of turbine has existed (monopteros), but is not manufactured anymore.

10.4 Adaptation to a Wind Regime

Figure 10.12 is a histogram with 1 m/s velocity classes, representative for the wind at 50 m height at a location on the West European seashore. The histogram shows the probability $P(v_i)$ that a certain wind speed occurs. The yearly average wind speed is about 7 m/s. Also shown is a distribution indicated with $P(v_i^3)$, which is the distribution of the energy density (energy flux through a unit surface). The meaning of the distributions is

$$P(v_i) = \frac{number\ of\ hours\ with\ wind\ velocity\ v_i\ in\ a\ year}{number\ of\ hours\ in\ a\ year\ (8760)},$$

$$P(v_i^3) = \frac{yearly\ energy\ in\ the\ wind\ at\ velocity\ v_i}{total\ yearly\ energy\ in\ the\ wind}.$$

The shape of the wind energy distribution is very different from the shape of the wind velocity distribution, since the cubic power of the velocity has a strong effect. The energy average wind speed is about 12 m/s.

Fig. 10.12 Distribution of wind speed and wind energy flux

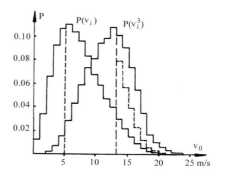

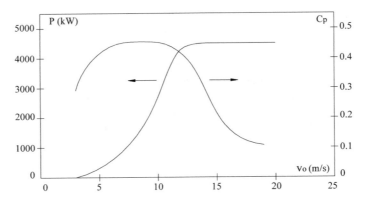

Fig. 10.13 Power coefficient and power of a former test wind turbine with variable speed and variable pitch (*Courtesy* Enercon)

The following wind speeds may be determined for a wind turbine functioning in this wind speed distribution.

- *Cut-in speed.* This is the wind speed below which the wind turbine is not operated, because the energy yield is too low to justify wearing the system. 5 m/s is taken as an example. The unused energy is small, although the time that the turbine does not function is quite significant.
- *Rated speed.* The power is limited to the value corresponding to this wind speed. From this speed onwards, the generator produces at its maximum power, called rated power, and the wind energy flux is not completely used.

  In Fig. 10.12, the rated speed is set at 13 m/s. A dashed line indicates the energy captured above the rated wind speed, approximated by reducing the energy proportional to the third power of the velocity. The unused energy is not negligible, but is not very large.
- *Cut-out speed.* This is the wind speed above which the turbine is stopped. This is done in order to avoid damage, because very high wind speeds occur in storm conditions. The turbine power is gradually reduced to zero over some speed range starting at the cut-out speed. In the example, the cut-out speed is 21 m/s. The available energy above this speed is very small.

Figure 10.13 shows the variation of the power coefficient and the power of a 4.5 MW turbine with variable speed and variable blade pitch angle. The characteristics are those of a test turbine of the early 2000s (Enercon).

The cut-in wind speed and the rated wind speed are 3 and 13 m/s. Within the 3–6 m/s wind speed range, the rotational speed is the minimum generator speed (-30%) and the pitch angle changes with the wind velocity. The speed ratio and pitch angle are optimal within the 6–10.5 m/s velocity range, where the turbine speed varies from -30 to $+30\%$. From 10.5 m/s onwards, the rotation speed is the maximum generator speed ($+30\%$) and the C_P value decreases with increasing wind velocity by blade pitch angle adjustment. From 13 m/s wind speed onwards, the power is limited to

4500 kW. The power during the cut-off phase is not drawn (C_P evolves gradually to zero). From the C_P-curve, one sees that the design wind speed is about 8.5 m/s. The ratio of rated wind speed to design wind speed of the example is thus about 1.50. The turbine had a diameter of 115 m and was tested on a shore location with 8.5 m/s average wind speed on the hub height.

The power coefficient of Fig. 10.13 may be combined with a wind histogram with the shape of that of Fig. 10.12, but with enlarged wind speeds such that the yearly averaged wind speed at hub height becomes 8.5 m/s (factor about 1.20). The observation is then that somewhat more wind energy is left unused than in Fig. 10.12, but that nevertheless a large part of the available energy is captured. Obviously, with a higher rated wind speed, the captured amount could even be larger. But the cost of the turbine then increases due to a larger generator, larger gearbox (if present) and larger wind forces on the rotor. Conversely, with a lower rated wind speed, the cost of the turbine is lower. The reasoning shows that there is an economic optimum rated wind speed.

The rated wind speed of the example of Fig. 10.13 is the economic optimum one for a high value of the produced electricity or a low investment cost. These were the circumstances up to around 2015, when there were installation subsidies or high feed-in tariffs for produced electricity.

Nowadays (2021) wind energy systems have to be economically justified on their own, which means that the economic value of the produced energy is much lower. Therefore, the rated power has to be lower. The specific power of the example of Fig. 10.13, which is the generator power divided by the rotor area, is 433 W/m^2 (also $\frac{1}{2}C_p\rho v^3$ with $C_p \approx 0.32$ and $v = 13$ m/s). Nowadays (2021), a wind turbine for a location with an average wind speed of 8.5 m/s at hub height in the 3–5 MW class has a specific power around 250 W/m^2 (derived from examples on the web sites of manufactures; see also Exercise 10.5.1). The corresponding rated wind speed is about 10.5 m/s, thus about 25% above the yearly mean wind speed. Typical power density values are nowadays about 325 W/m^2 for off-shore use (average wind speed at hub height about 10 m/s) to around 200 W/m^2 for inland use.

With a lower specific power, the capacity factor is higher. The capacity factor is the ratio of the yearly produced electricity (MWh) to the product of the rated power (MW) and the number of hours in a year (8760 h). With modern wind turbines it is about 60%, while it was only around 35% with older types. This means that the fraction of the time that wind turbines produce is much higher than in the past. This is also much better for the operation of electricity grids.

10.5 Exercises

10.5.1 Wind regimes are classified in three basic types by the IEC (International Electrotechnical Commission): class I for yearly average wind speed at hub height of 10 m/s, II for 8.5 m/s and III for 7.5 m/s. A letter is then added for the turbulence level: A for turbulence level 16%, B for 14% and C for 12%. Manufacturers define wind turbines as designed for one of these classes and may define also a specific class, then denoted by S. For use at the West European coast, the representative classes are IIA and IIB for hub height around 100 m. Verify the main properties of some wind turbines in these classes in the portfolio of a manufacturer. Estimate the specific power, the rated wind speed and the capacity factor. Take 0.35 as approximation of the power coefficient at rated wind speed.

A: In 2021, there were 2 turbines on the website of manufacturer Vestas (Denmark) in the class IEC IIB with rated power around 2 MW (the lowest offered rated power): V100-2.0 MW and V120-2.2 MW. A turbine in the class IEC IIA on the website of Siemens Gamesa Renewable Energy (Spain) was: SG 2.1-114. A comparable turbine in the class S was SG 2.9-129. The specified data are listed hereafter in a first table and the derived characteristics in a second table. The specific power is the quotient of the rated power and the rotor area. The rated wind speed is derived from this value using a power coefficient of 0.35 and the air density 1.225 kg/m^3. The wind speed ratio is the quotient of the rated wind speed and the yearly averaged wind speed (8.5 m/s).

Type	V100-2.0 MW	V120-2.2 MW	SG 2.1-114	SG 2.9-129
Rated power	2000 kW	2200 kW	2100 kW	2900 kW
Rotor diameter	100 m	120 m	114 m	129 m
Yr. Prod. @ 8.5 m/s	9.8 GWh	11.4 GWh	–	15.0 GWh

Type	V100-2.0 MW	V120-2.2 MW	SG 2.1–114	SG 2.9–129
Specific power	255 W/m^2	195 W/m^2	206 W/m^2	222 W/m^2
Rated wind speed	10.60 m/s	9.68 m/s	9.86 m/s	10.12 m/s
Wind speed ratio	1.25	1.14	1.16	1.19
Capacity factor	56.0%	59.2%	–	59.1%

The specific power range is 200–250 W/m^2. The capacity factor varies from about 56% (wind speed factor 1.25) to about 59% (wind speed factor 1.15).

10.5.2 Perform a similar search as in the previous exercise on wind turbines for onshore or offshore use, yearly wind speed at hub height 10 m/s, and rated power in the order of 5–6 MW.

A: In 2021, the turbines V150-6.0 MW and V162-6.0 MW were on the website of manufacturer Vestas (Denmark), the SG 5.8-155 on the website of Siemens Gamesa Renewable Energy (Spain) and the Haliade 150-6 MW on the website of General

Electric Renewable Energy (France). The specified data are listed hereafter in a first table and the derived characteristics in a second table.

Type	V150-6.0 MW	V162-6.0 MW	SG 5.8-155	Haliade 150-6 MW
Rated power	6000 kW	6000 kW	5800 kW	6000 kW
Rotor diameter	150 m	162 m	155 m	150 m
Drive train	Geared 2St	Geared 2St	Geared 3St	Direct
Yr. Prod. @ 10 m/s	29.0 GWh	31.0 GWh	–	–

Type	V150-6.0 MW	V162-6.0 MW	SG 5.8-155	Haliade 150-6 MW
Specific power	340 W/m^2	291 W/m^2	307 W/m^2	340 W/m^2
Rated wind speed	11.66 m/s	11.07 m/s	11.28 m/s	11.66 m/s
Wind speed ratio	1.17	1.11	1.13	1.17
Capacity factor	55.2%	59.0%	–	–

The specific power is in the order of 300 W/m^2 (290–340 W/m^2), thus higher than observed in the previous exercise. The wind speed ratio is about 1.15.

10.5.3 In Sect. 10.3.3 on blade-element analysis, it is stated that the expressions of the interference factors by Eq. (10.28) can be combined into a relation which reduces to the relation (10.18), by setting the drag coefficient to zero. Verify the correctness of this statement by taking the quotient of the expressions (10.28) and expressing $\tan \phi$ by Eq. (10.23). The consistency on this aspect between blade-element analysis and multiple-streamtube analysis is remarkable, because there is no reasoning on the force components on a blade element in absence of drag in the multiple-streamtube analysis. The expression of the axial force by Eq. 10.26 is the same in both analyses. The expression of the moment by Eq. 10.27 is also the same, because it is equivalent to the power expression by Eq. 10.19. The use of the blade-element force components in the blade-element analysis is thus replaced by the work balance by Eq. 10.16 in the streamtube analysis. This equation expresses work extraction without loss. It thus supposes that the intervening force is without dissipation, which is indirectly the expression that the intervening force is perpendicular to the relative velocity, thus without drag (see Sects. 2.2.5 and 2.2.7 in Chap. 2 on work by a lift force and by a drag force).

A: The combination of the expressions results in:

$$C_L\left[a(1-a) - b(1+b)\lambda_r^2\right] = C_D(a+b)\lambda_r. \tag{10.30}$$

This expression becomes Eq. (10.18) for $C_D = 0$.

10.5.4 Consider the changes of the velocity components (absolute and relative) by $-av_0$ in axial direction and $-bu$ in tangential direction between the oncoming free velocity and the velocity at the position of the rotor. Verify that the relation (10.18) expresses that the velocity change is in the direction of the lift force, but in opposite sense. It is thus also perpendicular to the relative velocity. This is a property that applies to a straight vortex line in a uniform flow. This property suggests that the effect of the rotor on the flow may be seen as analogous to the velocity field induced by vortices. The induced velocity by a straight vortex line at large downstream distance is twice the induced velocity at the position of the vortex itself. This analogy may thus be seen as a justification for the representation of the axial velocity by Eq. (10.14). But the representation of the flow through the rotor by vortices requires also the tip vortices and the vortices in the wakes of the blades and the hub. The vortex representation is thus much more complex than that of a single straight vortex. Verify that the expression (10.30), derived in the previous exercise, also expresses that the velocity change is in the direction of the force on the blade element. This property can be derived directly from the expressions (10.25) to (10.27). But the effect of a drag force cannot be represented by that of a vortex. This observation leads to the argumentation that it may be better justified to construct the interference factors from the relations (10.26) and (10.27) by only using the lift contribution in the expressions (10.25). There are thus several possible formulations of a blade-element analysis.

10.5.5 Estimate with the formulae (10.23), (10.28) and (10.29) the local power coefficient of a blade segment annulus with local speed ratio $\lambda_r = 3$ (thus on the mid radius of a turbine with $\lambda_T = 6$) and local solidity $\sigma_r = 0.08$. Calculate first for lift- and drag coefficients of the profile $C_L = 1$ and $C_D = 0$ and calculate then for $C_L = 1$ and $C_D = 0.02$. Use the iterative procedure as described in Sect. 10.3.3.

A: The local power coefficient is $4b\,(1 - a)\lambda_r^2$.

for $C_L = 1$ and $C_D = 0$: $a = 0.2613$, $b = 0.02101$ and $(C_P)_{\text{local}} = 0.5586$;

for $C_L = 1$ and $C_D = 0.02$: $a = 0.2619$, $b = 0.01923$ and $(C_P)_{\text{local}} = 0.5111$.

One may compare with the Betz limit equal to 0.5926.

10.5.6 The choice of the solidity is not optimal in the previous exercise. The optimal solidity is somewhat larger. Calculate the local power coefficient of a blade segment annulus with local speed ratio $\lambda_r = 3$, $C_L = 1$ and $C_D = 0.02$ for the local solidities $\sigma_r = 0.085, 0.090, 0.095$. Observe that the optimum solidity is 0.090. Observe also that a solution is not possible for large solidity. The blade-element analysis needs modification for large solidity.

A: for $\sigma_r = 0.085$: $a = 0.2887$, $b = 0.02034$ and $(C_P)_{\text{local}} = 0.5209$;

for $\sigma_r = 0.090$: $a = 0.3194$, $b = 0.02142$ and $(C_P)_{\text{local}} = 0.5249$;

for $\sigma_r = 0.095$: $a = 0.3561$, $b = 0.02245$ and $(C_P)_{\text{local}} = 0.5204$.

References

1. Sørensen JN (2016) General momentum theory for horizontal axis wind turbines. Springer. ISBN 978-3-319-22113-7
2. Hunt VD (1981) Wind power: a handbook on wind energy conversion systems. Van Nostrand Reinhold. ISBN 0-442-27389-4
3. Schaffarczyk AP (2020) Introduction to wind turbine aerodynamics. Springer. ISBN 978-3-030-41028-5
4. Shen WZ, Mikkelsen R, Sørensen JN (2005) Tip loss corrections for wind turbine computations. Wind Energy 8:457–475

Chapter 11
Power Gas Turbines

Abstract A gas turbine is a turbomachine composed of a compressor part, a part with heat supply to the compressed gas and a turbine part in which the hot gas expands. The present chapter discusses gas turbines for mechanical power generation. These are machines with an outward shaft, which drives a load. The largest market sector of such machines is electric power generation, but machines for driving compressors and pumps in industrial plants and for driving large vehicles and ships are also examples. We discuss the working principles of the components of power gas turbines in the present chapter. As electric power generation is the largest sector of application, we choose components of such machines for illustrations. The main purpose of the chapter is the discussion of the overall performance of power gas turbines. Performance analysis is a matter of thermodynamic modelling and is not strongly linked to a particular application.

11.1 General Concept and Components

11.1.1 Definition of a Gas Turbine

The general definition of a gas turbine is given above. Figure 11.1 is a schematic of a gas turbine with an outward shaft driving a load. The machine is thus intended for mechanical power generation. The general term for such a machine in the present book is *power gas turbine*, but this term does not universally refer to the broad category of gas turbines with an outward shaft. The most likely is that the term is understood as a gas turbine driving a generator in an electric power plant.

Gas turbines for power plants form the largest market sector of gas turbines with an outward shaft. They are also the largest machines with power up to about 600 MW. There is no unique term to denominate gas turbines for power plants. They are sometimes called land-based or stationary gas turbines, but the term *industrial gas turbine* is used as well. The term industrial gas turbine for power generation is sometimes used with the purpose to distinguish from industrial gas turbines for mechanical drive, which form a second subcategory. This term refers to a gas turbine

Fig. 11.1 Components of a
power gas turbine with
closed cycle

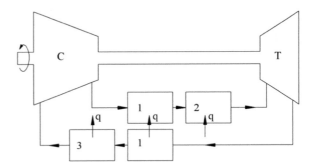

in an industrial plant driving a rotating machine requiring large power. Mostly this
is a compressor. Less common is drive of a large pump.

Power gas turbines are also used to drive vehicles, ships, propeller aircraft and
helicopters. Gas turbine driven vehicles are large-power vehicles, as off-road trucks,
excavators, tanks and train locomotives. Also ships requiring large propulsive power
in limited available space are gas turbine driven. Examples are fast container ships,
fast ferries, coast guard patrol boats and military ships like frigates and destroyers.

Gas turbines are also applied for aircraft propulsion. These are commonly called
aero gas turbines. Gas turbines for larger aircraft do not have an outward shaft. The
turbine part drives a compressor part and a part for thrust generation by acceleration
of air, termed *fan*. The combustion gas leaving the turbine part generates thrust as
well. Aero gas turbines are discussed in Chap. 12. The term aero gas turbine is
commonly used for all gas turbines for aircraft propulsion, including outward shaft
machines for driving the propeller of a propeller aircraft or a helicopter rotor.

Figure 11.1 represents a closed-cycle gas turbine. The gas delivered by the
compressor is heated by two heat exchangers. Heat exchanger 1 extracts heat from the
exhaust gas of the turbine. Heat from an external source is added in heat exchanger
2. The gas leaving the turbine part is further cooled by heat exchanger 3, using an
external cooling source. The closed cycle is rarely applied. The cycle is mostly open,
which then means that the compressor takes air from the atmosphere and that the
exhaust gas of the turbine part, having passed through heat exchanger 1, is discharged
into the atmosphere. In that case, there is no heat exchanger 3. In an open cycle, heat
exchanger 2 is always a combustion chamber, where the air is heated by internal
combustion. The atmosphere is a component in an open cycle. The gas passing
through the cycle is converted from air into a combustion gas. It is supposed that the
exhaust gas is cooled and regenerated in the atmosphere.

In many gas turbines, there are no heat exchangers. The gas turbine consists then
of a compressor part, a combustion chamber and a turbine part. The cycle is then
termed a *simple cycle*. Heat exchanger 1, if present, may take two forms. The term
recuperator means a heat exchanger with heat transfer through metal walls. The term
regenerator refers to a rotating tube matrix of ceramic material, with the hot and cold
gas alternately flowing through it.

Figure 11.1 suggests a machine with a single shaft. Most power gas turbines for electricity generation are built this way, but multi-shaft machines exist as well. A typical gas turbine for transport (vehicle, ship, propeller aircraft) has two turbine parts. The high-pressure (HP) part drives the compressor and the low-pressure (LP) part drives the external load. The machine is split into a *gas generator*, encompassing the compressor, the combustion chamber and the HP turbine and a *power turbine*, with different rotational speeds of both shafts. The gas generator turns at a rather high rotational speed, whereas the power turbine turns at a lower speed, adapted to the load. With a fixed fuel flow rate, the power supplied by the gas generator to the power turbine is approximately constant and so is the product of the torque and the rotational speed on the outward shaft. For transport applications, such a torque characteristic is advantageous, namely decreasing torque for increasing speed.

Some machines have a gas generator with two parts (the reason is explained later): compressor split into two parts and turbine split into three parts. The machine then has three shafts. A compressor part connected to a turbine part is called a spool. With multi-shaft machines, the shafts are mostly concentric, with the outward shaft, as on Fig. 11.1, at the compressor side. This is no general rule, but it is mostly advantageous that the outward shaft is at the cold side of the machine.

11.1.2 Comparison with Other Thermal Engines

Gas turbines show similarities with reciprocating internal combustion engines and steam turbines. Within a reciprocating engine, the gas completes also a cycle of compression, heating by combustion and expansion, but the difference is that the stages of the cycle occur within the same space, but at different times. By alternating cold and hot stages, the thermal load on the walls is lower with the same combustion temperature. Reciprocating engines thus allow higher combustion temperatures. Further, the combustion happens for a large part under a constant volume. A high combustion temperature is then advantageous for efficiency. The combustion in a gas turbine is at constant pressure. The consequence is that the simple-cycle efficiency does not depend much on the combustion temperature (see Sect. 11.3.3). Both machine types reach comparable efficiencies. For a modern advanced simple-cycle gas turbine, it is about 44% (shaft power). This is only a little lower than the efficiency of a diesel engine, which is about 46–48%.

With both machine types, a high combustion temperature is important for a large power density, or power per volume occupied. For this aspect, gas turbines have a significant advantage over reciprocating engines, as flow through a gas turbine is continuous and at high speed, with a through-flow Mach number at compressor entry of the order of 0.5 (the axial velocity at entry is about 160 m/s).

The power of large modern land-based gas turbines is about 500–600 MW, reached with a temperature of about 1500 °C at the entry to the turbine part. The dimensions are approximately: diameter 6 m and length 16 m. A diesel engine with these dimensions only attains about 10 MW.

A high temperature at the combustion chamber exit creates big technological problems for the combustor and the turbine. The melting temperature of the materials for the combustor and the turbine blades and vanes is about 1350 °C. With regard to sufficient strength and avoidance of creep (elongation under stress due to softening), the highest allowable metal temperature of turbine blades is about 900 °C. Combustor and turbine parts thus need intensive cooling and protection against the hot gas. The technological gas turbine design is strongly determined by cooling and heat protection aspects, which we discuss below.

Gas turbines share similarities with steam turbines, with similar cycles for both machines. The main difference is that the boiler feed pump of the steam turbine replaces the compressor of the gas turbine. With regard to efficiency potential, this is very advantageous for the steam turbine. A pressure increase of 300 bar (typical pressure rise in a large steam turbine) in water requires, with 90% efficiency, about 33 kJ/kg. A compression with the same efficiency and pressure ratio 20 (typical for a gas turbine for electricity generation) requires about 450 kJ/kg for air at 288 K at compression start (one may verify with the formulae of Sect. 11.2: $R \approx 288$ J/kgK, $c_p \approx 1045$ J/kgK, $\eta_\infty \approx 0.90$). An expansion with the same pressure ratio starting at 1450 °C yields about 1000 kJ/kg ($R \approx 288$ J/kgK, $c_p \approx 1250$ J/kgK, $\eta_\infty \approx 0.90$). The pump power is almost negligible with a steam turbine. The power of the compressor with a gas turbine is about half the power of the turbine part. This proportion is nearly the same for other pressure ratios. The global gas turbine efficiency is thus very sensitive to the efficiency of the components.

11.1.3   Example of a Power Gas Turbine

Figure 11.2 shows a large gas turbine for electric power generation. It is called a *heavy-duty type* machine, which means a very robust design meant for long production time per year (continuous operation is possible) and long periods between overhaul (typically 20,000 h). These machines are built for large power (from about 200 MW to about 600 MW). They are heavy, but this is no drawback for stationary land-based use. The compressor and turbine components are axial. The type shown is an SGT6-5000F (Siemens Energy). The power is 260 MW at 3600 rpm (for use at 60 Hz), with compressor pressure ratio 19.5 realised with 13 stages (the average pressure ratio per stage is about 1.26). The turbine has 4 stages, with 3 of them cooled. For examples, we refer to websites of manufacturers (large manufacturers are Siemens Energy, General Electric Power and Mitsubishi Power; smaller manufacturers are Ansaldo Energia and MAN Energy Solutions).

The compressor rotor is composed of discs and the compressor blades are attached to the discs by foots inserted in slots (see the later Fig. 11.4). The discs are connected to each other by teethed annular zones and tightened with bolts. The construction of the turbine rotor is similar (see the later Fig. 11.7). The compressor and turbine rotors are connected with a hollow tube.

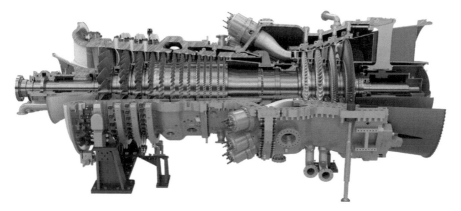

Fig. 11.2 Power gas turbine (Siemens SGT6-5000F; *Courtesy* Siemens Energy)

A second category of machines for electricity production is the *aero-derivative type*. These machines are derived from aero gas turbines for thrust generation, by replacing the fan (front of the machine) by a compressor stage and by replacing the end part of the low-pressure turbine (rear of the machine) by adapted turbine stages. Such machines are offered by aero gas turbine manufacturers as a second product line. The machines are light (required for aero use) and have a pressure ratio around 40, so much larger than with a typical heavy-duty gas turbine (see Chap. 12).

They also have multiple shafts (2 or 3). The compressor is split into a low-pressure (LP) and a high-pressure (HP) part. In two-shaft machines, the turbine is split into two parts, with the HP turbine driving the HP compressor and the LP turbine driving the LP compressor and the external load at the cold side of the machine (compressor side). In three-shaft machines, the turbine is split into three parts: the HP part drives the HP compressor, the intermediate-pressure part (IP) drives the LP compressor and the LP part drives the external load, either at the cold side or at the hot side of the machine. Typical manufacturers are Rolls Royce and General Electric (see websites; search for the name of the manufacturer combined with the term aero-derivative).

Aero-derivative gas turbines have limited power (from about 5 MW to about 50 MW; there is one type with 115 MW). They are used for electricity generation, but typically not in power stations, but very often in industrial plants requiring large electric power (order of several tens of MW). Typical are off-shore oil or gas platforms, desalination plants and refineries. These machines are also used for mechanical drive of compressors and pumps in such plants.

There are aero-derivative turbines for ship propulsion (see websites; e.g. Rolls Royce or General Electric, in combination with the term marine gas turbines).

Manufacturers of gas turbines for power generation typically also offer heavy-duty machines of smaller power for electricity generation in industrial plants or mechanical drive (search for small heavy-duty gas turbines) and there are manufacturers who only make such smaller machines (e.g. MAN Energy Solutions).

11.1.4 Compressor Part

Figure 11.3 sketches the blade profiles and the velocity triangles on the mean radius of an axial compressor. The shaft direction is horizontal in the figure. Figure 11.4 is a view on the rotor of the compressor of the SGT6-5000F. As usual, we employ the term *blade* in a rotor and the term *vane* in a stator, but the term blade refers to both in a general sense. A compressor stage consists of a rotor followed by a stator, but, typically, the inflow direction deviates from the axial direction. This means that the compressor has an inlet guide vane ring. The deviation is mostly about 15° in the rotation sense. This is normally needed to reduce the Mach number of the relative velocity at the entry to the first stage.

A major constraint is that the velocity reduction in the rotor (ratio w_2/w_1) and the stator (v_1/v_2 for a repeating stage) have to stay above 0.7 (approximately) in order to avoid flow separation. The consequence is a limitation to the flow turning in both stator and rotor, which limits the work coefficient ($\psi = \Delta W / u^2$, thus $\psi =$

Fig. 11.3 Blade and vane profiles and velocity triangles of an axial compressor (mean radius); full lines: design flow rate; dashed lines: reduced flow rate

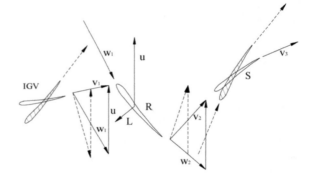

Fig. 11.4 Compressor rotor of the SGT6-5000F (*Courtesy* Siemens Energy)

$\Delta w_u/u = \Delta v_u/u$) to about 0.4 on the mean radius. Therefore, the blade speed is chosen as high as possible. In a heavy-duty machine, it is about 300 m/s on the mean radius.

The precise value depends on the size of the machine and the rotational speed (300 m/s corresponds to a diameter of 1.91 m at 3000 rpm). The blade speed is near to the speed of sound in the first stage, but below it (in air at 288 K, the speed of sound is 340 m/s). The relative velocity at the rotor entry of the first stage is also approximately 300 m/s (Mach number around 0.85). So, the limitation of the blade speed comes from keeping inflow velocities just subsonic and is not caused by strength limitations (for more detailed discussion, see Chap. 13).

Usually, the angular position of the inlet guide vanes of a compressor can be adjusted to adapt the machine to a smaller mass flow rate than the design mass flow rate. With many compressors, the stator vane angles of a number of stages at the front are also adjustable, up to about one fourth of the total number of stages (typically 3 or 4 stages). Such stages are mostly called *variable-geometry stages*. The term variable-geometry is used because adjustment is possible during operation.

The SGT6-5000F shown in Fig. 11.2 has a variable guide vane ring and 3 rows of variable stators. On each vane, there is a pivot, linked by a lever arm to a ring, allowing collective adjustment of the vane angles. The movable rings and level arms are visible on Fig. 11.2. With variable-geometry front stages, the compressor may be adapted to a smaller mass flow rate than the design value. We can easily understand the need for adjustable inlet guide vanes. It is somewhat more difficult to understand the benefit of a number of front stages with adjustable stator vanes for reduced load operation of the gas turbine (reduced mass flow rate).

During starting up a compressor, the density rise from the first stage to the last stage is less than at design conditions. For a constant-density fluid, the shape of the velocity triangles of the stages does not change when the stages maintain their flow coefficient and their work coefficient. So, at a reduced rotational speed, the shape is maintained for an operating point that obeys kinematic similitude. With a compressor, due to equal mass flow rate in successive stages, the reduced density rise causes deviations with increased through-flow velocity in the rear stages and decreased through-flow velocity in the front stages.

Figure 11.3 shows that by decreased through-flow velocity, with unchanged flow direction at the exit of a rotor (w_2) and unchanged flow direction at the exit of a stator (v_1), because these directions are imposed by the blades and the vanes, the incidences at the entry of a downstream stator or rotor increase. Flow turnings in rotor and stator increase, which means that the stage work increases with diminished flow rate. But, a large incidence may cause flow separation at the suction side of the blades. A flow with separation is commonly called a stalled flow. The term *stall* expresses that the flow does not behave well anymore. When this happens, the tangential lift diminishes and the work diminishes, which then typically leads to impossibility to bring the gas turbine to full speed.

There are several techniques for avoiding stall in the first stages of a compressor during starting up. First, part of the flow after a few stages may be blown off through valves and led to the turbine part. A second way is splitting the compressor in parts and so making several spools. The objective is limiting the number of connected stages in a compressor part. The compressor has to be split into 2 or 3 parts, depending on the pressure ratio. Splitting the compressor is typically done with aero-derivative gas turbines, but is not typical with heavy-duty machines. With large gas turbines for electricity generation, it is advantageous with respect to the precision of the rotational speed control to use only one shaft. Mostly, stall is avoided by variable stator vanes. Figure 11.3 shows that by turning the stator vanes more to the tangential direction (hatched profiles), the incidences at stator and rotor entries decrease. This allows starting up of the gas turbine.

With only the inlet guide vane ring variable, starting up of the gas turbine becomes possible, because avoiding stall in the first compressor stage is the critical action. However, good compressor efficiency cannot be reached at reduced mass flow rate for unchanged rotational speed, as incidence angles are then too large for good efficiency. When a compressor has multiple variable-geometry front stages, it means that part-load operation of the gas turbine has been optimised. The output power of a gas turbine is diminished by reducing the fuel flow rate. In order to understand the consequence for the mass flow rate, one has to know that the turbine part normally operates with a mass flow rate very near to the choked value or even at the choked value (for a detailed discussion, see Chap. 15). The choking mass flow rate is proportional to the density and the velocity of sound at the turbine entry. The product of these quantities is proportional to the pressure and inversely proportional to the square root of the temperature. A reduced fuel flow rate causes, in first instance, lowering of the turbine inlet temperature (TIT) and thus an increased mass flow rate through the turbine. Increased mass flow rate means for the compressor, at fixed rotational speed as in a single-shaft machine, reduced pressure ratio (the characteristic is decreasing work with increasing flow rate).

This result is illustrated in Fig. 11.5 with sketches of turbine and compressor characteristics of pressure ratio as a function of mass flow rate, ignoring the small increase of mass flow rate and small decrease of pressure in the combustion chamber. The larger air flow rate at part load, obtained by lowering the TIT (operation point 1) reduces the efficiency of the compressor. The smaller fuel flow rate and the larger air flow rate at part load reduce the fuel–air ratio in the combustion chamber. The fuel–air ratio is always small in a gas turbine, in the sense that full mixing of fuel with air leads to a mixture that is much too lean for stable combustion. This makes

Fig. 11.5 Operating points with lower turbine inlet temperature (1) or more tangential compressor stator vanes (2)

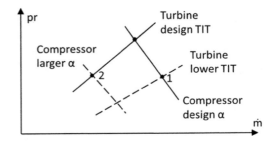

the operation of the combustion chamber delicate (see discussion in a later section). Variation of the fuel–air ratio with the load makes good combustion, which means keeping low the noxious combustion products (CO and NOx), more difficult. So, it is better for compressor efficiency and for combustion chamber performance that the mass flow rate is reduced at part load and that the fuel–air ratio stays approximately constant. This then means approximately constant turbine inlet temperature, but reduced pressure ratio [5].

The mass flow rate may be reduced by variable stator geometry in the front stages of the compressor. Figure 11.3 shows that the mass flow rate is reduced by turning the stator vanes more towards the tangential direction. This is called closing the stator vanes. By the same action, the stage work may also be reduced (smaller flow turnings in rotor and stator). This leads then to a reduced pressure ratio. By doing this in a number of front stages, the mass flow rate through the compressor is reduced, but also the density rise is reduced. This makes that in further stages the discrepancy between the through-flow velocity in design and reduced-load conditions may become sufficiently small so that variable stator geometry becomes unnecessary. In principle, the part-load efficiency benefits from more stages with variable stator vanes, but leakage in the stator parts increases which diminishes the efficiency, also in design conditions. So, a practical optimum is with not many variable-geometry stages. Figure 11.5 illustrates that reduced mass flow rate may be obtained at constant turbine inlet temperature by closing compressor stator vanes (operating point 2 by larger α).

11.1.5 Turbine Part

Figure 11.6 sketches the blade and vane profiles and the velocity triangles at the hub of an axial turbine. A turbine stage consists of a stator (with *vanes*) followed by a rotor (with *blades*). As with steam turbines, we may use the term *nozzle* for the channels formed by the stator vanes. Figure 11.7 is a view on the turbine rotor of the SGT6-5000F. Vanes and blades in a heavy-duty gas turbine are similar to those in the first stages of an LP part of a steam turbine. But, work and flow coefficients are larger (see Chap. 15). The reason is realisation of large power per unit of volume and limitation of blade and vane surfaces that need cooling and hot gas protection. The work coefficient is typically above 2 at the hub ($\psi \approx 2.20$).

In Fig. 11.6, the kinematic parameters are: $\psi = 2$, $\phi = v_a/u = 0.6$, R (degree of reaction) $= 0.4$. The blade speed is typically around 350 m/s at the hub (this value comes from strength limitations). So, the stage work is about 250 kJ/kg ($\psi \approx 2.20$). With a compressor pressure ratio until 20, four stages suffice in the turbine part (the enthalpy drop is about 1000 kJ/kg). With a compressor pressure ratio 40, five stages are needed (verify with the formulae of Sect. 11.2).

The interior parts of the combustion chamber and the first stages of the turbine part are exposed to the hottest combustion gas. These parts therefore need cooling and shielding from the hot gas. Heat shielding is done with a *thermal barrier coating*

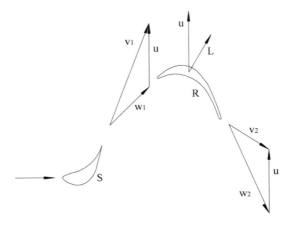

Fig. 11.6 Vane and blade profiles and velocity triangles of an axial turbine (hub section), ($\psi = 2$, $\phi = 0.6$, $R = 0.4$)

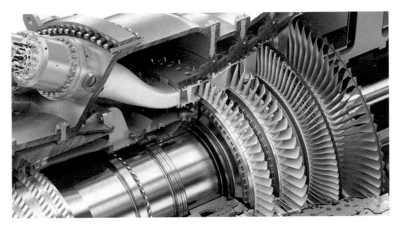

Fig. 11.7 Turbine part and combustion chamber of the SGT6-5000F (*Courtesy* Siemens Energy)

(TBC). This is a layer of 100–400 μm thickness of ceramic material, formed by vapour deposition of a mixture of zirconium and yttrium. It can withstand very high temperatures and has an extremely small heat conduction coefficient (in the order of 1 W/mK). This layer creates a temperature difference of 100–300 °C.

The ceramic layer is attached to a metal bond layer of about 100 μm thickness, formed by an alloy of nickel, cobalt, chromium, aluminium and yttrium, which itself covers the blade metal. The intermediate layer, or bond coat, serves as an elastic layer between the blade metal and the ceramic top coat. The white surface material of the blades of the first three stages in Fig. 11.7 is the ceramic layer.

The materials for combustors, turbine blades and vanes are alloys of nickel, chromium and cobalt (nickel-based alloys; melting temperature about 1350 °C). The blades of the first stage are typically cast with temperature controlled cooling such that a blade forms a *single crystal*. This very expensive technique is necessary for obtaining maximum strength. The blades of the second stage are subjected to lower temperatures. These blades are typically cast with a less delicate cooling technique in a *directionally solidified* form, which means that the blade is composed of several crystals, but all in the longitudinal direction. Blades of non-cooled stages may be cast without controlled cooling, leading to a large number of crystals with borders in all directions.

Cooling of the first stages of the turbine part of a gas turbine (stator and rotor parts of stages 1 and 2 and the stator of stage 3 of the SGT6-5000F in Fig. 11.2) is normally realised by air bleeds on the compressor. This air is led to the interior of the hollow stator vanes and rotor blades of the turbine, typically by external conduits in a land-based gas turbine and by internal perforations in an aero gas turbine. For rotor cooling, the air has to be brought into perforations of the rotor discs through slip ring sealed chambers. Three cooling principles are often combined.

Figure 11.8 (left) sketches sections through a rotor blade. First, there is *convection cooling* by air flowing through the internal passages. Heat transfer is enhanced by ribs in the wider passages and by pins in the narrower ones. A large fraction

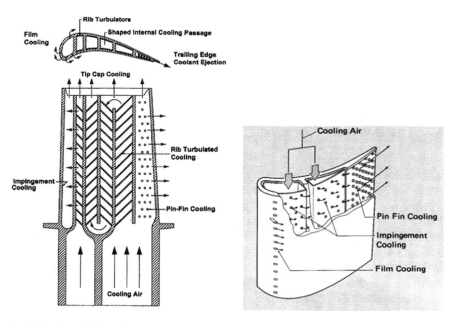

Fig. 11.8 Rotor blade with convection cooling, impingement cooling and film cooling (left: from [2], permission by ASME); stator vane with impingement cooling by inserts and film cooling (right, *Courtesy* Mitsubishi Power)

Fig. 11.9 Film cooling
holes in stator vanes
(*Courtesy* Siemens Energy)

of the air is exhausted at the tip of the blade for cooling of the casing. A small
fraction is exhausted at the trailing edge of the blade. By perforations between the
front convection cooling channel and a channel at the leading edge, jets are formed
impinging at the internal side of the leading edge. This way of cooling is called
impingement cooling. Impinging jets perpendicular to the surface realise a much
larger heat transfer than flow aligned to the surface, as with convection cooling. This
more intense cooling is very advantageous at the leading edge, where the heat load
is the largest. *Film cooling* may be added. The air is led to the blade surface by holes,
typically with an angle of about 30° to the surface, forming a protective film. With
stator vanes, the same principles may be used as for rotor blades, but more intense
cooling is reached by using inserts, as shown on Fig. 11.8 (right), so that impingement cooling is realised over a large part of the vane surface. Inserts are often used
with first row stator vanes (e.g. SGT6-5000F). Figure 11.9 shows the film cooling
holes in a row of stator vanes.

11.1.6   Combustion Chamber

A combustion chamber has an outer shell, mostly cylindrical, as on Fig. 11.2, not
exposed to the heat of the combustion, and an inner part which confines the combustion gas. The outer part has to withstand the pressure difference between the interior
of the gas turbine and the atmosphere. The inner part is generally called the *liner*,
which means that it functions as a lining or an inner surface. The inner part has
to withstand the heat of the combustion and therefore must be cooled. With large
machines, there are two types of combustion chambers. With the *can-annular* form,
as on Fig. 11.2, the inner part is composed of a number of approximate cylindrical
cans (16 for the SGT6-5000F on Fig. 11.2) placed in the outer annular space. The
individual liners are connected with tubes so that one liner can ignite another. A
second possible form is *annular*, which then means that the inner part is annular.
Small or medium size gas turbines often have combustion chambers of can type,
which means a number of cylindrical cans, each with a cylindrical liner.

The typical fuel for a gas turbine for electricity generation is natural gas. For this fuel, the cans of a can-annular combustion chamber typically contain 6–9 burners. The annular combustion chamber has a large number of burners. A first advantage of the annular combustion chamber is less friction area, resulting in a smaller pressure drop in the combustion chamber. The second advantage is a more homogeneous temperature profile at turbine entrance. The disadvantage is that is more difficult to obtain stable combustion.

Land-based gas turbines typically have reverse flow combustion chambers. This means that the flow from the compressor inverses direction, as on Fig. 11.2, before it enters the internal parts of the combustion chamber. The through-flow velocity in the rear part of the compressor is of the order of 120 m/s. The allowable velocity at the position of the combustion itself is only of the order of 10 m/s. Realisation of the strong velocity reduction is the easiest with reversal of the flow direction. Gas turbines for propulsion always have a straight through-flow combustion chamber, because the frontal area has to be minimised.

The cooling and the heat protection of liners follow the same principles as with turbine blades. The most common is convection cooling by air flowing over the exterior liner surface and a thermal barrier coating on the interior surface. But impingement cooling of cans, requiring then a shell with holes around the cans, is also employed. With annular liners, other systems are film cooling of the inner surfaces and heat protection by ceramic tiles covering the metal surface.

Figure 11.10 is a sketch of a longitudinal section of a somewhat simplified version of the burner system used, with some variants, in can-annular combustion chambers. A number of equal burners are positioned around a more complex central one, called the *pilot burner*. Figure 11.11 is a picture of eight peripheral burners, without the pilot burner. Burners are constructed for low production of noxious combustion

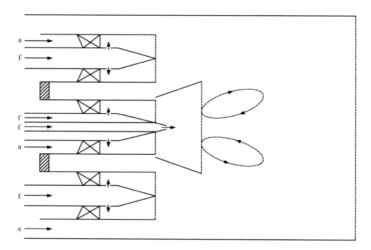

Fig. 11.10 Longitudinal section of a simplified dry low emissions burner system (DLE) for a can-annular combustion chamber

Fig. 11.11 Peripheral
burners of a DLE system
(*Courtesy* Siemens Energy)

products, which are nitrogen oxides (NOx) and carbon monoxide (CO). NOx forms
by reactions at high temperature between nitrogen and oxygen of the air. CO forms
by incomplete combustion due to lack of oxygen. For low production of both, the
combustion principles are *lean premixing* and *staged combustion*.

The basic organisation of combustion is injecting pure fuel into air. The (gaseous)
fuel then diffuses into the air and a flame front forms where the gas-air composition
is about stoichiometric. The flame is then very stable but the combustion temperature
is high and much NOx is formed.

By mixing fuel and air prior to combustion into a lean mixture, the combustion
temperature is much lower, so that much less NOx is formed. The drawback is
then that much CO is formed due to incomplete combustion. For completing the
combustion, air is added in a second stage to the combustion gas. The difficulty with
premixed combustion is that the combustion may be quite unstable for a mixture
with composition near to the lean flammability limit (there is also a rich limit, but
this one is not relevant here).

The burner system of Fig. 11.10 is started by injecting pure fuel through the central
pipe of the pilot burner. The mixture that forms by diffusion is ignited by a spark and
a diffusion flame is formed, which is very stable. From a low load on, the pure fuel
stream is stopped and replaced by a premixed flow, but with a composition sufficiently
richer than the lean flammability limit, so that stable combustion is possible. The
premixing is realised by swirling the air flow and injection of the fuel through holes
in the hollow swirl vanes (or holes downstream of the vanes as in Fig. 11.10). The
mixture passes a diverging cone. Due to the swirl of the flow, a low pressure zone
forms in the centre of the flow. By the low pressure, combustion gas is sucked
from downstream so that a circulating flow pattern is formed (Fig. 11.10). The hot
combustion gas ignites the oncoming mixture. The stabilisation and ignition process
is called *swirl stabilisation* and is used in many industrial burners.

The peripheral burners with premixing of fuel and air into a lean mixture, near to
the lean flammability limit, are started in stages. With eight peripheral burners, the
stages may be a first group of four burners, followed by another four, as load increases.
This mixture burns as it reaches the stable pilot flame. Air is added downstream of the
primary combustion zone through slots or holes in the liner surface in order to allow
the further combustion of the formed CO (second stage of the combustion). More air
is added further downstream (dilution air). With can liners, there are transition tubes

which bring the combustion gas to the turbine entry (Fig. 11.7). Also these transition tubes have to be cooled. The air used for convection cooling of these tubes is mixed in just before the turbine entry.

The same principles are used in other combustion systems, but the technical realisation may be different from described above. In particular, with an annular combustion chamber, all burners are equal. A possible realisation is then a burner in the style of the pilot burner of Fig. 11.10, with a second ring of swirl vanes around it with injection of fuel forming a very lean mixture. The basic ingredients are always: gas injection in a swirling air flow, swirl-stabilisation of the flame, and staging of the combustion air. This technology is commonly called dry low NOx (DLN) or dry low emissions (DLE). Alternative techniques for reducing NOx are injection of water or steam in the combustion chamber. These wet techniques are sometimes used with smaller gas turbines, but not with large turbines that run almost continuously.

11.2   Thermodynamic Modelling

At the entry and exit of a compressor or expander, the flow is nearly homogeneous and steady. This is not the case at the entry and exit of a stage. For a stage, we take averages over the entry and exit areas as representative states. As unsteadiness originates from the turning of the rotor, such averages are steady for constant rotor speed. We wish to assess the stage efficiency only using these average states. Hereafter, we discuss the efficiency of the work exchange with the flow that goes from the stage entry to the exit and passes through the rotor, i.e. the internal efficiency, as defined in Chap. 1 (Sect. 1.6) and Chap. 3 (Sect. 3.5). Similar discussions can be found in books on fundamentals of gas turbines [7].

For brevity, we use the term efficiency without any further specification. A volumetric efficiency is never used with gas turbine components. Losses due to clearances at blade tips or blade foots are included in the internal efficiency, because the leakage flows mix with the main flow.

11.2.1   Isentropic Efficiency with Adiabatic Compression or Expansion

For assessing the efficiency of a compressor or a turbine stage, we consider steady flow between the average states at entry and exit. To a streamline in the absolute frame applies:

Energy equation:

$$dW + dq = d\frac{1}{2}v^2 + dh + dU, \tag{11.1}$$

Work equation:

$$dW = d\frac{1}{2}v^2 + (1/\rho)dp + dU + dq_{irr},\qquad(11.2)$$

where dW is the elementary work done on the fluid and dq the elementary heat supplied to it. The gravitational potential energy is denoted by U and dq_{irr} is the heat internally produced by irreversibility (dissipation). First, we consider adiabatic processes, i.e. $dq = 0$.

The energy equation is a total differential and thus can be integrated. The work equation forms a total differential if density depends only on pressure. A special case meeting this condition is $\rho =$ constant. The reversible part of the work done on the fluid or by the fluid is then the change of the *mechanical energy*, the sum of kinetic energy, pressure potential energy and gravitational potential energy (see Chap. 1): $E_m = \frac{1}{2}v^2 + p/\rho + U$. Moreover, comparison of the energy and work balances (11.1 and 11.2) learns that work cannot be produced from internal energy. The definition of the internal efficiency is then quite straightforward. With work done on the fluid, as in a pump stage, it is defined as the ratio of the increase of the mechanical energy in the fluid to the work done by the rotor. With work done by the fluid, as in a hydraulic turbine, it is the ratio of the work done on the rotor to the decrease of the mechanical energy in the fluid.

Assessment of the efficiency is more difficult with a compressible fluid. The traditional procedure consists in considering a lossless flow and comparing its performance with the flow with losses. From the difference between the work equation and the energy equation (for $dq = 0$) it follows that $dq_{irr} = dh - (1/\rho)dp = Tds$. The lossless flow is thus isentropic. For an isentropic flow, ρ is a unique function of p and integration of the work equation is possible.

We first notice that changes of gravitational potential energy are negligible for a gas with the processes that we analyse. E.g., for $\Delta z = 1$ m is $\Delta U = g\,\Delta z = 9.81$ J/kg. Converted into enthalpy, this means for $c_p = 1005$ J/kgK (air): $c_p\,\Delta T = \Delta U$ or $\Delta T \approx 0.01$ K. Further we apply the concept of *total state* or *stagnation state*. The stagnation state is the state that would be attained by bringing a fluid particle to zero velocity in a reversible adiabatic way. We note this state with the subscript 0; especially: $h_0 = h + \frac{1}{2}v^2$. The total state is then attained by additionally bringing (in mind) the fluid particle to a reference height for the gravitational potential energy. But, since the effect of gravitational potential energy is negligible, the stagnation and total states are identical. The common practice is to use the term *total state*, which we follow.

With the concept of total enthalpy and ignoring the small changes in gravitational potential energy, Eqs. (11.1) and (11.2) become

$$dW = dh_0 = dh + d\frac{1}{2}v^2,\qquad(11.3)$$

Fig. 11.12 h–s diagram for compression and expansion

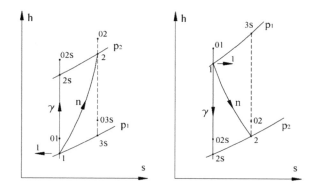

$$dW = (1/\rho)dp + d\frac{1}{2}v^2 + dq_{irr}. \qquad (11.4)$$

Figure 11.12 sketches a compression and an expansion (with work) in the h–s diagram. With a compression with the same starting point and the same final pressure, the isentropic process requires less work. The efficiency can thus be defined by

$$\eta_s = \frac{h_{02s} - h_{01}}{h_{02} - h_{01}} = \frac{\Delta W_s}{\Delta W}. \qquad (11.5)$$

For an expansion, Eqs. (11.3) and (11.4) are written as

$$-dh_0 = -dW, \quad -(1/\rho)dp - d\frac{1}{2}v^2 = -dW + dq_{irr}.$$

The isentropic expansion with the same starting point and the same final pressure extracts more work from the fluid and the efficiency can be defined as

$$\eta_s = \frac{h_{01} - h_{02}}{h_{01} - h_{02s}} = \frac{-\Delta W}{-\Delta W_s}. \qquad (11.6)$$

The total state points 02 and 02s are obtained by adding the kinetic energy to the static enthalpy in the state points 2 and 2s. But, the total points 02 and 02s are often positioned on the isobar p_{02}. This is, strictly spoken incorrect, as isobars in the h–s diagram diverge with increasing entropy.

Expressions (11.5) and (11.6) define the *isentropic total-to-total efficiency*. The work is compared to the total enthalpy change of an isentropic process with the same starting point and the same final pressure. The isentropic total enthalpy change is termed *isentropic head*. It is assumed that the isentropic head is the change of *mechanical energy*, meaning the maximum available enthalpy difference for work generation by the fluid (see Sect. 1.4.4 in Chap. 1).

For an expansion, this is correct, but for a compression the available enthalpy difference for work generation at the end of the compression is larger. In Fig. 11.12

(left), it is $h_{02} - h_{03s}$, which is larger than $h_{02s} - h_{01}$, used in Eq. 11.5, due to the divergence of the isobars. The isobars diverge in an h–s diagram, since $(dh/ds)_p = T$. Therefore, it seems appropriate to define an isentropic efficiency of a compression process based on the re-expansion isentropic head $h_{02} - h_{03s}$.

We denominate this efficiency by the term *isentropic re-expansion efficiency*:

$$\eta_{sre} = \frac{h_{02} - h_{03s}}{h_{02} - h_{01}}. \tag{11.7}$$

When the kinetic energy at the end of a compression is not useful, a *total-to-static efficiency* is defined by

$$\eta_s = \frac{h_{2s} - h_{01}}{h_{02} - h_{01}}. \tag{11.8}$$

Analogously, for a turbine, the total-to-static efficiency is

$$\eta_s = \frac{h_{01} - h_{02}}{h_{01} - h_{2s}}. \tag{11.9}$$

The isentropic efficiencies are no efficiencies in the strict thermodynamic sense. Efficiency is, by strict definition, the ratio of the useful part of the work to the total work with a given process. But, a compressible fluid does not allow a precise definition of the useful part, as this would require detailed knowledge of the real process path in the h–s diagram. This path is not defined since the real process is unsteady and, moreover, the objective is to define efficiency solely using the states at the start and the end of the process. Efficiencies defined here are *quality numbers*, as they compare the work from two processes, namely that of a lossless steady process and that of the real one. Strictly spoken, such a quality number is termed *effectiveness*, but the two terms are not distinguished in practice, so that, henceforth, we use the term isentropic efficiency.

11.2.2 Reheat Effect

A problem with the efficiency assessment of a compression or an expansion is the internal heating of the fluid by dissipation. This causes a shift of a state point in the h–s diagram to the right side, for a given pressure. The divergence of the isobars thus increases the enthalpy difference available for work, in the same way as an enthalpy difference increase by external heating. This implies that part of the work dissipated is recoverable. On the other hand, the work required to continue compression or to recompress after expansion increases. This effect is termed *reheat effect*.

In further analyses, we assume that the change of the kinetic energy is negligible in the overall process, so that we may state that $d\frac{1}{2}v^2 = 0$. The total-to-total efficiency may then be noted by comparing static states. To a compression and an expansion respectively apply

$$\eta_s = \frac{h_{2s} - h_1}{h_2 - h_1}, \quad \eta_{sre} = \frac{h_2 - h_{3s}}{h_2 - h_1} \quad \text{and} \quad \eta_s = \frac{h_1 - h_2}{h_1 - h_{2s}}.$$

We then speak of isentropic efficiency, without any further specification.

In order to study the reheat effect, we first consider the series operation of two compressions with the same isentropic efficiency, as sketched in Fig. 11.13, left. The isentropic efficiencies are equal:

$$\eta_{s1} = \frac{h_{2s} - h_1}{h_2 - h_1} = \eta_{s2} = \frac{h_{3s} - h_2}{h_3 - h_2}.$$

The isentropic efficiency of the entire compression is lower:

$$\eta_s = \frac{h_{3ss} - h_1}{h_3 - h_1} = \frac{h_{3ss} - h_{2s} + h_{2s} - h_1}{h_3 - h_2 + h_2 - h_1}.$$

The reason is the divergence of the isobars: $h_{3s} - h_2 > h_{3ss} - h_{2s}$. The heating by energy dissipation in the first compression increases the work required in the second compression.

An opposite effect occurs with series operation of two expansions with the same isentropic efficiency (Fig. 11.13, right):

$$\eta_{s1} = \frac{h_1 - h_2}{h_1 - h_{2s}} = \eta_{s2} = \frac{h_2 - h_3}{h_2 - h_{3s}}.$$

The isentropic efficiency of the entire expansion is larger:

$$\eta_s = \frac{h_1 - h_3}{h_1 - h_{3ss}} = \frac{h_1 - h_2 + h_2 - h_3}{h_1 - h_{2s} + h_{2s} - h_{3ss}}.$$

Fig. 11.13 Series operation of compressions and expansions

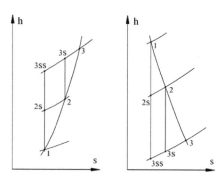

This follows from the divergence of the isobars: $h_2 - h_{3s} > h_{2s} - h_{3ss}$. The internal heating by energy dissipation in the first expansion increases the work potential of the second expansion in the same way as external heating does. So, part of the energy dissipation in the first expansion is recovered in the second expansion.

Series arrangement of compressor stages with equal isentropic efficiency results in an isentropic efficiency of the entire compression that is lower than the isentropic efficiency of the components. For the isentropic re-expansion efficiency, the result is higher efficiency of the series arrangement. With series operation of turbine stages the result is also higher isentropic efficiency of the series arrangement. So, isentropic efficiency is unsuitable for the study of the effects of changing the pressure ratio in gas turbine cycles.

11.2.3   Infinitesimal Efficiency; Polytropic Efficiency

The problem caused by the reheat effect can be avoided by defining efficiency on an infinitesimal base. The relations for a compression are (with $d\frac{1}{2}v^2 = 0$):

$$dW = dh, \quad dW = (1/\rho)\ dp + dq_{irr}.$$

For an infinitesimal process part, efficiency, in strict thermodynamic sense, is

$$\eta_\infty = \frac{dh_s}{dh} = \frac{(1/\rho)dp}{dh}.$$

This is termed *small stage efficiency* or *infinitesimal efficiency*. The subscript ∞ indicates that the complete process is thought as being composed of an infinite number of infinitesimal parts. For an ideal gas, with $p = \rho RT$ and $dh = c_p\ dT$, it follows that

$$\eta_\infty = \frac{RT}{p}\frac{dp}{c_p dT} = \frac{\gamma - 1}{\gamma}\frac{dp/p}{dT/T}. \qquad (11.10)$$

The specific heat capacity at constant pressure is denoted by c_p and we use the terms *specific heat* and *heat capacity* as equivalent abbreviations of the full term. With constancy of the heat capacity ratio (perfect gas) and η_∞ over the entire compression, the process is polytropic with exponent n:

$$p \sim \rho^n \quad \text{or} \quad p \sim T^{\frac{n}{n-1}}, \quad \text{thus} \quad \eta_{\infty,comp} = \left(\frac{n}{n-1}\right)\Big/\left(\frac{\gamma}{\gamma - 1}\right). \qquad (11.11)$$

An efficiency lower than unity requires $n > \gamma > 1$. This can graphically be deduced from the h–s diagram (Fig. 11.12, left). For example: $\gamma = 1.4$, $\eta_\infty = 0.9$, $n = 1.465$. Integration of (11.10) results in:

$$\frac{\gamma - 1}{\gamma} \ln(p_2/p_1) = \eta_\infty \ln(T_2/T_1).$$

The similar result for an isentropic compression is

$$\frac{\gamma - 1}{\gamma} \ln(p_2/p_1) = \ln(T_{2s}/T_1).$$

So:

$$\eta_\infty = \frac{\ln(T_{2s}) - \ln(T_1)}{\ln(T_2) - \ln(T_1)} = \frac{\ln(h_{2s}) - \ln(h_1)}{\ln(h_2) - \ln(h_1)}.$$

The obtained efficiency is termed *polytropic efficiency*. By convention it may also be applied to entire stages and the entire compressor by

$$\eta_p = \frac{\ln(h_{02s}) - \ln(h_{01})}{\ln(h_{02}) - \ln(h_{01})}. \tag{11.12}$$

Assuming that the compression may be represented as polytropic, we calculate for an ideal and perfect gas (Fig. 11.12, left) the following quantities.
Isentropic work (*isentropic head*):

$$\Delta h_s = \int_1^{2s} \frac{1}{\rho} dp = \frac{p_1}{\rho_1} \int_1^2 \left(\frac{p}{p_1}\right)^{-\frac{1}{\gamma}} d\left(\frac{p}{p_1}\right) = \frac{\gamma}{\gamma - 1} \frac{p_1}{\rho_1} \left[\left(\frac{p_2}{p_1}\right)^{\frac{\gamma-1}{\gamma}} - 1\right].$$

Reversible part of the work (*polytropic head*):

$$\Delta h_r = \int_1^2 \frac{1}{\rho} dp = \frac{p_1}{\rho_1} \int_1^2 \left(\frac{p}{p_1}\right)^{-\frac{1}{n}} d\left(\frac{p}{p_1}\right) = \frac{n}{n - 1} \frac{p_1}{\rho_1} \left[\left(\frac{p_2}{p_1}\right)^{\frac{n-1}{n}} - 1\right].$$

Work available with isentropic re-expansion (*isentropic re-expansion head*):

$$\Delta h_{sre} = \int_2^{3s} -\frac{1}{\rho} dp = -\frac{\gamma}{\gamma - 1} \frac{p_2}{\rho_2} \left[\left(\frac{p_1}{p_2}\right)^{\frac{\gamma-1}{\gamma}} - 1\right]; \quad \text{thus with } \rho_2 = \rho_1 \left(\frac{p_2}{p_1}\right)^{\frac{1}{n}}:$$

$$\Delta h_{sre} = \frac{\gamma}{\gamma - 1} \frac{p_1}{\rho_1} \left(\frac{p_2}{p_1}\right)^{1-\frac{1}{n}} \left[1 - \left(\frac{p_1}{p_2}\right)^{\frac{\gamma-1}{\gamma}}\right]$$

$$= \frac{\gamma}{\gamma - 1} \frac{p_1}{\rho_1} \left(\frac{p_2}{p_1}\right)^{\frac{1}{\gamma} - \frac{1}{n}} \left[\left(\frac{p_2}{p_1}\right)^{\frac{\gamma-1}{\gamma}} - 1\right].$$

Total work:

$$\Delta h = c_p(T_2 - T_1) = c_p T_1 \left(\frac{T_2}{T_1} - 1\right) = \frac{\gamma}{\gamma - 1} \frac{p_1}{\rho_1} \left[\left(\frac{p_2}{p_1}\right)^{\frac{n-1}{n}} - 1\right].$$

Table 11.1 Comparison of efficiency definitions of a compression with $\eta_\infty = 0.9$ and $\gamma = 1.4$

	$r = p_2/p_1 = 2$	$r = 10$	$r = 40$
$\eta_s = \Delta h_s / \Delta h$	0.8898	0.8641	0.8398
$\eta_\infty = \Delta h_r / \Delta h$	0.9000	0.9000	0.9000
$\eta_{sre} = \Delta h_{sre} / \Delta h$	0.9096	0.9296	0.9442

Results for $\eta_\infty = 0.9$ and $\gamma = 1.4$ are listed in Table 11.1. For a given infinitesimal efficiency, the difference with the other efficiencies increases with increasing pressure ratio.

The infinitesimal efficiency may be defined in the same way for an expansion. We note (with $d\frac{1}{2}v^2 = 0$):

$$-dh = -dW, \quad -(1/\rho)dp = -dW + dq_{irr}.$$

The infinitesimal efficiency is

$$\eta_\infty = \frac{-dh}{-dh_s} = \frac{-dh}{-(1/\rho)dp} = \frac{-c_p dT}{-RT(dp/p)} = \frac{\gamma}{\gamma - 1}\frac{(-dT/T)}{(-dp/p)}.$$

So:

$$\eta_{\infty,exp} = \left(\frac{\gamma}{\gamma - 1}\right) / \left(\frac{n}{n - 1}\right). \tag{11.13}$$

An efficiency lower than unity requires $\gamma > n > 1$, which can graphically be deduced from the h–s diagram (Fig. 11.12, right). For example: $\gamma = 1.4$, $\eta_\infty = 0.9$, $n = 1.346$.

Further:

$$\eta_p = \frac{\ln(h_{01}) - \ln(h_{02})}{\ln(h_{01}) - \ln(h_{02s})}. \tag{11.14}$$

Assuming that the expansion may be represented as polytropic, we calculate for an ideal and perfect gas (Fig. 11.12, right) the following quantities.

Isentropic work (*isentropic head*):

$$\Delta h_s = \int_1^{2s} -\frac{1}{\rho}dp = \frac{p_1}{\rho_1}\int_1^2 -\left(\frac{p}{p_1}\right)^{-\frac{1}{\gamma}}d\left(\frac{p}{p_1}\right) = \frac{\gamma}{\gamma - 1}\frac{p_1}{\rho_1}\left[1 - \left(\frac{p_2}{p_1}\right)^{\frac{\gamma-1}{\gamma}}\right].$$

Reversible part of the work (*polytropic head*):

$$\Delta h_r = \int_1^2 -\frac{1}{\rho}dp = \frac{n}{n - 1}\frac{p_1}{\rho_1}\left[1 - \left(\frac{p_2}{p_1}\right)^{\frac{n-1}{n}}\right].$$

Table 11.2 Comparison of efficiency definitions of an expansion with $\eta_\infty = 0.9$ and $\gamma = 1.4$

	$r = p_2/p_1 = 1/2$	$r = 1/10$	$r = 1/40$
$\eta_s = \Delta h/\Delta h_s$	0.9087	0.9269	0.9405
$\eta_\infty = \Delta h/\Delta h_r$	0.9000	0.9000	0.9000

Total work:

$$\Delta h = c_p(T_1 - T_2) = \frac{\gamma}{\gamma - 1} \frac{p_1}{\rho_1} \left[1 - \left(\frac{p_2}{p_1} \right)^{\frac{n-1}{n}} \right].$$

Results for $\eta_\infty = 0.9$ and $\gamma = 1.4$ are listed in Table 11.2. The conclusion is similar to that of a compression.

The practical application of polytropic efficiency has two minor difficulties. This efficiency cannot be read directly from an h–s diagram and may only be calculated directly for an ideal and perfect gas with formulae (11.11) and (11.13). The polytropic efficiency may be calculated for a general gas with the logarithmic formulae (11.12) and (11.14), which then implicitly represent the gas in an averaged way as ideal and perfect. There is no real problem when studying gas turbine cycles. Air is an ideal gas ($p = \rho RT$, $R = 287$ J/kgK for dry air). Air is no perfect gas. Its heat capacity does not depend on pressure, but it increases with temperature. The same applies to a combustion gas, as we discuss in the next section.

Taking variable c_p and R into account with numerical simulations is no problem. A polytropic representation is not used then, but only the definition of infinitesimal efficiency. With an expansion this is $-dh = -\eta_\infty (1/\rho) dp$. The expansion path is divided into small intervals. For each interval, the relation is integrated with average values of c_p and R. This has to be done iteratively, starting with the values at the beginning of the interval. From the provisional value of h at the end of the interval follows T with a tabulated relationship $h = h(T)$, and similarly c_p and R. Iteration is done a few times. As c_p and R only slightly vary, two or three iterations suffice. This allows a simple determination of the end point of an expansion with given η_∞.

The opposite operation, with a given end point and seeking the infinitesimal efficiency is more difficult. It requires iteration on the value of the infinitesimal efficiency. Calculations for compressions run analogously.

11.2.4 Thermodynamic Properties of Air and Combustion Gas

Dry air is a mixture of the following components in mass fractions:

$$0.7553 \, N_2 + 0.2314 \, O_2 + 0.0128 \, Ar + 0.0005 \, CO_2.$$

Table 11.3 Gas constant and heat capacity of gases relevant with combustion

	R (J/kgK)	C_p (J/kgK)			C_p/R	
		0 °C	500 °C	1500 °C	0 °C	1500 °C
N_2	295.31	1026	1057	1150	3.47	3.89
O_2	259.83	908.4	979.1	1071	3.50	4.12
H_2O	461.52	1858	1976	2302	4.03	4.99
CO_2	188.92	820.5	1016	1195	4.34	6.33

At 1 bar, 15 °C, 60% relative humidity, air contains a 0.0064 mass fraction of water. The air composition is then

$$0.7505\,N_2 + 0.2299\,O_2 + 0.0127\,Ar + 0.0064\,H_2O + 0.0005\,CO_2.$$

Henceforth we assume this air composition with detailed simulations.

Fuel is a mixture of carbon and hydrogen, spread over several chemical compounds. O_2 in air is consumed by combustion with generation of H_2O and CO_2. The ratio between the Ar and N_2 fractions in the air remains constant. Therefore, it is customary to combine the Ar with the N_2 and denote it then as N_2 from air. We so note the air composition as

$$0.7632\,N_2 + 0.2299\,O_2 + 0.0064\,H_2O + 0.0005\,CO_2.$$

Gas properties are listed in Table 11.3, with N_2 from air [1]. The specific heat (heat capacity) C_p is the integral value (denoted with big C), which means that the enthalpy difference between 0 °C and the temperature indicated is this value of C_p, multiplied with the temperature difference.

For air:

$$R = 0.7632\,R_{N_2} + 0.2299\,R_{O_2} + 0.0064\,R_{H_2O} + 0.0005\,R_{CO_2},$$
$$C_p = 0.7632\,C_{pN_2} + 0.2299\,C_{pO_2} + 0.0064\,C_{pH_2O} + 0.0005\,C_{pCO_2}.$$

The result is $R = 288.2$ J/kgK; $C_p = 1004.2, 1045.0, 1139.2$ J/kgK, at 0 °C, 500 °C, 1500 °C. The gas constant is somewhat larger than for dry air, which is 287.1 J/kgK. The heat capacity increases with temperature. For N_2 and O_2, $C_p/R = \gamma/(1-\gamma)$ is about 3.5 at low temperature (0 °C) and 4 at high temperature. These values are 4–4.3 and 5–6.3 for respectively H_2O and CO_2. So C_p/R increases with temperature and increases also by the generation of combustion products. For air in the range 0–500 °C, C_p is about 1045 J/kgK. The corresponding value of γ is about 1.38. So, we may assume this value of γ, together with $R = 288$ J/kgK for hand calculations with compressors.

The chemical combustion equations are

$$C + O_2 \rightarrow CO_2,$$

$$12.011 \text{ kg C} + 31.999 \text{ kg O}_2 \rightarrow 44.010 \text{ kg CO}_2,$$

$$H_2 + \frac{1}{2}O_2 \rightarrow H_2O,$$

$$2.0159 \text{ kg H}_2 + \frac{1}{2} \times 31.999 \text{ kg O}_2 \rightarrow 18.015 \text{ kg H}_2O.$$

So, 1 kg C forms 3.664 kg CO_2 and consumes 2.664 kg O_2; 1 kg H_2 forms 8.937 kg H_2O and consumes 7.937 kg O_2.

For a fuel with mass fractions c carbon and h hydrogen and a fuel–air ratio f (kg/kg), the composition of the combustion gas for $(1 + f)$ kg is

$$0.7632\,N_2 + [0.2299 - (h \times 7.937 + c \times 2.664)f]O_2$$
$$+ (0.0064 + h \times 8.937 \times f)H_2O + (0.0005 + c \times 3.664 \times f)CO_2.$$

The gas constant follows from

$$(1 + f)R = 288.2 - (h \times 7.937 + c \times 2.664) \times 259.83 \times f$$
$$+ h \times 8.937 \times 461.52 \times f + c \times 3.664 \times 188.92 \times f$$
$$= 288.2 + (h \times 2062.33 + c \times 0.016) \times f.$$

With $h = 0.13974$, $c = 0.86026$, this becomes $(1 + f)\,R = 288.2 + 288.2f$.

The composition of a liquid fuel is always near to $h = 0.14$ and $c = 0.86$. So, the gas constant of the combustion gas is approximately that of air, independent of the fuel–air ratio.

The heat capacity follows analogously, for $h = 0.14$, $c = 0.86$, from

$$(1 + f)C_p(0\,^\circ C) = 1004.2 + 1819.6f,$$
$$(1 + f)C_p(500^\circ C) = 1045.0 + 2342.7f,$$
$$(1 + f)C_p(1500^\circ C) = 1139.2 + 3002.0f.$$

E.g., for $f = 0.03$: $C_p = 1027.9$, 1082.8, 1193.5 J/kgK, at 0, 500, 1500 °C.

We take, as an example, combustion in air following a compression with pressure ratio 20 and $\eta_\infty = 0.9$, starting from $T_a = 288$ K ($\gamma = 1.38$).

For air:

$$\frac{\gamma}{\gamma - 1} = \frac{C_p}{R} = 3.632. \quad \text{Thus:} \quad \frac{n}{n - 1} = \eta_\infty \frac{\gamma}{\gamma - 1} = 3.268.$$

$$\left(\frac{T_2}{T_1}\right)^{\frac{n}{n-1}} = \frac{p_2}{p_1}, \quad from\ which\ T_2 = 720 \text{ K} = 447\,^\circ C.$$

The thermal combustion equation is:

$$C_{pa}(T_2 - T_r) + f\,C_{pf}(T_f - T_r) + f\,H_L(T_r) = (1 + f)C_{pg}(T_3 - T_r). \quad (11.15)$$

The subscripts a, f and g refer to air, fuel and combustion gas. H_L is the lower heating value of the fuel, determined at the reference temperature T_r, which often is $T_r = 288$ K, the standard temperature at sea level (sometimes 25 °C is used). The lower heating value of a liquid fuel is around $H_L = 43{,}000$ kJ/kg. The temperatures T_2 and T_3 are upstream and downstream of the combustion chamber. The heat content of fuel may be ignored. The values of C_{pa} and C_{pg} are average values over the temperature ranges. It follows that

$$f = \frac{C_{pg}(T_3 - T_r) - C_{pa}(T_2 - T_r)}{H_L(T_r) - C_{pg}(T_3 - T_r)}.$$

With $R = 288$ J/kgK, $\gamma_a = 1.38$, $\gamma_g = 1.32$, $T_3 = 1773$ K (1500 °C), $T_2 = 700$ K and $T_r = 288$ K, if follows that $f \approx 0.033$. The example demonstrates that the fuel–air ratio is at maximum about 0.03 in practice, as a gas temperature 1500 °C is about the maximum turbine inlet temperature used nowadays.

For $f = 0.03$, the average heat capacity of the combustion gas over the range 500–1500 °C is

$$C_p = (1193.5 \times 1500 - 1082.8 \times 500)/1000 = 1248.9\,\text{J/kg K}.$$

The corresponding value of γ is 1.30. So, we may use $\gamma = 1.30$ together with the gas constant $R = 288$ J/kgK with hand calculations in the turbine part of a gas turbine.

For $f = 0.03$, the average C_p in the range 0–1500 °C is 1193.5 J/kgK. The corresponding value of γ is 1.318. So, we may use $\gamma = 1.32$ together with $R = 288$ J/kgK for the combustion gas in the thermal balance equation of the combustion chamber (11.15).

With gaseous fuels such as natural gas (CH_4), the gas constant of the combustion gas differs somewhat from the value of air. For pure CH_4, 12.011 kg C goes together with 2×2.0159 kg H, which means $c = 0.749$; $h = 0.251$. Then:

$$(1 + f)R = 288.2 + 517.7\,f.$$

The lower heating value of methane is about 50,000 kJ/kg. This implies a fuel–air ratio of about 0.027 for $T_3 = 1500$ °C. With $f = 0.027$: $R = 294.0$ J/kgK. Real natural gas also contains higher hydrocarbons, such as C_2H_6, and also N_2 and CO_2. The lower heating value is somewhat smaller, typically about 46,000 kJ/kg. Compared to pure methane, the fuel–air ratio is somewhat larger, but the gas constant of the combustion gas stays about the same, thus somewhat larger than that of air.

Table 11.4 Polynomial fitting of integral heat capacity (coefficients in J/kgK)

	A	B	C	D
N_2	1026.3	26.5	91.2	-36.3
O_2	906.8	155.2	-18.3	-8.4
H_2O	1856.6	156.4	199.5	-70.7
CO_2	821.0	502.5	-254.3	57.1

11.2.5 Heat Capacity Representation

The integral specific heat capacity of a gas as a function of temperature may be expressed, with good accuracy, by a polynomial, according to

$$C_p = A + B\left(\frac{T}{1000\,K}\right) + C\left(\frac{T}{1000\,K}\right)^2 + D\left(\frac{T}{1000\,K}\right)^3.$$

Mean square fitting of the integral heat capacity between 0 °C and the temperature T, in °C, for values between 0 °C and 1500 °C per 100 °C [1], leads to the results in Table 11.4. The error of the fitted polynomials is lower than 2 ‰. The differential specific heat capacity $c_p = dh/dT$, with $h = C_p(T - T_r)$, is

$$c_p = A + 2B\left(\frac{T}{1000\,K}\right) + 3C\left(\frac{T}{1000\,K}\right)^2 + 4D\left(\frac{T}{1000\,K}\right)^3.$$

Henceforth, we distinguish between a differential value of the heat capacity (small c) and an averaged value over a temperature interval (big C), if necessary.

11.2.6 Cooled Expansion

Vanes and blades are cooled within the HP part of the turbine. This has to be taken into account when describing the expansion. Two phenomena intervene, namely cooling of the through-flow and mixing of the cooling air into the expanding gas. In order to model the cooled expansion we use here a method with the cooling distributed along the expansion, analogous to the distribution of the energy dissipation in the polytropic description. This way of simulation is, with many variants, commonly applied in the literature. There is an enormous literature on the topic. We refer to a few examples (El-Masri [3], Sanjay et al. [6], Wilcock et al. [8]). The simulation methodology used here contains some simplifications with respect to what typically is done.

For an infinitesimal expansion, the work equation (with no change of kinetic energy) reads

$$-(1/\rho)dp = -dW + dq_{irr}. \tag{11.16}$$

The infinitesimal efficiency is defined by

$$-dW = \eta_\infty(-(1/\rho)dp). \tag{11.17}$$

The heat removed from the main flow by cooling may be noted for an entire stage as

$$-\Delta Q = -\dot{m}_g \Delta q = A_b \alpha\,(\overline{T}_g - \overline{T}_b), \tag{11.18}$$

where $\dot{m}_g$ is the gas mass flow rate (kg/s), $-\Delta q$ the heat removed per unit of mass of gas, A_b the blade surface, α the heat transfer coefficient (W/m²K) and $(\overline{T}_g - \overline{T}_b)$ the average temperature difference between the gas and the blades. The work of a stage is similarly $-\dot{m}_g \Delta W$, with $-\Delta W$ the work done per unit of mass of gas. The gas flow rate may be written as $\dot{m}_g = \rho\, v\, A_g$, with ρ and v representative values of density and through-flow velocity and A_g the through-flow area. The ratio of the heat removed to the work done is

$$\frac{-\Delta q}{-\Delta W} = \frac{A_b}{A_g}\,\frac{\alpha}{\rho\, v\, \overline{c}_{pg}}\,\frac{\overline{c}_{pg}}{(-\Delta W)}\,(\overline{T}_g - \overline{T}_b).$$

The surface ratio is typically around 8. The Stanton number $St = \alpha/(\rho\, v\, \overline{c}_{pg})$ is around 0.005 for convective heat transfer in a flow along a flat plate, with $\overline{c}_{pg}$ the specific heat at $\overline{T}_g$ [3]. A coefficient $\kappa = \frac{A_b}{A_g} St \frac{\overline{c}_{pg}}{(-\Delta W)}$ appears.

With $\overline{c}_{pg} \approx 1250$ J/kgK and $-\Delta W \approx 250$ kJ/kg, it follows that $\kappa \approx 0.20$ J/kJK.

By distributing the work and the cooling along the expansion, we may assume the differential equation:

$$\frac{-dq}{-dW} = \kappa\,(T_g - T_b), \tag{11.19}$$

with $-dq$ and $-dW$ heat removed and work done per mass unit for an infinitesimal part of the expansion and T_g and T_b local gas and blade temperatures.

With state-of-the-art machines, the tolerable blade temperature $T_b \approx 900$ °C. The value of the heat transfer coefficient κ is lower than estimated up to now. The resistance to heat transfer increases by a protective ceramic coating on the blades (TBC: *Thermal Barrier Coating*) and by using film cooling. It is obvious that the coating constitutes an additional thermal resistance. Some reasoning is required to understand that a cooling film may be considered as an additional resistance too. The heat transferred from the gas to the blades (11.18) may, with an adapted definition of the heat transfer coefficient, be noted as

$$-\Delta Q = A_b\,\alpha\,(\overline{T}_{aw} - \overline{T}_b), \tag{11.20}$$

with $\overline{T}_{aw}$ the mean adiabatic wall temperature. This is the blade temperature that would be attained in absence of heat transfer. Without a cooling film, it is approximately the gas temperature (but there is some temperature increase by friction and leading edge stagnation).

The adiabatic film cooling effectiveness may be defined by

$$\eta_f = \frac{\overline{T}_g - \overline{T}_{aw}}{\overline{T}_g - \overline{T}_e}. \tag{11.21}$$

This effectiveness then expresses how much the adiabatic wall temperature decreases compared to the gas temperature, due to the protection by the film, as a fraction of the temperature difference between the gas $\overline{T}_g$ and the cooling air $\overline{T}_e$ leaving the ejection holes. It is a realistic approximation that the temperature of the cooling air entering the film is near to the blade temperature, due to the intense contact between cooling air and blade within the ejection holes. We so may assume $\overline{T}_e = \overline{T}_b$ and thus combine Eqs. (11.20) and (11.21) into

$$-\Delta Q = A_b \alpha \left(\overline{T}_g - \eta_f(\overline{T}_g - \overline{T}_b) - \overline{T}_b\right) = A_b \alpha \left(1 - \eta_f\right)(\overline{T}_g - \overline{T}_b).$$

From this reasoning emerges that film cooling decreases the heat transfer between gas and blade. The film cooling effectiveness may be up to 30%, so that film cooling then reduces the heat transfer to about 70%. In combination with a thermal barrier coating, κ may be halved. Henceforth, $\kappa = 0.15$ J/kJK is used, taking into account that heat transfer is larger at a leading edge due to flow stagnation.

Figure 11.14 (left) represents the h–s diagram of the expansion of a cooled stage. The $12'$ part is the expansion with distributed cooling. The $2'2''$ part renders the temperature drop by mixing and the $2''2$ part is the pressure drop by mixing.

In order to calculate the expansion through a turbine, the entire pressure drop is divided into a large number of small intervals. In Fig. 11.14 a pressure interval corresponds to the pressure difference $p_1 - p_2$. This interval is considered as a stage. This fictitious stage has a pressure drop that is much smaller than a stage in reality. With a first iteration, the pressure difference $p_1 - p_2'$ may be set equal to $p_1 - p_2$.

Fig. 11.14 Expansion with cooling and mixing of cooling air and gas; left: simultaneous cooling and expansion on the $12'$ path; right: adiabatic expansion followed by cooling

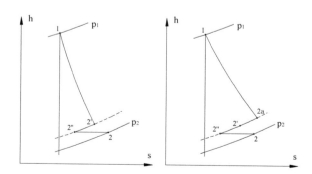

On the 12′ path, the energy equation is written as

$$-dh = -dW - dq = -dW\big(1 + \kappa(T_g - T_b)\big), \tag{11.22}$$

and

$$-c_{pg} dT_g = \eta_\infty\left(-\frac{RT_g dp}{p}\right)\big(1 + \kappa(T_g - T_b)\big). \tag{11.23}$$

Equations (11.17) and (11.19), together with the gas law, are applied to obtain (11.22) and (11.23). c_{pg} is the differential value of the specific heat. Equations (11.22) and (11.23) may be integrated exactly for constant coefficients. But, the resulting expressions are rather complex. For gaining insight, approximate integration with an average value of the factor $1 + \kappa(T_g - T_b)$ is therefore more convenient. This factor changes only little over a small pressure interval. Approximate integration of Eq. (11.23) with averaged values of c_{pg} and T_g results in

$$\frac{T_{2'}}{T_1} = \left(\frac{p_{2'}}{p_1}\right)^{\frac{R\,\eta_\infty}{\overline{c}_{pg}}\left(1+\kappa(\overline{T}_g-T_b)\right)}. \tag{11.24}$$

The relation is polytropic and the exponent increases by cooling. The increased exponent expresses that, for a given pressure interval, the final temperature with cooling is lower than without cooling. With (11.24) follows the enthalpy difference over the pressure interval:

$$-\Delta h = \overline{c}_{pg}(T_1 - T_{2'}) = \overline{c}_{pg}T_1\left(1 - \frac{T_{2'}}{T_1}\right). \tag{11.25}$$

With an approximate integration, (11.22) results in

$$\frac{-\Delta W}{-\Delta h} = \frac{1}{1 + \kappa(\overline{T}_g - T_b)} \quad \text{and} \quad \frac{-\Delta q}{-\Delta h} = \frac{\kappa(\overline{T}_g - T_b)}{1 + \kappa(\overline{T}_g - T_b)}. \tag{11.26}$$

With (11.24)–(11.26), the work expression is

$$-\Delta W = \overline{c}_{pg}T_1\frac{1 - \left(\frac{p_{2'}}{p_1}\right)^{\frac{R\,\eta_\infty}{\overline{c}_{pg}}\left(1+\kappa(\overline{T}_g-T_b)\right)}}{1 + \kappa(\overline{T}_g - T_b)}. \tag{11.27}$$

We take as an example $p_1/p_{2'} = 2$, which is a large ratio (approximately the value of an entire stage in reality). With $\overline{T}_g - T_b = 1300 - 900\,°C$ and $\kappa = 0.15\,\text{J/kJK}$, it follows that $\kappa(\overline{T}_g - T_b) \approx 0.060$. For $R\,\eta_\infty/\overline{c}_{pg} = 0.200$, the factor of $\overline{c}_{pg}T_1$ at the right-hand side of (11.27) then is 0.1289. This value may be compared to the result of an adiabatic flow, obtained with $\kappa = 0$, being 0.1295. There is only about 0.5%

difference. The same applies to the exact integration of (11.22) and (11.23). It means that there is almost complete compensation between the increase of the enthalpy drop on the 12' path in Fig. 11.14 (left) by the cooling and the consequence that the work is a fraction of this enthalpy drop.

So, for almost the same final result, a cooled expansion may be represented by an adiabatic expansion with work, followed by a cooling without work. These processes are sketched in Fig. 11.14 (right). The 1-2a path is the adiabatic expansion. The 2a-2' path is the temperature decrease by blade cooling. The 2'-2'' path is the temperature decrease by mixing the ejected cooling air with the gas and the 2''-2 represents the pressure drop by mixing. It is not necessary to determine the intermediate temperatures. The final temperature T_2 results directly from a mixing equation, for which the cooling air mass flow rate must be known. The cooling air mass flow rate follows from a heat balance with the blade seen as a heat exchanger:

$$-\Delta Q = -\dot{m}_g \Delta q = \dot{m}_c c_{pc}(T_e - T_c), \qquad (11.28)$$

with $\dot{m}_c$ the mass flow rate and c_{pc} the specific heat of the cooling air (average value over the $T_e - T_c$ range). T_c is the temperature of the cooling air at entrance of the blade and T_e the end temperature of the cooling air.

The temperature T_e follows from $T_e - T_c = \varepsilon (T_b - T_c)$, with ε the effectiveness of the cooling. The effectiveness is about 0.5 with convection cooling and about 0.7 with combined cooling by convection and impingement. With the assumption $T_e = T_b$, already used above for film cooling, the effectiveness of film cooling is set to 1. Equation (11.28) is completed by $-\Delta Q$ from Eq. (11.19) into

$$-\Delta Q = -\dot{m}_g \Delta q = \dot{m}_g \kappa (\overline{T}_g - T_b)c_{pg}(T_1 - T_{2a}) = \dot{m}_c c_{pc}(T_e - T_c), \quad (11.29)$$

with $\overline{T}_g$ and c_{pg} average values over the $T_1 - T_{2a}$ temperature range. With the effectiveness ε, the cooling air mass flow rate follows from

$$\frac{\dot{m}_c \, c_{pc}}{\dot{m}_g \, c_{pg}} = \left(\frac{\kappa}{\varepsilon}\right) \frac{(\overline{T}_g - T_b)\,(T_1 - T_{2a})}{(T_b - T_c)}. \qquad (11.30)$$

The thermal mixing equation determining T_2 reads

$$\dot{m}_g c_{pg}'(T_{2a} - T_2) + \dot{m}_c c_{pc}'(T_c - T_2) = 0. \qquad (11.31)$$

The result is

$$T_2 = \frac{T_{2a} + \delta \, T_c}{1 + \delta}, \quad \text{with} \quad \delta = \frac{\dot{m}_c \, c_{pc}'}{\dot{m}_g \, c_{pg}'}. \qquad (11.32)$$

The heat capacity ratios in Eqs. (11.30) and (11.32) differ somewhat. The ratios are determined iteratively (see next section).

Assuming that the cooling air does not contribute any momentum to the main flow, the decrease of the momentum flow rate within the main flow, thus the decelerating force, is $v\,\dot{m}_c$, with v the average through-flow velocity just before mixing. The corresponding pressure drop is $(p_{2'} - p_2)A_g$, with A_g the through-flow area. The gas mass flow rate is $\dot{m}_g = \rho\,v\,A_g$.

So:

$$p_{2'} - p_2 = \frac{\dot{m}_c}{\dot{m}_g}\rho\,v^2 = \frac{\dot{m}_c}{\dot{m}_g}\gamma M^2 p_{2'}, \tag{11.33}$$

where M is the Mach number of the flow just before mixing. The mixing occurs nearby the trailing edge, i.e. where the Mach number is close to 1. We may assume $M \approx 0.8$, so that $\mu = \gamma M^2 \approx 0.85$. In reality, the cooling air supplies some momentum to the main flow. This reduces the loss coefficient, with a realistic value about $\mu \approx 0.7$ [3].

11.2.7 Compression with Extraction

The calculation of the cooled turbine cannot be separated from the calculation of the compressor from which the cooling air is extracted. In practice, there are only a few extractions (2–3) from the compressor and a few injections (4–5) into the turbine. But, in the modelling, these are distributed. This is achieved by dividing the entire pressure interval in the compressor and the turbine into an equal number of subintervals, which form fictitious stages. This is done according to a power series, so that the pressure ratio of the stages is always the same. We choose, e.g., as many stages as the overall pressure ratio is (40 stages at pressure ratio 40). Due to the pressure drop in the combustion chamber, corresponding points on the compressor path are at a somewhat higher pressure than on the turbine path. We take extraction from the compressor and injection into the turbine at corresponding points and consider the pressure difference as necessary to bring the cooling air from the compressor to the turbine.

It is, e.g., assumed in a first iteration that 95% of the air aspirated by the compressor is supplied to the combustion chamber and 5% is used as cooling air. For a given air flow rate into the combustion chamber, the fuel flow rate required to attain a specified turbine inlet temperature (TIT) can then be determined, followed by the calculation of the specific heat and the gas constant for the first expansion stage. Cooling air is mixed with the gas at the end of each stage. The gas composition changes and adapted values of specific heat and gas constant are determined. The procedure is repeated for the further fictitious turbine stages until the gas temperature reaches the allowable blade temperature. From that point, calculations are performed without cooling, i.e. with $\kappa = 0$. When the expansion has been fully calculated, the air flow rate to be extracted from the compressor at the various pressure levels is known. The air flow rate supplied to the combustion chamber is recalculated and the foregoing procedure is repeated until convergence is attained. Thereafter, the power input to the compressor and the power output by the turbine are calculated, taking the mass flow rates in the various stages into account.

11.3 Performance of Simple-Cycle Power Gas Turbines

11.3.1 Idealised Simple Cycle

In a simple-cycle gas turbine, the flow passes through a compressor, a combustion chamber and a turbine. The main loss with this cycle is thermal in the sense of unused heat in the exhaust gas of the turbine. Thermodynamic losses range in second order. They encompass work dissipated into heat within the compressor and the turbine and work loss by cooling of the turbine. Mechanical losses in the work transfer from the turbine to the compressor and to the external load range in third order.

To estimate the effects of the various losses, we first analyse a cycle with only thermal loss, i.e. $\eta_{\infty c} = \eta_{\infty t} = \eta_m = 1$ and $\kappa = 0$ and with unchanged fluid in the cycle.

Atmospheric conditions: p_0, T_0.

Compressor ($1 \rightarrow 2$):

$$r = p_{02}/p_{01} = p_{02}/p_0,$$

$$W_c = C_p(T_{02} - T_{01}) = C_p T_{02}\left(1 - \frac{T_{01}}{T_{02}}\right) = C_p T_{02}\left(1 - r^{-\frac{\gamma-1}{\gamma}}\right).$$

Combustion chamber ($2 \rightarrow 3$):

$$\Delta Q = C_p(T_{03} - T_{02}).$$

We adopt the simplification that there is no change in mass flow rate and in heat capacity of the gas.

Turbine ($3 \rightarrow 4$):

$$W_t = C_p(T_{03} - T_{04}) = C_p T_{03}\left(1 - r^{-\frac{\gamma-1}{\gamma}}\right).$$

The work output is

$$\Delta W = W_t - W_c = C_p(T_{03} - T_{02})\left(1 - r^{-\frac{\gamma-1}{\gamma}}\right). \tag{11.34}$$

The *thermal efficiency* is the ratio of the work produced to the heat supplied by the fuel:

$$\eta_t = \frac{\Delta W}{\Delta Q} = 1 - r^{-\frac{\gamma-1}{\gamma}}.$$
(11.35)

This theoretical efficiency does not depend on the turbine inlet temperature. The result is significantly different from the efficiency of real simple-cycle gas turbines. For instance, with $r = 20$ and $r = 30$, for $\gamma = 1.40$ (dry air), the theoretical efficiencies are 0.575 and 0.622. Gross electrical efficiencies (alternator output) of real machines are, on average, about 40% (e.g. Siemens SGT6-5000F, $r \approx 20$, representative for a heavy-duty machine) and 42% (e.g. General Electric LM6000, $r \approx 30$, representative for an aero-derivative machine). The net electrical efficiency (plant output) is somewhat less than one percentage point lower and the thermal efficiency (shaft output) is somewhat less than one percentage point higher.

The much lower efficiency in reality is mainly due to the change of the gas composition in the combustion chamber and to the compressor and turbine efficiencies, as we analyse in the next section.

For later discussions, it is important to remark the quite important effect of the exponent in the efficiency expression (11.35). From now on, we use the term exponent as the exponent in the polytropic relation of a pressure ratio as a function of the corresponding temperature ratio.

Thus:

$$\frac{p_b}{p_a} = \left(\frac{T_b}{T_a}\right)^e.$$

For isentropic flow, the exponent is $\gamma/(\gamma - 1) = C_p/R$. For dry air at 0 °C, $\gamma = 1.40$ and the exponent is 3.5. For a combustion gas $\gamma \approx 1.30$ and the exponent is around 4.333. The theoretical efficiencies for $r = 20$ and $r = 30$ become for $\gamma = 1.30$: 0.499 and 0.544. These values are significantly lower than for $\gamma = 1.40$. This observation is already an indication that the efficiency of the simply cycle decreases by the conversion of air into combustion gas. This is one of the aspects that we study in the next section.

11.3.2 Simple Cycle with Component Efficiencies and Different Heat Capacities of Air and Combustion Gas

Extending the previous analysis for the efficiencies of the compressor and turbine parts is quite simple. It is also easy to take into account the larger heat capacity of the combustion gas than that of air, if constant values of heat capacities are used.

The adapted relations are listed hereafter. The subscripts a and g refer to air and combustion gas. The subscripts c and t refer to the compressor and the turbine.
    Compressor $(1 \rightarrow 2)$:

$$W_c = C_{pa}(T_{02} - T_{01}) = C_{pa}T_{01}\left(\frac{T_{02}}{T_{01}} - 1\right) = C_{pa}T_{01}\left(r^{\frac{R}{\eta_{\infty c}C_{pa}}} - 1\right). \quad (11.36)$$

Turbine $(3 \rightarrow 4)$:

$$W_t = C_{pg}(T_{03} - T_{04}) = C_{pg}T_{03}\left(1 - r^{\frac{-\eta_{\infty t}R}{C_{pg}}}\right). \quad (11.37)$$

Combustion chamber $(2 \rightarrow 3)$:

$$\Delta Q = C_{pg}(T_{03} - T_{01}) - C_{pa}(T_{02} - T_{01}). \quad (11.38)$$

We adopt the simplification that there is no change of mass flow rate. The total inlet temperature of the compressor is taken as the reference temperature for the heat input $(T_{01} = T_0)$, which comes from the point of view that heating is meant for increasing the work capacity of the fluid, additionally to the work capacity already realised by the compression.
    The thermal efficiency is

$$\eta = \frac{C_{pg}\frac{T_{03}}{T_{01}}\left(1 - r^{\frac{-\eta_{\infty t}R}{C_{pg}}}\right) - C_{pa}\left(r^{\frac{R}{\eta_{\infty c}C_{pa}}} - 1\right)}{C_{pg}\left(\frac{T_{03}}{T_{01}} - 1\right) - C_{pa}\left(r^{\frac{R}{\eta_{\infty c}C_{pa}}} - 1\right)}. \quad (11.39)$$

In principle, the values of the heat capacities of the combustion gas in the combustion chamber equation and the turbine equation have to be different, as these are averages over different temperature ranges. For the combustion chamber $\gamma \approx 1.32$ and for the turbine $\gamma \approx 1.30$. We take here $C_{pg}/R = 4.25$ as an average. For the compressor, we take $\gamma = 1.40$, so that $C_{pa}/R = 3.50$.
    Table 11.5 illustrates the effects of the different parameters. The temperature ratio T_{03}/T_{01} is set at 5 and 6, corresponding to T_{03} equal to 1440 K $(= 1167 \,°C)$ and 1728 K $(= 1455 \,°C)$ with T_{01} equal to 288 K.
    The first part of Table 11.5 reproduces the results obtained by the analysis of the previous section. The second part illustrates that the efficiency decreases with a factor of about 0.85 by the increase of the heat capacity due to the combustion. The results also show that the efficiency becomes somewhat dependent on the temperature ratio. The efficiency benefits from a higher turbine entry temperature. The third part of Table 11.5 illustrates the combined effects of the component efficiencies and the gas composition change. The thermal efficiency decreases considerably due to the component efficiencies. The values obtained are near to the values for real machines (e.g. the SGT6-5000F and the LM6000 used in the previous section). This observation means that supplementary changes of the efficiency due to variability of the gas

Table 11.5 Effects on thermal efficiency of the simple cycle of gas properties and component efficiencies according to the simplified expression (11.39)

$C_{pa}/R = 3.50,\ C_{pg}/R = 3.50;\ \eta_{\infty c} = \eta_{\infty t} = 1$	$r = 20$	$r = 30$	$r = 40$
$T_{03}/T_{01} = 5$	0.575	0.622	0.651
$T_{03}/T_{01} = 6$	0.575	0.622	0.651
$C_{pa}/R = 3.50,\ C_{pg}/R = 4.25;\ \eta_{\infty c} = \eta_{\infty t} = 1$	$r = 20$	$r = 30$	$r = 40$
$T_{03}/T_{01} = 5$	0.490	0.529	0.533
$T_{03}/T_{01} = 6$	0.494	0.535	0.561
$C_{pa}/R = 3.50,\ C_{pg}/R = 4.25;\ \eta_{\infty c} = \eta_{\infty t} = 0.9$	$r = 20$	$r = 30$	$r = 40$
$T_{03}/T_{01} = 5$	0.387	0.403	0.405
$T_{03}/T_{01} = 6$	0.409	0.435	0.448
$C_{pa}/R = 3.50,\ C_{pg}/R = 4.25;\ \eta_{\infty c} = \eta_{\infty t} = 0.8$	$r = 20$	$r = 30$	$r = 40$
$T_{03}/T_{01} = 5$	0.239	0.202	0.144
$T_{03}/T_{01} = 6$	0.295	0.291	0.274

properties and due to cooling remain quite limited (see next section). The fourth part of Table 11.5 illustrates the very low thermal efficiency with low efficiencies of the compressor and turbine parts. The strong reduction of the global efficiency proves the need for high component efficiencies.

11.3.3 Simple Cycle with Component Efficiencies, Cooling and Variable Gas Properties

Figure 11.15 shows the efficiency of the simple cycle obtained with the simulation methodology explained in Sect. 11.2. The ambient conditions are 288 K and 1 bar. The infinitesimal efficiencies of the compressor and the turbine are set to 90%. Further

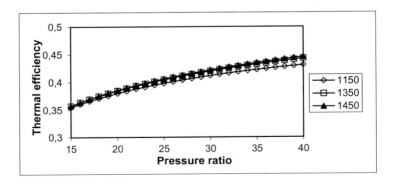

Fig. 11.15 Efficiency of a simple-cycle gas turbine (CH$_4$, TIT = 1150–1450 °C)

parameters are: pressure drop in the combustion chamber: 4%; thermal efficiency of the combustion chamber: 98%; mechanical efficiency of the work transfer of the turbine to the compressor and the external load: 99%. The fuel is pure CH_4 with lower heating value 50 MJ/kg. The energy balance of the combustion chamber is written in the same way as Eq. (11.38):

$$\Delta Q = f\, H_L = (1 + f)C_{pg}(T_{03} - T_{01}) - C_{pa}(T_{02} - T_{01}). \qquad (11.40)$$

This means that the reference temperature of the lower heating value is the total temperature at the entry to the compressor and that any energy that has been supplied to the gaseous fuel by heating it and compressing it is absorbed in the definition of the lower heating value. The heat exchange coefficient in the turbine is set to $\kappa = 0.15$ J/kJK. Film cooling is assumed with heat exchanger effectiveness $\varepsilon = 1$. This means that we consider advanced thermal protection and cooling.

The results in Fig. 11.15 are realistic, but the efficiency is somewhat too low for the lower pressure ratios. This may be verified with the two machines that we already used (SGT6-5000F and LM6000). The thermal efficiency depends strongly on the efficiencies of the compressor and turbine components, as already became clear in the previous section. Modern gas turbines have component efficiencies above 90%, certainly for lower pressure ratios.

The most advanced and largest gas turbines reach nowadays (2021) gross electrical efficiencies of about 43–44% for a pressure ratio of 24. E.g., SGT5-9000HL (Siemens Energy): pressure ratio 24, 593 MWe, 43.0% efficiency; 9HA.02 (GE Power): pressure ratio 23.8, 571 MWe, 44.0% efficiency. The high efficiencies are mainly reached by high efficiencies of the compressor and turbine parts.

The thermal efficiency depends somewhat on the fuel composition. Simulations with kerosene, the fuel of aero-engines, represented as $C_{12}H_{23}$ (diesel oil is also near to $C_{12}H_{23}$), show an efficiency decrease of about 2% (results not shown). The effects are by the larger fuel–air ratio as a consequence of the smaller heating value of 43 MJ/kg and the higher fraction of CO_2 in the combustion gas.

The efficiency increases with the pressure ratio in the indicated range ($r = 15$ to 40), and it reaches a maximum above $r = 40$. The obtained efficiencies are somewhat lower than obtained in the previous section. This is due to the turbine cooling (this was verified by simulations with $\kappa = 0$; results not shown).

The small improvement of the efficiency with increasing turbine inlet temperature (TIT) in absence of cooling, observed in the previous section, is almost completely neutralised by the cooling. This means that TIT has almost no influence on obtainable efficiency. For this result, the heat transfer in the turbine has to be sufficiently small. It was verified by simulations with a larger value of the heat exchange coefficient, namely $\kappa = 0.3$ J/kJK, that the efficiency at higher TIT becomes smaller than at lower TIT (results not shown). So, high TIT is not essential for high efficiency. The benefit of high TIT is mainly large work output, as we illustrate with a further figure.

The efficiency results of Fig. 11.15 are obtained with cooling air extracted from the compressor, as in most machines. Further, it is assumed that there is no pre-cooling of this air by an external heat exchanger before supplying it to the turbine.

In some machines such pre-cooling is implemented. Cooling may also be realised with an externally provided fluid. In particular, machines have been built with *steam cooling*. The choice for this cooling fluid comes from applications in combined-cycle systems (see Sect. 11.4.4). The heat capacity of steam is larger than that of air, which, principally, lowers the necessary cooling flow rates. Further, the heat absorbed by the steam is used in the steam cycle. However, there are drawbacks with steam. Steam has to be led into the machine, but also led out of it. This is possible, but is complicated for rotor parts. Further, film cooling cannot be used. This increases the heat loading of the turbine part ($\kappa = 0.20$ J/kJK instead of 0.15 J/kJK) and decreases the heat exchanger effectiveness of the blades and the vanes ($\varepsilon = 0.7$ instead of 1).

Simulations with cooling by an external fluid have as a result that the thermal efficiency of the simple cycle is slightly lower than with cooling by air from the compressor (results not shown). But these are results assuming that the heat absorbed by the cooling fluid is useless. Taking into account that the heat may contribute to a second cycle, one may conclude that there is a small advantage on efficiency with steam cooling in a combined-cycle application.

Steam cooling was used in the past but has been abandoned for cooling of the turbine part (see Sect. 11.4.4). The reason is that the role of combined gas and steam turbine power plants has moved from base-load plants to mid-load plants, and even to peak-load plants. Mid-load plants follow approximately the variation of the electricity consumption on the time scale of the day and peak-load plants allow fast power changes. The changed operation comes from the growing fraction of electricity produced by wind turbines and solar cells. Combined steam and gas turbines with steam cooling of the gas turbine have a large thermal inertia. This is avoided with air cooling. In the past years, much effort has been made by manufacturers to decrease the time for starting up. With modern machines, the time needed for reaching full power of the gas turbine (not including the steam turbine), starting from zero power but with the gas turbine ready, slowly rotating on the turning gear, and all auxiliaries ready, has been reduced to about 10 min, which was about 30 min up to 10 years ago. This is not possible with large thermal inertia.

Figures 11.16, 11.17, 11.18 and 11.19 show specific power, turbine outlet temper-

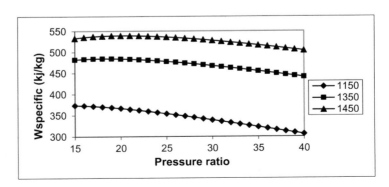

Fig. 11.16 Specific power of a simple-cycle gas turbine (CH$_4$)

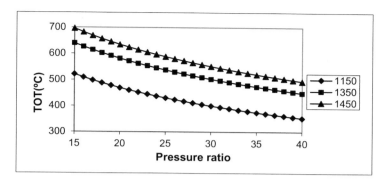

Fig. 11.17 Turbine outlet temperature with the simple cycle (CH$_4$)

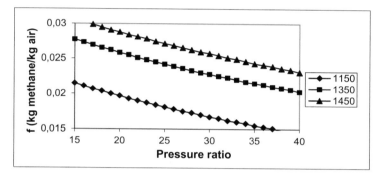

Fig. 11.18 Fuel flow rate to inlet air flow rate with the simple cycle (CH$_4$)

ature (TOT), fuel flow rate and cooling flow rate obtained by simulation with pure CH$_4$ as fuel. The results with kerosene differ slightly. E.g., the specific power is about 4–5% lower. The specific power is expressed as the power per unit of mass flow rate of the exhaust gas. The increase of the specific power, the turbine outlet temperature, the fuel flow rate and the cooling flow rate with increasing turbine inlet temperature is obvious.

One may verify the results with the data of the Siemens SGT6-5000F, producing 260 MW at $r = 19.5$, with 586 kg/s exhaust mass flow, TIT around 1350 °C, TOT = 592 °C and gross efficiency about 40%. The specific power is thus about 445 kJ/kg. The predicted TOT is nearly correct, which is an indication that the TIT in reality and simulation are about the same. The predicted specific power is about 480 kJ/kg and thus somewhat too high. The overprediction is very likely due to the use of pure CH$_4$ as fuel, with a higher heating value than natural gas and a lower fraction of CO$_2$ in the combustion gas. The predicted efficiency is about 38% and thus somewhat too low, very likely due to higher efficiencies of the compressor and turbine parts in the real machine.

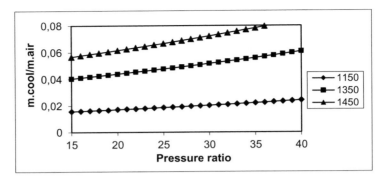

Fig. 11.19 Cooling air flow rate with the simple cycle (CH_4)

The specific power has a weak maximum as a function of the pressure ratio, around pressure ratio 15–20. With respect to specific power, it is thus not useful to choose the pressure ratio too high. Gas turbines for electric power stations have typically a pressure ratio in the order of 20. The simple-cycle efficiency (Fig. 11.15) is then not at maximum, but these machines are almost always used in combined-cycle power stations and the optimum efficiency with combined cycles is for a lower pressure ratio (see Sect. 11.4.4).

The fuel–air ratio (Fig. 11.18) is in the order of what we verified earlier. The ratio of the coolant flow rate to the air flow rate at the entry to the compressor (Fig. 11.19) is of the order of 4–6%, but the simulations are done with advanced values of heat transfer coefficient κ and effectiveness ε. Doubling the value of κ/ε causes, in principle, doubling the value of the cooling air flow rate. So, a precise prediction of the coolant mass flow rate is not possible.

The values of efficiency and specific power, obtained by the simulations and by the hand calculations in the previous section do not differ much (one may verify for some combinations of TIT and PR; see exercise 11.5.2). This may seem somewhat surprising, but with the study of the cooled expansion, we remarked already that cooling does not change much the turbine work per unit of mass.

11.4 Performance of Power Gas Turbines with Enhanced Cycles

11.4.1 Compression with Intercooling

Intercooling reduces the compressor work. But, on the other hand, air led to the combustion chamber needs more heating. An additional effect is reduced cooling flow rate, as the cooling air becomes colder.

Intercooling at an early stage in the compression is advantageous for efficiency. The positive effect may be understood by considering a cycle with intercooling as a superposition of two cycles, as sketched in Fig. 11.20. The same figure also shows a similar interpretation of reheat within the turbine.

Intercooling may be considered as adding a cycle with a lower pressure ratio than that of the original cycle, but with better component efficiencies. The compression and the expansion paths run almost parallel, which implies that the additional cycle approximately reaches the efficiency with ideal components. The exponent of the pressure–temperature relation is $\eta_{\infty c} C_{pa}/R$, which is about 3.20. For pressure ratios 20 and 40, the efficiency of the added cycle is about 0.61 and 0.68, according to formula (11.35) with an adapted exponent. With these high values, an efficiency gain is possible if the pressure ratio of the added cycle is not much lower than that of the main cycle.

With the simulation methodology used here, the optimum pressure ratio for placing the intercooling comes at about 1.7 for compressor pressure ratio 15 and at about 2.2 for compressor pressure ratio 35 (results not shown in a figure). The optimum intercooling pressure ratio depends somewhat on TIT. The efficiency gain is about 1.5 percentage points for compressor pressure ratio 15 and about 5 percentage points for pressure ratio 35 (results not shown in a figure). These results are obtained with neglect of pressure losses in the intercooler. So, the efficiency gain is smaller in reality. In particular, intercooling is attractive for a gas turbine with a large compressor pressure ratio.

There is only one intercooled gas turbine for power generation, the LMS100 by GE Power (see website). It is an aero-derivative with pressure ratio about 42 and 43.5% net electrical efficiency. We may compare with the simple-cycle aero-derivative LM6000 of GE, with pressure ratio about 30 and 41.0% net electrical efficiency. The higher efficiency of the LMS100 is principally due to the higher pressure ratio. So, the efficiency gain due to intercooling, if any, is very small. This means that the main benefit of the intercooling is the larger specific work.

The much smaller efficiency gain in practice compared to the simulation result is due to pressure losses in collecting the air with a volute after the LP compressor, passing the air through a cooler and leading the air to the HP part of the compressor by a volute. The LMS100 has an LP compressor with 6 stages and a HP compressor with 14 stages. The turbine is split into 3 parts: 2 stages HP, 2 stages IP and 5 stages LP. There are 3 shafts. The LP part of the turbine is the power turbine driving the load

Fig. 11.20 Cycle with intercooling in the compressor and reheat in the turbine

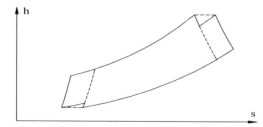

at the hot end side of the machine. Assuming that the compressor stages all have the same pressure ratio, the pressure ratio of the LP compressor would be about 3, which is higher than the result of the simulation methodology for the optimum intermediate pressure. This higher ratio is also an indication that the target of the intercooling in the LMS100 is mainly increase of the specific work.

11.4.2   Expansion with Reheat

Adding reheat to the turbine, which means expanding the combustion gas in one or more turbine stages and reheating it to the original TIT, is principally advantageous for efficiency, if the intermediate heating occurs at an early stage in the expansion. The reason is similar to that with compressor intercooling, as is clear from Fig. 11.20. The exponent of the pressure–temperature relation is $C_{pg}/R\,\eta_{\infty t}$, which is about 4.80. For pressure ratios 20 and 40, the efficiency of the added cycle is about 0.464 and 0.536 according to formula (11.35). These values are much lower than with intercooling, but are still larger than these of the basic cycle.

Reheat causes, of course, a higher need for cooling flow, which degrades somewhat the resulting efficiency. Further, there are pressure losses in the second combustion chamber. With the simulation methodology used here and ignoring the pressure drop in the second combustion chamber, a small efficiency gain is obtained for larger pressure ratios. There is no gain for compressor pressure ratios lower than about 20 and the gain is about 2.5 percentage points for compressor pressure ratio 35 (results not shown in a figure). Moreover, the gain increases somewhat with lower TIT. From these simulation results, we conclude that reheat is feasible with high pressure ratio and preferably with lower TIT. Turbine reheat may be combined with compressor intercooling. The effects approximately add. But there are no gas turbines with combined intercooling and reheat.

At present, there is only one gas turbine with reheat: the GT26 (370 MW) by Ansaldo Energia (see website). The pressure ratio is about 35. The gross electrical efficiency is about 42%. The TIT is not communicated by the manufacturer, but it is known that it is rather low. Clearly, the efficiency is not improved by the reheat, but the primary objective is to reach good efficiency in combined-cycle mode, which requires that the turbine outlet temperature is high enough (see Sect. 11.4.4). With reheat, the TOT comes at about 625 °C, which is much higher than would be attained without reheat (around 475 °C according to Fig. 11.17, with an estimated TIT of 1350 °C).

Reheat forms, of course, a technical complication. In the GT26, the turbine has five stages and the second combustion chamber is placed after the first stage. The flow leaving the first stage is led through rectangular orifices with vortex generators placed in these. The fuel is injected by lances and surrounded by injected air. The fuel and the air mix through the vortex motion and combustion takes places in the second annular combustion chamber.

11.4.3   Recuperator

Adding a recuperator, i.e. a heat exchanger where the turbine exhaust gas heats the air delivered by the compressor, is advantageous for efficiency with lower pressure ratios. For efficiency gain, the exit temperature of the compressor must be lower than exit temperature of the turbine. With a recuperator, it is always advantageous to provide intercooling on the compressor as well. The compressor work is then reduced and the lower temperature at the compressor exit is compensated by heating by the recuperator, with no need for an increase of the fuel flow rate. Hereby, we also understand that the most efficient intercooling is reached at the minimum of the compressor work, which means equal pressure ratios before and after the intercooling.

Figure 11.21 shows the efficiency by combined recuperator and intercooler. The air is cooled halfway the compression to the ambient temperature (minimum compressor work). The recuperator has effectiveness 0.7. The pressure loss is set to 0.02 bar on the air side of the intercooler and the same value is used on the air side and combustion gas side of the recuperator. The efficiency comes to about 47% for TIT above 1350 °C, almost independent of the pressure ratio.

Reheat, enhancing the efficiency, could be added to a cycle with recuperator. Intermediate cooling in the compressor combined with intermediate heating in the turbine is advantageous, in principle. The reason is that an infinite number of intermediate coolers and intermediate heaters results in a Carnot cycle. But, reheat increases the TOT and the recuperator must be able to withstand the higher combustion gas temperature. This is a technical problem. There are no gas turbines with recuperator and reheat.

The cycle with intercooler and recuperator is frequently used in marine applications. Implementation of intercooling is easily realised due to the availability of cooling water. Good efficiency is already attained with a lower TIT and low pressure ratio. For completeness, we have to mention that many marine gas turbines are aero-derivatives and that these have a large pressure ratio. The smaller gain in efficiency

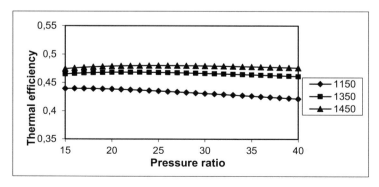

Fig. 11.21 Efficiency of a cycle with intercooler and recuperator (fuel CH_4)

with intercooling for large pressure ratio is less attractive and the intercooler requires space. Military ships are therefore often equipped with simple-cycle gas turbines.

11.4.4 Combined Gas and Steam Cycles

In a combined-cycle plant, the exhaust gas of a gas turbine is used for steam generation in a recovery boiler. The steam is fed to a steam turbine. For good efficiency of the steam cycle, the temperature at the entry to the steam turbine must be sufficiently high.

Figure 11.22 sketches the cooling of the combustion gas and the heating of the steam, with subcritical pressure, in the T-s diagram of the gas, with rescaled entropy for steam such that gas and steam temperatures correspond to each other, as on the surfaces of the steam generator.

The dashed line is the steam path for a single-pressure steam generator. A pinch point forms, which is the smallest temperature difference at the beginning of the evaporation of the water. Due to the pinch point, the temperature difference is quite large between the steam leaving the steam generator and the gas entering it. A similar phenomenon occurs at the outlet side of the gas. These differences may be reduced by using two pressure levels in the steam generator, as shown by the full lines. In practice, even three pressure levels with reheat between the high pressure part and the intermediate pressure part are used.

With a triple-pressure reheat cycle, with pressure levels 117 bar, 15.2 bar and 3.7 bar, the steam cycle efficiency is about 40.8% for steam temperature at the entrance of the recovery boiler 647 °C [4] and the efficiency decreases to 40%, 38.5% and 34% for the entry temperatures 577, 527 and 477 °C. These efficiencies

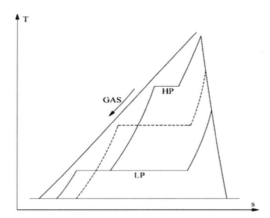

Fig. 11.22 Single pressure level (dashed line) and dual pressure level (full line) steam generation in a recovery boiler

are based on the heat available in the gas by cooling it down to 20 °C, excluding the condensation heat. The importance of a high entry temperature at the gas side is obvious. This effect is similar as with a steam cycle with combustion.

The cited efficiencies are much lower than the typical values of a fossil fuel plant, due to much lower gas temperature in the boiler, much lower pressure levels and absence of regenerative feed water preheating and combustion air preheating with combined cycles. Feed water preheating by extraction steam could, in principle, be applied, but would result in a higher exit temperature at the gas side. Since there is no combustion, the remaining heat cannot be used for preheating combustion air. This makes feed water preheating useless. The cited efficiency can be improved with about 4 percentage points for the entry temperature 647 °C by optimisation of the pressure levels [4], but the highest level becomes then much higher. The optimum levels are 217 bar (just below the critical pressure of 220 bar), 28.2 bar, 2.2 bar, which means a larger cost of the steam generator and the steam turbine.

Figure 11.23 shows the efficiency obtained by simulation of the combined cycles, with the efficiencies of the steam part as given above, without optimisation of the steam pressure levels. A 96% efficiency of the recovery boiler is assumed. The obtained efficiency is in the order of 57% for pressure ratio above 20 and TIT around 1350 °C. This efficiency level is typical for first-generation gas turbines of the F-class, but the efficiency of modern machines (2021) is much larger than shown on Fig. 11.23, due to better efficiencies of the compressor and turbine parts of the gas turbine and better efficiency of the steam turbine part (see Sect. 11.5): about 64% net combined-cycle efficiency. The steam turbine adds about 50% to the power by the gas turbine(s). E.g. SGT5-9000HL (Siemens Energy): 593 MWe; two gas turbines + one steam turbine: 1760 MWe.

The figure shows that the efficiency does not depend much on the pressure ratio. So, a large pressure ratio is not very useful, but it is not harmful either. The efficiency improves by a higher turbine inlet temperature. This is different from the simple cycle (Fig. 11.15), where the efficiency does not benefit from higher TIT. This observation demonstrates the effect of the better efficiency of the steam cycle with higher inlet temperature. With this observation, we also understand the benefit by reheat as with the GT26.

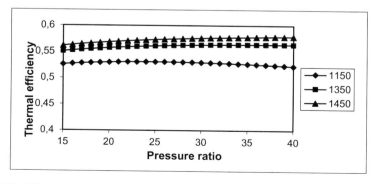

Fig. 11.23 Efficiency of combined gas and steam cycles (CH_4)

11.4.5   Steam Injection

In a combined-cycle system, steam is generated in the recovery boiler and supplied to a steam turbine. But steam injection into the gas turbine is possible as well. Injection of steam at the entrance of the combustion chamber is the most efficient way, as the mixture of combustion gas and steam then enters the turbine at the original TIT. Of course, this requires increase of the fuel flow rate. The efficiency with steam injection is much lower than with combined cycles, because the steam cycle is less favourable, being identical to the gas cycle. The highest pressure in the cycle is far too low for the steam (30 bar compared to 300 bar), the lowest pressure is far too high (1 bar compared to almost vacuum). The much higher temperature does not compensate for this.

Figure 11.24 shows the thermal efficiency with the same assumed characteristics of the recovery boiler as with combined cycles, with application of one pressure level within the recovery boiler at a pressure three bar above the gas pressure in the combustion chamber and assuming 10 °C temperature difference at the pinch point. The gain in efficiency with respect to the simple cycle (Fig. 11.15) is about eight percentage points, independent of pressure ratio and TIT.

Difficulties with steam injection are that the steam flow rate is rather large (of the order of 15% of the air flow rate at pressure ratio 30 and TIT = 1350 °C) and that the required cooling air flow also becomes rather large (also of the order of 15% for pressure ratio 30 and TIT = 1350 °C). Providing a simple-cycle gas turbine for steam injection with these steam and cooling air flow rates is not possible. Gas turbines allowing steam injection are limited to steam injection flow rates in the 12% order. Operation with larger steam flow rates is technically possible, but requires turbine adaptation to a larger flow rate, and thus efficiency loss when running without steam injection.

Some enhancements are possible. E.g., with a gas turbine running continuously in steam-injected mode, steam cooling of the turbine part is possible. Extraction of cooling air from the compressor becomes then unnecessary and the mass flow

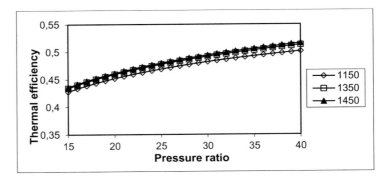

Fig. 11.24 Thermal efficiency of a steam-injected gas turbine (CH$_4$)

rate of the steam injected into the combustion chamber is reduced. Moreover, the cooling steam may be released in the gas path. So, contrary to closed-loop steam cooling, there is no technical difficulty. It is also easy to recuperate the boiler feed water by cooling the exhaust gas, because supplementary water is produced by the combustion, especially with natural gas (CH_4) as fuel. Such enhancements were studied in the past, but machines with these features were never produced. Steam-injected gas turbines are very rare. The complication is the addition of the boiler. With large-power turbines, it is more economical to opt for combined-cycle use. There is then the supplementary cost of the steam turbine, but the efficiency is much larger.

Steam-injected gas turbines are sometimes used in cogeneration plants (CHP: combined heat and power). The gas turbine drives an electric generator and steam for process heating is produced by the exhaust gas. Steam is only injected into the gas turbine when heat is not required. The steam injection is then rather exceptional and, therefore, the machine is not optimised for steam injection operation. The steam-loaded exhaust gas is discharged into the atmosphere for the same reason.

11.5 Classification of Power Gas Turbines

With respect to technology steps, it has become customary to denote gas turbine types with subsequent letters of the alphabet. Current classes are E, F, G, H and J (A, B and C are older types, but some still exist, and I is not used).

Class E are machines with convection and impingement cooling, originally with a turbine inlet temperature (TIT) around 1100 °C and reaching about 50% combined-cycle efficiency. Modern versions of these machines reach a TIT of 1250 °C due to thermal barrier coatings (TBC) and directionally solidified blades. Class E machines are very robust and are typically used with crude fuels, which cause deposits in the turbine part, making film cooling impossible.

Class F are machines with film cooling, TBC and, originally, directionally solidified blades. Originally, these allowed a TIT of about 1350 °C and reached 58–59% combined-cycle efficiency.

Class H meant up to some years ago machines with closed-loop steam cooling of the combustion chamber and stationary and rotating parts of the turbine, advanced TBC and single-crystal blades in the first stage of the turbine part. Such machines allow a TIT up to 1500 °C and reach combined-cycle efficiency above 60%. As earlier said, steam cooling of turbine parts has been abandoned. It is possible to obtain combined-cycle efficiency above 60% with air-cooled machines with an advanced technology level. Improved efficiency of the gas turbine comes from better efficiency of the compressor and turbine components, due to improved blade design and reduced clearance losses. The combined-cycle efficiency is raised by a higher TIT (up to 1600 °C) obtained with better TBC, single-crystal blades in the first and second stages of the turbine part, improved cooling and optimisation of the triple-pressure reheat steam cycle. These machines are called H-class by many manufacturers (see websites of Siemens Energy, General Electric Power, Ansaldo Energia), although there is no steam cooling, to express the advanced technology level. The net electrical simple-cycle efficiency (plant output) is nowadays (2021) 42.8–43.8% and the net combined-cycle efficiency is 63.5–64.3%.

A recent evolution is use of H-technology in F-machines, as e.g., single crystal blades in the first turbine stage and improved blade shapes. This brings the simple-cycle efficiency above 41% and the combined-cycle efficiency above 61%. The essential difference between F and H machines is then only the level of TIT.

Mitsubishi Power still produces gas turbines with steam cooling, but the steam cooling is only applied to the combustor and there is air cooling in the turbine part. Such machines are classified as G-types (TIT about 1500 °C) and J-types (TIT about 1600 °C). Mitsubishi also offers machines with air cooling only, comparable to H-machines and calls these JAC (J with air cooling).

11.6 Exercises

11.5.1 Verify with the formulae of Sect. 11.3.2 the statements in Sect. 11.1.2 that compression of air with pressure ratio 20 starting at 288 K requires about 450 kJ/kg (take $R = 288$ J/kgK and $\gamma = 1.38$) and that expansion of combustion gas with pressure ratio 20 starting at 1723 K (1450 °C) yields about 1000 kJ/kg (take $R = 288$ J/kgK and $\gamma = 1.31$). Assume adiabatic flow with 90% polytropic efficiency. Calculate the corresponding values for pressure ratio 40. Compare the ratios of compressor work to turbine work. Ignore the small difference in mass flow rate of compressor and turbine and ignore the small pressure drop in the combustion chamber.

A: $r = 20$: $Wc/Wt \approx 0.46$; $r = 40$: $Wc/Wt \approx 0.55$.

11.5.2 The observation in Sect. 11.3.2 is that representation of a gas turbine by adiabatic compression and expansion, thus ignoring cooling in the turbine part, ignoring cooling flow extractions in the compressor part and cooling flow injections in the turbine part, and also ignoring the increase of mass flow rate due to fuel addition, the heat loss and the pressure drop in the combustion chamber, results in realistic estimates of the efficiency.

Refine the representation somewhat by multiplying the turbine work with a mechanical efficiency of 0.99 for the transfer of work to the compressor and the external load and by multiplying the shaft work by the efficiency of the electric generator of 0.985. This way, an estimate of the gross electrical efficiency is obtained.

Take gas properties ($\gamma = 1.38$ and $\gamma = 1.31$) and polytropic efficiency of compressor and turbine parts (0.90) as in the previous exercise. Calculate for TIT = 1350 °C, for pressure ratios 20, 30 and 40: the gross electric efficiency, the specific work, the turbine outlet temperature and the fuel–air ratio. Assume, for this last result, a lower heating value of 46,000 kJ/kg for natural gas.

Judge on the realism of the results by comparing with data of the SGT6-5000F (pressure ratio = 19.5, gross electrical efficiency 40%, 260 MW, 586 kg/s exhaust flow rate, exhaust temperature 592 °C) and the LM6000 (pressure ratio $\approx$ 30, net electrical efficiency 41.2%, 44 MW, 125.6 kg/s exhaust flow rate, exhaust temperature 461 °C) and by results on the Figs. 11.15, 11.16, 11.17 and 11.18.

A: r = 20, eff = 0.395, work = 463.2 kJ/kg, TOT = 584.5 °C, f = 0.0255;
r = 30, eff = 0.417, work = 447.2 kJ/kg, TOT = 513.5 °C, f = 0.0233;
r = 40, eff = 0.430, work = 427.6 kJ/kg, TOT = 466.8 °C, f = 0.0216.

The efficiencies accord well with these of the SGT6-5000F and the LM6000 and accord even better than these by the simulations in Fig. 11.15. Specific work, turbine outlet temperature and fuel–air ratio accord well with results on Figs. 11.16, 11.17 and 11.18. Specific work and turbine outlet temperature accord also quite well with the data of the SGT6-5000F. The good correspondence of the results shows that the simple calculation method with adiabatic compressor and turbine components is sufficiently accurate for the estimation of the main performance characteristics. Specific work and turbine outlet temperature of the LM6000 are lower than calculated, which is an indication that the TIT of this gas turbine is lower than assumed.

11.5.3 A simple-cycle gas turbine, fired by natural gas, features: $r_c = 24$, TIT = 1550 °C, $\eta_{\infty c} = \eta_{\infty t} = 0.9$. Ambient conditions are T = 15 °C, p = 1 bar. Take as gas constant $R = 288$ J/kgK combined with $\gamma_{air} = 1.38$ in the compressor, $\gamma_{gas} = 1.32$ in the combustor and $\gamma_{gas} = 1.30$ in the turbine. Take as lower heating value of the fuel 46,000 kJ/kg. Ignore cooling in the turbine. Assume 4% pressure drop in the combustor, 98% combustor thermal efficiency, 99% mechanical efficiency and 98.5% generator efficiency.

Determine the fuel–air ratio, the turbine outlet temperature, the specific work (based on exhaust mass flow rate) and the gross electrical efficiency. Compare the results with these of the SGT5-9000HL, which are: pressure ratio 24, electrical output power 593 MW, gross electrical efficiency 43.0%, exhaust temperature 670 °C, exhaust mass flow 1050 kg/s, thus specific work equal to 565 kJ/kg. Observe that the predicted values are quite good, but that the predicted exhaust temperature and specific work are somewhat higher than these of the SGT5-9000HL. The data can thus be better matched by lowering somewhat the turbine inlet temperature (but see next exercise). Since this temperature is never precisely known, such matching is a practical way of tuning a mathematical representation, which then may be used for analysis of changes to the cycle (we study some changes in the next exercises). Compare with the results of the more simplified method used in the previous exercise. The predictions are somewhat less precise with the simplified method.

A: $\dot{m}_{fuel}/\dot{m}_{air} = 0.0307$, TOT = 677 °C, specific work = 589 kJ/kg, gross electrical efficiency = 43.0%; simplified method: TOT = 650 °C, specific work = 576 kJ/kg, gross electrical efficiency = 42.0%.

11.5.4 Estimate the necessary coolant mass flow rate for the turbine in the previous exercise. Assume 900 °C as admissible metal temperature. Approximate the heat extracted from the turbine flow based on a heat transfer coefficient $\kappa = 0.15$ J(heat)/kJ(work)K, the work of the adiabatic turbine between turbine inlet (1550 °C) and the position where the gas temperature becomes the admissible metal temperature (900 °C), and the average temperature difference between gas and admissible metal temperature in this turbine part.

Further assume that all air for cooling is extracted at the compressor exit and that there is a pressure drop of 4% by bringing this air to the turbine part (same drop as in the combustor). Assume that this extracted air can be heated to the admissible metal temperature, which means heat exchanger effectiveness of the blades and vanes equal to unity (possible for very efficient film cooling). Model the cooling by mixing the coolant air flow with the combustion gas just before turbine entry (at the same pressure), ignoring a pressure drop caused by mixing. Take into account that a consistent mixing equation requires that the heat capacity of the mixture is the mass weighted average of the heat capacities of the components.

Observe that by the chosen approximation of the cooling, the turbine outlet temperature decreases (temperature drop is taken into account) but that the efficiency remains almost the same (pressure drop is not taken into account). Observe also the decrease of the specific work by the cooling. The turbine exit temperature and the specific work are now lower than with the SGT5-9000HL. This means that the turbine entry temperature of 1550 °C, considered as producing a quite good match for a turbine part ignoring cooling cannot be the value in reality and that 1600 °C is more realistic.

A: $\dot{m}_{cool}/\dot{m}_{air} = 0.0947$, $C_{p,mix} = 1229.4$ J/kgK, $T_{0,mix} = 1466.8$ °C, TOT = 624.9 °C; specific work = 534.6 kJ/kg, gross electric efficiency = 43.0%.

11.5.5 Modify the simple-cycle gas turbine of exercise 11.5.3 by adding intercooling in the compressor half-way the compression ($r = \sqrt{24}$) lowering the air temperature to the compressor entry temperature (15 °C) and by adding a recuperator with 70% effectiveness. Ignore pressure drops in the heat exchangers. Calculate the gross electrical efficiency and compare with Fig. 11.21.

A: $\dot{m}_{fuel}/\dot{m}_{air} = 0.0296$, efficiency = 0.530. The efficiency gain is 10 percentage points. The efficiency is higher than in Fig. 11.21 due to neglect of cooling. With the numerical simulation, the efficiency gain is from 0.40 (Fig. 11.15) to 0.485 (Fig. 11.21), thus 8.5 percentage points.

11.5.6 Combine the simple-cycle gas turbine with adiabatic components of exercise 11.5.3 with a steam turbine. Assume 1100 J/kgK as heat capacity of the exhaust gas of the gas turbine ($\gamma = 1.355$). Ignore pressure losses at the combustion gas side in the recovery boiler. Assume 96% thermal efficiency of the boiler and 40% thermal efficiency of the steam turbine, based on the heat in the exhaust gas of the gas turbine made available by cooling down to the ambient temperature (excluding the latent heat of the water fraction). Determine the gross electrical efficiency of the combined gas and steam turbine system. Observe that the obtained estimate is realistic (combined-cycle efficiency of SGT5-9000HL > 63%).

A: efficiency = 63.4% (is much higher than in Fig. 11.23).

References

1. Baehr HD, Kabelac S (2009) Thermodynamik, 14th edn. Springer. ISBN 978-3-642-00555-8
2. Chen AF, Wu HW, Wang N, Han JC (2018) Heat transfer in a rotating cooling channel (AR = 2:1) with rib turbulators and a tip turning vane. J Heat Transfer 140:102007
3. El-Masri MA (1986) On thermodynamics of gas turbine cycles: part 2—a model for expansion in cooled turbines. J Eng Gas Turbines Power 108:151–159
4. Franco A, Casarosa C (2002) On some perspectives for increasing the efficiency of combined cycle power plants. Appl Therm Eng 22:1501–1518
5. Razak AMY (2007) Industrial gas turbines; performance and operability. Woodward Publishing and CRC Press. ISBN 978-1-84569-205-6
6. Sanjay Y, Singh O, Prasad BN (2008) Influence of different means of turbine blade cooling on the thermodynamic performance of combined cycle. Appl Therm Eng 28:2315–2326
7. Saravanamuttoo HIH, Rogers GFC, Cohen H, Straznicky PV, Nix AC (2017) Gas turbine theory, 7th edn. Pearson. ISBN 978-1292093093
8. Wilcock RC, Young JB, Horlock JH (2005) The effect of turbine blade cooling on the cycle efficiency of gas turbine power cycles. J Eng Gas Turbines Power 127:109–120

Chapter 12
Thrust Gas Turbines

Abstract Due to high power compared to weight and volume, gas turbines are very suitable components of aircraft propulsion systems. Almost all modern aircraft propulsion is gas turbine based. Only with low power (<300 kW), as for very light aircraft, reciprocating engines are used. Aircraft propulsion systems exist in a wide variety of types. The basic principle is always that the propulsive force (thrust) is a reaction to the acceleration of an air flow. In this chapter, we discuss the different systems and we analyse the performance of the core part, which is a gas turbine, and the performance of double-jet engines with mixed and unmixed jets, which are the most common types. The chapter is concluded by a discussion of some technological aspects.

12.1 Thrust Generation

There are three basic propulsion systems: propeller, reactor and rocket. The first two systems are applied in conventional aircraft.

12.1.1 Propeller

A propeller is a rotor with twisted blades. It provides acceleration to a part of the air around the aircraft. A force is generated as reaction to this acceleration. With an airplane, the propeller generates the propulsive force, called *thrust*, while lift force is produced by the wings. Helicopter rotors generate both thrust and lift force. Helicopters therefore are sometimes called rotary wing aircraft, to distinguish them from fixed wing aircraft. The acceleration of the surrounding air is generated by forces exerted on the blades by the relative air flow. Thrust may thus also be considered as a result of the blade forces. Figure 12.1 represents a blade section and the forces on this section.

As an absolute coordinate system, we take a frame bound to the aircraft (moving with respect to the surrounding air) and as a relative system, a frame rotating with the

Fig. 12.1 Force generation
on a propeller blade segment

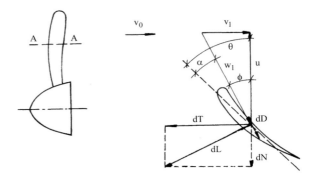

propeller. The inflow velocity in the absolute frame is v_0, the flight speed. The flow accelerates due to the propeller force. Within the plane of the propeller, the velocity in the absolute frame v_1 exceeds v_0. Blade speed is denoted with u, relative velocity with w_1.

The relative velocity forms an angle of attack α to the blade reference line (e.g., the zero-lift line). The stagger angle of the blade is θ (in Fig. 12.1, with respect to the tangential direction). The lift force, dL, is perpendicular to the relative velocity. The drag force is along this velocity. Lift may be resolved into an axial component dT, contributing to the thrust and a tangential component dN. The latter causes the power taken from the shaft: $dP = -\overrightarrow{dN}.\vec{u}$. The drag force dD has an axial component which diminishes the thrust and a tangential component which enlarges the power consumed.

In Fig. 12.1 the velocity in the absolute frame at the position of the blade is in the axial direction. This is a simplification. In reality, there is also a tangential velocity component as a reaction to the tangential component of the blade force. This component has two principal consequences.

First, the power transferred by the lift increases for the same thrust (dT). The power exchanged by the lift force is

$$-\overrightarrow{dL}.\vec{u} = -\overrightarrow{dL}.\overrightarrow{v_1} \quad (\overrightarrow{dL}.\overrightarrow{w_1} = 0).$$

The expression demonstrates that the power taken from the shaft is completely transferred to the flow as mechanical energy (displacement work) and thus lift is a pure active force (see Chap. 2). The tangential component of v_1 is in the sense of the blade speed and so increases the power exchanged for given thrust dT.

The second consequence of the tangential component is a change of the active and dissipative parts of the power exchanged by the drag force:

$$-\overrightarrow{dD}.\vec{u} = -\overrightarrow{dD}.\overrightarrow{v_1} + \overrightarrow{dD}.\overrightarrow{w_1}.$$

Fig. 12.2 Through-flow of a propeller in one-dimensional representation

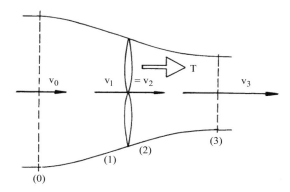

The $\overrightarrow{dD}.\overrightarrow{w_1}$ part is the dissipated part of the power exchanged (see Chap. 2). This part reaches the flow as heat. The $-\overrightarrow{dD}.\overrightarrow{v_1}$ part causes exchange of mechanical energy. This part is negative with the velocity triangle drawn in Fig. 12.1. Mechanical energy is thus taken from the flow by the drag force. The tangential component of the velocity v_1 changes somewhat both components of the power exchanged by the drag force.

We further analyse the flow through the propeller with the simplified velocity triangle in Fig. 12.1, so with velocity v_1 in axial direction. The through-flow in one-dimensional representation is sketched in Fig. 12.2. The propeller is represented by a plane perpendicular to the flow. Based on continuity of the mass flow, the axial velocity in the flow through the propeller ($v_1 = v_2$) remains constant (for unchanged density across the propeller). Sections (0) and (3) lay so far upstream and downstream that there is atmospheric pressure in these sections. A control volume is formed with the envelope of a streamtube through the rotor and the inlet and outlet sections (0) and (3). The thrust generated by the rotor is denoted by T. The force denoted by T in Fig. 12.2 represents the force exerted on the flow, so in the sense opposite to the thrust.

From the assumptions follows that thrust is (see Chap. 1, Exercise 1.9.4):

$$T = \dot{m}(v_3 - v_0), \tag{12.1}$$

where $\dot{m}$ is the mass flow rate through the propeller. This result requires the assumption that the resulting force by the pressure and the shear stress on the control volume surface equals zero. In absence of losses (see also Sect. 12.1.2), this result is obtained by considering a second streamtube, concentric around the streamtube drawn in Fig. 12.2, with an envelope so far from the propeller that the pressure on it equals atmospheric pressure. As there is no work within the second streamtube, the inflow and outflow velocities thus equal v_0. The consequence is that there is no change of momentum flux in this streamtube. The resulting force on the entire surface of the second streamtube thus must equal zero. This demonstrates that the resulting force on the envelope of the streamtube through the propeller in Fig. 12.2, in the absence of losses, equals zero.

In absence of losses and ignoring the tangential component of v_1, the power transferred to the flow is

$$T\, v_1 = \dot{m}\, (v_3 - v_0)v_1. \tag{12.2}$$

The power corresponding to the mechanical energy rise within the flow, termed *dynamic power* (because there is no pressure rise), is

$$P_{dyn} = \dot{m}\, (\frac{1}{2}\, v_3^2 - \frac{1}{2}\, v_0^2). \tag{12.3}$$

Equating (12.2) and (12.3) gives

$$v_1 = \frac{1}{2}(v_0 + v_3).$$

The dynamic power (12.3) is the theoretical power required to generate the thrust (12.1). In reality, the power extracted from the propeller shaft is larger due to the tangential velocity component, non-uniform axial velocity over the propeller and loss mechanisms. These effects may be studied by a multiple streamtube analysis, analogous to the analysis in Chap. 10 on wind turbines. But, for the study of the gas turbine as a driving motor, a detailed analysis of the propeller is not needed.

Expression (12.1) gives a theoretical value of the velocity increase to be provided by the propeller to the through-flowing air for the generation of the thrust T. Expression (12.3) represents the corresponding theoretical power to be supplied to the shaft by a motor. The real power is larger and the *propeller efficiency* may be defined as the ratio of the dynamic power to the shaft power. With analysis of the gas turbine, we assume a certain value of the propeller efficiency. This form of efficiency is near to 90% with optimally designed propellers.

In propeller theory, the velocity difference $\Delta v = v_1 - v_0$ is called the induced velocity at the propeller and similarly, $\Delta v = v_3 - v_0$ the induced velocity far downstream. The dynamic power (12.3) in propulsion theory is mostly called induced power in propeller theory. The propeller is driven by a motor, which is normally a gas turbine, because of the high ratio of power to weight and power to volume, as already mentioned. A propeller combined with a gas turbine engine is called a *turbo-propeller* or simply *turboprop*. The gas ejected by the gas turbine contributes to the thrust by about 5–10%. So, strictly spoken, a propeller driven by a gas turbine is a double-jet propulsion system (propeller + reactor, see next section). Double-jet thrust generation is discussed later in this chapter.

12.1.2 Reactor or Jet Engine

Figure 12.3 sketches the through-flow of a reactor. The term reactor indicates an air-breathing machine that aspirates air and ejects it, partly or totally altered into a combustion gas, as a jet with a higher velocity and possibly a pressure exceeding the atmospheric pressure. *Jet engine* is an equivalent term. Henceforth we often just use the term *engine*.

The internal part of the reactor in Fig. 12.3 represents a gas turbine without any net shaft power. Its turbine part just drives a compressor. Due to the energy addition by heating in the combustion chamber, the pressure drop through the turbine is smaller than the pressure rise through the compressor. The remaining pressure difference to the atmosphere at the turbine exit is supplied to a nozzle to generate a thrust jet.

In most engines, the air delivered by the compressor only partially flows through the combustion chamber and the turbine. Part of the compressor air bypasses the combustion chamber. Mostly, this air flows through a second concentric nozzle. There is then a hot jet and a cold jet. Further, we will derive that for optimum efficiency, the cold-jet velocity has to be somewhat lower than the hot-jet velocity. In other engines, the bypass flow is mixed with the flow leaving the turbine part and the mixed flow feeds a single nozzle. There is then only one thrust jet. So, we distinguish single-jet and double-jet engines. Figure 12.3 suggests a single-jet engine. We first analyse the single-jet engine.

The mass flow passing through the reactor in Fig. 12.3 is denoted with $\dot{m}$. We ignore the mass flow increase by fuel injection. We consider a control volume with an inlet section far upstream of the reactor and an outlet section at the exit plane of the nozzle. The nozzle exit pressure and velocity are denoted with p_j and v_j (the subscript j indicates the jet at the nozzle exit). As the pressure distribution on the envelope of the reactor is not known a priori, we consider besides the streamtube through the reactor, a second concentric streamtube with an envelope so far from the reactor that the pressure on it equals atmospheric pressure. The control volume is formed by the envelope of the outer streamtube. We assume that pressure equals atmospheric pressure in the inlet and outlet planes of the outer streamtube.

No energy is added to the flow that does not pass through the reactor. In the outlet plane of the outer streamtube, pressure is atmospheric. So, in absence of losses, the

Fig. 12.3 Through-flow of a reactor (single-jet engine variant)

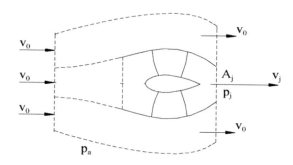

outflow velocity equals the inflow velocity and the momentum fluxes at the inlet and outlet of the outer streamtube are equal.

The momentum balance is:

$$\dot{m}\,(v_j - v_0) = T - A_j(p_j - p_a),$$

or

$$T = \dot{m}\,(v_j - v_0) + A_j(p_j - p_a) = \dot{m}\,\Delta v + A_j\,\Delta p. \tag{12.4}$$

Thrust T thus consists of a momentum part and a pressure part. We notice that the thrust of a propeller (12.1) follows from expression (12.4) for $\Delta p = 0$. In order to absorb the pressure thrust in the momentum thrust, an *effective jet velocity* is defined by

$$v_{je} = v_j + A_j\,\Delta p/\dot{m}, \quad \text{thus} \quad T = \dot{m}\,(v_{je} - v_0). \tag{12.5}$$

The nozzle exit pressure can only exceed atmospheric pressure when choking occurs. In practice, this may happen with a converging nozzle with a supercritical pressure ratio. The nozzle exit is then almost sonic (sonic speed is not fully attained because of losses). Free expansion after the nozzle occurs with losses by oblique shocks. Therefore, controlled expansion in an added diverging nozzle part may be more efficient.

Determination of the possible benefit is rather simple. Isentropic expansion from temperature T_j and pressure p_j to atmospheric pressure results in the temperature at the end of the expansion:

$$T_e = T_j\left(\frac{p_a}{p_j}\right)^{\frac{\gamma-1}{\gamma}} = T_j\left(\frac{p_j - \Delta p}{p_j}\right)^{\frac{\gamma-1}{\gamma}} = T_j(1 - \varepsilon)^{\frac{\gamma-1}{\gamma}}, \quad \text{with} \ \ \varepsilon = \Delta p/p_j.$$

The attainable velocity is

$$v_e^2 = v_j^2 + 2\,C_p(T_j - T_e) = v_j^2 + 2\,C_p T_j\left[1 - (1 - \varepsilon)^{\frac{\gamma-1}{\gamma}}\right],$$

Thus:

$$v_e = v_j\sqrt{1 + \frac{2}{\gamma - 1}\frac{1}{M_j^2}\left[1 - (1 - \varepsilon)^{\frac{\gamma-1}{\gamma}}\right]}.$$

With a choked nozzle, the effective exit velocity is

$$v_{je} = v_j + \frac{p_j}{\rho_j v_j}\varepsilon = v_j\left(1 + \frac{1}{\gamma}\frac{1}{M_j^2}\varepsilon\right).$$

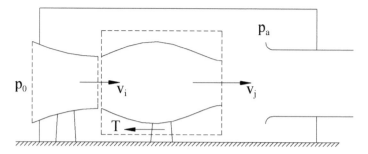

Fig. 12.4 Measurement of uninstalled thrust of a jet engine at sea level

The gain is $\Delta v_j = v_e - v_{je}$. For $\gamma = 1.33$ and $M_j = 1$, this leads to

$$\varepsilon = 0.2 : \frac{\Delta v_j}{v_j} = 0.0013; \quad \varepsilon = 0.5 : \frac{\Delta v_j}{v_j} = 0.0232.$$

The estimated gain is in absence of nozzle losses. Since the gain is small, the conclusion is that, unless the pressure ratio significantly exceeds the critical value, designing the nozzle with a diverging part is inefficient.

Henceforth we calculate with the effective jet velocity and consider the pressure at the exit of the reactor as atmospheric pressure. In aero-engine practice, choked and non-choked nozzles are both used.

The thrust according to (12.5) is called the uninstalled thrust. The thrust transferred to the aircraft is the installed trust, which is lower due to the drag on the nacelle (the envelope of the engine). The difference is the installation drag. Uninstalled engine thrust may be measured on a stationary test bench, as sketched in Fig. 12.4. The figure sketches the arrangement for measurement at sea level.

The inlet nozzle guiding the aspirated air is not connected to the engine. It is used for mass flow measurement based on inlet and outlet pressures. Momentum exchange of the thrust jet lowers the pressure in the chamber around the engine. So, for maintaining atmospheric pressure, air has to be supplied to the chamber (not represented in the figure). Comparison of the control volumes in Figs. 12.3 and 12.4 demonstrates that uninstalled thrust is measured. With uninstalled thrust, it is assumed that there are no losses between the inlet plane of the control volume in Fig. 12.3 and the engine entry.

The arrangement is somewhat more complex for thrust measurement at altitude flight conditions. The inlet nozzle then takes air from a plenum chamber in which the air is lowered in pressure and, possibly, also cooled.

A correction to the measured thrust is necessary because it is not possible to realise the correct inflow and outflow conditions of the engine. The installation drag, which depends on the flight speed, is normally calculated (measurement requires a very big wind tunnel). For calculation, the engine nacelle is divided into a front part (inlet) and a rear part (nozzle). Inlet drag results from ram effect (pressure rise due to the obstruction by the engine) and friction on the front engine part. Nozzle drag results from friction and from non-uniformity of velocity and pressure due to curvature of streamlines in the flow around the nozzle.

12.1.3   Rocket

With a rocket, no air is aspirated. The ejected gas is generated in a combustion chamber by means of a fuel and an oxidiser. This makes a rocket engine suitable for use outside the atmosphere. The fuel and the oxidiser are the sources of the thrust gas and are therefore called propellants. Propellants may be liquid or solid. With fluid propellants, the fuel and the oxidiser are separate liquids. A solid propellant is generally a mixture of a fuel and an oxidiser. A solid fuel may also be combined with a liquid oxidiser.

The thrust formula follows from the reactor formula with $v_0 = 0$:

$$T = \dot{m}v_{je} \text{ with } v_{je} = v_j + \frac{A_j \Delta p}{\dot{m}}.$$

The ambient pressure is zero outside the earth atmosphere. The nozzle pressure ratio is thus always strongly supercritical. The nozzle is thus converging–diverging. The pressure thrust is nevertheless an important component of the overall thrust.

12.2   Overview of Aircraft Gas Turbine Engines

12.2.1   Turbojet

Figure 12.5 represents the simplest form of a jet engine, called *turbojet*. There is one internal flow successively passing through a compressor, a combustion chamber, a turbine and a nozzle. There is a single propulsive jet. This type, historically the first one, is not used anymore, as it is only efficient with flight Mach numbers above 2.5, as we will derive later. There are no such aircraft at present.

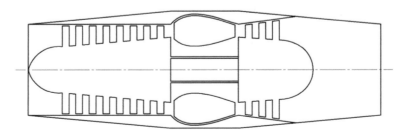

Fig. 12.5 Turbojet

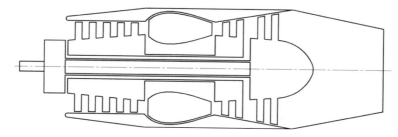

Fig. 12.6 Turboprop and turbo-shaft

12.2.2 Turboprop and Turbo-Shaft

Figure 12.6 is a sketch of the gas turbine part of a turbo-propeller, commonly called a *turboprop*. At the front side of the engine, there is an outward shaft that drives the propeller through a gearbox. The figure suggests an epicyclic gearbox, but parallel-shaft gearboxes are used as well. The gas turbine has a nozzle contributing to the thrust. In the example, the compressor and turbine components are axial and there are two shafts. With small engines, the compressor may be composed of axial stages followed by a radial one. There are also single-shaft engines. With two shafts, it is typical that there is no compressor part on the outward shaft (as in the figure). The turbine that drives the outward shaft is then a power turbine running at a much lower rotational speed than the core part, i.e. the part around the hollow shaft.

A *turbo-shaft* is similar, but, in principle, no thrust is generated by a nozzle. So, basically, a turbo-shaft does not differ from a power gas turbine, except that a gearbox is integrated. A typical turbo-shaft application is helicopter rotor drive (but some thrust may be generated with an outlet nozzle). Such gas turbines are also applied to drive heavy vehicles such as off-road trucks.

12.2.3 Bypass Turbojet

In a bypass turbojet, as sketched in Fig. 12.7, a part of the compressor air is diverted before the final compressor pressure is attained. This air does not pass through the combustion chamber and the turbine parts. It is mixed with the combustion gas that leaves the turbine parts. There are two flows inside the engine: a cold flow and a hot one. The mixed flows both feed the nozzle, generating a single jet. This engine is called *bypass turbojet* or bypass jet engine or just *bypass engine*. The bypass ratio, i.e. the ratio of the mass flow rates of the cold and hot flows, is typically between 0.5 and 1. The reason for organising the engine with two internal flows is discussed in Sect. 12.5.

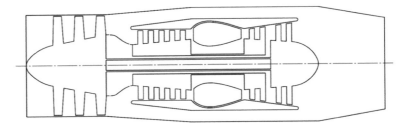

Fig. 12.7 Bypass turbojet

12.2.4 *Turbofan*

Figure 12.8 represents a *turbofan*. Two jets are generated by separate nozzles. The compressor part feeding the cold-flow nozzle is called a fan, as its pressure ratio is quite moderate (1.5 to 1.7). The central part of the fan delivers air passing through compressor parts, the combustion chamber, turbine parts and the hot-flow nozzle. In many current (2021) turbofan engines the ratio of the mass flow rates of the cold and the hot flows is 5 to 7. As with the bypass engine, this ratio is called the bypass ratio. The practical optimum is however about 12–15 for a typical airliner. There is thus a reason to increase the bypass ratio. The newest turbofan types feature a bypass ratio of 10–12.

In some turbofan engines, the cold and hot jets partially mix in a common nozzle. The term turbofan is still applied for these types, expressing that the bypass ratio is much larger than unity. Further, there is a tendency nowadays to use the term fan also for the low-pressure compressor part in a bypass engine.

This makes the distinction between a bypass engine and a turbofan rather vague. The fundamental difference is, indeed, not very large. Both are jet engines with two

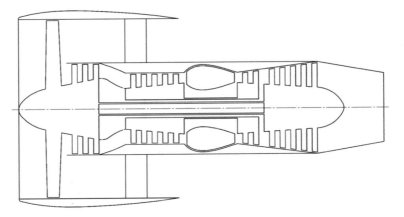

Fig. 12.8 Turbofan

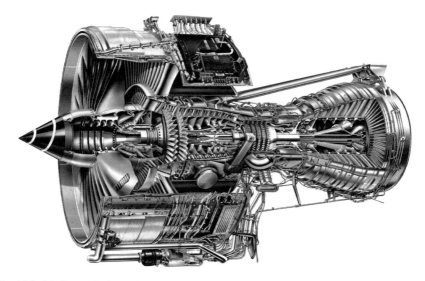

Fig. 12.9 Modern turbofan engine: Trent800 (*Courtesy* Rolls-Royce)

internal flows. The variants are that the flows feed separate nozzles or that the flows partially or fully mix at the entrance to a common nozzle. We derive in Sect. 12.5 that mixing of the flows is principally beneficial for efficiency, but efficient mixing is only possible for lower values of the bypass ratio.

Figure 12.9 shows a modern turbofan engine: a Tent800 from Rolls-Royce. It is a three-shaft engine. The internal shaft bears the fan and the low-pressure turbine with five stages. The fan diameter is 2.80 m. There are eight stages in the intermediate-pressure compressor and six stages in the high-pressure compressor. The inlet guide vanes and the vanes of the first two stages of the IP compressor are variable. The HP turbine and IP turbine have one stage. Sea level thrust is 420 kN and bypass ratio is 6. Similar engines exist with two shafts. There is then a small IP compressor part, called a *booster*, running together with the fan, as in Fig. 12.8.

12.2.5 Prop-Fan and Unducted Fan

A turboprop may be considered as an extreme form of turbofan, namely with a very large bypass ratio, typically around 50. There are propulsive systems that are classified in between the turboprop and the turbofan. They are denominated *prop-fan* or *unducted fan* (UDF). The thrust comes from two *contra-rotating propellers* (two propellers in row, with the same axis of rotation, rotating in opposite senses).

The reason for their existence is that the optimum bypass ratio of a turbofan for flight Mach number 0.85 is between 12 and 15 and that this optimum ratio increases with lowering of the flight Mach number (see Sect. 12.5).

Turbofan engines with high bypass ratio have a large fan diameter. The diameter of the largest current (2021) turbofans is around 3 m: e.g., Rolls-Royce Trent1000 (diameter 2.85 m, bypass ratio 10), General Electric GE90 (diameter 3.40 m, bypass ratio 9), Pratt and Whitney PW4000-112 (diameter 2.85 m, bypass ratio 6) (see websites of these manufacturers) with thrust at sea level of 350 to 500 kN.

A turbofan engine is mounted under the wings and there must be sufficient ground clearance with the airplane on the ground and during take-off or landing. The problem is similar with smaller engines on smaller aircraft. So, further increase of the bypass ratio and thus the fan diameter becomes problematic.

A cruise flight Mach number 0.85 is typical for airliners, as with this flight Mach number shocks in the flow over the suction side of the wings remain rather weak. The flight speed is around 900 km/h with speed of sound about 300 m/s ($T \approx 225$ K at 10 km height). But fuel saving may be realised with a somewhat lower flight Mach number (0.70–0.75). Then, shock losses disappear and a larger part of the wings can be operated with laminar boundary layers (shocks create the risk of separation of a laminar boundary layer). With lower flight speed, the optimum bypass ratio is larger (for flight Mach number 0.70 it may be around 20; see Sect. 12.5) and the diameter of a turbofan may become impractically large. On the other hand, aircraft with conventional turbo-propellers are limited in flight speed to about 550–650 km/h. The reason is limitation of the Mach number of the relative flow attacking the propeller tip.

An example is the Q400 turboprop aircraft (Bombardier; 68 to 78 passengers; production has been stopped, now a variant by De Havilland) with two Dowty-R408 six-bladed propellers (GE Aviation). The cruise altitude is 7500 m ($T \approx 240$ K, speed of sound ≈ 310 m/s). The long-range cruise speed is 150 m/s (540 km/h) and the fast-travel cruise speed is 180 m/s (648 km/h), both with propeller tip speed of 180 m/s. The oncoming flow speed at the propeller tip, based on the velocity triangle with these speeds, is 235–255 m/s (the actual speed is somewhat larger due to acceleration of the flow). The corresponding Mach number is 0.76–0.82. With this Mach number, shock losses at the propeller tip and noise generation stay limited. For the same tip Mach number, a larger flight speed requires a lower tip speed of the propeller. For the same axial velocity increase through the propeller, the solidity then has to be larger. This means more and broader propeller blades and larger post-whirl. The larger post-whirl diminishes the propeller efficiency. There is thus a practical limitation to the flight speed.

There are single-row propellers for aircraft with flight Mach number 0.70–0.75, cruising at 10–11 km altitude. An example is the A400M (Airbus) medium-size military transport aircraft with four TP400-D6 eight-bladed propellers (Europrop = RR + MTU + Safran + ITP) with cruise flight speed 780 km/h and cruise altitude 9500 m. The blade tip speed is about 235 m/s (propeller diameter is 5.30 m).

The Mach number of the relative flow approaching the propeller tips is slightly above unity (about 1.10). Although blade tips are adapted to this Mach number through backsweep and appropriate blade profiles, shock losses are quite high at the tip and noise production is also quite high. For a military aircraft, this is easier to accept than for a commercial airliner.

Contra-rotating propellers may be a solution for commercial aircraft. The tip speed can be lower than with a single-row propeller and the work transferred can be larger. The flow turning per propeller is larger, but the contra-rotation concept avoids post-whirl. The solidity is larger and so is the number of blades. The corresponding velocity increase in the flow through the propeller is then larger than in a conventional propeller, but smaller than in a conventional turbofan.

Figure 12.10 shows an engine, generally called an *unducted fan*, studied already for a long time by GE Aviation. The propeller blades are mounted on contra-rotating turbines in the rear part of the engine. The blades are swept back at the tips, in order to lower the Mach number of the velocity component perpendicular to the leading edge, as with backswept wings of an airplane. The blades have a compensating forward sweep at the hub. The diameter of the forward propeller is larger in order to avoid interference of the tip vortices with the rearward rotor. The pitch angle of the blades can be varied for adjustment of the thrust.

The engine in Fig. 12.10 is a pusher, meant for mounting on the rear part of an aircraft fuselage, at a position much higher than the wings. The tip speed of the propellers is to be limited to about 200 m/s at flight speed 200 m/s (720 km/h). The consequence is low tip speed of the turbine carrying the propeller blades. So, the stage work is low and many stages are required. For instance, the GE36 UDF has 14 stages in the LP-turbine driving the propellers (a stage is one blade row; there are no stationary parts).

An alternative for the low-speed power turbine of an unducted fan is a high-speed power turbine with a low number of stages, driving the propellers over a reduction gearbox, as with a conventional propeller. The gearbox is then epicyclic with two concentric contra-rotating driving shafts. A pusher type (propeller downstream of the gas turbine) and a tractor type (propeller upstream of the gas turbine) are both possible. The concept is called a *prop-fan*. Both have been studied since long time by Rolls-Royce (RB3011). The tractor type is meant for mounting on the upside of the wings, as with many conventional propellers.

Fig. 12.10 High-bypass unducted fan (*Courtesy* GE Aviation)

A tractor type, the D27 by Ivchenko-Progress in Ukraine, is used since the 1980s on the Antonov An-70, which is a medium-size military transport aircraft. The front propeller with diameter 4.5 m has eight blades and provides the largest part of the thrust. The somewhat smaller rear propeller has six blades. The power through the gearbox is about 10 MW and the thrust is about 100 kN at sea level.

The main problem with the unducted fan and the prop-fan is high noise due to interference between the two rotors, actually not much lower than by single-row propellers with high tip speed. This impedes airliner use. Further, the prop-fan is limited to medium-size aircraft, requiring a sea-level thrust per engine of about 100 kN. The reason is the power limitation on the gearbox, currently to about 10 MW (the same limitation applies to single-row propellers).

The benefit of reduced fuel consumption would be more interesting with the larger class of engines with sea-level thrust of the 400 kN order for large airliners. The problem of noise emission has been under study since long time, but a satisfactory solution has not been obtained yet. The noise problem can be weakened by shrouding the propellers. This has been studied in the past, but the concept is then not essentially different from a turbofan.

12.2.6   Geared Turbofan

A high bypass ratio necessitates a large fan diameter and, in case of direct drive of the fan, a low tip speed of the blades of the driving LP-turbine. The bypass ratio of the newest large turbofan engines is 10, as already mentioned. But, larger bypass ratios are wanted. There is not only the problem of the large fan diameter, but also of the number of stages, and thus the weight, of the LP-turbine driving the fan. The LP-turbine may be compacted by letting it run faster and driving the fan over a reduction gearbox, as with a propeller. The concept is called a *geared turbofan*.

Such engines are manufactured by Pratt and Whitney: the PW1000G series. The gearbox is epicyclic with a speed ratio 3:1. The LP-turbine has three stages. The bypass ratio is 12 and the thrust at sea level is 100–150 kN. The power passing through the gearbox is about 13.5 MW. Further development aims at increasing the speed ratio of the gearbox (larger bypass ratio) and the power passing through it (larger engine). The technological challenges of the gearbox are development of light, high-strength materials and reduction of the friction heat.

12.3 Performance Parameters of Aircraft Propulsion Systems

We first define some concepts of power and efficiency for single-flow turbojets. Afterwards, these concepts are extended to double-flow systems. With double-flow systems, we treat turbo-propellers, bypass engines, turbofan engines, prop-fan engines and unducted fan engines all with the same form of analysis.

12.3.1 Specific Thrust

For a single-jet engine, thrust is

$$T = \dot{m} \, \Delta v,$$

with Δv the effective velocity increase. In the formula, the mass flow increase due to fuel injection is neglected. *Specific thrust* is thrust divided by the mass flow rate aspirated by the propulsion system. With the simplified formula for thrust, this is

$$T_s = T/\dot{m} = \Delta v.$$

The extension to double-jet propulsion systems is immediate. It is then the total thrust divided by the sum of the aspirated mass flows. With turbofan engines, the order of magnitude is 300 m/s at start of take-off and 125–150 m/s at cruise. With propellers, the flow acceleration Δv is around 100 m/s at start of take-off and 10–15% of the flight speed in cruise flight.

12.3.2 Dynamic Power

Dynamic power is the power corresponding to the kinetic energy increase in the flow through the propulsion system:

$$P_{dyn} = \dot{m} \frac{1}{2} \left(v_j^2 - v_0^2 \right). \tag{12.6}$$

Again, extension of the notion to double-jet systems is immediate.

12.3.3   Gas Power and Dynamic Efficiency

In a single-jet engine, the dynamic power originates from the flow acceleration in the nozzle. The gas entering the nozzle is delivered by a gas turbine, which thus functions as a *gas generator*. The nozzle may then be defined as the *jet generator*. *Gas power* denominates the power that would be delivered to the propulsive jet formed in the nozzle in absence of losses, thus by an isentropic expansion. The gas power is thus the ideal value of the dynamic power. It may be interpreted as the output from the gas generator and the input to the jet generator. The efficiency of the jet generator, or the *dynamic efficiency* (η_d), is then defined as the ratio of dynamic power to gas power.

12.3.4   Thermal Power, Thermodynamic and Thermal Efficiencies

Thermal power is the power released by the combustion. *Thermodynamic efficiency* (η_{td}) is the ratio of gas power to thermal power. Thermodynamic efficiency thus characterises the efficiency of the gas generator. *Thermal efficiency* (η_t) of the propulsion system is defined as the ratio of dynamic power to thermal power.

12.3.5   Propulsive Power and Propulsive Efficiency

Propulsive power or *thrust power* is the power consumed by the aircraft for its propulsion:

$$P_{prop} = T \, v_0.$$

The propulsive power differs from the dynamic power by

$$P_{dyn} - P_{prop} = \dot{m}\frac{1}{2}(v_j^2 - v_0^2) - \dot{m}(v_j - v_0)v_0 = \dot{m}\frac{1}{2}(v_j - v_0)^2.$$

The difference is the flux of kinetic energy in the propulsive jet dissipated in the atmosphere behind the aircraft. The power difference is called *residual power* and the kinetic energy is called the *residual kinetic energy*. The residual power is a part of the dynamic power that contributes to the generation of the propulsive jet, but that is not useful for the propulsion of the aircraft. *Propulsive efficiency* is therefore defined as the ratio of the propulsive power to the dynamic power:

$$\eta_p = \frac{\dot{m}\left(v_j - v_0\right)v_0}{\dot{m}\frac{1}{2}\left(v_j^2 - v_0^2\right)} = \frac{2\,v_0}{v_j + v_0} = \frac{2}{1 + \left(v_j/v_0\right)} = \frac{1}{1 + \frac{1}{2}\frac{\Delta v}{v_0}}. \tag{12.7}$$

The propulsive efficiency can never be near to unity, as it is essential for propulsion that the effective jet velocity exceeds the flight speed. At this stage, we may use the formula for propulsive efficiency in a loose sense also for double-jet engines, in order to obtain an order of magnitude. For a turbofan engine with $\Delta v = 150$ m/s and $v_0 = 250$ m/s, $\eta_p \approx 0.77$. For a propeller with $\Delta v/v_0 = 0.15$, $\eta_p \approx 0.93$.

12.3.6 Overall Efficiency

The ratio of propulsive power to thermal power is the overall efficiency:

$$\eta = \frac{P_{prop}}{P_{therm}} = \frac{P_{dyn}}{P_{therm}}\frac{P_{prop}}{P_{dyn}} = \eta_t\eta_p. \tag{12.8}$$

It is obvious that the overall efficiency has to be optimised.

12.3.7 Generalisation for Double-Flow Engines

In order to understand the way of generalisation, we consider the transformation of a turbojet into a turboprop. This is achieved by mounting a power turbine at the exit of the gas generator of the turbojet. The propeller is driven by the power turbine, normally over a reduction gearbox.

It is obvious that the conversion is of no use regarding the thermal efficiency of the propulsive system. The gas power, available with a turbojet, can optimally be converted into dynamic power through a nozzle. The efficiency of a nozzle (ratio of dynamic power to gas power) is very high, about 0.98. Dynamic power generation through the propeller involves the power turbine efficiency ($\eta \approx 0.92$), the transmission efficiency ($\eta \approx 0.95$ with a reduction gearbox) and the propeller efficiency (propeller efficiency considered as the ratio of the dynamic power to the mechanical power: $\eta \approx 0.9$). The efficiency of the conversion is only about 0.80.

The difference between both systems is that the jet velocity is high (≈ 1050 m/s; we derive the value later) with a turbojet, whereas the jet velocity can be made very low with a propeller. This mainly has significance for the propulsive efficiency.

Examples:

$$T_0 = 225\,\text{K}(10\,\text{km height}), M_0 = 0.85 \rightarrow v_0 \approx 250\,\text{m/s}(900\,\text{km/h}),$$
$$\text{for } v_j = 1050\,\text{m/s} \rightarrow \eta_p = 0.385.$$

$T_0 = 225\,\text{K}, M_0 = 2.5 \rightarrow v_0 \approx 750\,\text{m/s}(2700\,\text{km/h})$,

for $v_j = 1050\,\text{m/s} \rightarrow \eta_p = 0.833$.

Achieving good propulsive efficiency with a single-jet engine requires a flight Mach number of about 2.5 or above. The purpose of the second jet is thus adaptation to a lower flight speed. The gas power supplied to the hot nozzle is decreased by converting part of it into mechanical power and by generating a second jet through a compressor, a fan or a propeller so that dynamic power becomes available at a lower jet velocity and with a larger mass flow rate. The higher the bypass ratio, the better the engine is adapted to low flight speed. Of course, the higher the bypass ratio, the lower the final thermal efficiency is. But clearly, the product of thermal efficiency and propulsive efficiency (12.8) must be optimised.

Precise performance analysis requires splitting of a gas turbine based propulsion system into two parts. We discern a *gas generator* and a *jet generator*. First, we discern a hot flow and a cold flow. The hot flow is the flow that passes through the combustion chamber. So, in a turbofan engine, the fan has to be thought as composed of an inner part processing the hot flow and an outer part processing the cold flow. The splitting is similar for the LP-compressor in a bypass engine. The gas generator is the part of the engine that processes the hot flow until a section in the turbine such that the part upstream of it drives the hot-flow compressor parts.

The compressor and turbine parts of the gas generator are not always easily discernible. The high-pressure compressor and the high-pressure turbine are part of it in a double-shaft engine. In a triple-shaft engine, it encompasses the high-pressure and intermediate-pressure compressor and turbine parts. In a bypass engine and a turbofan, the low-pressure turbine drives the low-pressure compressor or the fan. These contribute both to the cold and the hot flows. For the gas generator, we have to think of the LP turbine as split into a front part belonging to the gas generator and a rear part belonging to the jet generator. In practice, the term *core engine* indicates the engine part exclusively processing the hot flow (normally the HP and IP parts). So, the term core engine is more restrictive than the term gas generator.

The gas generator is characterised by the thermodynamic efficiency. This is the ratio of gas power to thermal power. The jet generator is characterised by the dynamic efficiency. This is the ratio of dynamic power (sum of the two jets) to gas power. The thermal efficiency of a propulsive system is the product of the thermodynamic efficiency and the dynamic efficiency ($\eta_t = \eta_{td}\eta_d$). The overall efficiency is the product of three efficiencies:

$$\eta = \eta_{td}\eta_d\eta_p. \tag{12.9}$$

Optimisation of the propulsive system is obtained by designing the gas generator for maximum thermodynamic efficiency and the jet generator for maximum product of dynamic efficiency and propulsive efficiency. The lower the flight speed, the higher the bypass ratio must be for good propulsive efficiency. This inevitably decreases the dynamic efficiency, so that the product of both goes through a maximum. This description is a simplification, as it overlooks that an increase of the bypass ratio causes increased engine dimensions, resulting in additional drag on the aircraft and added weight. An overall evaluation must take engine drag and engine weight into account.

The efficiency terminology with propulsion systems is inevitably more complicated than with power gas turbines. With power generation, the thermal efficiency is defined as the ratio of shaft power to thermal power. Thermodynamic efficiency may also be introduced, but is not strictly necessary. With power gas turbines, a gas generator and a power turbine may be distinguished. The thermodynamic efficiency is then the gas generator efficiency in the sense discussed. The thermal efficiency is the product of the thermodynamic efficiency and the isentropic efficiency of the power turbine.

12.3.8 Specific Fuel Consumption

Thrust specific fuel consumption (TSFC) denominates the fuel consumption per unit of thrust. In technical literature, it is often expressed in pound mass per hour per pound force (1 lbm = 0.4536 kg, 1 lbf = 4.448 N). Thrust specific fuel consumption by modern airliner turbofans in these imperial units is around 0.55 in cruise flight. The value at take-off is about 0.35. In international units, thrust specific consumption may be expressed in (g/s)/kN. Departing from imperial units, multiplication by $10^6/(9.8065 \times 3600) = 28.325$ is required, leading to about 15.5 (g/s)/kN in cruise flight. With about 43,000 kJ/kg lower heating value of the fuel, this is approximately 670 kW fuel flow rate per kN thrust.

For an engine driving a propeller, the specific fuel consumption is the fuel mass flow rate per unit of power. The technical literature often uses pound mass per hour per horsepower (1 HP = 745.7 W). In these imperial units, the specific fuel consumption of a turbo-shaft engine is about 0.50 in cruise flight, so similar to the value of the TSFC of a turbofan. With a lower heating value of 43,000 kJ/kg, this is about 3.65 kW fuel flow rate per kW shaft power. With a turboprop, the residual thrust, which is the thrust of the nozzle, is converted into mechanical power according to the convention that 1 HP of power is equivalent to 2.6 lbf of thrust (1 kW is about 15.5 N), which is a value appropriate for static sea-level conditions (see Exercise 12.7.1). The precise value of the conversion factor is not very important as residual thrust is only about 5 to 10% of the total thrust. The entire mechanical power achieved is called equivalent shaft horsepower = ESHP.

The concept of thrust specific fuel consumption with jet engines is rather peculiar. It does not strictly express efficiency. The efficiency is given by

$$\eta = \frac{T\,v_0}{\dot{m}_f\,H_L}, \quad \text{thus} \quad f_s = \frac{\dot{m}_f}{T} = \frac{v_0}{H_L\eta}.$$

Thrust specific fuel consumption thus depends on the flight speed. Values must be compared with the same flight Mach number. This is not a problem in practice. With airliners, the cruise flight Mach number is always about 0.85. With fighter aircraft, the design flight Mach number is typically around 2. But, there are also applications with other flight Mach numbers. An example is paratrooper transport aircraft with a flight Mach number around 0.5. Note that thrust specific fuel consumption has meaning in static conditions (start of take-off at sea level) where efficiency has no meaning.

12.4 Performance of the Gas Generator and the Single-Jet Engine

The gas generator in an aero gas turbine is analogous with a simple-cycle power gas turbine. Good performance therefore requires a high pressure ratio and a high temperature ratio. A high pressure ratio is important for high thermodynamic efficiency. The temperature ratio mainly determines the specific power. Compared to power gas turbines, aero gas turbines have an advantage concerning pressure ratio and temperature ratio. The compression is partially obtained aerodynamically by the ram effect of the incoming air within the engine inlet. The ram effect is the spontaneous deceleration due to the obstruction by the engine.

With an airliner, the aerodynamic compression achieves a very high efficiency. Further, it does not require turbine work. Of course, the ram effect causes drag for the engine and for the aircraft. The through-flow Mach number (Mach number of the average meridional velocity component) in an aero-engine compressor is around 0.5 in the front stages. In a fan, it is somewhat higher, around 0.6. The flight Mach number of an airliner is 0.8 to 0.85. So, the deceleration at the engine entrance is very moderate. Hence, the isentropic efficiency of the inlet is very high, of the order of 0.98. For analysis purposes, we assume it equal to unity. For a supersonic fighter aircraft, the efficiency of the inlet is much lower due to formation of shock waves. The atmosphere temperature is low at cruising altitude: about 225 K at 10 km height. With a given combustion temperature, the temperature ratio is thus better than with a power gas turbine.

A turbojet engine consists of a gas generator with a directly added nozzle. Nozzle efficiency is about 0.98, being very close to 1. The gas generator may thus be described as a single-jet engine with an ideal nozzle. Gas power and dynamic power are then identical. Henceforth we analyse the single-jet engine with an ideal nozzle. The results strictly apply to the gas generator. Thermal efficiency and specific power values are about 2% lower with a real turbojet engine. As with a power gas turbine, the major loss in a turbojet engine is thermal, in the sense of remaining heat in the combustion gas after expansion in the nozzle. In second order, there are thermodynamic losses due to dissipation in the compressor and the turbine parts and due to cooling of turbine parts. As with the power gas turbine it is therefore relevant to first analyse with neglect of thermodynamic losses.

12.4.1 Analysis with Loss-Free Components

The cycle without component losses is sketched in Fig. 12.11. The parts are.

  0–01: inlet
  01–02: compressor
  02–03: combustion chamber
  03–04: turbine
  04–4e: nozzle (the symbol e denotes the end of an ideal expansion).

States 0 and 4e are static. The other states are total (stagnation with respect to the engine frame).

The total parameters of the oncoming air are

$$h_{00} = h_0 + \frac{v_0^2}{2}, \quad T_{00} = T_0 + \frac{v_0^2}{2C_p}, \quad \frac{p_{00}}{p_0} = \left(\frac{T_{00}}{T_0}\right)^{\frac{\gamma}{\gamma-1}}.$$

(a) **Inlet**

$$p_{01} = p_{00}, \quad T_{01} = T_{00},$$

$$T_{01} = T_0\left(1 + \frac{\gamma-1}{2}M_0^2\right), \quad \frac{p_{01}}{p_0} = \left(1 + \frac{\gamma-1}{2}M_0^2\right)^{\frac{\gamma}{\gamma-1}} = r_a.$$

The p_{01}/p_0 ratio is denominated aerodynamic compression ratio.

(b) **Compressor**

$$r_c = p_{02}/p_{01}, \quad W_c = C_p(T_{02} - T_{01}).$$

(c) **Combustion chamber**

Fig. 12.11 Cycle of a single-jet engine without component losses

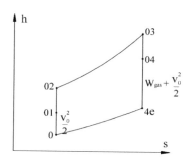

$$\Delta Q = C_p(T_{03} - T_{02}).$$

Here, we simplify by the assumption of constant mass flow rate and constant specific heat.

(d) **Turbine**

$$W_t = C_p(T_{03} - T_{04}) = W_c.$$

(e) **Jet generator**

$$\frac{p_{03}}{p_0} = r_a r_c, \quad \frac{T_{03}}{T_{4e}} = (r_a r_c)^{\frac{\gamma-1}{\gamma}},$$

$$W_{gas} = C_p(T_{04} - T_{4e}) - C_p(T_{01} - T_0) = C_p(T_{03} - T_{4e}) - C_p(T_{02} - T_0).$$

$$W_{gas} = C_p T_{03}\left(1 - (r_a r_c)^{-\frac{\gamma-1}{\gamma}}\right) - C_p T_{02}\left(1 - (r_a r_c)^{-\frac{\gamma-1}{\gamma}}\right).$$

The thermodynamic efficiency is: $\eta_{td} = 1 - r^{-\frac{\gamma-1}{\gamma}}$, with $r = r_a r_c$.

The expression is the same as with a stationary gas turbine, when the overall pressure ratio is applied, being the product of the aerodynamic (r_a) and mechanical (r_c) pressure ratios. The mechanical pressure ratio of many current turbofan engines is around 40 [1]. Due to the product of the aerodynamic and mechanical pressure ratios, the efficiency potential is very high. The two examples hereafter illustrate the high efficiency potential.

$M_0 = 0.85 : r_a = 1.60(\gamma = 1.40)$; with $r_c = 40 : r = r_a r_c = 64$;

$\eta_{td} = 0.695(\gamma = 1.40)$, $\eta_{td} = 0.644(\gamma = 1.33)$.

$M_0 = 3 : r_a = 36.7$; with $r_c = 1 : \eta_{td} = 0.643(g = 1.40)$, $\eta_{td} = 0.591(\gamma = 1.33)$.

The last result demonstrates that jet propulsion can be efficient at a sufficiently high flight Mach number, without any compressor and turbine parts. This is called a ramjet (see Exercise 12.7.10).

12.4.2 Analysis with Component Losses

The foregoing analysis may be adapted for adiabatic compressor and turbine efficiencies and change of the heat capacity of the working fluid in the combustion chamber (but with constant values for air and for combustion gas). Taking into account cooling of the turbine and variable gas properties requires a numerical simulation technique. But, with the analysis of power gas turbines (Chap. 11), we observed that efficiency and power results of simplified hand calculations (Sect. 11.3.2) are quite near to real values and results of numerical simulations (Sect. 11.3.3). Therefore, we consider here only analysis by hand calculations.

We set the polytropic efficiencies of the compressor and the turbine parts at 0.9. We set the heat capacity ratios for air and combustion gas at 1.40 and 1.33, combined with the gas constant 287 J/kgK (these are the typical values in the literature; air is dry in cruise conditions). We consider flight at Mach number 0.85, so that $r_a = 1.60$. We still treat the inlet and the nozzle with efficiency equal to 1. As in Sect. 11.3.2, we ignore the pressure drop in the combustion chamber and the mass flow rate increase due to fuel injection.

The design aim of an aircraft engine is optimum efficiency for cruise conditions, but the engine also has to function during take-off and climb. The operating points of the fan, which provides the largest part of the thrust in a turbofan engine (the thrust ratio of the two flows is about the bypass ratio), therefore, must be close to each other for all flight conditions in its non-dimensional performance chart [1–3]. By the non-dimensional performance chart is meant pressure ratio as a function of non-dimensional mass flow function $\dot{m}\sqrt{RT_{01}}/(p_{01}D^2)$, at constant values of non-dimensional rotational speed $\Omega D/\sqrt{RT_{01}}$, where D is a representative diameter and T_{01} and p_{01} are total temperature and total pressure at fan entry (see Chap. 15, Sect. 15.3).

The necessity for having the non-dimensional operating parameters during take-off and climb near to these for cruise comes from the requirement of stable operation of the fan in other flight conditions than cruise (sufficient margin to choking and stall; see Chap. 13). In reality, there are more conditions, as all compressor and turbine components must function in all flight conditions of the aircraft. But compressors parts typically have front stages with variable stator vane angles, so that the operation characteristic can be changed. The fan of a turbofan engine has no variable parts.

The analysis of the equilibrium between the components of an aero gas turbine for different flight conditions and the determination of optimal characteristics of these components is quite complicated. We refer to specialised books on aircraft gas turbines [3]. For basic understanding, as intended here, such detailed analyses are not necessary, as will become clear hereafter.

The previous requirements make that the total temperature ratio T_{03}/T_{01} (T_{03} = TIT) of the cycle and the mechanical pressure ratio PR = p_{02}/p_{01} have to be similar for cruise, start of take-off and climb (ratios are somewhat higher for top of climb). Limitations come to the maximum turbine inlet temperature (TIT) and maximum pressure ratio (PR) during take-off from maintaining sufficient strength of the turbine blades in the HP section and sufficient strength of the outer confinement of the combustion chamber.

Cumpsty and Heyes [3] estimate common take-off values around 1750 K (about 1475 °C) for TIT and 40 for PR. The newest engines have somewhat larger values of TIT and PR. We take here 1775 K (about 1500 °C). At static sea level conditions ($T_{01} = T_{00} = T_0 = 288$ K), the total temperature ratio is then about 6.16. The ratio during cruise must be comparable, which means that the TIT at cruise must be much lower than at take-off. Manufacturers do not communicate values of TIT in cruise conditions and only specify values of maximum TIT.

Cumpsty and Heyes [3] estimate common cruise values of TIT around 1500 K (about 1225 °C). Again, this level of TIT is somewhat low for the newest engines. We take 1575 K (about 1300 °C). For a flight Mach number of M = 0.85, corresponds a total temperature $T_{01} = T_{00} = 257.5$ K to the static temperature $T_0 = 225$ K. The total-to-total temperature ratio T_{03}/T_{01} is then about 6.12, thus comparable to the ratio at start of take-off. The total-to-static temperature ratio is $T_{03}/T_0 = 7$. We take this value for analysis, but it will become clear during the analysis that the turbine inlet temperature can be optimised, once the bypass ratio is chosen. Remark that the cruise TIT is much lower than typical for power gas turbines (Chap. 11).

The relations of the previous section are adapted.

(a) **Compressor**

$$W_c = C_{pa}(T_{02} - T_{01}) = C_{pa}T_0 \frac{T_{01}}{T_0}\left(\frac{T_{02}}{T_{01}} - 1\right) = C_{pa}T_0\, r_a^{\frac{R}{C_{pa}}}\left(r_c^{\frac{R}{\eta_{\infty c} C_{pa}}} - 1\right).$$

(b) **Turbine**

$$W_t = C_{pg}(T_{03} - T_{04}) = C_{pg}T_0 \frac{T_{03}}{T_0}\left(1 - \frac{T_{04}}{T_{03}}\right) = C_{pg}T_0\frac{T_{03}}{T_0}\left(1 - r_t^{-\frac{\eta_{\infty t} R}{C_{pg}}}\right).$$

Equating compressor and turbine work determines the pressure ratio of the turbine for a chosen pressure ratio of the compressor.

(c) **Combustion chamber**

$$\Delta Q = C_{pg}(T_{03} - T_0) - C_{pa}(T_{02} - T_0)$$

$$= C_{pg}T_0\left(\frac{T_{03}}{T_0} - 1\right) - C_{pa}T_0\left(\frac{T_{02}}{T_{01}}\frac{T_{01}}{T_0} - 1\right),$$

$$\frac{\Delta Q}{C_{pa}T_0} = \frac{C_{pg}}{C_{pa}}\left(\frac{T_{03}}{T_0} - 1\right) - \left(r_c^{\frac{R}{\eta_{\infty c} C_{pa}}}\, r_a^{\frac{R}{C_{pa}}} - 1\right).$$

Here, we still simplify by the assumption of constant mass flow rate, but the change of specific heat is taken into account. Further, we take the reference temperature for the heat input equal to T_0.

(d) **Jet generator**

$$W_{gas} = C_{pg}(T_{04} - T_{4e}) - C_{pa}(T_{01} - T_0)$$

$$= C_{pg}T_{04}\left(1 - \frac{T_{4e}}{T_{04}}\right) - C_{pa}T_0\left(\frac{T_{01}}{T_0} - 1\right),$$

$$\frac{W_{gas}}{C_{pa}T_0} = \frac{C_{pg}}{C_{pa}}\frac{T_{03}}{T_0}(r_t)^{-\frac{\eta_{\infty t} R}{C_{pg}}}\left(1 - \left(\frac{r_a r_c}{r_t}\right)^{-\frac{R}{C_{pg}}}\right) - \left((r_a)^{\frac{R}{C_{pa}}} - 1\right).$$

Results for three values of pressure ratio are listed in Table 12.1 for $r_a = 1.60$ ($M_0 \approx 0.85$, $v_0 \approx 225$ m/s). The jet velocity is the ideal value of the single-jet engine.

Table 12.1 Gas generator parameters for $T_{03}/T_0 = 7$ and $r_a = 1.60$

r_c	r_t	$\frac{W_{gas}}{C_{pa}T_0}$	$\frac{\Delta Q}{C_{pa}T_0}$	η_{td}	v_{j1}/v_0
30	4.2426	2.4962	4.5421	0.5496	4.2859
40	5.4699	2.3757	4.2201	0.5629	4.1869
50	6.8533	2.2503	3.9493	0.5698	4.0814

The obtained values of thermodynamic efficiency are very high. The real values are a few percentage points lower, but still above 50%. As expected, the thermodynamic efficiency increases with increasing pressure ratio. So, in principle, it is useful to enlarge the pressure ratio above the common values of around 40. E.g., the most recent (2021) large turbofan engine by Roll-Royce, the Trent1000, has pressure ratio 50 and bypass ratio 10. The newest large turbofan engine by GE Aviation, the GE9X, has pressure ratio 60 and bypass ratio 10. On the other hand, the specific power lowers somewhat with increasing pressure ratio. But this is not a drawback, as will become clear hereafter. For pressure ratio 40, and flight speed 255 m/s, the ideal jet velocity of a single-jet engine is about 1067 m/s. The actual value is then about 1050 m/s. We used this value in Sect. 12.3.7.

12.5 Performance of Double-Flow Engines

A double-flow engine consists of a gas generator with an added turbine that drives the cold-flow compressor, fan or propeller. As already discussed, it is often impossible to distinguish physically the gas generator and jet generator parts in a double-flow engine. The distinction is made here thermodynamically, enabling us to study the effect of the second flow. Additional parameters are the bypass ratio b and the pressure ratio of the cold flow. Furthermore, there is a difference between engines with both flows mixed before a common nozzle (bypass engine) and engines with separate nozzles (turbofan).

We assume a given gas generator and a given flight speed. Engine optimisation implies then obtaining the highest possible thrust per unit of mass flow rate in the hot flow. In the following derivations, the small flow rate increase by adding fuel is not taken into account.

12.5.1 Unmixed Flows (Double-Jet Engine: Turbofan, Turboprop)

A single-jet engine has the dynamic power

$$P_{dyn} = \dot{m}\frac{1}{2}\left(v_{j1}^2 - v_0^2\right).$$

Neglecting nozzle loss, this power also equals the gas power:

$$P_{gas} = \dot{m}\frac{1}{2}\,v_0^2\left(s_1^2 - 1\right), \quad s_1 = \frac{v_{j1}}{v_0}, \tag{12.10}$$

where the symbol s denotes a speed ratio.
Thrust is

$$T_1 = \dot{m}\left(v_{j1} - v_0\right) = \dot{m}\,v_0(s_1 - 1). \tag{12.11}$$

Figure 12.12 sketches the cycle of a double-flow engine in the h–s diagram. Some parts are:

01–02: hot-flow compressor;
01–05: cold-flow compressor;
03–04: hot-flow turbine;
04–06: cold-flow turbine.

The state point 07 is not relevant for the analysis of an engine with separate cold and hot jets, as we perform now (we use it in the next section on mixed flows).

Fig. 12.12 Cycle of a double-flow engine with component losses

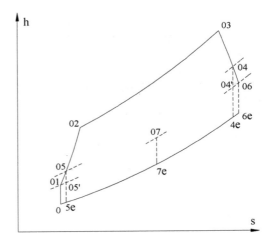

With the conversion of a single-flow engine into a double-flow engine, the mass flow rate of the hot flow (flow through the combustion chamber) stays $\dot{m}$ and the mass flow rate of the cold flow becomes $b\,\dot{m}$, with b the bypass ratio.

The work balance of the cold flow is

$$b\,C_{pa}(T_{05} - T_{01}) = \eta_{trans} C_{pg}(T_{04} - T_{06}), \tag{12.12}$$

where η_{trans} is the efficiency of the mechanical transmission between the added turbine and the cold-flow compressor part. The work extracted from the flow by the cold-flow turbine may be written as

$$C_{pg}(T_{04} - T_{06}) = \eta_t \frac{1}{2}\left(v_{j1}^2 - v_{j2hot}^2\right), \tag{12.13}$$

where η_t is the isentropic turbine efficiency, v_{j2hot} is the hot jet velocity of the double-jet engine and where we neglect the isentropic enthalpy difference between the expansions 04′–4e and 06–6e. The work done on the flow by the cold-flow compressor may be written as

$$C_{pa}(T_{05} - T_{01}) = \frac{1}{\eta_c}\frac{1}{2}\left(v_{j2cold}^2 - v_0^2\right), \tag{12.14}$$

where η_c is the isentropic re-expansion efficiency of the compressor, v_{j2cold} is the cold-jet velocity of the double-jet engine and where we neglect the isentropic enthalpy difference between the expansions 05′–5e and 01–0.

Combination of Eqs. (12.12), (12.13) and (12.14) results in

$$b\frac{1}{2}\left(v_{j2cold}^2 - v_0^2\right) = \eta_c \eta_{trans} \eta_t \frac{1}{2}\left(v_{j1}^2 - v_{j2hot}^2\right), \tag{12.15}$$

which may be written as a balance of dynamic power as

$$\frac{1}{2}\left(v_{j1}^2 - v_0^2\right) = \frac{1}{2}\left(v_{j2hot}^2 - v_0^2\right) + \frac{b}{\eta_b}\frac{1}{2}\left(v_{j2cold}^2 - v_0^2\right), \tag{12.16}$$

where η_b is the product of the efficiencies involved, which we may call the *bypass efficiency*. Henceforth we take $\eta_b = 0.85$, which is appropriate for a turbofan engine. For a turboprop the value is lower due to the reduction gearbox. Remark that the balance of dynamic power may be written directly in the form (12.16), but the role of the intervening efficiencies is clearer with the derivation in parts.

With velocity ratios, Eq. (12.16) is

$$s_{2hot} = \sqrt{s_1^2 - \frac{b}{\eta_b}\left(s_{2cold}^2 - 1\right)}. \tag{12.17}$$

The ratio of the thrust of the double-jet engine to the thrust of the single-jet engine is

$$\frac{T_2}{T_1} = \frac{\dot{m}\left(v_{j2hot} - v_0\right) + \dot{m}\, b\left(v_{j2cold} - v_0\right)}{\dot{m}\left(v_{j1} - v_0\right)} = \frac{(s_{2hot} - 1) + b(s_{2cold} - 1)}{s_1 - 1}.$$

$$(12.18)$$

With a chosen bypass ratio b, the hot jet velocity may be determined as a function of the cold-jet velocity, so that the thrust attains a maximum with the double-jet engine. This requires.

$$\frac{d\, s_{2hot}}{d\, s_{2cold}} + b = 0.$$

Furthermore from Eq. (12.17):

$$\frac{d\, s_{2hot}}{d\, s_{2cold}} = \frac{1}{2\sqrt{(-)}}\left(-2\frac{b}{\eta_b}\, s_{2cold}\right) = -\frac{b}{\eta_b}\frac{s_{2cold}}{s_{2hot}}.$$

Maximum thrust is attained for

$$s_{2cold} = \eta_b\, s_{2hot}.$$

$$(12.19)$$

Then

$$s_{2hot}^2 = s_1^2 - b\,\eta_b\, s_{2hot}^2 + \frac{b}{\eta_b} \quad \text{or} \quad s_{2hot}^2 = \frac{s_1^2 + (b/\eta_b)}{1 + b\,\eta_b}.$$

The resulting overall efficiency is

$$\eta = \eta_{td}\frac{T_2 v_0}{P_{gas}} = \eta_{td}\left(\frac{T_2}{T_1}\right)\frac{T_1 v_0}{P_{gas}} = \eta_{td}\left(\frac{T_2}{T_1}\right)\eta_{p1}, \quad \text{and thus} \quad \eta_d\eta_p = \left(\frac{T_2}{T_1}\right)\eta_{p1},$$

where η_{p1} is the propulsive efficiency of the single-jet engine.

We take the velocity ratio of the single-jet engine $s_1 = 4.20$, which is approximately the value in Table 12.1 for flight Mach number 0.85, $T_0 = 225$ K, TIT = 1575 K and pressure ratio of the hot-flow compressor equal to 40. The corresponding propulsive efficiency is $\eta_{p1} = 2/(1 + s_1) \approx 0.385$. The results for the double-jet engine for varying bypass ratio are listed in Table 12.2. The thrust ratio T_2/T_1 is the factor by which the global efficiency increases. In Table 12.2, this ratio keeps increasing for increasing bypass ratio and the limit value of $\eta_d\eta_p$ for bypass ratio tending towards infinity is approximately the bypass efficiency. That the highest possible bypass ratio comes out as the best is because increased drag and weight of the engine, which decrease the aerodynamic performance of the aircraft fuselage, are not taken into

Table 12.2 Double-jet engine (turbofan) parameters for $s_1 = 4.20$ and $\eta_b = 0.85$

b	s_{2hot}	s_{2cold}	r_{cold}	T_2/T_1	$\eta_d \, \eta_p$
1	3.189	2.711	6.984	1.219	0.469
5	2.117	1.799	2.311	1.598	0.615
10	1.759	1.495	1.636	1.786	0.687
20	1.512	1.285	1.310	1.944	0.748
50	1.326	1.127	1.122	2.085	0.802
100	1.254	1.066	1.060	2.145	0.825

account in the analysis. Taking these effects into account, the optimum bypass ratio is likely around 15.

For pressure ratio 40 and bypass ratio 10, the overall efficiency becomes $0.563 \times 0.687 = 0.387$. For $b = 10$, the mass averaged speed ratio v_j/v_0 is 1.519. With $v_0 = 255$ m/s, the specific thrust is 0.519×255 m/s $= 132.3$ m/s. The dynamic efficiency is 0.868 and the propulsive efficiency is 0.790 (Formula 12.7, which forms an approximation for a double-jet engine gives 0.794).

For $b = 5$, the mass averaged velocity ratio is 1.852. With $v_0 = 255$ m/s, the specific thrust is 217.3 m/s. For $b = 5$, the specific thrust is then much too high for reaching a good propulsive efficiency.

The specific thrust may be lowered by lowering the TIT. This lowers somewhat the thermodynamic efficiency, but not very much. So, obviously, there is an optimum TIT for each bypass ratio and TIT $= 1575$ K, as used here, is certainly too high for $b = 5$. It is even somewhat too high for $b = 10$.

We do not attempt here to determine the optimum TIT, but, of course, results may be recalculated for a different value of TIT. Remark that the specific thrust may also be lowered by increasing the pressure ratio. The results show that TIT $= 1575$ K or higher is only justified for a sufficiently high bypass ratio (higher than 10), combined with a sufficiently high pressure ratio (higher than 40).

The efficiency gain by the conversion of a single-jet engine into a double-jet engine is, as already discussed before, due to increased propulsive efficiency by spreading the energy provided by the gas generator over a larger mass flow rate and thus reducing the mass-averaged jet velocity.

The optimum velocity ratio of the jets, for chosen bypass ratio, (12.19), is a result that can be understood by inspection of the character of the optimisation problem. The relation (12.16) expresses that a weighted sum of the increases of the kinetic energies in the two flows is a fixed quantity imposed by the energy in the flow produced by the gas generator. The relation for thrust ratio (12.18) is a weighted sum of increases of velocities. So, the optimisation problem is of the type that the maximum of a sum, say $x + y$, is sought for a fixed sum of the squares of these variables $x^2 + y^2$. The optimum is obtained for $x = y$. One remarks that equal jet velocities are obtained for $\eta_b = 1$.

The cold-flow pressure ratio may be calculated from the h–s diagram in Fig. 12.12. Again, we neglect losses in the engine inlet and in the cold-flow nozzle.

$$\frac{T_{01}}{T_0} = r_a^{\frac{R}{C_{pa}}}, \quad \frac{T_{05}}{T_{01}} = (r_{cold})^{\frac{R}{\eta_{\infty c}C_{pa}}}.$$

$$\frac{v_{j2cold}^2}{2} = C_{pa}(T_{05} - T_{5e}) = C_{pa}T_{05}\left(1 - (r_a r_{cold})^{-\frac{R}{C_{pa}}}\right),$$

$$\frac{v_{j2cold}^2}{2} = C_{pa}T_0(r_a)^{\frac{R}{C_{pa}}}(r_{cold})^{\frac{R}{\eta_{\infty c}C_{pa}}}\left(1 - (r_a r_{cold})^{-\frac{R}{C_{pa}}}\right). \tag{12.20}$$

Determination of r_{cold} from Eq. (12.20) for given v_{j2cold} requires iteration. The results are listed in Table 12.2.

The fan pressure ratio is about 2.3 for $b = 5$ and about 1.6 for $b = 10$ according to Table 12.2. The choking pressure ratio of a nozzle is about 1.9. Thus for large bypass ratio, the bypass flow nozzle is not choked at start of take-off, but it becomes choked in cruise flight, because the pressure ratio of the nozzle is the product of the aerodynamic pressure ratio and the fan pressure ratio and the aerodynamic pressure ratio is about 1.60 for flight Mach number 0.85. The core flow nozzle is never choked due to the high temperature. Choking decreases the efficiency of a nozzle. This effect is not taken into account in the previous analysis.

The pressure ratio attainable within a stage of an axial compressor is at maximum about 1.75, when flow velocity and blade speed are set to the maximally attainable values (see Chap. 13). The bypass ratio has thus to be at least 10 for optimal use of a single-stage fan. With bypass ratio 5, the pressure ratio of a single-stage fan is too low for obtaining optimum functioning (but the necessary value lowers with lower TIT). With pressure ratio 1.75, the efficiency may be calculated with the above formulae.

The results are:

$$b = 5; \ r_{cold} = 1.75; \ s_{2cold} = 1.556; \ s_{2hot} = 3.027; \ \frac{T_2}{T_1} = 1.509.$$

The thrust ratio is lower than the optimal value for $b = 5$ in Table 12.2. We further note that the cold flow requires several compressor stages for low bypass ratio.

12.5.2   Mixed Flows (Bypass Engine)

Pressures and velocities of two flows to be mixed have to be equal for zero mixing loss. This implies comparable total pressures before mixing. Therefore, for the purpose of the analysis, we take equal total pressures. The h–s diagram is as sketched in Fig. 12.12, but now with the condition that the total pressure p_{05} at the exit of the cold-flow compressor is equal to the total pressure p_{06} at the exit of the turbine that drives the cold-flow compressor. This condition implies a relation between the bypass ratio and the common pressure level $p_{05} = p_{06}$. The state point 07 is the mixed flow entering the common nozzle.

To illustrate the effect of mixing, we take once more: $r_a = 1.60$, $r_c = 40$, $T_{03}/T_0 = 7$ and we choose $b = 10$. First, we calculate separate hot and cold jets. The hot and cold flow compressions are determined by

$$\frac{T_{01}}{T_0} = (r_a)^{\frac{R}{C_{pa}}}, \quad \frac{T_{02}}{T_{01}} = (r_c)^{\frac{R}{\eta_{\infty c} C_{pa}}}, \quad \frac{T_{05}}{T_{01}} = (r_{cold})^{\frac{R}{\eta_{\infty c} C_{pa}}}.$$

With a chosen cold-flow compression ratio r_{cold}, follows T_{05}.
The power balance of the hot flow, with transmission efficiency equal to unity, is

$$C_{pg}(T_{03} - T_{04}) = C_{pa}(T_{02} - T_{01}).$$

This balance determines T_{04}.
The power balance of the cold flow, with transmission efficiency equal to unity, is

$$C_{pg}(T_{04} - T_{06}) = b\, C_{pa}(T_{05} - T_{01}).$$

For a chosen bypass ratio, this balance determines T_{06}.
From the temperature ratio T_{04}/T_{06} follows the pressure ratio p_{04}/p_{06}. The jet velocities v_{j1}, v_{j2hot} and v_{j2cold} can then be determined. The results are listed in Table 12.3 ($s_1 = 4.1869$). The bypass efficiency is calculated from (12.15). The product $\eta_d \eta_p$ is maximum for $r_{cold} = 1.60$ to 1.65. The values of optimum r_{cold} and optimum $\eta_d \eta_p$ corresponds approximately with the values in Table 12.2 (there are small differences because s_1 and η_b are not exactly the same).
From Table 12.3 follows that equal values of total pressure at the exits of the cold-flow compressor and the-cold flow turbine are approximately reached for $r_{cold} = 1.52$. With the values for this pressure ratio, we may calculate the performance by mixing the hot and cold flows before entry to a common nozzle. The energy balance of the nozzles, ignoring mixing loss, is

$$\dot{m}(b+1)\frac{1}{2}v_{j2}^2 = \dot{m}\,\frac{1}{2}v_{j2hot}^2 + \dot{m}\,b\frac{1}{2}v_{j2cold}^2. \tag{12.21}$$

Table 12.3 Double-jet engine (turbofan) parameters for $r_a = 1.60$, $r_c = 40$, $T_{03}/T_0 = 7$ and b = 10

r_{cold}	p_{06}/p_{05}	s_{2hot}	s_{2cold}	T_2/T_1	η_b	$\eta_d \eta_p$
1.50	1.086	2.410	1.415	1.745	0.856	0.673
1.52	1.007	2.322	1.428	1.757	0.855	0.677
1.55	0.898	2.185	1.446	1.771	0.855	0.683
1.60	0.742	1.940	1.475	1.786	0.854	**0.689**
1.65	0.612	1.664	1.503	1.787	0.853	**0.689**
1.70	0.504	1.338	1.530	1.770	0.853	0.683
1.75	0.414	0.910	1.556	1.718	0.852	0.662

With the values obtained, the results are $s_2 = 1.531$ and $T_2/T_1 = 1.833$. This thrust ratio is larger than the maximum value in Table 12.3.

The better result, compared to separate flows, even with optimised velocity ratio, comes from the higher efficiency of the energy transfer from the hot flow to the cold flow. The efficiency of the mechanical energy transfer (turbine, transmission, fan) is the bypass efficiency, which is significantly lower than unity (about 0.85 in Table 12.3). The efficiency of the thermal energy transfer by mixing has been set to unity in the previous analysis. This means that the efficiency of the combined mechanical and thermal transfers is larger than solely with mechanical transfer. So, it is clearly advantageous to mix, even if there is some loss during mixing.

Not too high bypass ratios, as in military engines ($b = 0.5$ to 1), enable a complete mixing of the hot and the cold flows. A lobed mixer surface is sometimes applied to enhance mixing (see the later Fig. 12.17). Mixing of the flows is then always favourable. Complete mixing cannot be realised with higher bypass ratios, because of the limited length of the mixing zone in practice. Some engines with moderate bypass ratio have two versions. An example is the V2500 (IEA = P&W + MTU + JAEC). The bypass ratio is about 5. It exists with separate nozzles, but there is also a version with the cold flow brought to the rear of the engine through a long duct. This is favourable for noise reduction, as the engine is then encapsulated. The cold flow is partially mixed with the hot flow without mixer surface. Both versions have approximately the same efficiency. With a high bypass ratio, e.g. 10, mixing of the cold and hot flows is technically not realisable.

12.5.3 Intercooling and Recuperation

Figure 12.13 is a sketch of a turbofan engine with intercooling between the IP compressor (booster) and the HP compressor. The cooling air comes from a part

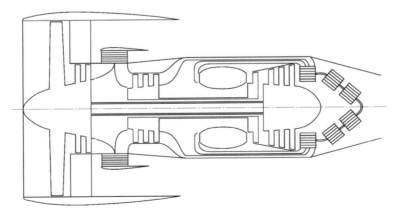

Fig. 12.13 Intercooled recuperated turbofan engine

of the exit flow of the fan. There is recuperation of outlet heat by a series of heat exchangers preheating the compressor air before entry to the combustion chamber. The thermodynamic benefits of the system are similar as with power gas turbines, but it is, of course, crucial that pressure losses in the heat exchangers be low. Such engines are studied nowadays, but not used in practice yet.

12.6 Technological Aspects of the Turbofan Engine

12.6.1 Discs and Shafts

In Fig. 12.9 it is visible that rotor drums of compressor and turbine parts are made from connected discs (welded or bolted together) with a large hole at the centre. Figure 12.14 is a meridional section of the turbine part of the Trent800. The rotor discs are clearly visible. One of the discs in the LP turbine and the single discs of the HP and IP turbines are connected to concentric hollow shafts. Discs in the LP and IP compressors and in the LP turbine are typically forged from a titanium alloy (alloy with aluminium as most important added component), due to the low weight of titanium. Discs of the HP compressor and the HP and IP turbines are made from

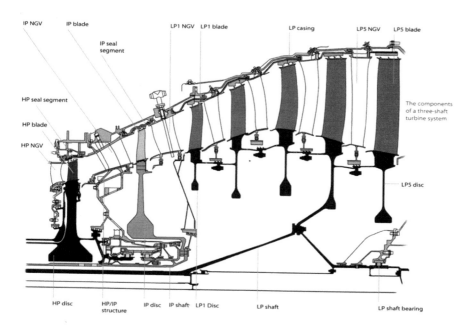

Fig. 12.14 Turbine part of the Tent800 turbofan engine (from the Jet Engine 2005 [5]; permission by Rolls-Royce)

nickel-base alloys, because titanium alloys cannot withstand a temperature higher than about 540 °C. Shafts are mostly made from steel.

12.6.2 Vanes and Blades

Fan blades are typically hollow structures produced from two titanium alloy sheets, shaped by plastic deformation, and a corrugated plate or a honeycomb structure of the same metal in between them. All parts are connected by diffusion bonding. A very modern technology is fan blades in composite materials: graphite fibres and epoxy resin. Vanes and blades in cold parts of the engine (LP and IP compressor, LP turbine) are made from titanium alloys, while nickel-base alloys are used for the hot parts (HP compressor, HP and IP turbines). HP turbine blades are normally cast as a single crystal and IP blades are cast directionally solidified. Vanes and blades in HP and IP turbines have a thermal barrier coating (TBC).

Cooling of vanes and blades is similar as with stationary gas turbines, but inserts are not used in stator vanes. Convection cooling, impingement cooling and film cooling are combined. Figure 12.15 sketches examples of cooling channels of nozzle vanes and blades. In the vane example, cooling air is fed to two internal channels (convection cooling), which both deliver air for film cooling on the pressure surface. The second channel delivers air through holes in an internal wall, forming impinging jets on the inner side of the suction surface. This air leaves through trailing edge slots passing over so-called pedestals, which are staggered cylinders creating a high level of turbulence and thus a high heat transfer. The rotor blade has three cooling channels. In the middle one, air passes three times.

Fig. 12.15 Cooling channels in vanes (left) and blades (right)

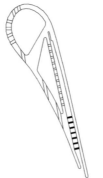

12.6.3 Combustion Chamber

Common modern combustion chambers are of annular type with several injectors. Compared to the can-annular type frequently used earlier, the main advantages are a smaller length and a smaller frontal area, which are vital properties for an aero-engine. The smaller wall area results in lower pressure drop and lower cooling air flow rate. The disadvantage of the annular type is that it is more difficult to manufacture and that it is more difficult to reach perfectly stable combustion resulting in a homogeneous exit temperature.

Figure 12.16 is a sketch of a longitudinal section through the liner of a combustion chamber (the inner part that confines the flame) and a fuel injector. The axial Mach number range in a compressor of an aero-engine is about 0.5 in front stages, but towards the exit this level is reduced to about 0.3 (the axial velocity is about 150 m/s throughout the compressor). The exit velocity is much too high for a stable combustion. The air is slowed down by diffuser blades and by dump diffusion.

A part of the air is supplied centrally to the combustion chamber through the injectors (primary air). This flow is split into two parts. Flow through the centre of the injector is forced into swirl. The outside flow is forced into swirl as well, but in the opposite sense. Fuel is injected as a liquid sheet between both contra-rotating air flows. The high shear on the liquid flow atomises it into small drops. This disintegrating mechanism enables variation of the fuel–air ratio over a large range, maintaining stable combustion. The swirl of the inner air flow causes a low pressure zone such that hot combustion products are drawn from downstream and brought to the injector by a circulating flow pattern. This means that swirl stabilisation is used (see Chap. 11). Part of the air is added around the injector and serves as secondary air (combustion of the CO). This air also cools a part of the liner. Downstream of the combustion zone, more air is added as dilution air.

Figure 12.16 is a schematic of a thin-wall liner made from a nickel-base alloy, covered with a thermal barrier coating (TBC) and cooled by injection of air through

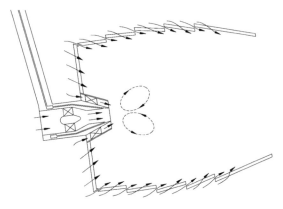

Fig. 12.16 Schematic of a liner of a combustion chamber and injector

arrays of a large number of small holes, either at about 30° in the wall (called effusion cooling) or perpendicular in rings (called diffusion cooling). Heat shielding and cooling are by the films formed at the inner side of the liner, but the liner is also cooled by heat removal within the holes. Such liners are nowadays used for smaller engines and are still in development phase for larger ones. More common is a double-wall liner being a metal shell covered at the inside with ceramic tiles cooled by a combination of convection, impingement and film cooling.

A common principle for keeping emissions of NO_x and CO low is a sequence of rich burning, quick quenching of the flame and lean burning. The mixture in the primary zone is very rich so that the combustion temperature is rather low (the combustion temperature is the highest for stoichiometric combustion). This ensures that not much NO_x is formed, but, of course, production of CO and soot (carbon) is high. With addition of the secondary air, the mixture passes quickly from rich to lean. This process is called quenching. The mixture passes through a stoichiometric zone with high production of NO_x but this zone is kept very short. In the lean zone, CO and soot are burnt.

With this combustion principle, emissions of CO and NO_x are higher than with Dry Low Emission (DLE) combustors of power gas turbines. Staged combustion systems with pilot burners ensuring stable combustion (diffusion flame) combined with main burners optimised for low NO_x production (lean mixture) are developed by all manufacturers nowadays. The combustors are similar to the DLE combustors of power gas turbines: either rows of pilot burners next to rows of lean premixed burners or rows of burners with a pilot core and a lean premixed part around it. Staged combustors are used in the newest aero-engines.

12.6.4   Mixer, Nozzles and Thrust Reverser

Figure 12.17 (left) shows a lobed surface for partial mixing of hot and cold flows. Figure 12.17 (right) demonstrates how, by pushing back the cold-flow nozzle, the

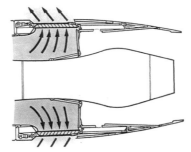

Fig. 12.17 Left: lobed mixer for mixing cold and hot flows; right: reverse thrust nozzle position (from The Jet Engine 1986 [4]; permission by Rolls-Royce)

Fig. 12.18 Serrated nozzle edges (*Courtesy* GE Aviation)

cold flow is deflected to generate reverse thrust for braking immediately after landing. Modern nozzles mostly have a serrated outlet edge for noise reduction (Fig. 12.18).

12.7 Exercises

12.7.1 The military transport aircraft A400M (Airbus), used as an example in Sect. 12.2.5, is equipped with four TP400-D6 eight-bladed propellers with diameter 5.30 m (Europrop = RR + MTU + Safran + ITP). The cruise speed is 780 km/h at cruise altitude 9500 m (the ceiling is 11300 m). The normal shaft power at take-off and the maximum continuous shaft power in cruise flight are both about 8000 kW per propeller (the maximum take-off power is about 8250 kW and the maximum take-off weight is about 140 tonnes). The propeller speed is 864 rpm for a normal take-off and 846 rpm in cruise flight.

Calculate with the formulae of Sect. 12.1.1 the flow acceleration, the thrust per propeller and the Mach number of the relative flow approaching the blade tips at the start of a normal take-off. Take $T = 288.2$ K and $p = 101.33$ kPa at sea level. Assume 0.825 as propeller efficiency, defined as the ratio of dynamic power to shaft power. Repeat the calculations for cruise flight with an assumed propeller efficiency of 0.875 and take $T = 226.5$ K and $p = 28.60$ kPa at 9500 m altitude. Take for air $R = 287$ J/kgK and $\gamma = 1.4$. A remark is that the thrust of the whole turboprop is about 10% larger due to thrust by the gas turbine nozzle. Observe that the ratio of thrust to power is near to 15.5 N/kW at start of take-off, which is the usual value for conversion of residual thrust into equivalent shaft power at static sea level conditions.

A: Take-off: $\Delta v = 99.2$ m/s, $M_{tip} = 0.72$, $T = 133.0$ kN; Cruise: $\Delta v = 14.4$ m/s, $M_{tip} = 1.075$, $T = 31.3$ kN.

12.7.2 Repeat the calculations of the previous exercise for the Q400 turboprop civil aircraft (Bombardier) with two Dowty-R408 six-bladed propellers (GE Aviation) with diameter 4.11 m. The cruise altitude is 7500 m and the maximum cruise speed (fast-travel) is 646 km/h. The normal shaft power at take-off is 3370 kW and the maximum continuous shaft power in cruise flight is 2943 kW per propeller. The propeller speed is 1020 rpm for a normal take-off and 850 rpm in cruise flight. Assume again 0.825 and 0.875 as propeller efficiency for take-off and cruise, and take $T = 239.5$ K and $p = 38.33$ kPa at 7500 m altitude.

A: Take-off: $\Delta v = 88.1$ m/s, $M_{tip} = 0.66$, $T = 63.1$ kN; Cruise: $\Delta v = 10.2$ m/s, $M_{tip} = 0.84$, $T = 14.0$ kN.

12.7.3 The EC145 helicopter (Airbus) has a maximum take-off weight of 3585 kg. The rotor diameter is 11 m. Calculate the rotor shaft power required to take off with maximum weight, ignoring ground effect, in other words, assuming that the helicopter is free in the air, and ignoring the downward force on the fuselage by the rotor flow. Take into account that the incoming velocity of the air is zero with hovering flight. This means that the through-flow velocity of the rotor is the induced velocity. First, calculate the theoretical flow acceleration through the rotor and the theoretically required induced power (=dynamic power). Then, determine the shaft power, assuming that the rotor efficiency is 0.70 (ratio of dynamic power to shaft power). This rotor efficiency is low compared to that of a propeller because the flow through the rotor of a helicopter is very inhomogeneous in hovering flight. Due to ground proximity the required induced power is lower (the flow partially thrusts on the ground). The reduction of the induced power is called ground effect. On the other hand, there is a downward drag on the fuselage by the flow through the rotor. At start of take-off, the required power is about 90% of the theoretically calculated value. The ground effect stops in hovering flight at a flight height of about one rotor diameter.

A: $\Delta v = 24.58$ m/s, $P = 617.5$ kW.

12.7.4 With a single-flow jet engine, a single thrust flow is generated by a thermodynamic cycle of compression, heating and expansion. Calculate the thrust per unit of mass flow rate for $T_0 = 225$ K (10 km height), $M_0 = 0.85$ (flight Mach number), pressure ratio of the compressor: $p_{02}/p_{01} = 40$ and combustion gas temperature at turbine inlet (TIT) $T_{03} = 1500$ K (thus 75 K lower than used in the text). Properties of atmospheric air: $C_p = 1005$ J/kgK, $R = 287$ J/kgK. Combustion gas properties $C_p = 1150$ J/kgK, $R = 287$ J/kgK. Assume $\eta_{\infty c} = \eta_{\infty t} = 0.90$ for compressor and turbine and assume $\eta_{\infty} = 0.97$ for the nozzle. Ignore the pressure drop in the combustion chamber (in practice about 4%) and heat losses to the surroundings (in practice about 1%). Ignore the flow rate increase within the turbine and nozzle by adding fuel (in practice about 2%). Ignore losses at the engine inlet.

Determine the thermal power, which is the power released by the fuel (take T_0 as reference temperature for the heat input). Determine gas power, dynamic power and propulsive power. Calculate the thermodynamic, thermal, dynamic, propulsive and global efficiencies. Compare with the results in Table 12.1 (the obtained thermodynamic efficiency is somewhat lower than in Table 12.1 due to the lower TIT). Observe that the propulsive efficiency it is very low for the aero-engine studied. The reason is the large difference between the jet flow speed and the flight speed so that the kinetic energy dissipated in the atmosphere is large.

A: $\eta_{td} = 0.557$, $\eta_t = 0.544$, $\eta_p = 0\ 407$, $\eta_g = 0.222$.

12.7.5 Transform the engine of the previous exercise into a turbofan engine with separate flows. This is achieved by mounting upstream of the hot nozzle an additional turbine, driving the fan. Take the bypass ratio equal to 5, the pressure ratio of the fan equal to 1.7 and the infinitesimal fan efficiency equal to 0.90.

First, determine the power required to drive the cold-flow part of the fan. Then, determine the pressure ratio of the additional turbine, assuming that the mechanical efficiency of the transmission between the turbine and the fan equals 1. Determine the jet velocities of the cold and the hot flows, their ratio, and the resulting specific thrust. Determine the dynamic, propulsive and overall efficiencies.

Compare with the results of the previous exercise. Compare also with the results in Table 12.2. The global efficiency is much better than in the previous exercise due to the bypass flow, but the efficiency is lower than in Table 12.2 for $b = 5$ because the pressure ratio in the bypass flow is lower than the optimum value.

A: $v_{2c}/v_{2h} = 0.545$, $\eta_d = 0.891$, $\eta_p = 0.661$, $\eta_d\eta_p = 0.589$, $\eta_g = 0.328$.

12.7.6 The ratio of the cold and hot flow jet velocities obtained in the previous exercise is too low for optimum operation of a bypass engine with separate cold and hot flows. It should be around 0.85 (see Sect. 12.5.1). Such a ratio could be obtained for $b = 5$ by enlarging the pressure ratio of the cold flow. Another possibility is to enlarge the bypass ratio. Verify that for $b = 8$ the velocity ratio becomes approximately optimal for pressure ratio of the cold flow equal to 1.70.

Calculate the product of the dynamic and propulsive efficiencies. Remark the large improvement with respect to the value obtained in the previous exercise. Remark also that the value is near to the interpolated value for $b = 8$ in Table 12.2.

A: $\eta_d\eta_p = 0.657$.

12.7.7 Verify that for $b = 6.35$, the pressure ratios of the cold and hot nozzle flows become approximately equal for the turbofan engine of Exercise 12.7.5. Efficient mixing of the flows before entry to a common nozzle is then possible.

Calculate the product of the dynamic and propulsive efficiencies for mixed flows fed to a single nozzle for $b = 6.35$, ignoring mixing loss. Do not use the energy balance (12.21) but calculate the mixed-out state 07 in Fig. 12.12 and the expansion in the common nozzle. Remark that for thermodynamic consistency, the heat capacity and the total enthalpy in state 07 are mass flow averaged values. Take also a mass flow averaged value of the slightly different total pressures. Verify afterwards the approximate validity of the energy balance (12.21). Observe the small improvement with respect to the result of the previous exercise.

A: $\eta_d\eta_p = 0.667$.

12.7.8 The military bypass engine EJ200 (Eurojet $=$ RR $+$ MTU $+$ ITP $+$ Avio) for fighter aircraft is designed for cruise flight at Mach number 1.60 at altitude 12 km (T $= 216.7$ K). The bypass ratio is 0.4, TIT is 1750 K (thus much higher than with civil aircraft), the total compressor pressure ratio is 26 and the fan pressure ratio (LP-part of the compressor) is 4.20. The cold and hot flows are mixed before the entry to a common nozzle.

Calculate the specific thrust and the thermodynamic, dynamic, propulsive and global efficiencies. Estimate the total pressure loss in the engine inlet according to the empirical formula $\Delta p_0/p_0 = -0.075 \, (M_0 - 1)^{1.35}$ for an inlet with deceleration by a system of oblique shocks. Ignore the pressure drop in the combustion chamber, heat losses to the surroundings and the flow rate increase by adding fuel. Take $R =$

287 J/kgK, $\gamma_a = 1.40$, $\gamma_g = 1.33$. Take 0.90 as polytropic efficiency for compressor and turbine parts and 0.97 for the nozzle. Calculate the mixing of the cold and hot flows as in the previous exercise.

   A: $T_s = 831.3$ m/s, $\eta_{td} = 0.621$, $\eta_d = 0.967$, $\eta_p = 0.614$, $\eta_g = 0.368$.

12.7.9 The EJ200 has as rated values of the mass flow rate 75 kg/s and 60 and 90 kN of the dry and wet thrust. With dry and wet thrust is meant the thrust without and with activated afterburner (reheat). With the previous values, the specific thrust without reheat is 800 m/s, which is close to the value obtained in the previous exercise.

   Calculate with the data of the previous exercise the temperature after reheat, necessary for augmenting the thrust to 150%. Take 1.30 as the specific heat ratio for the combustion gas after reheat. Calculate the additional heat input by taking T_0 as reference temperature for the heating, and calculate the resulting global efficiency. Observe the strong decrease of the global efficiency with afterburning activated.

   With military engines (bypass engines with small bypass ratio and mixed flows) afterburning is switched on at take-off. Due to increased thrust, the aircraft can take off faster and can reach flight height faster. Afterburning is also switched on during combat to increase thrust and thus manoeuvrability.

   A: $T_{0,reheat} = 1581.8$ K, $\eta_g = 0.265$.

12.7.10 For supersonic flight at very high Mach number, e.g. flight Mach number around 5, engines that can be switched from turbojet configuration to ramjet configuration have been studied for some time. With these, the gas turbine is switched off at a sufficiently high flight Mach number. Aerodynamically compressed air runs then from the inlet through a combustion chamber to the nozzle.

   Calculate the specific thrust and the overall efficiency for flight Mach number 5 at ambient temperature 225 K and nozzle inlet temperature 1800 K. Take $R = 287$ J/kgK, $\gamma_a = 1.40$, $\gamma_g = 1.33$. Take 0.90 as polytropic efficiency of the aerodynamic compression (losses due to shockwaves during supersonic flight) and take 0.97 as polytropic nozzle efficiency. Ignore, as in the previous exercises, pressure loss in the combustion chamber, heat loss to the surroundings and flow rate increase due to fuel injection. Calculate the heat input with the ambient temperature as reference temperature for the heating. Conclude that the overall efficiency is quite high.

   A: $T_s = 255.5$ m/s, $\eta_g = 0.555$.

References

1. Cumpsty NA (2010) Preparing for the future: reducing gas turbine environmental impact—IGTI scholar lecture. J Turbomach 132:041017
2. Crichton D, Xu L, Hall CA (2007) Preliminary fan design for a silent aircraft. J Turbomach 127:184–191
3. Cumpsty N, Heyes A (2015) Jet propulsion. Cambridge University Press. ISBN 978-1-107-51122-4
4. The Jet Engine (1986) Rolls-Royce. ISBN 0-902121-04-9
5. The Jet Engine (2005) Rolls-Royce. ISBN 0-902121-2-35

Chapter 13
Axial Compressors

Abstract The chapter starts with the analysis of the circumferentially averaged flow on the mean radius of a stage of an axial compressor. Efficiency and loss representations are discussed together with the diffusion factor concept for estimating the loading capacity of blade rows. A further step is analysis of the radial variation of flow parameters, but without taking into account the effect of boundary layer flows on the hub and casing. This flow is called the primary flow. A next aspect is the secondary flow, which is the difference between the complete flow and the primary flow. The different vortex patterns in the secondary flow are described. All considerations together allow a conclusion on the optimal radial distribution of the flow parameters. Some aspects of three-dimensional blade shaping conclude the flow study. Next, blade profile shapes for subsonic, supercritical, transonic and supersonic cascades are studied. The chapter concludes with a discussion on the performance characteristics and the operating limits due to stall and choking.

13.1 Mean Line Analysis

A *mean line analysis* is an analysis of the circumferentially averaged flow on the mean radius. Tangential variations are filtered out by the circumferential averaging and the averaged parameters are steady for constant rotational speed. A further added assumption is absence of radial variations. It is thus assumed that the simplified flow is representative for the entire blade span and thus for the entire orthogonal section of the blade ring. Secondary simplifications of zero radial velocity components and constant axial velocity through the stage are often added. The analysis is one-dimensional in the sense that changes of parameters only occur with the axial coordinate: tangential velocity component, pressure and density. In a real compressor, the axial velocity decreases somewhat from entry to exit. There is also a decrease of the radial height, causing small radial velocity components. Periodic variations occur in the circumferential direction and the real flow is unsteady.

© The Author(s), under exclusive license to Springer Nature Switzerland AG 2022
E. Dick, *Fundamentals of Turbomachines*, Fluid Mechanics and Its Applications 130,
https://doi.org/10.1007/978-3-030-93578-8_13

13.1.1   Velocity Triangles

Figure 13.1 sketches the mean velocity triangles for a repeating stage (drawn with constant axial velocity component). A stage is a rotor followed by a stator.

The x, y and z-directions are the axial, circumferential and radial directions. The coordinate system is right-handed, but the machine is left-turning (real machines can also be right-turning). The symbol u represents the blade speed. Velocities are denoted by v in the absolute frame (angles are α) and by w in the relative frame (angles are β). The inlet velocity usually deviates from the axial direction ($\alpha_1 \approx 15°$) (see Sect. 13.3). So, a compressor normally starts with a stator vane ring, with *inlet guide vanes*. In most literature sources, the analysis is with the axial direction drawn vertically with the through-flow from top to bottom. We follow this custom in later figures.

The flow decelerates in both the rotor and stator parts. The velocity reduction ratio that can be withstood in subsonic flow is about 0.7. With an intended stronger reduction, normally, boundary layer separation occurs. As a consequence, the possible flow turning is limited, typically to about 40° in subsonic flow. The flow turning is further limited by the Mach number level and becomes smaller when this level is higher. The angle changes $\Delta\alpha = \alpha_2 - \alpha_1$ and $\Delta\beta = \beta_2 - \beta_1$ ($\alpha > 0$, $\beta < 0$) may amount to about 40° with subsonic cascades (all velocities lower than sonic) and supercritical cascades (both inlet and outlet subsonic, but with a supersonic zone in the interior). With transonic cascades (inlet supersonic, outlet subsonic and a strong shock) and supersonic cascades (inlet supersonic, outlet supersonic and weak shocks) small losses are only attainable if the flow turning is limited to about 20° (see Sect. 13.4).

Fig. 13.1 Velocity triangles of a repeating axial compressor stage

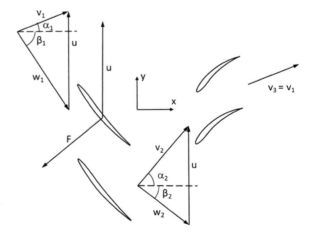

13.1.2 Fundamental Equations

We make the analysis for the rotor (it is similar for the stator). Figure 13.2 sketches a control volume for determination of momentum balances.

Axial momentum

$$\dot{m}(w_{2a} - w_{1a}) \approx A_m(p_1 - p_2) - F_a.$$

F_a is the axial component of the force exerted on the blades and A_m is the area of the average orthogonal section (perpendicular to the shaft). $-F_a$ is the axial force exerted on the flow. The approximation implies a linear representation of the pressure distribution on the envelope of the control volume.

$$w_a = cst \;\rightarrow\; F_a = A_m(p_1 - p_2) = -A_m \Delta p. \tag{13.1}$$

Since $p_2 > p_1$: $F_a < 0$ (see Fig. 13.1).

Tangential momentum

As we reason with $r = $ constant, the equation of moment of momentum around the shaft equals the equation of tangential momentum:

$$\dot{m}(w_{2u} - w_{1u}) = -F_u, \quad F_u = -\dot{m}\,\Delta w_u, \tag{13.2}$$

$$\Delta w_u = w_{2u} - w_{1u} = v_{2u} - v_{1u} > 0.$$

Thus: $F_u < 0$ (see Fig. 13.1).

Rotor work

The work done on the flow by the blades, per mass unit (or the power per mass flow rate unit) is given by Euler's formula:

Fig. 13.2 Control volume
for momentum

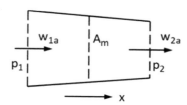

$$\Delta W = \frac{-u \, F_u}{\dot{m}} = u \, \Delta w_u. \tag{13.3}$$

Energy

In the absolute frame, the total enthalpy rise equals the work done (for adiabatic flow):

$$\Delta W = H_2 - H_1, \ \text{with} \ H = e + \frac{p}{\rho} + \frac{1}{2} v^2 + U.$$

The contribution of the gravitational potential energy is negligible, resulting in:

$$\Delta W = h_{02} - h_{01}, \tag{13.4}$$

where $h_0 = h + \frac{1}{2} v^2$ is the *stagnation enthalpy*. Because the gravitational potential energy is negligible, hereafter, we use the term *total enthalpy* to indicate h_0, by which thus is meant the sum of static enthalpy and kinetic energy. From the velocity triangles (Fig. 13.1), it follows that

$$
\begin{aligned}
w_1^2 &= v_1^2 + u^2 - 2u \, v_{1u} \\
w_2^2 &= v_2^2 + u^2 - 2u \, v_{2u} \\
\hline
w_1^2 - w_2^2 &= v_1^2 - v_2^2 + 2u \, \Delta v_u \quad = \quad v_1^2 - v_2^2 + 2 \, \Delta W
\end{aligned}
$$

Thus:

$$\frac{w_1^2}{2} - \frac{w_2^2}{2} = h_2 - h_1, \ or \ h + \frac{1}{2} w^2 = h_{0r} = \text{constant}. \tag{13.5}$$

The *total relative enthalpy* h_{0r} is constant in a rotor. This result stays valid for $v_a \neq$ cst, but $u = $ cst is a requirement. We express the above result by stating that no work is done in the relative frame. The total enthalpy in the absolute frame is constant in the stator part of a compressor stage as no work is done in a stator.

13.1.3   Kinematic Parameters

The degree of reaction is the ratio of the static enthalpy rise to the total enthalpy rise in the rotor. With Eqs. (13.3), (13.4) and (13.5), it follows that

$$R = \frac{h_2 - h_1}{h_{02} - h_{01}} = \frac{\frac{1}{2} w_1^2 - \frac{1}{2} w_2^2}{u(w_{2u} - w_{1u})}$$

Fig. 13.3 Dimensionless
velocity triangles (R = 0.6,
$\psi = 0.4$, $\phi = 0.5$)

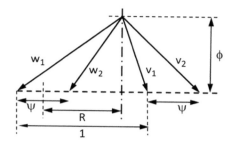

$$= \frac{\frac{1}{2}w_{1u}^2 - \frac{1}{2}w_{2u}^2}{u(w_{2u} - w_{1u})} = -\frac{w_{mu}}{u}.$$

Further:

work coefficient: $\psi = \frac{\Delta W}{u^2} = \frac{w_{2u} - w_{1u}}{u}$, flow coefficient: $\phi = \frac{w_a}{u}$.
Thus:

$$\left.\begin{array}{l} \frac{w_{2u}}{u} - \frac{w_{1u}}{u} = \psi \\ -\frac{w_{2u}}{u} - \frac{w_{1u}}{u} = 2R \end{array}\right\} \quad -\frac{w_{1u}}{u} = R + \frac{\psi}{2}; \quad -\frac{w_{2u}}{u} = R - \frac{\psi}{2}.$$

The parameters R, ϕ and ψ determine the shape of the velocity triangles. Therefore, they are denominated *kinematic parameters*. Figure 13.3 represents the velocity
triangles with velocity components made dimensionless by dividing them by the
blade speed. The figure is made with $R = 0.6$, $\psi = 0.4$ and $\phi = 0.5$.

The triangles for $R = 0.8$, $\psi = 0.3$, $\phi = 0.4$; $R = 0.7$, $\psi = 0.4$, $\phi = 0.5$; $R = 0.5$,
$\psi = 0.7$, $\phi = 0.65$ are sketched in Fig. 13.4. These combinations are representative

Fig. 13.4 Velocity triangles
for R = 0.8, $\psi = 0.3$, $\phi =$
0.4 (top); R = 0.7, $\psi = 0.4$,
$\phi = 0.5$ (middle); R = 0.5, ψ
= 0.7, $\phi = 0.65$ (bottom),
representative for near
casing, mean radius and near
hub of a compressor with
radius ratio 1.20/0.75

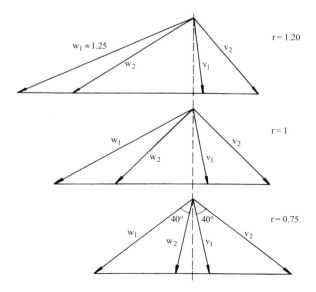

for the flow near the casing, at the mean radius and near the hub of a compressor with radius ratio 1.20/0.75 (see Sect. 13.3). With $R = 0.50$, there is symmetry between the rotor and the stator (inlet and outlet velocities are equal). For $R > 0.5$, rotor velocities are larger than stator velocities and the turning angle in the stator is larger than in the rotor.

13.1.4   Total-To-Total Efficiency and Loss Representation

Figure 13.5 represents the h–s diagram of a compressor stage, drawn with the kinematic parameters of Fig. 13.3. The total enthalpy levels, h_{01} and h_{03}, and the relative total enthalpy level, $h_{0r} = h_{0r1} = h_{0r2}$, are indicated. The slanted dashed lines represent isobars (all drawn with the same slope; which is a simplification) on which points with the same entropy as in the points 1 and 2 are marked.

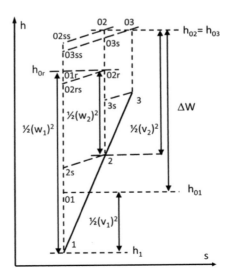

Fig. 13.5   h–s diagram of an axial compressor stage (R = 0.60)

Efficiencies and losses are conventionally defined by comparing the real process with an isentropic process. The isentropic total-to-total efficiency of the stage is defined as the ratio of the isentropic total enthalpy increase that realises the same total pressure increase as the actual compression, $h_{03ss} - h_{01}$, to the actual total enthalpy increase, which is equal to the work transferred, $h_{03} - h_{01}$:

$$\eta_{tt} = \frac{h_{03ss} - h_{01}}{h_{03} - h_{01}}. \tag{13.6}$$

In the stator, the available kinetic energy for increasing the pressure is $\frac{1}{2}v_2^2$. The corresponding enthalpy difference is $h_{02} - h_2$. The velocity magnitude is reduced to the target value (Fig. 13.1), but meanwhile, due to irreversibility, the total pressure decreases (Fig. 13.5). For reaching the same total pressure, p_{03}, the enthalpy difference $h_{03s} - h_2$ would be sufficient in a flow without irreversibility.

The *isentropic deceleration efficiency* and the *enthalpy loss coefficient* of the stator may thus be defined by

$$\eta_{sd,stat} = \frac{h_{03s} - h_2}{h_{02} - h_2} \quad \text{and} \quad \xi_{d,stat} = (h_{02} - h_{03s})/\left(\frac{1}{2}v_2^2\right), \tag{13.7}$$

where $\Delta h_{loss,stat} = h_{02} - h_{03s}$ is the conventional loss. A *pressure loss coefficient* corresponding to the *enthalpy loss coefficient* is often used in practice (see Exercise 13.6.3), defined by $\omega = (p_{02} - p_{03})/(p_{02} - p_2)$. With constant density, both coefficients are equal. For the rotor, with similar definitions, the quantities are:

$$\eta_{sd,rot} = \frac{h_{02rs} - h_1}{h_{01r} - h_1} \quad \text{and} \quad \xi_{d,rot} = (h_{01r} - h_{02rs})/\left(\frac{1}{2}w_1^2\right), \tag{13.8}$$

where $\Delta h_{loss,rot} = h_{01r} - h_{02rs}$ is the conventional loss.

With neglect of the slope changes of the isobars in the h–s diagram (as in the figure) due to changed temperature, the relation between the quantities used for the definition of the total-to-total efficiency is

$$h_{03ss} - h_{01} \approx h_{03} - h_{01} - \Delta h_{loss,rot} - \Delta h_{loss,stat}. \tag{13.9}$$

The isentropic total-to-total efficiency may thus be approximated by

$$\eta_{tt} = \frac{h_{03ss} - h_{01}}{h_{03} - h_{01}} \approx 1 - \frac{\Delta h_{loss,rot} + \Delta h_{loss,stat}}{\Delta W}. \tag{13.10}$$

Obviously, with neglect of the slope changes of the isobars, a loss may also be expressed by the static enthalpy difference corresponding to the used total enthalpy difference, e.g. $h_2 - h_{2s}$ instead of $h_{02r} - h_{02rs}$ in Fig. 13.5. This replacement is very often done in analytic derivations. A loss is even sometimes directly defined with a static enthalpy difference.

13.1.5 Average Density

The conventional loss may be approximated using a pressure difference and an average density. We derive the relation for the rotor. The derivation for the stator is identical. Fundamentally, loss follows from

- Work equation (in the rotor): $dW = 0 = d\frac{1}{2}w^2 + \frac{1}{\rho}dp + dq_{irr}$,
- Energy equation (in the rotor): $dW = 0 = d\frac{1}{2}w^2 + dh$.

The difference of these equations is

$$0 = dh - \frac{1}{\rho}dp - dq_{irr} \text{ or } dq_{irr} = dh - \frac{1}{\rho}dp.$$

For $\rho = $ constant:

$$dh_s = \Delta h_s = \frac{1}{\rho}\Delta p \text{ and } q_{irr} = \Delta h - \frac{1}{\rho}\Delta p = \Delta h - \Delta h_s.$$

For constant density, the slope of isobars is constant. The conventional loss is thus the loss the flow would be subjected to, if density were constant.

With a compressible fluid, a static isentropic enthalpy difference may be written, using a *process average density*, as

$$\Delta h_s = \frac{\Delta p}{\rho_m} \text{ or } \rho_m = \frac{\Delta p}{\Delta h_s}.$$

This average density differs only slightly from the arithmetic average density. With polytropic representation of the compression path (exponent n), it follows that:

$$\frac{\rho_2}{\rho_1} = \left(\frac{p_2}{p_1}\right)^{1/n}, \ \rho_a = \frac{1}{2}(\rho_1 + \rho_2) = \frac{1}{2}\rho_1(1 + r^{1/n}), \text{ with } r = p_2/p_1.$$

From $h_{2s} - h_1 = h_1\left(\frac{h_{2s}}{h_1} - 1\right) = h_1\left(r^{(\gamma-1)/\gamma} - 1\right)$, it follows that $\rho_m = \frac{p_1(r-1)}{c_p T_1(r^{(\gamma-1)/\gamma} - 1)} = \rho_1 \frac{\gamma-1}{\gamma} \frac{r-1}{r^{(\gamma-1)/\gamma} - 1}$.

Further $\eta_\infty = \frac{n/(n-1)}{\gamma/(\gamma-1)}$.

Example: for $r = 2$ (so very large),

$$\eta_\infty = 0.9, \ \gamma = 1.4 \rightarrow n = 1.465,$$
$$\rho_a = \rho_1 \, 1.302, \ \rho_m = \rho_1 \, 1.305.$$

We may thus consider ρ_m as the arithmetic average density. With a good approximation, the conventional loss, using static enthalpy differences, may be written as

$$\Delta h_{loss} \approx \Delta h - \Delta h_s \approx \Delta h - \frac{\Delta p}{\rho_m}. \tag{13.11}$$

It is remarkable that the term $\Delta p/\rho_m$ constitutes a good approximation of the isentropic enthalpy rise. It may be verified that this applies for a wide range of pressure ratio values and infinitesimal efficiency values (see Exercise 13.6.1). The term is sometimes a slight underestimation and sometimes a slight overestimation.

With the average density, the analysis of the blade force of a compressor cascade may be performed analogously to the analysis of a constant-density fluid cascade, as done in Chap. 2 (Sect. 2.2).

13.1.6 Force Components

With the approximate loss expression (13.11), the following relations apply:

$$\Delta p = \rho_m(\Delta h - \Delta h_{loss}),$$
$$F_a = -A_m \Delta p = -\rho_m A_m (h_2 - h_1 - \Delta h_{loss}),$$
$$F_a = -\rho_m A_m (1/2\, w_1^2 - 1/2\, w_2^2) + \rho_m A_m \Delta h_{loss} \quad (h_{0r} = cst)$$
$$= \rho_m A_m (1/2\, w_{2u}^2 - 1/2\, w_{1u}^2) + \rho_m A_m \Delta h_{loss} \quad (w_a = cst)$$
$$= \rho_m A_m w_{mu} \Delta w_u + \rho_m A_m \Delta h_{loss},$$
$$F_u = -\dot{m}\, \Delta w_u = -\rho_m A_m w_a \Delta w_u,$$

with $w_{mu} = \frac{1}{2}(w_{2u} + w_{1u})$ the tangential component of the average velocity.

Without loss, it follows that

$$w_a F_a + w_{mu} F_u = 0.$$

This implies that the resulting force is perpendicular to the average velocity. In presence of losses we may define lift as the component of the force perpendicular to the average velocity and drag as the component in the direction of the average velocity. This result is identical to the one obtained in Chap. 2 with the analysis of constant-density fluid flow through blade cascades. For a compressible fluid, the lift coefficient and the drag coefficient may be defined in the same way. The only difference is the need to use the average density.

By denoting by w_m the magnitude of the average relative velocity, the drag is

$$D = (w_a F_a + w_{mu} F_u)/w_m = \rho_m A_m \Delta h_{loss} \frac{w_a}{w_m} \quad \text{or } D\, w_m = \dot{m}\, \Delta h_{loss}.$$

A drag coefficient related to inlet kinetic energy thus is:

$$C_{D1} = \frac{D}{\frac{1}{2}\rho_m w_1^2 A_b},$$

where $A_b = c.\ell$ is an approximation for the blade surface, with c the chord on the average radius and ℓ the average blade span. Further $A_m = s.\ell$, with s being the spacing (or pitch). Solidity is defined by $\sigma = c/s$.

It follows that

$$C_{D1} = \frac{\Delta h_{loss}}{\frac{1}{2} w_1^2} \frac{s}{c} \frac{w_a}{w_m} = \frac{\xi_1}{\sigma} \cos \beta_m \text{ or } \sigma\, C_{D1} = \xi_1 \cos \beta_m. \tag{13.12}$$

A lift coefficient may be defined analogously. This is formally possible, but not useful. The solidity is mostly very high ($\sigma \approx 1.5$ to 2.0) so that it is not possible to derive the lift coefficient of a cascade from that of an isolated aerofoil.

Moreover, the lift coefficients for optimum operation and for stall do not have approximate universal values with compressor cascades. A lift coefficient is thus not meaningful for expressing the loading of a cascade and for selecting a cascade that realises a desired flow. In practice, semi-empirical correlations are applied (classic correlations are by Howell and by Lieblein) for determining, with given velocity triangles ($w_1, w_2, \beta_1, \beta_2$), the cascade that realises the desired flow turning (e.g., see Lewis [11]). The notion of diffusion factor is used to determine the solidity required to realise a target flow turning, as discussed hereafter.

13.1.7   Diffusion Factor and Loss Correlations

The enthalpy loss coefficient or the pressure loss coefficient can be correlated to a factor called *diffusion factor DF*, expressing the cascade loading. The diffusion factor replaces the lift coefficient and the relation $\xi_1 = f\,(DF)$ or $\omega_1 = f\,(DF)$ is equivalent to the polar $C_D = f\,(C_L)$ of an aerofoil.

There are variants of the diffusion factor definition. In Chap. 3 we used the local diffusion factor D_{loc}. An expression better related to external parameters is the diffusion factor DF (name without any further specification).

$$D_{loc} = \frac{w_{\max} - w_2}{w_{\max}}, \ DF = \frac{w_{\max} - w_2}{w_1}. \tag{13.13}$$

In these expressions, w_{max} is the maximum velocity on the suction side of the blade. Figure 13.6 sketches the velocity distribution on a compressor blade (velocity at the boundary layer edge). The flow on the pressure side is represented as uniformly accelerating, but in reality a zone with slightly decelerating flow may occur (see Sect. 13.4). The strongly decelerating part in the flow on the suction side corresponds to an adverse pressure gradient. With a too large adverse pressure gradient, the boundary layer separates.

The diffusion factors are measures for the pressure gradient in the decelerating part of the boundary layer on the suction side. They determine the maximum velocity on the suction side and so the minimum pressure and thus the obtainable tangential force

Fig. 13.6 Velocity distribution on a compressor blade (rotor)

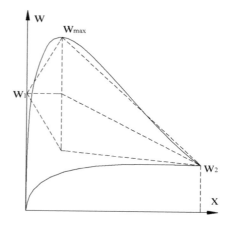

on a compressor blade. We reason upon this relation intuitively here. A justification is given by Lieblein et al. [12, 13], based on boundary layer theory.

The diffusion factor can be related to the velocity triangles. This may be understood by expressing the circulation around a blade in two ways as sketched in Fig. 13.7. A first way is along a contour composed by periodic lines in between two neighbouring blades and tangential parts upstream and downstream of the cascade. This contour cuts the wake at large distance from the trailing edge. The second way is along a contour following the edges of the boundary layers and cutting through the wake near to the trailing edge. With this second way, the circulation is equal to the surface integral of the velocity plot of Fig. 13.6, but with a curved abscissa following the profile surface. The two expressions of the circulation are equal for lossless flow.

The two expressions of the circulation result in

Fig. 13.7 Calculation of the circulation around a profile in two ways

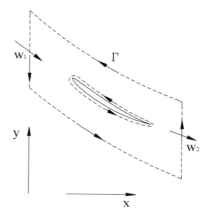

$$\Gamma = \int_C \overrightarrow{w} \cdot \overrightarrow{d\ell} = s(w_{2u} - w_{1u}) = f_1 f_2 \, 2(w_{max} - w_1)\frac{c}{2}.$$

The factor f_1 takes into account that the integration path length is larger than the chord in Fig. 13.7 and the factor f_2 that the integrated area is larger than the sum of the triangles drawn in Fig. 13.6. The relation thus implies

$$w_{max} - w_1 = \frac{1}{f_1 f_2}\frac{\Delta w_u}{\sigma}.$$

The quantity $\Delta w_u / \sigma$ was called the circulation parameter by Lieblein et al. [12, 13] and they proved that the acceleration from the inlet velocity w_1 to the maximum velocity w_{max} can be very well correlated, for a large set of compressor cascade data for attached flow, by

$$w_{max} - w_1 = 0.5\frac{\Delta w_u}{\sigma} + 0.1\, w_1.$$

This allows, by dropping the constant factor 0.1, writing the diffusion factor as

$$DF = 1 - \frac{w_2}{w_1} + \frac{\Delta w_u}{2\sigma w_1}. \qquad (13.14)$$

The diffusion factor can be seen as the sum of two terms, one indicating the deceleration in the suction side boundary layer due to the general deceleration in the cascade and one as the supplementary deceleration due to the blade loading, i.e. the pressure difference between pressure and suction sides. Expressed as a function of the angles (see Fig. 13.1), the diffusion factor is

$$DF = 1 - \frac{\cos\beta_1}{\cos\beta_2} + \frac{\cos\beta_1}{2\sigma}(\tan\beta_2 - \tan\beta_1). \qquad (13.15)$$

Fig. 13.8 Boundary layer zones and wake

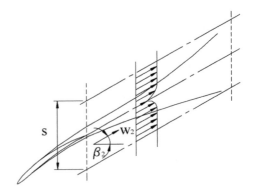

Figure 13.8 sketches the boundary layers and the wake of a blade, together with a velocity profile across the wake. The momentum loss flow rate (defect of momentum flux due to presence of boundary layers) at the trailing edge of a blade may be expressed by

$$\rho_2 w_2 w_2 (\delta_{2s} + \delta_{2p}).$$

The thickness δ_2 is called the momentum loss thickness and subscripts s and p indicate the suction and pressure sides, with w_2 being the average velocity in the core flow immediately downstream of the cascade. With core flow is meant the zone outside the boundary layers. There is no loss in the core flow between entry and exit of the cascade, because the loss by friction on the blades stays confined to the boundary layers. Downstream of the cascade, the wakes and the core flow mix.

The total loss after complete mixing can be derived from a momentum balance between a section just downstream of the cascade and a section far downstream with an assumed uniform flow.

The momentum balance, with the subscript m indicating the mixed-out flow, is

$$\rho_m A w_m w_m - (\rho_2 A w_2 w_2 - \rho_2 w_2 w_2 \delta_2) = (p_2 - p_m) A,$$

with $A = s \cos \beta_2$ and $\delta_2 = \delta_{2s} + \delta_{2p}$ the total momentum loss thickness. This thickness is small with respect to the blade spacing. So, we may ignore the difference between w_2 and w_m (w_2 is somewhat larger than w_m) and ignore the difference between ρ_2 and ρ_m (ρ_2 is somewhat smaller than ρ_m). A consistent approximation of the total pressure loss is then (the core flow total pressure is constant between entry and exit):

$$\Delta p_0 s \cos \beta_2 = \rho_2 w_2^2 \delta_2,$$

so that

$$\omega_2 = \frac{\Delta p_0}{\frac{1}{2} \rho_2 w_2^2} = \frac{2}{\cos \beta_2} \frac{\delta_2}{s} = \frac{2\sigma}{\cos \beta_2} \frac{\delta_2}{c}. \tag{13.16}$$

The boundary layer momentum thickness at the trailing edge, relative to the chord, δ_2/c, is strongly linked to the cascade loading, in other words, the diffusion factor. So, on the basis of the relation (13.16) we expect a strong correlation between

$$\omega_1' = \frac{\Delta p_0}{\frac{1}{2} \rho_1 w_1^2} \frac{w_1^2 \cos \beta_2}{w_2^2 \ 2\sigma} = \omega_1 \left(\frac{\cos \beta_2}{\cos \beta_1} \right)^2 \frac{\cos \beta_2}{2\sigma} \tag{13.17}$$

and the diffusion factor. The strong correlation, shown in Fig. 13.9, was proven by Lieblein [13] for flows at low Mach number ($\rho_1 \approx \rho_2$).

Fig. 13.9 Transformed
pressure loss coefficient ω_1'
as a function of diffusion
factor with compressor
cascades, according to
Lieblein [13]

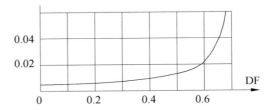

The line in Fig. 13.9 is an average through many experimental data, with a small spreading around the line. The loss coefficient rises sharply above $DF = 0.6$. This value may be considered as the value above which stall occurs (boundary layer separation on the suction side). $DF \approx 0.45$ is mostly accepted as the optimum value. Henceforth we apply this value to estimate the optimum solidity. However, compressor cascades may have a larger design diffusion factor, up to $DF = 0.50$ in stator cascades and $DF = 0.60$ in rotor cascades. The reason is that a deceleration is also realised at a higher inflow Mach number by weak shock structures, without the need for large flow turning (see Sect. 13.4). To conclude, we remark that the loss coefficient has a rather good correlation with D_{loc} as well, with a value for strong loss increase around 0.50 [13].

13.1.8   Secondary Flow: Principle

The flow in a turbomachine is generally subdivided into two flows. The purpose is to discern between losses on the blades and losses on the hub and the casing. The flow that would be obtained in absence of viscous effects on the hub and the casing and in absence of clearances between the rotor blades and the casing and between the stator vanes and the hub is called the *primary flow*. So, it is the flow through the blade passages, confined without clearances by the *end walls* (hub and casing), assuming that slip occurs at these surfaces. The difference between the full flow and the primary flow is called the *secondary flow*. The secondary flow thus describes the effects of the end wall boundary layers and the clearances.

The primary flow is sometimes defined more restrictively as the flow on the average radius. The radial variation of the primary flow is then ignored. This simplification is not necessary for a good understanding, as will become clear from the discussion in Sect. 13.1.10 about flow optimisation. Some basics of secondary flow and of radial flow variation are required to understand this optimisation. A detailed discussion of these subjects follows later. Here, we briefly discuss secondary flow. The next section is a brief discussion of radial variation.

The end wall boundary layers interact with the core flow or primary flow. This generates a series of flow patterns, all of them with a vortex character. We therefore refer to these as *secondary vortices*. The dominant vortex is the passage vortex, caused by the turning of the primary flow. A centrifugal force that is balanced by the pressure

Fig. 13.10 Passage vortex
and trailing vortices in wake
zones

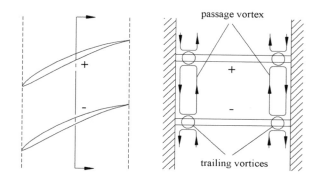

difference between the pressure and the suction sides of the blades corresponds with this turning. But the velocity is lower in the end wall boundary layers than in the core flow. For the same turning, the centrifugal force and the pressure difference are weaker there. The consequence is generation of two vortex motions, as sketched in Fig. 13.10, similar to the vortices in a bend of a pipe (Chap. 2, Sect. 2.3). These are together denominated the *passage vortex*.

The figure also shows that passage vortices from neighbouring channels may trigger vortex motions in the wakes of the blades. These vortices are called *trailing vortices*.

A passage vortex changes the direction of the flow near an end wall. With Fig. 13.10, one sees that the turning increases near an end wall. The phenomenon is called overturning and the enlarged turning brings the centrifugal force due to flow turning near the wall in equilibrium with the centrifugal force more to the middle of the blade passage, where the turning is smaller. The creation of the passage vortex may thus also be explained by the necessity of this radial equilibrium.

Another consequence is a supplementary loss, caused by the kinetic energy associated to the vortices. This energy is taken from the primary flow and almost completely dissipated downstream of the blade row. The sum of the losses due to the secondary flow (sum over several vortices) may be very important. For smaller aspect ratio (say around two), the secondary flow loss may be comparable to the loss caused by the primary boundary layers (the boundary layers on the blades) for optimal incidence (minimum primary loss). The loss in the end wall boundary layers itself is typically about half as large as the primary loss with optimum incidence. In other words, end wall effects, i.e. viscous effects, clearances and secondary vortices, may increase the sum of losses with a factor of about 2.5 compared to the losses of the primary flow.

13.1.9 Radial Variation of Flow: Principle

At the entry and exit of a compressor stage, the flow in the absolute frame is close to the axial direction (small tangential velocity component; see Fig. 13.1). So, there is only little centrifugal force in the direction from hub to casing within the inlet and the outlet flows. As a consequence, the flow is almost homogenous. The flow has a significant tangential velocity component in the space between the rotor and the stator. The associated centrifugal force causes a higher pressure at the casing and a lower one at the hub.

A stage is normally designed with the same work on all radii. Because of the larger pressure rise through the rotor at the casing and the lower pressure rise at the hub, the degree of reaction increases from the hub to the casing. For example, with $R = 0.7$ on the mean radius, $R \approx 0.8$ at the casing and $R \approx 0.5$ at the hub may be expected at ratios for outer and inner radii $r_{or}/r_{mr} = 1.2$ and $r_{ir}/r_{mr} = 0.75$ (a detailed analysis follows in Sect. 13.3). With $\psi \sim 1/u^2$, the velocity triangles are as sketched in Fig. 13.4. We note that the rotor blades have a strong twist and that the flow turnings in rotor and stator are the largest at the hub.

13.1.10 Optimisation of a Stage

Not only primary losses, but also losses within the end wall boundary layers and secondary losses intervene. Primary losses are not even dominant. With $R = 0.5$, the velocity levels within rotor and stator are the same and flow turnings are the same. So, primary losses in rotor and stator are equal. Thus, with regard to primary losses, $R = 0.5$ on the average radius is optimal: minimum sum of losses when both terms are equal, because losses are proportional to velocity squared.

With increased R, the absolute flow is nearer to the axial direction, especially at the casing. Figure 13.11 (left) sketches the entry part of the low-pressure compressor of a bypass aero-engine. The figure is also representative for the entry part of a compressor of a power gas turbine. The rotor blades have a clearance at the casing. The boundary layer flow on the casing occurs within the absolute frame, but, of course, the boundary layer gets distorted by the passing of rotor tips. Boundary layer flow nearer to the axial direction aligns the momentum better with the adverse

SHROUDED VANES

Fig. 13.11 Compressor hub and casing (left: from [2], permission by SAGE Publications; right: from [15], permission by Rolls-Royce)

pressure gradient. This reduces the growth of the end wall boundary layer on the casing, causing a decrease of losses in this boundary layer and a decrease of the intensity of the secondary flow. A high degree of reaction at the casing is therefore favourable for reduction of secondary losses and boundary layer losses at the casing.

Vanes are mostly connected to segments of circular bands, as illustrated in Fig. 13.11 (right). The bands are stationary and the one on the hub side is sealed with labyrinth seals. The end wall boundary layer at the hub then alternately flows in the absolute frame and the relative frame. Therefore, with regard to minimisation of the sum of the losses, 50% degree of reaction at the hub is optimum (symmetry between rotor and stator).

Because of both effects, a degree of reaction on the average radius above $R = 0.5$ is better. Typical values for optimum flow are (see Sect. 13.3):

average radius	$R = 0.7$, $\psi = 0.4$, $\phi = 0.5$
casing	$R = 0.8$, $\psi = 0.3$, $\phi = 0.4$
hub	$R = 0.5$, $\psi = 0.7$, $\phi = 0.65$.

The corresponding dimensionless velocity triangles for constant axial velocity (see the later Table 13.2) are sketched in Fig. 13.4. Due to the limitation of the deceleration ratios w_2/w_1 and v_1/v_2 to about 0.7 at the hub, the flow turning on the mean radius is limited. The consequence is that the value of ψ on the average radius cannot be higher than about 0.4, when combined with $\phi = 0.5$.

The value of ϕ should, in principle, be as large as possible in order to allow the largest possible flow rate. But $\dot{V} \sim \phi u$ and $\Delta W = \psi u^2$. Thus, with regard to stage work it is advantageous to choose the blade speed u as high as possible and to moderate somewhat ϕ, taking into account that the Mach number of w_1 should be limited at the casing. In the entry part of a compressor, the mean radius blade speed is normally chosen near to the speed of sound, so that the Mach number corresponding to w_1 at the casing is about 1.25 with the radius ratio of Fig. 13.4. This can hardly be increased without large losses by shock waves. These considerations explain the values of degree of reaction, work coefficient and flow coefficient, given above.

The cited values are not universal however. The optimum depends on the boundary layer status on the end walls, determined by losses in previous stages, and the relative importance of secondary losses and end-wall boundary losses, determined by the aspect ratio of the blades (height to chord ratio) and the radius ratio of the annulus (casing radius to hub radius). The cited values are typical for stages with significant secondary losses (aspect ratio around 2 or lower), not an entry stage of the entire compressor, but a high ratio of casing radius to hub radius. Examples are the second and third stages of the low-pressure compressor of a bypass aero-engine, as shown in Fig. 13.11, or such stages of the compressor of a power gas turbine. Other examples are the front stages of the HP-compressor of a two-shaft turbofan engine or the IP-compressor of a three-shaft turbofan engine.

The degree of reaction $R = 0.7$ on the mean radius is quite universal, but the entry stage of a compressor typically has a lower degree of reaction, 50% to 60%, because this lowers the tip Mach number at the rotor entry, and an exit stage may have a higher degree of reaction because this brings the outlet direction nearer to the axial direction. Rear stages have normally lower values of flow and work coefficients and are thus less loaded ($\phi = 0.4$ and $\psi = 0.3$ to 0.35 for a last stage), because of thicker end-wall boundary layers due to losses in previous stages and larger influence of end-wall boundary layers due small blade height relative to the mean radius. A front stage is usually more loaded ($\phi = 0.6$ and $\psi = 0.45$).

A particular case is the fan stage of a turbofan engine. It is the first stage of the engine and the aspect ratio is extremely high. Secondary losses do not have a big role. So, aiming at a degree of reaction of 50% at the hub has no meaning. Further, the variation of the degree of reaction from hub to casing is extreme, which means that the degree of reaction at the hub must be small. The Mach number limitation at the rotor entrance at the tip of the rotor blade is crucial. This means that the velocity triangles at the tip are still close to the triangles shown on the top of Fig. 13.4, but with a somewhat larger degree of reaction such that the entrance velocity v_1 is in the axial direction (also a somewhat larger Mach number of the relative flow at rotor inlet). The hub has a low degree of reaction. So, rotor flow turning becomes large near the hub and in order to limit it, work must be lowered near the hub (see Exercise 13.6.5). Thus, work cannot be constant along the radius.

13.1.11 Blade Shape

With the velocity triangles at the mean radius of Fig. 13.4 (see also the later Table 13.2), the diffusion factors for rotor and stator are:

$$DF_r = 1 - \frac{w_2}{w_1} + \frac{\Delta w_u}{2\sigma_r w_1} = 0.313 + \frac{0.194}{\sigma_r},$$

$$DF_s = 1 - \frac{v_1}{v_2} + \frac{\Delta v_u}{2\sigma_s v_2} = 0.279 + \frac{0.283}{\sigma_s}.$$

For $DF = 0.45$ it follows that $\sigma_r = 1.42$ and $\sigma_s = 1.65$. These are rather high solidities. Note also that $w_2/w_1 = 0.69$ and $v_1/v_2 = 0.72$. So, rather strong velocity reductions are required. About 50% of the kinetic energy at the inlet has to be converted into static enthalpy increase. This is about the possible maximum. We note that the velocity reductions at the hub in Fig. 13.4 are 0.62 and 0.64. Such strong decelerations are actually unrealisable. A solution is discussed in Sect. 13.3.6.

Figure 13.12 shows a rotor example and the computed flow. Grid lines of the computational grid are shown on the left and Mach contour lines on the right. The solidity is about 1.33 on the mean radius, 1.70 at the hub and 1.20 at the casing.

At the hub, the flow is supercritical, which means subsonic inlet and outlet flows, but with a supersonic zone at the suction side ended by a normal shock. The shock is weak with a Mach number just upstream of the shock about 1.1. On the mean radius, the flow is transonic, which means supersonic inlet flow (the inlet Mach number is just above 1) and subsonic outlet flow with velocity reduction through a normal

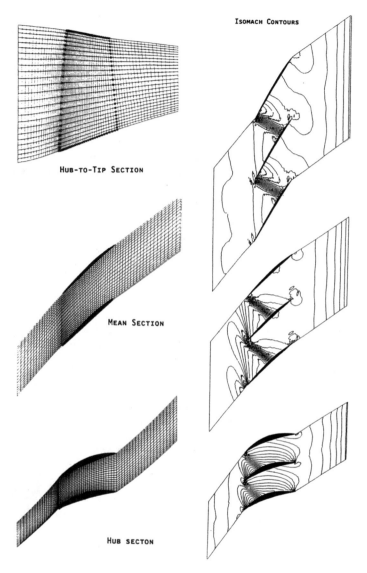

Fig. 13.12 Mach number distribution in the rotor of a transonic compressor stage (from [10], permission by Springer)

shock. The flow is transonic as well at the casing with inlet Mach number about 1.2 and outlet Mach number about 0.9.

A compressor stage is called transonic when the relative flow entering the rotor is supersonic in the tip region and subsonic in the hub region. With a subsonic or a supersonic stage, the relative flow at rotor inlet is subsonic or supersonic over the whole span.

13.1.12   Attainable Pressure Ratio

As a first example, we take the cold-flow part of a fan of a turbofan engine. The tip speed is about 450 m/s at take-off. The radius ratio of the cold-flow part is about 0.5. So, at the equal area mean radius of the cold flow (radius ratio about 0.79), the blade speed is $u \approx 355$ m/s. For $\psi = 0.45$ (which is a rather high value), the stage work is $\Delta W \approx 57$ kJ/kg. The temperature rise is then about 56.5 K ($R = 287$ J/kgK, $\gamma = 1.40$). From this follows, for $T_{01} = 288$ K (sea level):

$$\frac{p_{03}}{p_{01}} = \left(\frac{T_{03}}{T_{01}}\right)^{\frac{\gamma}{\gamma-1}\eta_\infty} \approx \left(1 + \frac{56.5}{288}\right)^{3.15} \approx 1.75.$$

This level of pressure ratio is attained in the outer part of the fan (see Exercises 13.6.5 and 13.6.6). With $u = 355$ m/s, the flow at the rotor inlet is supersonic ($w_1 > u$: $w_1 \approx 400$ m/s; $T_1 < T_{01}$: $T_1 \approx 268$ K, $c_1 \approx 328$ m/s: Mach number is about 1.20). At the tip, the rotor inlet Mach number is much higher, about 1.5.

The pressure ratio and Mach numbers are somewhat smaller in cruise flight. The tip speed is lower, about 400 m/s. The atmospheric temperature is also lower, about 225 K at 10 km height. For a flight Mach number of 0.85, the total temperature at fan entry is about 257.5 K. The tip speed corresponding to perfect similitude would be about 425 m/s (proportional to the square root of the total inlet temperature).

As a second example, we take a stage in the middle of the compressor of a three-shaft turbofan engine. For start of take-off at sea level, $T_{01} = 288$ K and for pressure ratio 40, the total exit temperature of the compressor is about 890 K ($\gamma = 1.38$). The temperature at the exit of the fan may be estimated at about 300 K, with corresponding speed of sound about 350 m/s ($\gamma = 1.40$). This value is the allowable blade speed at the entry of the IP part of the compressor, if we accept that the Mach number of the relative entry flow is around unity. The blade speed that can be allowed in the HP part of the compressor is also about 350 m/s due to strength limitations associated to the high temperatures. So, a blade speed of 350 m/s is typical for both IP and HP parts. With $\psi = 0.4$, the stage work is 49 kJ/kg. The corresponding temperature rise is about 47 K ($\gamma = 1.38$). In the middle of the compressor, the stage entry total temperature is approximately 570 K. The pressure ratio is then

$$r = \left(1 + \frac{47}{570}\right)^{\frac{\gamma}{\gamma-1}\eta_\infty} \approx 1.30.$$

In present-day airliner aero-engines, the pressure ratio per stage is about this value, e.g. Tent800 (Rolls-Royce): $r = 40$ in 15 stages, thus 1.28 per stage on average (see Exercise 13.6.7 for a more detailed analysis). The average pressure ratio for compressors of power gas turbines is similar, e.g. the SGT6-5000F (Chap. 11): $r = 18.9$ in 13 stages, thus about 1.25 per stage. A higher pressure ratio is used in military engines (bypass), e.g. EJ200 (Eurojet): $r = 26$ in 8 stages, thus 1.50 per stage. The main consequence is higher Mach number levels.

The conclusion is that the attainable pressure ratio per stage may be strongly different in different applications and that it is largely determined by the allowable Mach number at the rotor inlet. The allowable Mach number depends strongly on the necessary operating range of the compressor (see Sects. 13.4 and 13.5).

13.2 Secondary Flow

13.2.1 Definition of Secondary Flow

Secondary flow was defined in Sect. 13.1.8. The secondary flow is the difference between the full flow and the primary flow. The primary flow is the flow that would be obtained in absence of boundary layers and clearances on the end walls. The interaction between end wall boundary layer flows and clearance flows with the primary flow generates a series of vortices called *secondary vortices*.

13.2.2 Passage Vortex and Trailing Vortices

The passage vortex was already discussed in Sect. 13.1.8. Figure 13.10 is a sketch of the passage vortex within a linear cascade with added end walls. In a real compressor, the passage vortex is distorted compared to the sketch, because the blade spacing increases with the radius. A significant effect of the passage vortex is the change of the outlet direction of the flow. The flow turning by the blades is increased in the boundary layers at the hub and the casing and reduced some distance away from the end walls. The trailing vortices by interaction between the passage vortices, sketched in Fig. 13.10, are very weak. But, there are similar vortices, called *trailing shed vortices* (TSV), shed from a trailing edge due to the lower pressure difference across a blade near an end wall (see Fig. 13.14, right).

13.2.3   Corner Vortices

In principle, corner vortices may form, driven by the passage vortex, in all corners of a blade passage. These corner vortices interact however with other vortices, as discussed below.

13.2.4   Horseshoe Vortex

Figure 13.13 demonstrates the higher stagnation pressure at a blade leading edge in the core flow than in the end-wall boundary layer (the static pressure is constant on a line perpendicular to parallel streamlines). The pressure difference generates a vortex that wraps around the leading edge, with forming of a pressure side leg and a suction side leg. Because of its shape, this vortex is denominated *horseshoe vortex*.

Figure 13.14 (left) demonstrates how the hub passage vortex (HPV) drives the horseshoe vortex leg at the pressure side (HSP) towards the suction side of the neighbouring blade (rotor blade view from downstream; see Fig. 13.1). The rotation sense of this horseshoe vortex leg is the same as that of the passage vortex. Both vortices develop together and usually merge before the cascade exit. The horseshoe

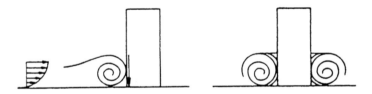

Fig. 13.13   Horseshoe vortex generation; left: side view; right: back view

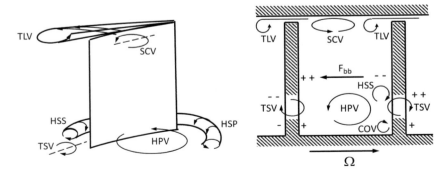

Fig. 13.14   Secondary vortices in a rotor blade passage with clearance at the casing: tip leakage vortex (TLV), scraping vortex (SCV), trailing shed vortex (TSV), horseshoe vortex (HSS and HSP), hub passage vortex (HPV), corner vortex (COV); blade-to-blade force F_{bb}

vortex leg at the suction side of the blade (HSS) is driven by the passage vortex towards the mid-span of the blade. The rotation sense of this horseshoe vortex leg is the same as that of the trailing shed vortex (TSV). The two vortices usually merge downstream of the cascade exit (the merging is not represented in Fig. 13.14).

A corner vortex (COV) may form under the passage vortex in the corner of the suction side of the blade and the end wall. In particular, this may happen with separated flow in this corner. Typically, no corner vortex forms in the corner at the pressure side, as no separation occurs.

Horseshoe vortex legs and corner vortices are similar with stator vanes near the casing and also similar near the hub in absence of clearances (typically: end bands; see Fig. 13.11). With clearances, a leakage vortex and a scraping vortex are generated, as discussed next.

13.2.5 Leakage Vortex and Scraping Vortex

With rotor blades, there is a clearance at the casing. The leakage flow from the pressure side to the suction side forms a *leakage vortex* (TLV) and the motion of the blade with respect to the casing causes a *scraping vortex* (SCV) (Fig. 13.14). The leakage vortex and the scraping vortex have the same sense of rotation and merge downstream of the blade passage.

Somewhat downstream of a rotor blade passage, three vortex structures are typically visible near the hub (Fig. 13.14, right): a hub passage vortex (which includes the horseshoe pressure leg), a trailing shed vortex (which includes the horseshoe suction leg; merging is not yet reached in Fig. 13.14, right), often called the *counter vortex*, and a corner vortex (in the corner of the suction side of the blade and the hub). At the casing, there is typically only a leakage vortex, as this is the dominant phenomenon (merged leakage and scraping vortices). A remark is that the real flow is unsteady by the blade rotation and strong turbulence. But the vortices are typically well visible in the time-averaged flow.

13.2.6 Loss Assessment

The secondary flow generates a loss that may be expressed by an additional drag coefficient. According to Howell's measurements on slender rotor blades (high aspect ratio), it is:

$$C_{D,\text{sec}} = 0.018\, C_L^2.$$

With a low aspect ratio, Vavra finds (c is chord, h is height):

$$C_{D,\text{sec}} = 0.04\, C_L^2 \frac{c}{h}.$$

Both formulae give the same result for $c/h = 0.45$, which means that Vavra's formula must be applied with aspect ratio h/c lower than 2.22.

The end boundary layers themselves generate a loss as well. The associated drag coefficient is approximately $0.02\, s/h$, with s the blade spacing. The typical magnitude of the loss coefficients ($c/s \approx 1.33$, $h/c \approx 2$) is

$$C_{D,prim} = 0.02,\ \ C_{D,\text{sec}} = 0.02,\ \ C_{D,end} = 0.01.$$

This means that the primary loss is mostly not dominant within the total loss.

13.3   Radial Flow Variation

13.3.1 S_1–S_2 Decomposition

Analytical study of the real three-dimensional flow in a turbomachine is impossible and only numerical solution of the flow equations is achievable.

An often used simplification is a *quasi-three-dimensional representation of the flow*. It is assumed then that flow within blade passages occurs on circumferential streamsurfaces (surfaces of revolution). The flow is represented two-dimensionally on such surfaces, denominated S_1-surfaces or *blade-to-blade surfaces*. The S_1-surfaces are streamsurfaces of the circumferentially averaged flow, which is a two-dimensional flow in a meridional plane, denominated flow on the S_2-surface or the *hub-to-shroud surface*.

The equations describing the flow on the S_2-surface contain terms expressing the average effect of the blade forces. The S_1–S_2 decomposition requires iteration between calculations on the S_2-surface (through-flow) and calculations on a series of S_1-surfaces (blade-to-blade flows). No analytical solution is possible for either of these two-dimensional flow types. So the technique has to be a numerical one.

The grid for the three-dimensional calculation is a Cartesian-type in Fig. 13.12. The projection of the quasi-radial and quasi-axial grid lines on a meridional plane is shown. This hub-to-shroud surface could be an S_2-surface and the surfaces of revolution obtained with the quasi-axial grid lines could be S_1-surfaces.

The S_1–S_2 decomposition technique was proposed by C. H. Wu in 1952 as an exact decomposition of a steady three-dimensional flow into flows on a series of S_1-surfaces and a series of S_2-surfaces. The surfaces are all streamsurfaces of the 3D flow and they get twisted in the flow through a blade cascade. The exact decomposition leads to a time-consuming iterative calculation with, nowadays, no advantage compared to an entirely three-dimensional flow calculation. Nowadays, flow analysis typically starts with a quasi-three-dimensional flow representation, followed by a fully three-dimensional simulation.

13.3.2 *Radial Equilibrium*

In the spaces in between rotor blade rows and stator blade rows, there are no blade forces in the circumferentially averaged S_2-flow. Therefore, the circumferential S_1-streamsurfaces may be approximated by interpolation between the shroud and hub surfaces. The momentum equation in the direction perpendicular to the streamsurfaces is then reduced to equilibrium between the centrifugal force associated to the tangential velocity component and a pressure gradient in the normal direction. A further simplification comes with the assumption of cylindrical stream-surfaces. Both forces are in the radial direction then and the equilibrium is called *radial equilibrium*. From radial equilibrium follows an equation of the variation in the radial direction of the velocity components. This allows an approximate representation of the real three-dimensional flow.

The equation of radial equilibrium is

$$\frac{1}{\rho}\frac{dp}{dr} = \frac{v_u^2}{r}.$$

The total enthalpy is $h_0 = h + \frac{1}{2}v_a^2 + \frac{1}{2}v_u^2$. By derivatives in the radial direction follows:

$$\frac{dh_0}{dr} = \frac{dh}{dr} + v_a\frac{dv_a}{dr} + v_u\frac{dv_u}{dr}.$$

The definition of entropy imposes

$$T\frac{ds}{dr} = \frac{dh}{dr} - \frac{1}{\rho}\frac{dp}{dr}.$$

Combination of the three differential equations results in

$$\frac{dh_0}{dr} - T\frac{ds}{dr} = v_a\frac{dv_a}{dr} + \frac{v_u^2}{r} + v_u\frac{dv_u}{dr}. \tag{13.18}$$

A stage is normally designed for constant work along the radius. This implies constant total enthalpy along the radius. Further, it is reasonable to assume constant losses along the radius, which then implies constant entropy along the radius. This approximation is adequate in the core part of the flow, but losses are obviously larger in the end wall boundary layers. With these simplifications, (13.18) is

$$v_a\frac{dv_a}{dr} + \frac{v_u^2}{r} + v_u\frac{dv_u}{dr} = 0, \text{ or } \frac{d\left(\frac{1}{2}v_a^2\right)}{dr} + \frac{v_u}{r}\frac{d}{dr}(rv_u) = 0. \tag{13.19}$$

Equation (13.18) is denominated the non-isentropic simple radial equilibrium equation (NISRE). Equation (13.19), where the term dh_0/dr may be kept, is denominated the isentropic simple radial equilibrium equation (ISRE). The term simple expresses the assumed absence of a radial velocity component.

13.3.3 Free-Vortex Blades

Equation (13.19) is met for constant values along the radius of rv_u and v_a. The corresponding blade shape is denominated *free-vortex blade*. The term means that the angular momentum within the flow is constant. No twisting is then exerted on circumferential streamsurfaces. The variation of the velocity triangles along the radius is then determined when velocity triangles on the average radius are chosen.

As a first example of kinematic parameters on the mean radius we take

$$\psi = 0.4, \ \phi = 0.5, \ R = 0.5.$$

We calculate the velocity triangles on the radii $r_{or}/r_{mr} = 1.20$ (outer radius of the flow annulus) and $r_{ir}/r_{mr} = 0.75$ (inner radius of the flow annulus). With this choice, the through-flow areas above and under the mean radius are equal. The rather large radius variation is typical for first stages in a real compressor. The reasoning hereafter does not apply to the fan stage of a turbofan engine, where the radius variation is much larger (see Exercise 13.6.5).

The results are listed in Table 13.1, with the blade speed on the mean radius taken as the unit value. The velocity components v_a, v_{1u} and v_{2u} on the mean radius result from the chosen kinematic parameters. Constant mass flow rate is approximately expressed by constant axial velocity along the axial direction on the mean radius. The velocity components on the inner and outer radii follow from the vortex distribution. With free-vortex blades, the axial velocity is constant along the radius and the tangential velocity components in the absolute frame are inversely proportional to the radius. All other values in the table follow from the velocity triangles.

The observations are:

Table 13.1 Flow parameter variation with free-vortex blades; $R = 0.50$ on the mean radius

u	v_a	v_{1u}	v_{2u}	v_1/v_2	w_{1u}	w_{2u}	w_2/w_1
0.75	0.5	0.40	0.933	**0.605**	− 0.350	0.183	0.872
1	0.5	0.30	0.7	0.678	− 0.7	− 0.3	0.678
1.20	0.5	0.25	0.583	0.728	− 0.950	− 0.617	0.740

R	α_1	α_m	$\Delta\alpha$	v_2	β_m	$\Delta\beta$	w_1
0.11	38.66	50.24	23.15	1.058	− 7.45	**55.09**	0.610
0.50	30.96	42.71	23.50	0.860	− 42.71	23.50	0.860
0.65	26.57	37.98	22.81	0.768	− 56.61	11.26	1.074

Table 13.2 Flow parameter variation with free-vortex blades; $R = 0.70$ on the mean radius

u	v_a	v_{1u}	v_{2u}	v_1/v_2	w_{1u}	w_{2u}	w_2/w_1
0.75	0.5	0.133	0.667	**0.621**	− 0.617	0.083	**0.638**
1	0.5	0.1	0.5	0.721	− 0.9	− 0.5	0.687
1.20	0.5	0.083	0.417	0.779	− 1.117	− 0.783	0.759

R	α_1	α_m	$\Delta\alpha$	v_2	β_m	$\Delta\beta$	w_1
0.47	14.57	33.92	38.70	0.84	− 30.10	42.02	0.80
0.70	11.31	28.16	33.69	0.71	− 52.98	15.95	1.03
0.79	9.09	24.56	30.94	0.65	− 61.64	8.60	**1.23**

- rotor blade twist is very large: β_m varies from $− 7.5°$ to $− 56.5°$: twist $\approx 50°$;
- flow turning at the rotor hub is very large: $\Delta\beta \approx 55°$;
- deceleration at the stator hub is very strong: velocity ratio ≈ 0.60;
- all inlet Mach numbers are moderate; as the blade speed on the mean radius is around the speed of sound, dimensionless velocities are approximately Mach numbers.

A higher degree of reaction on the mean radius reduces the observed disadvantages. With $\psi = 0.4$, $\phi = 0.5$ and $R = 0.7$ on the mean radius, the results are listed in Table 13.2.

Favourable findings are:

- rotor blade twist is acceptable: $\approx 31.5°$;
- flow turning at the rotor hub is acceptable: $42°$ with inlet Mach number 0.80;
- deceleration at the stator hub is less strong: velocity ratio ≈ 0.62.

Unfavourable findings are that the deceleration at the rotor hub becomes strong and that the Mach number at the rotor entry at the tip becomes high. Further raising the degree of reaction on the average radius causes a further decrease of the deceleration ratio at the rotor hub and a further increase of the inlet Mach number at the rotor tip, as demonstrated by the results in Table 13.3 for $\psi = 0.4$, $\phi = 0.5$ and $R = 0.8$ on the mean radius.

Free-vortex blades do not result in an entirely advantageous variation of the flow parameters, unless work is decreased (ψ lower than 0.40 on the mean radius) or the

Table 13.3 Flow parameter variation with free-vortex blades; $R = 0.80$ on the mean radius

u	v_a	v_{1u}	v_{2u}	v_1/v_2	w_{1u}	w_{2u}	w_2/w_1
0.75	0.5	0	0.533	0.684	− 0.75	− 0.217	**0.605**
1	0.5	0	0.4	0.781	− 1	− 0.6	0.698
1.20	0.5	0	0.333	0.832	− 1.20	− 0.867	0.770

R	α_1	α_m	$\Delta\alpha$	v_2	β_m	$\Delta\beta$	w_1
0.65	0	23.42	46.83	0.73	− 39.89	32.85	0.90
0.80	0	19.33	38.66	0.64	− 56.81	13.24	1.12
0.86	0	16.84	33.67	0.60	− 63.71	7.35	**1.30**

choice for a repeating stage is abandoned. The best flow for the radius ratio 1.20/0.75 is obtained with $R = 0.70$ on the mean radius (Table 13.2). The corresponding velocity triangles are shown on Fig. 13.4. Remark that the flow at the inlet and outlet of the stage deviates from the meridional plane ($\alpha_1 \approx 15°$ at the hub). This deviation from the meridional plane mainly has as benefit that the Mach number of the relative flow at the rotor entry lowers somewhat with respect to the value without pre-whirl. Further improvement at the rotor tip is possible by forcing the vortex distribution somewhat, as discussed in the next section.

13.3.4 Forcing of the Vortex Distribution

A more favourable variation of flow parameters is obtained by forcing the whirl distribution somewhat away from the free vortex. The following relations apply (m denotes the mean value between inlet and outlet of the rotor):

$$v_{1u} = v_{mu} - \frac{\Delta v_u}{2} = v_{mu} - \frac{\Delta W}{2u}, \quad v_{2u} = v_{mu} + \frac{\Delta W}{2u}.$$

ΔW is kept constant along the radius. The variation of v_{mu} with free-vortex blades is inversely proportional to the radius. This variation may be chosen differently, however. In order to analyse the influence of this variation, we assume $v_{mu}\ r^n$. With the exponent $n = -1$, free-vortex blades are recovered.

Hereafter, we calculate the variation of the flow parameters for $n = 0$. Velocities are written in dimensionless form by dividing them by the blade speed on the mean radius (mr denotes mean radius). The Eq. (13.19) stays valid with dimensionless quantities. The variation of tangential velocity components follows from

$$\overline{v}_{1u} = \frac{v_{1u}}{u_{mr}} = \frac{v_{mu}}{(v_{mu})_{mr}} \frac{(v_{mu})_{mr}}{u_{mr}} - \frac{1}{2} \frac{\Delta W}{u^2_{mr}} \frac{u_{mr}}{u},$$

or

$$\overline{v}_{1u} = (1 - R_{mr})\overline{r}^n - \frac{1}{2}\psi_{mr}/\overline{r}, \tag{13.20}$$

and analogously:

$$\overline{v}_{2u} = (1 - R_{mr})\overline{r}^n + \frac{1}{2}\psi_{mr}/\overline{r}, \tag{13.21}$$

with $R_{mr} = (-w_{mu}/u)_{mr}$ and $\psi_{mr} = \Delta W/u^2_{mr}$. On the mean radius we choose $\psi = 0.4$, $\phi = 0.5$ and $R = 0.7$. Constant mass flow rate through the stage is approximated by constant axial velocity along the axial direction on the mean radius. Hereafter, we omit the overbars indicating dimensionless values.

Table 13.4 Whirl forcing with $n = 0$; $R = 0.70$ on the mean radius

v_{1a}	v_{2a}	v_{1u}	v_{2u}	v_1/v_2	w_{1u}	w_{2u}	w_2/w_1
0.512	0.585	0.033	0.567	**0.63**	− 0.717	− 0.183	**0.70**
0.5	0.5	0.1	0.5	0.72	− 0.9	− 0.5	0.69
0.487	0.444	0.133	0.467	0.78	− 1.067	− 0.733	0.73

R	α_1	α_m	$\Delta\alpha$	v_2	β_m	$\Delta\beta$	w_1
0.50	3.69	23.90	40.41	0.82	− 35.92	37.10	0.88
0.70	11.31	28.16	33.69	0.71	− 52.98	15.95	1.03
0.80	15.28	30.87	31.17	0.64	− 62.14	6.67	**1.18**

For $n = 0$, Eq. (13.20) becomes

$$v_{1u} = 0.3 - 0.2/r \ \ or \ \ r\,v_{1u} = 0.3\,r - 0.2.$$

Equation (13.19) becomes

$$\frac{d}{dr}v_{1a}^2 = -\left(\frac{0.6}{r} - \frac{0.4}{r^2}\right)(0.3) \ \ or \ \ v_{1a}^2 = c - 0.18 \ln r - 0.12/r.$$

The value of v_{1a} on the mean radius determines the constant: $c = 0.37$. Analogously we obtain $v_{2a}^2 = 0.13 - 0.18 \ln r + 0.12/r$. The results are shown in Table 13.4.

The flow variation has become more favourable. The blade twist is moderate: $7°$ on the stator and $26°$ on the rotor. Maximal flow turnings are acceptable: $40°$ on the stator hub with Mach number 0.8 and $37°$ on the rotor hub with Mach number 0.88. The Mach number at the rotor blade tip is acceptable: 1.18. Even a somewhat higher Mach number would be acceptable. We further note that velocity reductions at the stator and rotor hub are 0.62 and 0.64 with a free-vortex blade. Actually, such strong decelerations are not realisable. Vortex forcing with $n = 0$ improves these reductions. The reduction values are 0.63 for the stator and 0.70 for the rotor. Only at the stator hub, the velocity reduction stays too strong.

Vortex forcing aims essentially at two improvements: increasing v_{mu} at the casing, which decreases the Mach number level at the rotor entry and decreasing v_{mu} at the hub, which decreases the flow turning within the rotor (turning within the stator increases somewhat). This softens the velocity reduction within the rotor at the hub to an acceptable value. There is still a problem with the stator, but it has not turned worse compared to free-vortex blades. With a blade design according to the above data, stall occurs in the corner of the suction side and the hub of the stator. In the past, many compressors were designed with this corner stall present. A solution is obtained by choosing a smaller radius ratio or by reducing work. But there is still another solution, discussed in Sect. 13.3.6.

The observed tendencies continue when increasing the exponent n above zero. But there is no benefit in doing this (results not shown). The major disadvantage is that the entry Mach number increases at the rotor hub. A vortex distribution with n

between -1 and 0 is optimum. The real optimisation, however, is more complex than studied here. The flow in the spaces outside the blade passages does not need to lay on cylindrical streamsurfaces. Further, the loss distribution should be taken into account. Enlarged work near the casing and the hub is desirable in order to homogenise the pressure rise (see next section).

13.3.5 Effect of End Wall Boundary Layers

Figure 13.15 sketches the velocity triangles at casing and hub with results for $n = 0$. The dashed lines represent the velocity triangles with axial velocity reduced to 75% and blade outlet angles kept unchanged. The triangles with reduced flow rate illustrate the changes in the end wall boundary layers. There is incidence increase, both on the rotor and the stator, leading to strong increase of the flow turning. Further, the velocity reduction factors become stronger. Based on these observations, one would fear strong separation, both at the hub and at the casing.

In order to restore the flow turnings to the original values, the stator stagger angle may be changed until the direction of w'_1 corresponds again to that of w_1. The flow turnings then approximate the original values, both in stator and rotor parts. The angle adjustment is similar to closing of adjustable stator vanes for adaption to a reduced flow rate. The stator vane then gets bended at the end parts. Such geometry modifications were used in the past, but did not prove to be effective.

The justification is purely two-dimensional and ignores the intense secondary flow phenomena in the end wall regions. Moreover, as already said, increased work

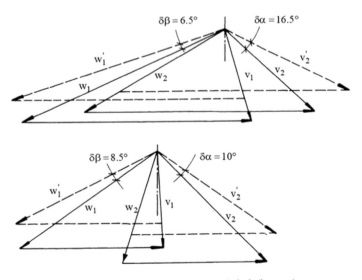

Fig. 13.15 Effect of end wall boundary layers; casing (top), hub (bottom)

near the end walls is beneficial for homogenisation of the pressure rise. In the next section, we discuss modern blade modifications in end wall regions.

13.3.6 Three-Dimensional Blade Design

The quasi-three-dimensional flow representation enables the determination of the profiles of the S_1-sections of a blade. The blade is obtained by stacking these S_1-sections. But, the quasi-three-dimensional flow representation does not provide a stacking strategy. One could be led by considerations about strength and stack the sections with their centres of gravity on a radial line, but shifting the profiles in the axial and tangential directions creates additional degrees of freedom.

The shifts are termed *sweep* and *lean*. In practice, three-dimensional blade shaping is done with three-dimensional flow simulations, and optimisation of the stacking is quite complex. The main objectives of the shifts can be understood from fundamental arguments, however, as discussed hereafter.

Sweep is displacement of the blade profile in the chord direction and lean is displacement perpendicular to the chord direction. The shifts generate radial forces on the flow due the pressure difference between pressure and suction sides, thus influencing the radial distribution of flow rate and blade loading. The effects are similar to those discussed with steam turbines in Sect. 6.9 of Chap. 6, but the terms sweep and lean were used there for inclinations in meridional and orthogonal planes.

A favourable result requires positive lean in the end wall parts, which means shifting blade profiles at the end walls in the sense of the lift. A blade then gets a bowed appearance, as sketched in Fig. 13.16 (left). By bowing, the blade force obtains a component in the radial direction near the end walls, pointing towards the mid of the blade passage. The flow in blade surface vicinity near the end walls is then influenced in a meridional plane in the same way as the flow over a two-dimensional aerofoil due to lift [8].

A streamline in the end wall boundary layer near the suction side is pushed away from the end wall at the entrance of a blade passage. In principle, it returns to the end wall in the outlet part of the passage, but this motion is hindered by the passage vortex. It may also be hindered by boundary layer separation in the corner of the suction side and the end wall (the adverse pressure gradient zone at the suction side).

Fig. 13.16 Stator cascade; left: bowing (view from downstream); right: sweep (side view with pressure side upside)

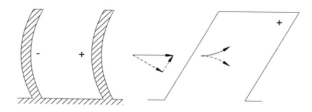

The result of the streamline shift is that the pressure minimum at the suction side is deepened somewhat away from the end wall and weakened at the end wall. A displacement of a streamline in the end wall boundary layer, in principle, occurs in the other sense near the pressure side, but the possibility for displacement is restricted by the sharp angle and changes in pressure are limited as the pressure in the leading edge zone is close to the stagnation pressure.

The result is increased blade loading somewhat away from the end wall and reduced loading in the end wall zone [6]. So the strongest boundary layers are loaded more and the weakest boundary layers are loaded less. For unchanged blade profiles, as in many experiments, losses decrease near the end walls, but increase away from the end walls and the average losses stay about the same. In order to benefit, blade profiles must be adapted to the changed loading.

The effect of sweep is illustrated in Fig. 13.16 (right). Sweep is called positive when the leading edge is moved parallel with the chord towards the oncoming flow. The figure demonstrates positive sweep at the hub and negative sweep at the casing. A first effect of sweep is torsion on the streamsurfaces. The flow velocity at the leading edge may be split in a component parallel to the leading edge and a component perpendicular to it.

The parallel component stays nearly unmodified in the flow over the profile. The perpendicular component increases near the suction side and decreases near the pressure side. With positive sweep, a streamline near the suction side moves towards the end wall and a streamline near the pressure side moves away from the end wall. The torsion on the streamsurface results in a vortex motion counteracting the passage vortex (Fig. 13.10).

So, secondary flow reduction requires combined sweep, positive at the hub and positive at the casing. But reduction of the secondary flow is not necessarily beneficial, as the secondary vortices move core flow fluid to boundary layer regions and vice versa. A particular beneficial effect of the secondary flow is the redistribution of energy between core flow and end wall layers. We note that the vortex induced by positive sweep at the hub supplies mass flow from the central part of a blade to the corner of the suction side and the hub. This supply decreases the corner stall. This is thus favourable for loss reduction in the hub region of a stator. Another favourable effect of sweep is, as with aircraft wings, Mach number lowering of the flow perpendicular to the leading edge, causing reduction of shock loss with supercritical and transonic profiles. This is favourable for rotor tip sections.

Figure 13.17 shows the loss coefficient measured on a stator vane with straight stacking and on a vane with the same profiles, but with a curved stacking line. Contours of isentropic Mach number (calculated from the pressure using isentropic formulae) and limiting streamlines on the suction surface are shown. Bowing is applied, i.e. positive lean both at the hub and at the casing. Positive sweep is applied at the hub and negative sweep at the casing. The reduction of the corner stall at the hub due to the positive sweep is obvious. Herewith, we also understand that positive sweep is not necessarily beneficial at the tip, because there is no corner stall there. The effect of the bowing is a redistribution of the loss with lower values at the end

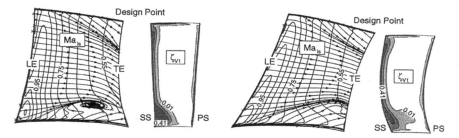

Fig. 13.17 Distribution of loss immediately downstream of a compressor stator cascade: straight stacking line (left); bowed and swept stacking line (right) (from [8], permission by ASME)

walls and higher values in the central part of the vane (but profiles are not adapted to the changed loading).

Also with rotor blades, positive lean and positive sweep at the hub are favourable. Sweep at the casing reduces the Mach number of the flow perpendicular to the blade. But this can be obtained both with backward and forward sweep. Further, realisation of a three-dimensional shape is more difficult from a structural point of view. At present, rotor blades have typically a slight general sweep (positive at the hub and negative at the casing), as in Fig. 13.17, and a slight positive lean at the hub. The sweep reduces the corner stall at the hub. The lean reduces the load on the end wall blade sections. Note that corner stall is less intense with a rotor than with a stator (Table 13.4). Back-sweep at the casing reduces the effective Mach number.

Fan blades in the most recent aero-engines have forward (positive) sweep at the casing. With forward sweep, the operating margin against stall increases because the shock wave in the blade passage is pushed backwards (see the later Fig. 13.24). Figure 13.18 illustrates the forward tip sweep of the fan of the GE9X. Forward sweep at rotor tips may also come into use in future compressor parts [9].

Fig. 13.18 Front view on the fan of the GE9X (courtesy by GE Aviation)

13.4   Compressor Blade Profiles

13.4.1   Subsonic and Supercritical Cascades

The optimum Mach number distribution (boundary layer edge velocity) and the corresponding profile shape for a stator are shown in Fig. 13.19. Also shown is the comparison between the optimum profile and a profile by a NACA-65 thickness distribution on a circular camber line. The parameters of the stator cascade are:

- inlet Mach number: 0.773 (supercritical)
- flow turning: 39.6°
- pitch-cord ratio: 0.465
- axial velocity-density ratio: $1.1 = \rho_2 v_{2a}/\rho_1 v_{1a} =$ streamtube contraction.

The features of the optimal Mach number distribution are:

- The acceleration at the leading edge of the suction side runs over about 30% of the chord. The objective of the acceleration is to lower the pressure in order to create pressure difference between both sides of the profile. The strength of the acceleration is tuned such that the boundary layer stays laminar within this zone. The boundary layer is kept in laminar state in order to minimise the friction and thus the contribution to the resulting drag.
- The deceleration on the suction side is optimised such that the boundary layer is near to separation at the trailing edge. The boundary layer evolution is made for maximum tangential force, but balanced against the drag contribution of this part of the profile. In order to withstand the adverse pressure gradient, the boundary layer must turn from laminar state to turbulent state at the maximum in the Mach number distribution.

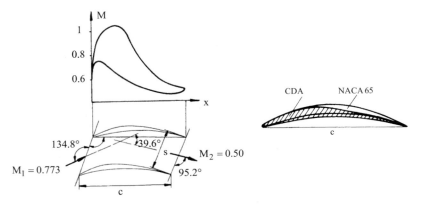

Fig. 13.19 Optimum Mach number distribution for a supercritical compressor cascade; comparison between a controlled diffusion profile and a NACA-65 thickness distribution on a circular camber line; adapted from [4]

- The Mach number just upstream of the shock wave on the suction side slightly exceeds 1 ($M \approx 1.05$) and the boundary layer is turbulent at the shock impact position. With the low upstream Mach number, shock loss is small.
- There is a slight acceleration in the front part of the pressure side caused by the rounding of the leading edge. The acceleration is followed by a deceleration that increases the pressure difference of the profile, but, as for the suction side, balanced against the drag contribution of this part of the profile. The optimised evolution ends with a weak acceleration zone at the trailing edge.

The phenomena that strongly determine the optimum profile are transition from laminar to turbulent flow in the boundary layers and possibility for boundary layer separation in the adverse pressure gradient zones.

Since the general flow in a compressor cascade is decelerating, the adverse pressure gradient zones are large on both the suction and pressure sides and these zones dominate the optimisation. Therefore, an optimised profile is called a *controlled diffusion aerofoil* (CDA).

The turbulence level in the core flow of a compressor is mostly very high. The degree of turbulence is the ratio of the magnitude of the fluctuating velocity to the average velocity. It is typically around 5%, except for an entry stage. The high free-stream turbulence comes from turbulence generated by the wakes of upstream blade rows. Boundary layer transition is then directly induced by the fluctuations in the main flow. This means that the transition is not caused by instability of the boundary layer flow with forming of Tolmien-Schlichting waves (see Fluid Mechanics). Transition caused by instability, called *natural transition*, is typical for aircraft wings in a very low turbulence level flow (about 0.01% turbulence level in cruise conditions). The directly induced transition is called *bypass transition*, meaning that the instability mechanism is bypassed.

A consequence is that the boundary layer thickness for transition, expressed with a Reynolds number based on momentum thickness, lowers considerably with increased free-stream turbulence level. So, it becomes harder to keep a boundary layer in laminar state in the presence of a high free-stream turbulence level. The acceleration of the boundary layer in the leading edge zone has to be sufficiently strong in order to keep the momentum thickness Reynolds number below the critical value. The optimisation of the accelerated zone thus depends strongly on the turbulence level of the incoming flow.

A major difference between a laminar and a turbulent boundary layer is that a turbulent boundary layer can withstand a much larger adverse pressure gradient before entering into separated state. The reason is turbulent exchange of momentum from the outer region of the boundary layer to the near-wall region. So, to avoid separation, it is best to let the boundary layer transition from laminar to turbulent state at the position where the boundary layer edge flow changes from accelerating flow to decelerating flow. On the other hand, by the turbulent momentum exchange, the shear stress in a turbulent boundary layer is larger for the same edge velocity.

Optimisation includes thus making a balance between the extent of the laminar part of the boundary layers with lower shear stress, and the turbulent part with higher resistance to separation. The optimum depends on the free-stream turbulence level and on the chord Reynolds number. The lower the chord Reynolds number is, the weaker is a laminar boundary layer against separation under an adverse pressure gradient. So, the optimum profile of a compressor blade in an aero-engine is different from that in a large heavy-duty gas turbine for electricity production because the global flow Reynolds number in an aero-engine is much lower.

A compressor cascade is called subsonic when the Mach number of the oncoming flow is low enough such that the entire flow over the profile is subsonic. For higher Mach number of the oncoming flow, a supersonic zone may form at the suction side, ended with a normal shock. The flow is then called supercritical (subsonic upstream flow, subsonic downstream flow, but a supersonic zone on the profile).

The shock at the suction side creates a sudden pressure rise which may lead to boundary layer separation. In order to avoid shock-induced separation, the Mach number just upstream of the shock should not be too high (a typical limit is 1.25) and the boundary layer state should be turbulent at the shock position (better resistance against separation). The consequence is that the optimum profile shape becomes dependent on the Mach number of the oncoming flow.

The optimisation strategy for supercritical rotor cascades is the same as for stator vanes, but with rotor cascades the Mach number is higher and the flow turning is lower (typically under 20° compared to 40°). The resulting blade shape depends on Mach number and flow turning.

Figure 13.20 is a sketch of the radial distribution of the Mach number in the oncoming flow of rotors and stators within a three-stage transonic axial compressor (LP-compressor of a bypass aero-engine). We repeat that a compressor stage is called transonic when the relative flow entering the rotor is supersonic in the tip region and subsonic in the hub region. The three stages are of this type. The stator cascades are supercritical near the hub and subsonic near the casing. The rotor cascades are supercritical near the hub and transonic near the casing. In compressor terminology, a cascade is transonic when the inlet flow is supersonic and the outlet flow is subsonic.

The design strategy of subsonic stator cascades near the casing is the same as with supercritical cascades. A supercritical rotor cascade only differs from a supercritical stator cascade by the higher Mach number level. With regard to shock strength, the maximum Mach number at the suction side should not exceed about 1.25. This

Fig. 13.20 Variation of the Mach number of the oncoming flow (rotor: R; stator: S) as a function of the radius (normalised span) with a three-stage axial compressor; adapted from [7]

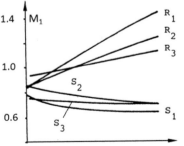

limitation is meant to prevent boundary layer separation due to the sudden shock pressure rise. For design conditions, blade shapes may be optimised with deceleration upstream of the shock so that the Mach number just before the shock is close to one. But away from the design operating point, the Mach number in front of the shock may be much higher. The main effect of limiting the maximum Mach number is that the optimum loading of a blade decreases with higher Mach number in the oncoming flow and that the solidity thus must increase.

13.4.2 Transonic Cascades

Figure 13.21 sketches the shock pattern within a transonic cascade with low or medium–high Mach number of the oncoming flow (up to Mach number about 1.2). Each blade has a shock just upstream of the leading edge with an oblique part upstream of the cascade, called the *bow shock* and a part normal to the adjacent blade surface, called the *blade passage shock*. The oblique shock is very weak, with negligible loss. The normal shock provides the main part of the compression. This shock is also the main loss source, not only because of its intrinsic dissipation, but also because of interaction between the shock and the boundary layer. For preventing separation, the boundary layer has to be turbulent in the impact zone.

Optimisation of a transonic cascade is strongly determined by minimising shock losses. Depending on the Mach number of the oncoming flow, at the suction side in front of the normal shock, either a slight acceleration is tolerable (convex shape), or no acceleration can be allowed (plane shape) or even a deceleration must be realised (concave shape). Downstream of the shock, the flow is subsonic. Within this part of the blade passage, further deceleration combined with flow turning occurs.

Figure 13.22 shows two transonic compressor cascade shapes for the middle section and the tip section of a transonic compressor rotor (a rotor is transonic when the inflow Mach number is subsonic at the hub and supersonic at the tip). For the lower inflow Mach number ($M_1 \leq 1.1$), the blade shape is similar to that of a supercritical cascade with higher inflow Mach number ($M_1 \geq 0.80$). The suction side is convex with a small acceleration upstream of the shock. For the higher inflow Mach number, there is a slight supersonic deceleration upstream of the shock (see next section), requiring a slightly concave form of the suction side.

Fig. 13.21 Shocks within a transonic compressor cascade for low or medium–high inflow Mach number

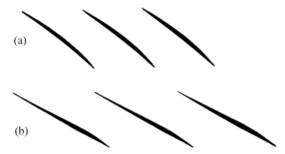

Fig. 13.22 Transonic compressor cascades with a single normal blade passage shock; a: low transonic ($M_1 \sim 1.1$), convex leading suction side b: medium transonic ($M_1 \sim 1.2$), slightly concave leading suction side

13.4.3 Supersonic Cascades and Transonic Cascades with High Inlet Mach Number

In compressor terminology, a cascade is supersonic when both inlet and outlet Mach numbers are above unity. Figure 13.23 sketches the ideal flow pattern. The suction side is concave at the leading edge. As a consequence, compression Mach lines form, which converge into a weak pre-compression shock. The main compression shock stands within the blade passage, but differently from the transonic cascades in Fig. 13.21, the blade passage shock is oblique.

The suction side shape downstream of the pre-compression and the pressure side shape are such that the shock penetrates far into the blade passage. This requires a sufficiently opened throat and some acceleration downstream of the pre-compression zone. With transonic cascades, the throat is sufficiently narrow so that the blade passage shock stands upstream of the throat and is strong. With supersonic cascades, the flow is called started, in analogy with starting of a supersonic wind tunnel or a supersonic engine intake by opening the throat and swallowing the shock wave.

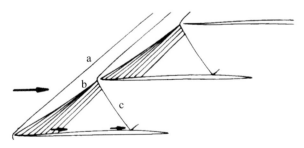

Fig. 13.23 Ideal flow pattern in a supersonic cascade; a: bow wave; b: pre-compression shock; c: oblique shock within the blade passage; adapted from [16]

Fig. 13.24 Flow patterns within a transonic cascade with high inlet Mach number ($M_1 \approx 1.4$); middle: design flow; left: stall; right: choking; adapted from [1]

The most striking feature of the optimum blade shape of a supersonic compressor cascade is the concave zone at the suction side, responsible for the pre-compression. The blade shapes are therefore denominated *pre-compression profiles*. The leading edge zone has a characteristic S-shape.

The ideal flow within a supersonic cascade has an oblique passage shock without reflection, followed by an almost uniform flow. So, the possible flow turning is small with supersonic cascades. In practice, the flow pattern is strongly determined by the boundary layer state on the suction side. Mostly, the boundary layer upstream of the shock impact zone is laminar and the oblique shock then causes boundary layer separation with forming of a lambda shock structure and downstream strong shock patterns, but finally acceleration into a supersonic outflow [16]. The losses of the separated flow are very high, with a loss coefficient in the order of 0.15. Therefore, supersonic cascades are not used in practice.

Pre-compression profiles, however, may be applied efficiently with transonic cascades with high inflow Mach number. From Fig. 13.22 it is clear that the principle of a single normal blade passage shock becomes impractical for a high inflow Mach number ($M_1 \approx 1.3$ to 1.4). The front part of the profile then becomes extremely thin due to the concave shape of the suction side. Figure 13.24 sketches shock patterns in design flow, near stall and near choking (stall and choking are discussed in Sect. 13.5) for pre-compression profiles in a transonic cascade.

In the design flow, there is a normal shock immediately following the oblique blade passage shock. With increasing backpressure, the normal shock is pushed forward and merges with the oblique shock, making the merged shock strong (pattern of Fig. 13.21). Boundary layer separation occurs on the blade suction side when the shock becomes sufficiently strong (*stall limit*). This happens when the shock gets expelled somewhat from the blade passage. Decreasing the backpressure moves the normal shock downstream. The mass flow rate attains a blocked value (*choking limit*), when the normal shock reaches the profile trailing edge at the suction side. With this example, we understand that the stall and choking limit flows are similar with the transonic cascades of Fig. 13.22. In particular, with reduced backpressure, the normal shock is swallowed into the blade passage (started flow) and becomes weak.

13.5   Performance Characteristics and Operating Range

13.5.1   General Shape of a Characteristic Curve

Figure 13.15 demonstrates that the rotor work of an axial compressor stage increases with decreasing flow rate. So, the general shape of the characteristic curve of pressure ratio as a function of mass flow rate at constant rotational speed is a descending line. Figure 13.15 is drawn assuming that the rotor outlet direction (w_2) and the stator outlet direction (v_1) stay unchanged with changing flow rate. On this assumption, which is only approximately satisfied in reality, work increases linearly with decreasing flow rate. But, incidences increase with decreasing flow rate. So, boundary layer separation occurs below a certain flow rate. The flow is then said to be *stalled*, meaning that the compressor does not function properly anymore. With stalled flow, the tangential force of the blade drops. This causes a maximum in the characteristic curve of pressure ratio as a function of flow rate. This maximum may be very sharp.

13.5.2   Rotating Stall

Stall never occurs in a compressor simultaneously on all blades over their full span. First, the incidence margin with respect to stall changes from stage to stage, depending on the operating point (see Sect. 13.5.5). Second, due to a higher Mach number level, the incidence margin at the tip of a rotor blade is smaller than at the hub (see Fig. 13.4). With decreasing flow rate, the flow thus first goes into stall at the tip of a rotor blade. With longer blades, as in the first stages of a compressor, stall zones therefore form at the casing, unless the flow rate reduction with respect to non-stalled flow is extreme. In the last stages of a compressor, where blade height is smaller, full-span stall is reached faster.

Whether stall is part-span or full-span, it is always organised in cells, rotating in the rotation sense of the compressor. Figure 13.25 sketches possible cell patterns. Zones with low axial velocity caused by stall are hatched. The typical speed of part-span cells is about 50% of the rotor speed and there are always several cells. With full-span stall, there is typically only one cell with speed 20% to 40% of the rotor speed. There is no simple explanation for the number of cells and their speed [3].

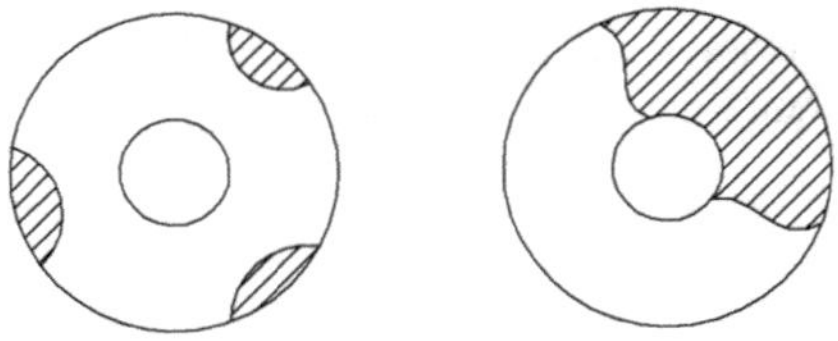

Fig. 13.25 Cell patterns
with rotating stall; part-span
and full-span stall

Fig. 13.26 Rotating stall
propagation mechanism

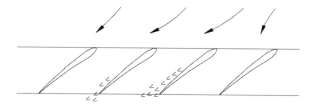

The cell propagation mechanism may be understood with Fig. 13.26. Assume that stall occurs on a specific blade. The flow near the blade with stall must then deviate somewhat. The incidence increases on the blade at the suction side of the blade with stall, forcing the flow into stall on this neighbouring blade. The incidence decreases on the blade at the pressure side of the blade with stall. So, the tendency to stall is suppressed on this neighbouring blade. It is typical with flow rate reduction that stall starts on only a few blades within a blade row, as there are small geometric differences between the blades. Cells are generated that way.

According to Fig. 13.26, the sense of propagation in a rotor blade row is opposite to the running sense in the relative frame. The propagation speed is lower than the blade speed, causing the stall cells to move in the rotation sense in the absolute frame. With the same figure one sees that the propagation sense of cells in a stator vane row is in the running sense of the rotor. Typically, rotating stall cells form in a number of consecutive stator vane rows and rotor blade rows.

13.5.3 Choking

By reduction of the backpressure of a compressor with subsonic flow, the flow rate increases and all velocities increase until sonic speed is attained in some through-flow section. With further reduction of the backpressure, the flow rate stays then blocked. This phenomenon is called *choking*. The choking flow rate depends on the rotational speed, both for choking in rotor blade passages and for choking in stator vane passages. This may be understood by the following, simplified, derivations.

Within the relative system it applies that

$$I = h + \frac{1}{2}w^2 - \frac{1}{2}u^2 = \text{constant.}$$

At the inlet of the rotor (see Fig. 13.1): $w_1^2 = u^2 + v_1^2 - 2uv_{1u}$, so that:

$$I = h_1 + \frac{1}{2}v_1^2 - uv_{1u} = h_{01} - uv_{1u}.$$

$M = 1$ is attained for $w^2 = c^2 = \gamma RT = (\gamma - 1)h$, so that then

$$I = h + \frac{1}{2}(\gamma - 1)h - \frac{1}{2}u^2 = \frac{1}{2}(\gamma + 1)h - \frac{1}{2}u^2.$$

Thus: $\frac{1}{2}(\gamma + 1)h = h_{01} - uv_{1u} + \frac{1}{2}u^2$, or

$$\frac{h}{h_{01}} = \frac{2}{\gamma + 1}\frac{h_{01} - uv_{1u} + \frac{1}{2}u^2}{h_{01}}$$

$$= \frac{2}{\gamma + 1}\left[1 + (\gamma - 1)\frac{(\frac{1}{2}u^2 - uv_{1u})}{c_{01}^2}\right]. \tag{13.22}$$

If sonic flow occurs, it happens near the rotor inlet, since the further the flow enters the rotor, the lower is the velocity w and the higher is the enthalpy h. Thus, the inlet flow is the most critical. The mass flow rate for rotor choking is

$$\dot{m}_c = A_1\rho c = A_1\rho_{01}c_{01}\left(\frac{\rho}{\rho_{01}}\right)\left(\frac{c}{c_{01}}\right) = A_1\rho_{01}c_{01}\left(\frac{h}{h_{01}}\right)^{\frac{1}{\gamma-1}+\frac{1}{2}}, \tag{13.23}$$

where A_1 is the through-flow area near the inlet. In all the relations above, isentropic flow is assumed, which means that losses in the rotor entry are ignored.

With an average value of the blade speed at the inlet, an approximation of the choking mass flow rate is obtained. The Mach number is higher at the tip than at the hub at the rotor entry. Thus, attaining $M = 1$ within the entire entry section is impossible. So, sonic flow can only occur somewhat more inside the rotor. This makes the estimation of the choking mass flow rate with (13.22) and (13.23) approximate. Nevertheless, a clear conclusion is that the mass flow rate at choking depends on the rotational speed and increases with increasing rotational speed.

At the inlet of the stator:

$$h_{02} = h + \frac{1}{2}v^2.$$

$M = 1$ is attained for $v^2 = c^2 = \gamma RT = (\gamma - 1)h$, so that then

$$h_{02} = h + \frac{1}{2}(\gamma - 1)h = \frac{1}{2}(\gamma + 1)h \quad \text{or} \quad \frac{h}{h_{02}} = \frac{2}{\gamma + 1},$$

with

$$h_{02} = h_{01} + \psi u^2 = h_{01}\left[1 + (\gamma - 1)\frac{\psi u^2}{c_{01}^2}\right]. \tag{13.24}$$

The corresponding mass flow rate is

$$\dot{m}_c = A_2 \rho c = A_2 \rho_{02} c_{02} \left(\frac{2}{\gamma + 1} \right)^{\frac{1}{\gamma - 1} + \frac{1}{2}},$$

$$\text{with } \frac{\rho_{02}}{\rho_{01}} = \left(\frac{h_{02}}{h_{01}} \right)^{\frac{1}{n-1}} \text{ and } \frac{c_{02}}{c_{01}} = \left(\frac{h_{02}}{h_{01}} \right)^{\frac{1}{2}},$$

where A_2 is the through-flow area near the inlet and n is the polytropic exponent of the total-to-total process in the rotor. As for the entry to the rotor, losses are ignored in the entry to the stator.

Thus:

$$\dot{m}_c = A_2 \rho_{01} c_{01} \left(\frac{2}{\gamma + 1} \right)^{\frac{1}{\gamma - 1} + \frac{1}{2}} \left[1 + (\gamma - 1) \frac{\psi u^2}{c_{01}^2} \right]^{\frac{1}{n-1} + \frac{1}{2}}. \tag{13.25}$$

The choking mass flow rate depends on the rotational speed as well.

The expressions (13.23) and (13.25) are similar so that discerning which is most critical for mass flow rate is difficult. For degree of reaction larger than 50%, it is the rotor because the Mach number level is the largest in the rotor ($A_2 < A_1$).

The above derivation is intended to illustrate that the choking mass flow rate depends on the rotational speed. The reasoning for subsonic cascades applies, in principle, to supercritical cascades as well. The choking condition for transonic cascades is different, as explained before. But, the dependence of the choking mass flow rate on the rotational speed is similar.

13.5.4 Surge

The term *surge* refers to flow varying in time between attached flow and separated flow. This periodic type of stall is caused by a dynamically unstable interaction between a compressor and its load. Figure 13.27 (left) is a sketch of a possible compressor characteristic together with a possible load characteristic.

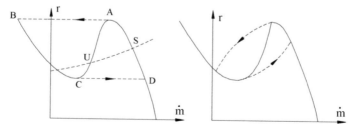

Fig. 13.27 Surge: flow cycle with very large reservoir (left) and with smaller reservoir (right); soft maximum (left) and sharp maximum (right)

Just as with pumps (Sect. 8.3 in Chap. 8), the intersection S of the load and compressor characteristics is a stable operating point and the intersection U is an unstable operating point. This may be verified by applying a small perturbation. We take as an example a compressor delivering to a reservoir and a short duration decrease of the flow rate to the reservoir by opening a bypass valve in the connecting pipe. The pressure in the reservoir then decreases somewhat and so does the backpressure of the compressor. The reaction of the compressor for operating point S is flow rate increase, which is a restoring reaction. The reaction of the compressor for operating point U is flow rate decrease, which is an aggravating reaction.

If the backpressure of the compressor is gradually enlarged, the load characteristic shifts upwards until there is only one tangent point with the compressor characteristic in the vicinity of the maximum. With further increase of the backpressure, there is no intersection anymore for positive mass flow rate and the operating point jumps to the point B on the negative branch. Point B is a stable operating point (this may be verified by a perturbation analysis). A cycle ABCD is run through, as sketched in Fig. 13.27 (left).

The cycle is ABCD drawn, assuming that the delivered pressure does not change during the transition from the operating point at the maximum of the characteristic A to the operating point on the negative branch B, and similarly for the transition from C to D. This can only be attained with a very large reservoir. With limited reservoir capacity, the pressure decreases during the transition from A to B and increases during the transition from C to D. So, a real surge cycle rather looks as sketched in the right-hand part of Fig. 13.27. The extension of the cycle depends on the buffer capacity of the reservoir. Negative flow rates may occur with large reservoir capacity. When the flow rate stays positive, the surge is called mild. When negative flow rates occur, the surge is called deep.

The above description makes clear that surge means that the flow runs alternatively through states with attached boundary layers and states with massively separated boundary layers. So, we may also describe the phenomenon as *pulsating stall*. The periodic forces on the blades are typically so high that blades cannot withstand them and surge causes destruction. So, surge is much more destructive than rotating stall and, therefore, surge is considered as not allowable.

Surge is only possible when there is a maximum in the compressor characteristic. As discussed before, this maximum comes from rotating stall. This means that both phenomena appear together, which makes flow patterns quite complex. In the right-hand part of Fig. 13.27 a sharp maximum is supposed. This occurs with full-span rotating stall, which is produced at high rotational speed (see next section). When the maximum is smooth, caused by part-span rotating stall at low rotational speed, the precise location of the start of the surge cycle depends on the shape of the load characteristic and the level of kinetic energy at the exit of the compressor. Often, with a large reservoir, surge is assumed to start at the maximum of the difference between the total pressure at compressor entry and the static pressure at compressor exit [3]. For a smooth maximum in the characteristic of total pressure ratio, this means a starting point somewhat at the left of the maximum.

13.5.5 Operating Range

Figure 13.28 is a field of characteristics at constant rotational speed of pressure ratio and efficiency as functions of the mass flow rate. *Choking* is attained with the larger rotational speeds. The characteristics are limited at the low flow rate side by a line called the *surge line* (line with short dashes in Fig. 13.28). But true surge only occurs with high rotational speed. The characteristic then has a sharp maximum and this maximum is the starting point of a possible surge cycle, independently from the capacity of the connected reservoirs and the compressor casing. At low rotational speed, the maximum is smooth and surge can only start from an operating point with lower flow rate than at the maximum. The limiting phenomenon is then rotating stall. But, conventionally, rotating stall is considered as a limiting instability phenomenon and the limiting instability line in the pressure ratio field is called the surge line, although surge only occurs at high rotational speed.

A possible operating line, imposed by the load, is drawn with short and long dashes in Fig. 13.28. An example of a unique operating line is with steady state operation of the compressor of an aero-engine. The compressor load is then by the downstream turbine and the downstream nozzle. For a given rotational speed, there is an operating margin against instability and choking. The *operating range* is expressed by the difference between the choking and surge mass flow rates divided by the choking mass flow rate at the design rotational speed.

Also, the notion of *surge margin* is used. There are several definitions of this margin. An example is the difference in pressure ratio between the instability and optimum efficiency values as a fraction of the optimum efficiency value for the design mass flow rate (vertical line in the chart). Expressed this way, the surge margin is typically around 25% for a compressor in an aero-engine [3].

Fig. 13.28 Chart of characteristics of an axial compressor; lines of pressure ratio and efficiency at constant rotational speed

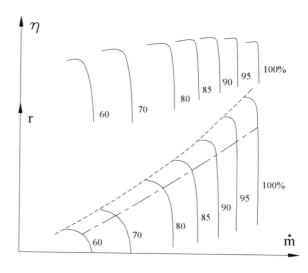

The operating line is not unique because of load transients. For instance, when an aero-engine is accelerated from lower speed to higher speed by fuel flow increase, the transient operating line comes closer to the surge line than the line of steady operating points. Actually, the maximum acceleration is limited by the margin between the line of steady operating points and the surge line.

A particular problem arises with series arrangement of stages. If, for design rotational speed and design backpressure, stages are matched in the sense that the operating point of each stage has a sufficient margin against instability and against choking, it is not guaranteed that a similar margin is kept at lower rotational speed or changed backpressure.

At a reduced rotational speed, the pressure rise is lower and the density in the rear stages is below the design value. For a constant-density fluid, the shape of the velocity triangles does not change when the stages maintain their flow coefficient and their work coefficient. So, stage matching is maintained for operating points that obey kinematic similitude. The reduced density rise causes a deviation with increased through-flow velocity in rear stages and decreased through-flow velocity in front stages. Thus, front stages are pushed towards stall (we discussed this phenomenon already in Sect. 11.1.4 of Chap. 11 for explaining the need of adjustable stator vanes). Rear stages are pushed towards choking, but choking is usually not attained due to lowered Mach number levels.

At design rotational speed, increased back pressure causes reduced through-flow velocities and stronger density rise. So, rear stages are driven towards stall. With increased backpressure, the following effects result: surge in the rear stages at design rotational speed or a somewhat enlarged or reduced speed, full-span rotating stall with one cell in the middle stages at reduced speed and rotating stall with several cells at the casing in the front stages at low rotational speed [5].

For lowered backpressure at design rotational speed, through-flow velocities increase in all stages. Mach number levels are high in front stages for reaching a high mass flow rate. Mach numbers are lower in rear stages due to enlarged velocity of sound by increased temperature and, possibly, reduced through-flow velocity for lowering exit velocity. Consequently, the margin with respect to choking is smaller in front stages. With reduced backpressure at design rotational speed, the density rise and the temperature rise decrease. Thus, in rear stages, velocities increase due larger mass flow rate and lower density. And, velocity of sound decreases due to lower temperature. This drives rear stages towards choking much faster than front stages. So, although the margin with respect to choking in rear stages is larger in design conditions, the usual result is that rear stages enter choking and not front stages [4].

High pressure ratio compressors (say $r = 20$ and above) cannot maintain stage matching at reduced speed without variable geometry, which means the possibility for stator vane angle adjustment. The dashed lines in Fig. 13.15 may also be seen as obtained with global mass flow rate reduction. The figure demonstrates that angle adjustment of stator vanes at reduced flow rate allows both the readjustment of inlet rotor velocity (w_1) and of inlet stator velocity (v_2) to the original direction. The adjustment is attained by a more tangential position of the stator vanes. This means *closing of the compressor*. Closing results in a comparable work, and thereby pressure ratio, but with a smaller flow rate than originally. The compressor characteristics are thus shifted towards lower flow rate this way.

With compressors operating with a largely changing load, many stages in the front part may be provided with variable geometry. This may be done with industrial compressors or compressors of stationary gas turbines for electric power generation. Aero-engines operate most of the time in cruise conditions. Actually, compressors of these gas turbines only have a problem during starting up. Often, only one or two stages in the front are provided with variable geometry.

Bleeds in the first stages is an alternative for variable geometry. Sufficient segmentation of the gas turbine is another method. E.g., some Rolls Royce Trent engine types have three shafts and no variable geometry (but the Tent800 taken as example in Chap. 12 has variable geometry in the front stages of the IP-compressor).

13.6 Exercises

13.6.1 Verify that the average density is well-suited to approximate the isentropic enthalpy change of a compression process. Represent the compression by a polytropic path. Calculate for air as an ideal gas ($\gamma = 1.40$) for pressure ratio values 1.2, 1.6, 2.0 and infinitesimal efficiency values 1, 0.95, 0.90: Δh_s, Δh_r, $\Delta p/\rho_m$, where $\Delta h_r = \int (1/\rho)\, dp$ is the reversible work. Observe that $\Delta p/\rho_m$ is a very good approximation of Δh_s (sometimes overestimation, sometimes underestimation) and a slight underestimation of Δh_r. With a compression, the ordering is:

$$\frac{\Delta p}{\rho_m} \approx \Delta h_s < \Delta h_r < \Delta h.$$

13.6.2 The average density concept is also applicable with expansion processes. The term $\Delta p/\rho_m$ constitutes an approximation of the isentropic enthalpy change, but is less accurate than with a compression. Calculate for air as an ideal gas ($\gamma = 1.40$) for pressure ratio values 1.2, 1.6, 2.0 and infinitesimal efficiency values 1, 0.95, 0.90: Δh_s, Δh_r, $\Delta p/\rho_m$. Observe that $\Delta p/\rho_m$ is a slight underestimation of Δh_s, and a somewhat stronger underestimation of Δh_r. With an expansion, the ordering is:

$$\Delta h < \frac{\Delta p}{\rho_m} \approx \Delta h_s < \Delta h_r.$$

13.6.3 An enthalpy loss coefficient cannot be determined accurately by measuring temperatures due to influence of heat transfer. For a stator (see Fig. 13.5), the enthalpy and pressure loss coefficients are $\xi = (h_{02} - h_{03s})/(h_{02} - h_2)$ and $\omega = (p_{02} - p_{03})/(p_{02} - p_2)$. The pressure loss coefficient follows directly from pressure measurements. Derive that the enthalpy loss coefficient also follows from pressure measurements by calculation of temperature ratios associated to pressure ratios. For constant density, both coefficients are identical, but with a compressible fluid, their difference increases with increasing Mach number.

Verify numerically that the enthalpy loss coefficient and the pressure loss coefficient are nearly equal for low inlet Mach number. Assume a compressor cascade with air ($\gamma = 1.40$) and infinitesimal efficiency 0.9. Consider inlet Mach numbers 0.3, 0.7 and 1.2. Assume, by way of simplification, zero outlet velocity. Observe that the enthalpy loss coefficient increases weakly with increasing inlet Mach number and that the pressure loss coefficient increases quite strongly.

Observe that ω is always larger than ξ. Remark that the Mach number influence on ω may be neutralised by taking the logarithm of pressure instead of pressure. Such a loss coefficient is 1 minus the infinitesimal efficiency (similarly as with expansions; see Exercise 6.10.1).

13.6.4 Formulate the relationship between the lift coefficient and the diffusion factor. Observe that, for a given diffusion factor value, a large variation of the lift coefficient is possible, strongly dependent on the general deceleration ratio w_2/w_1. This demonstrates that lift coefficient is an unworkable concept with compressor cascades.

13.6.5 The radius ratio chosen in Sect. 13.3 is 1.60 (1.20/0.75). This ratio is smaller with many compressor stages, but is much larger with a fan stage, where it may be up to 4. Take as an example the fan of the Tent800 (Fig. 12.9). Tip and hub radii are 1.40 m and 0.35 m. Consider the fan on the ground with $T_0 = 288$ K and $p_0 = 100$ kPa. Take as air properties $R = 287$ J/kgK and $\gamma = 1.40$. Take tip blade speed 450 m/s and inflow velocity (absolute system) in axial direction, constant over the radius, equal to 200 m/s (Mach number approximately 0.6).

Determine the velocity triangles at entry and exit of the rotor on the equal area mean radius (take $u = 328.0$ m/s) taking into account that the deceleration ratio in the rotor (w_2/w_1) cannot be lower than 0.7. Consider constant axial velocity through the rotor. Remark that deceleration in the stator is not useful on the mean radius since the exit flow enters the cold-flow nozzle.

Determine the work coefficient. Remark that it is quite high and that the inflow of the rotor is supersonic. But the flow is realisable since the turning in the rotor is small. Determine the total pressure ratio, assuming a polytropic efficiency of 0.9.

A: $\psi \approx 0.45$, $M_{w1} \approx 1.17$, $p_{03}/p_{01} \approx 1.63$.

Observe that at the casing, the same work as on the mean radius leads to a low work coefficient (**A:** $\psi \approx 0.24$). So, larger work is possible (see next exercise). Observe that it is not possible to realise the same work at the hub. Verify that for constant axial velocity through the stage, outlet of the stage at $+ 15°$ and strongest deceleration ratio of 0.7 in the stator, the rotor blade profile at the hub is nearly symmetric and the degree of reaction is approximately zero (verify the realism on Fig. 12.9). So, calculate with an exactly symmetric rotor blade profile.

A: $\psi \approx 2$, $\varphi \approx 1.78$, work ≈ 25 kJ/kg, $p_{03}/p_{01} \approx 1.30$.

13.6.6 Calculate the total pressure ratio of the fan rotor at the tip with the data of the previous exercise, assuming that the flow in the relative system is straight and that all compression comes from a normal shock. This forms an approximation. In reality, the flow passes through a bow shock, a compression fan, an oblique shock and a normal shock and experiences a small turning. Derive the flow parameters downstream of the shock from balances of mass, momentum and energy. Observe that a quadratic equation in velocity is formed from the energy equation by substitution of density as a function of velocity from the mass balance and substitution of pressure as a function of velocity from the momentum balance.

Determine the Mach number at the rotor inlet, the work coefficient and the isentropic re-expansion efficiency of the rotor. Observe the large work coefficient, the large pressure ratio and the high efficiency, but boundary layer losses in the rotor are not taken into account and neither are the losses in the post-connected stator. But, the conclusion remains that compression by a shock is quite efficient.

A: $M_{w1} \approx 1.50$, $p_{02}/p_{01} \approx 2.48$, $\psi \approx 0.46$, $\eta_{sre} \approx 0.935$.

13.6.7 Analyse the work in the stages of the compressor of the Trent800 for static sea level conditions ($T_0 = 288$ K). The compressor consists of a fan, 8 IP stages and 6 HP stages. From exercise 13.6.5 we know that the fan work at the hub is about 25 kJ/kg. Assume this value for the entire hot flow part of the fan. Assume the mean blade speed 350 m/s, both in the IP and HP compressor parts (see Sect. 13.1.12). Take $\gamma = 1.40$, 1.38 and 1.36 in the fan, the IP and HP parts and take $\gamma = 1.38$ as an average for the entire compressor. First, verify the work needed for the entire compression with pressure ratio 40 and polytropic efficiency 0.90 (**A:** about 630 kJ/kg).

Calculate then the work in the first and last stages of the IP part assuming $\phi = 0.5, \psi = 0.4$ for the first stage and $\phi = 0.45, \psi = 0.35$ for the last stage. Calculate similarly the work in the first and last stages of the HP part with $\phi = 0.45, \psi = 0.35$ for the first stage and $\phi = 0.4, \psi = 0.3$ for the last stage (flow coefficient and work coefficient diminish gradually from front to rear in a compressor).

Verify the total work in the compressor by summing the work contributions of the stages. Observe that the sum is near to the earlier determined value, which proves that the values of blade speeds and work coefficients are realistic. Calculate the pressure ratios in the first and last stages of the IP and HP parts (**A:** 1.58, 1.23, 1.22, 1.14). Observe the strong variation of the stage pressure ratio. Calculate the axial velocity at the entry of the IP part and the exit of the HP part (**A:** 175 m/s and 140 m/s).

13.6.8 Analyse the work transferred in the stages of the compressor of the SGT6-5000F for sea level conditions ($T_0 = 288$ K). The pressure ratio is 18.9 with 13 stages. Take $\gamma = 1.38$. First, verify the work needed for the entire compression with pressure ratio 18.9 and polytropic efficiency 0.90 (**A:** about 438 kJ/kg). Calculate then the work in the first and last stages assuming $\phi = 0.5, \psi = 0.4$ for the first stage and $\phi = 0.4, \psi = 0.35$ for the last stage. Assume the mean blade speed 300 m/s. Verify the total work in the compressor by summing the work contributions of the stages. Observe that the sum is near to the earlier determined value, which proves that the values of blade speed and work coefficients are realistic. Calculate the axial velocity at the entry and the exit (**A:** 150 m/s and 120 m/s).

References

1. Boyer KM, O'Brien WF (2003) An improved streamline curvature approach for off-design analysis of transonic axial compression systems. J Turbomachinery 125:475–481
2. Calvert WJ, Ginder RB (1999) Transonic fan and compressor design. J Mech Eng Sci IMechE 213:419–436
3. Cumpsty NA (1989) Compressor aerodynamics. Longman Scientific & Technical, ISBN 0-582-01364-X
4. Cumpsty NA, Greitzer EM (2004) Ideas and methods of turbomachinery aerodynamics: a historical view. J Propulsion Power 20:15–26
5. Day IJ, Freeman C (1994) The unstable behavior of low and high-speed compressors. J Turbomachinery 116:194–201
6. Fischer A, Riess W, Seume JR (2004) Performance of strongly bowed stators in a four-stage high-speed compressor. J Turbomachinery 126:333–338
7. Fottner L (1989) Review of turbomachinery blading design problems. Chapter 1 in AGARD LS 167. Blading design of axial turbomachines. AGARD, ISBN 92-835-0512-3
8. Gümmer V, Wenger U, Kau HP (2001) Using sweep and dihedral to control three-dimensional flow in transonic stators of axial compressors. J Turbomachinery 123:40–48
9. Hah C, Shin HW (2012) Study of near-stall flow behaviour in modern transonic fan with compound sweep. J Fluids Eng 134:071101
10. Happel HW, Stubert B (1988) Application of a 3D time-marching Euler code to transonic turbomachinery flow. In: 7th GAMM-conference on numerical methods in fluid dynamics. Vieweg Verlag, ISBN 3-528-08094-9, pp. 120–129
11. Lewis RI (1996) Turbomachinery performance analysis. Wiley, ISBN 0-470-23596-9
12. Lieblein S, Schwenk FC, Broderick RL (1953) Diffusion factor for estimating losses and limit blade loadings in axial-flow compressor blade elements. NACA RM E53D01
13. Lieblein S (1965) Experimental flow in two-dimensional cascades. In: Chapter 6 of Aerodynamic design of axial-flow compressors. NASA SP-36
14. Rechter H, Steinert W, Lehmann K (1985) Comparison of controlled diffusion airfoils with conventional NACA 65 airfoils developed for stator blade application in a multistage axial compressor. J Eng Gas Turbines Power 107:494–498
15. The Jet Engine (1986) Rolls-royce. ISBN 0-902121-04-9
16. Tweedt DL, Schreiber HA, Starken H (1988) Experimental investigation of the performance of a supersonic compressor cascade. J Turbomachinery 110:456–466

Chapter 14
Radial Compressors

Abstract Radial compressors (or centrifugal compressors) resemble radial fans and radial pumps for basic operation aspects. So, based on the study of fans and pumps, done in previous chapters, we can understand the basic operation of centrifugal compressors. In this chapter, we repeat the analysis of the working principles of centrifugal machines, but applied to centrifugal compressors. We discuss aspects that are particular for centrifugal compressors. These are the inducer part at the entry to a rotor for large work and the diffuser downstream of the rotor. We also analyse the operating characteristics with the limits by stall and choking.

14.1 Construction Forms and Applications

14.1.1 Rotor Types

Figure 14.1 shows a radial or centrifugal compressor for chemical plants and refineries and for transport of natural gas. The terms radial and centrifugal are equivalent with compressors. The successive rotors have increasingly smaller through-flow areas due to increasing density. A rotor is mostly called an *impeller*, which means the element that actuates the fluid. The rotors of Fig. 14.1 have a shroud, so are closed, and the blades are moderately swept backward at the exit.

There is no strong difference with rotors of a centrifugal pump. At the rotor entry, blades begin downstream of the orthogonal entry plane. Thus, as with pumps, there is a rotor eye. At the leading edge of the rotor blades, the velocity has both an axial and a radial component. With terminology of centrifugal pumps, the blades are thus extended into the rotor eye. The blades are doubly-curved, which means that the curvature differs in different orthogonal sections, but the axial variation of the curvature is moderate. Some compressors have bladed rotor parts with a purely radial entry (no axial velocity component), but these are rare.

Compared to pumps, the ratio of the mean diameter at rotor entry (bladed part) to the diameter at rotor exit is larger (this ratio lowers in the successive rotors), and the backward sweep of the blades at the rotor exit is less. Flow factors and work factors are thus higher. The tendency to higher values comes from the lower density of the

© The Author(s), under exclusive license to Springer Nature Switzerland AG 2022   539
E. Dick, *Fundamentals of Turbomachines*, Fluid Mechanics and Its Applications 130,
https://doi.org/10.1007/978-3-030-93578-8_14

Fig. 14.1 Multistage centrifugal compressor for chemical plants, refineries and gas transport; closed rotors (*Courtesy* MAN Energy Solutions)

fluid, which is a gas instead of a liquid, and the property of driven turbomachines that pressure rise and mass flow rate are proportional to the density at the machine entry.

Figure 14.2 shows the compressor rotor of a turbocharger for a large diesel engine. The rotor is open with an axial entry part: the leading edges of the longest blades are in the orthogonal entry plane of the rotor. The shorter blades, called splitter blades, start more downstream. Blades are moderately backward swept at the rotor exit (typical

Fig. 14.2 Open radial compressor rotor with axial inducer entry part and splitter blades; blades have slight backward sweep and slight lean at the rotor exit; compressor part of a diesel engine turbocharger (*Courtesy* ABB Turbocharging)

is about 30° backward sweep) and have a slight forward lean (the tip is advanced in the rotation sense).

Open rotors are intended for a large pressure ratio and a large flow rate. A high peripheral speed, 450–500 m/s, is used therefore. With a high peripheral speed, a shroud on the rotor would cause too high stress on the blades due to the centrifugal force. For limiting bending stress, a blade should not deviate too strongly from a surface formed by radial filaments (see Sect. 14.4.2). Therefore, the leading edge part of the blades is formed with near-radial filaments. The rotor entry part resembles an axial compressor. This part is termed an *inducer*.

Half of the blades are extended into the inducer, the other half are not. In other words, there are *primary blades* and *splitter blades*. Splitter blades divide the channel that is formed by primary blades into two parts, from a certain distance beyond the rotor entry plane. The reason for the splitter blades will be discussed in Sect. 14.4.2. The lean at the rotor exit brings the blade shape closer to one with radial filaments. The figure is also intended to demonstrate that even the primary blades do not have a very complex shape.

14.1.2 General Shape of a Radial Compressor

Figure 14.3 sketches a single-stage radial compressor with an open rotor and radial-end blades (but we discuss the effect of backsweep). The incoming flow is in the axial direction. Some machines feature a stationary blade row at the inlet. The inlet

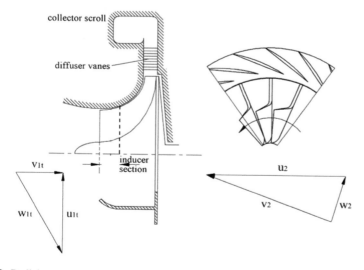

Fig. 14.3 Radial compressor: parts and velocity triangles

guide vanes are intended to give whirl to the flow at the rotor entry, so to deviate it from the axial direction.

The pre-whirl vanes change the operating characteristic (see Exercise 14.8.2). The entry velocity triangle is drawn at the inducer tip. Slip has been taken into account with the velocity triangle at the rotor exit, so that the relative outflow is inclined somewhat backward. With radial-end blades, the tangential outflow velocity component (v_{2u}) is somewhat lower than the peripheral speed (u_2).

The rotor work is $\Delta W = u_2 v_{2u}$.

For the same rotor work, the rotor exit velocity decreases with more backward sweep, but an increased peripheral speed is necessary. E.g., compare $u_2 = 100\%$ and $v_{2u} = 75\%$ to $u_2 = 125\%$ and $v_{2u} = 60\%$. But, high rotor work inevitably causes a high outflow velocity of the impeller. So, a diffuser becomes necessary.

A diffuser with vanes is sketched. Other diffuser shapes are possible and will be discussed in a later section. A scroll-shaped collector comes around the diffuser. Diffusers inherently cause large loss (see later discussion). Compressors with slightly backward swept blades thus have a lower efficiency than compressors with more inclined ones. Rotors with forward inclined blades are never used.

14.1.3   Comparison Between Radial and Axial Compressors

The fundamental difference between both types is that radial compressors, for equal rotor diameter and equal peripheral speed, handle a smaller flow rate, but realise a larger stage work. With an axial compressor, 75% of the frontal area may be available for through-flow (at a 0.5 radius ratio) with an axial velocity up to about 50% of the blade speed on the mean radius (about 40% of the tip speed). A flow factor defined with frontal area and peripheral speed may thus reach about 0.30.

The flow area with a radial compressor maximally covers about 45% of the frontal area with a through-flow velocity maximally about 50% of the blade speed at the inlet tip (about 35% of the peripheral speed of the rotor). The flow factor, according to the above definition, is maximally about 0.15. Shrouded rotors (Fig. 14.1) often feature a much lower flow factor. The flow factor of radial rotors varies between about 0.02–0.15, with the largest values obtained by open rotors.

The work coefficient of an axial compressor stage, related to the blade speed on the mean radius, is at maximum about 0.4 ($\psi = \Delta W/u^2$). Related to tip speed it is maximally about 0.3. With a radial compressor with radial-end blades, the work coefficient is about 0.9.

The peripheral speeds of open-rotor axial and radial compressors do not differ strongly. Front stages of axial compressors are mostly made of titanium alloys. Titanium alloys and high-strength aluminium alloys are used for rotors of radial compressors. Fabrication of shrouded radial rotors is then by precision casting, while for unshrouded rotors it is either precision casting or milling. High-strength Ni–Cr steel alloys are used as well. The hub with the blades is then cut as a single piece by flank-milling and a shrouded rotor is formed by welding the shroud, after turning it on a lathe. Welding is possible, because the rotor channels are rather open.

The maximum peripheral speed of a shrouded radial rotor is about 350 m/s, independent of the material. The peripheral speed of an axial compressor is generally until about 400 m/s, while for an open-rotor radial compressor it is generally until 500 m/s. Strength does not strictly limit this speed. With titanium alloys in axial compressors and unshrouded centrifugal compressors, 550 m/s is achievable with both types.

A speed limitation normally follows from Mach number limitation. With 400 m/s peripheral speed, the relative velocity on the mean radius at the rotor entry of an axial compressor is slightly supersonic and with 500 m/s, the relative velocity at the inducer tip and the absolute velocity at the rotor exit of a radial compressor with moderate outlet backsweep are slightly supersonic too (see Sect. 14.3).

But unshrouded radial rotors of high-strength stainless steel may run at peripheral speeds above 650 m/s, even with rather large outlet backsweep. The attainable pressure ratio is then up to 9 (see Sect. 14.3), but Mach numbers at inducer entry and diffuser entry are then high, which penalises the efficiency and the operating range. So, the pressure ratio of a centrifugal compressor stage can be much higher than that of an axial stage, where it is about 1.75 at maximum.

The efficiency of an axial compressor is significantly better than that of a radial one. The isentropic total-to-total efficiency of an axial compressor is about 0.92 at the maximum compared to maximally about 0.85 for a centrifugal compressor with open rotor and moderate backsweep. The efficiency improves with more backward sweep and closed rotors to maximally about 0.87.

The mean reason for lower efficiency is the inhomogeneous flow at the rotor exit of a radial compressor by jet-wake forming (Sect. 2.3.2 in Chap. 2; Sect. 14.4.3). The inhomogeneity causes a significant mixing loss downstream of the rotor. Further, rotors for high work require a significant velocity reduction in a diffuser downstream of the rotor. A strong deceleration inevitably creates large loss. A third reason is rather important leakage loss with an open impeller.

14.1.4 Examples of Radial Compressors

Some industrial centrifugal compressors resemble centrifugal pumps. Figure 14.1 is an example. Rotors have backward swept blades, as pumps do. After each rotor, there is a ring-shaped vaneless chamber, functioning as a diffuser, followed by a chamber with return channels leading the flow to the next stage. After the last stage, the gas is collected with a scroll.

The work per rotor is not very high in compressors as in Fig. 14.1. The stage work coefficient is maximally about 0.5. So, the rotor exit velocity is maximally about 50% of the peripheral speed. Therefore, a vaneless annular chamber is sufficient as diffuser (see Sect. 14.5). A compressor as in Fig. 14.1 is intended to allow a large flow rate variation. The range is limited by rotating stall or surge at low flow rate and choking at high flow rate (similarly as with axial compressors).

The operating range will be discussed in Sect. 14.7. With moderate backward sweep of the blades and vaneless diffusers, the design flow rate has a large margin against the limiting phenomena.

The air compressor shown in Fig. 14.4 features a first open rotor stage (but there are no splitter blades), followed by three shrouded ones. With this type, the first stage has an open rotor because the volume flow rate is the largest in the first stage. The compressor shown in Fig. 14.4 is intended for a relatively high flow rate. Similar machines exist with only shrouded rotors, intended for lower flow rates.

The successive radial stages feature increasingly smaller through-flow areas. The rotors have moderately backswept blades, resulting in a higher work coefficient than typical for pump rotors. This implies the need for more blades. Typically, rotors have 16–20 blades (see also Fig. 14.1).

The air is cooled after each stage (called an isothermal compressor). The coolers are integrated in the machine. The supply to the coolers requires a rather special vaned chamber, functioning as diffuser. A simpler vaned chamber guides the flow

Fig. 14.4 Radial air compressor with integrated coolers (*Courtesy* MAN Energy Solutions)

from the exit of a heat exchanger to the entry of a rotor. Figure 14.5 shows cross sections of the chambers.

Figure 14.6 shows parts of a radial compressor with ten stages and an *integrated gearbox* (integrally geared compressor). This arrangement is an alternative to an inline construction. The objective is a sufficiently high and approximately equal peripheral speed of the successive stages with decreasing diameter for compressors with rather small rotor dimensions.

Figure 14.7 shows a diesel engine turbocharger with a radial compressor for high work and an axial turbine. Further, a radial compressor is often used as the last stage of the compressor part of a small gas turbine for helicopter rotor drive. The general appearance is then somewhat similar to the turbocharger of Fig. 14.7, but with one or

Fig. 14.5 Cross sections of
the vaned chambers for inlet
to (left) and outlet from
(right) the radial rotors of
Fig. 14.4

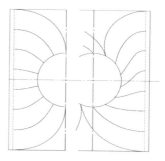

Fig. 14.6 Centrifugal
compressor with 10 stages
for CO_2 with integrated
gearbox (*Courtesy* MAN
Energy Solutions)

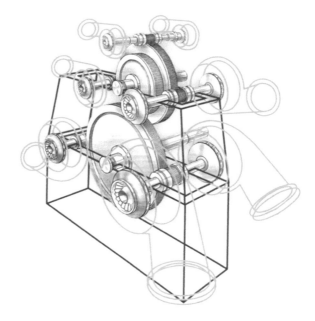

two axial stages upstream of the radial compressor and addition of a power turbine
stage.

14.2 Kinematic Parameters

We take a rotor with radial-end blades, as sketched in Fig. 14.3, as a first example,
but we discuss the effect of backsweep. We assume the absolute entry velocity in
the meridional plane and a constant meridional velocity component (v_m) through the
rotor.

The work coefficient is

$$\psi = \frac{\Delta W}{u_2^2} = \frac{u_2 v_{2u}}{u_2^2} = \frac{v_{2u}}{u_2}. \tag{14.1}$$

The degree of reaction is

$$R = \frac{h_2 - h_1}{h_{02} - h_{01}}, \tag{14.2}$$

with $h_{02} - h_{01} = u_2 v_{2u}$ and $h_2 - h_1 = u_2 v_{2u} - \left(\frac{1}{2} v_2^2 - \frac{1}{2} v_1^2\right)$.
With $v_2^2 = v_{2u}^2 + v_{2m}^2$ and $v_{2m} = v_1$: $h_2 - h_1 = u_2 v_{2u} - \frac{1}{2} v_{2u}^2$.
Thus:

$$R = 1 - \frac{1}{2} \frac{v_{2u}}{u_2} = 1 - \frac{1}{2} \psi. \tag{14.3}$$

This relation is the same as the one derived for radial fans in Chap. 3. For radial-end
blades, taking slip into account, v_{2u} is somewhat smaller than u_2. Typically: $\psi \approx 0.9$

Fig. 14.8 h–s diagram of a centrifugal compressor; slanted dashed lines represent isobars (R ≈ 0.65)

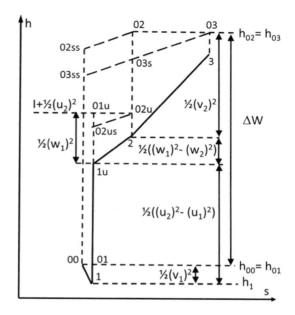

(see Sect. 14.3), thus $R \approx 0.55$. $\frac{1}{2}v_2^2 - \frac{1}{2}v_1^2$ is then about 45% of the work and $\frac{1}{2}v_2^2$ is about half of the work. Because the diffuser loss is then quite high, it is better to provide some backsweep. This reduces v_{2u}. It is also useful to reduce somewhat v_{2m}. We illustrate the tendencies with $u_1 = 0.45\, u_2$ (mean radius), $v_1 = 0.35\, u_2$, $v_{2u} = 0.75\, u_2$ and $v_{2m} = 0.3\, u_2$. The h–s diagram of Fig. 14.8 is constructed with these data. The degree of reaction is approximately 0.65.

The rothalpy, $I = h + \frac{1}{2}w^2 - \frac{1}{2}u^2$, is constant in the rotor. With axial inflow velocity, this implies:

$$I = h_1 + \frac{1}{2}w_1^2 - \frac{1}{2}u_1^2 = h_1 + \frac{1}{2}v_1^2 = h_{01}.$$

The static enthalpy rise in the rotor is

$$h_2 - h_1 = \frac{1}{2}\left(u_2^2 - u_1^2\right) + \frac{1}{2}\left(w_1^2 - w_2^2\right).$$

The representation of the deceleration in the diffuser is the same as with an axial compressor. For representation of the deceleration in the impeller, it has to be taken into account that the enthalpy increase due to the centrifugal force is free of losses. Therefore, a supplementary state is defined, with subscript *1u*, obtained by isentropic addition of the enthalpy $\frac{1}{2}\left(u_2^2 - u_1^2\right)$ to the point representing the static state with subscript *1*. By further isentropic addition of the enthalpy $\frac{1}{2}w_1^2$, a corresponding total state point, with subscript *01u*, is obtained. The resulting enthalpy level is a constant of the rotor flow:

$$h_{01u} = h_1 + \frac{1}{2}\left(u_2^2 - u_1^2\right) + \frac{1}{2}w_1^2 = I + \frac{1}{2}u_2^2.$$

The loss coefficients of the diffuser and the rotor and the corresponding isentropic deceleration efficiencies can be defined by:

$$\xi_{dif} = (h_{03} - h_{03s})/(1/2)v_2^2 \text{ and } \xi_{rot} = (h_{02u} - h_{02us})/(1/2)w_1^2. \quad (14.4)$$

$$\eta_{sd,dif} = (h_{03s} - h_2)/(h_{03} - h_2), \quad \eta_{sd,rot} = (h_{02us} - h_{1u})/(h_{02u} - h_{1u}). \quad (14.5)$$

The enthalpy increase in the rotor due to deceleration is quite small in Fig. 14.8. This is typical for any centrifugal compressor. Station 2 is the exit of the rotor. The real rotor outflow is very inhomogeneous. Mostly, a 1D representation is defined by mass flow rate averaged quantities. Losses downstream of the rotor are then accounted to the downstream stator part, encompassing a mixing space, a vaned or vaneless diffuser and a volute (or a return chamber). But, it is also possible to define a mixed-out state and account the mixing loss to the rotor (done in Fig. 14.8).

The stage efficiency is generally defined between the total state at the compressor entry, which is the rotor eye entry (state 00), and the total state at the compressor exit, which is the return chamber or volute exit (state 03). The total state at the compressor entry is generally taken as the reference state for determination of dimensionless quantities. The usual definition of isentropic total-to-total internal efficiency is (see Fig. 14.8):

$$\eta_{tt} = \frac{h_{03ss} - h_{00}}{h_{03} - h_{00}} = \frac{\Delta h_{0s}}{\Delta h_0} = \frac{\Delta h_{0s}}{\Delta W}. \quad (14.6)$$

The third kinematic parameter, the flow coefficient, follows from efficiency optimisation. It is commonly assumed, as a first approximation, that maximum efficiency is obtained by minimising the Mach number of the relative velocity at the inducer entry, at the casing position (tip): w_{1t} [4, 10]. The objectives are to minimise shock losses at the entry (the relative Mach number is always quite high), minimise friction loss in the rotor, but also make the rotor outflow in the absolute frame as tangential as possible. This minimises the radial dimensions of the diffuser and the volute, which reduces the losses within these components. Strong deceleration in the rotor helps to attain an absolute rotor exit velocity as tangential as possible. It is mostly assumed that the best is to use a deceleration that is near to the limit for avoiding boundary layer separation. The limit deceleration is usually expressed by the ratio w_{1t}/w_{2mix}, where w_{2mix} is the mixed-out value. There is no universally valid limit value of this ratio. An estimate of the maximum possible ratio is $w_{1t}/w_{2mix} \approx 1.75$, thus $w_{2mix}/w_{1t} \approx 0.57$ [3, 9, 10]. Minimising the Mach number at the rotor inlet also realises the largest possible margin against choking, if the rotor inflow is the most critical for choking (see Sect. 14.7).

The problem of minimising the inlet Mach number for imposed mass flow rate is equivalent to maximising the mass flow rate for imposed Mach number of the relative flow at the rotor tip. Without pre-whirl at the rotor entry, the following relations hold (see Fig. 14.3):

$$v_{1t} = w_{1t}\cos(\beta_{1t}) = M_{1t}\,c_1\cos(\beta_{1t}), \quad u_{1t} = w_{1t}\sin(\beta_{1t}) = M_{1t}\,c_1\sin(\beta_{1t}),$$

$$\frac{\rho_1}{\rho_{10}} = \left(\frac{T_1}{T_{10}}\right)^{1/(\gamma-1)}, \quad \frac{c_1}{c_{10}} = \left(\frac{T_1}{T_{10}}\right)^{1/2}, \quad \frac{T_{01}}{T_1} = 1 + \frac{\gamma-1}{2}M_{1t}^2\cos^2(\beta_{1t}).$$

The mass flow rate is:

$$\dot{m} = \rho_1 v_1 A_1 = \frac{\rho_1}{\rho_{01}}M_{1t}\cos(\beta_{1t})\frac{c_1}{c_{01}}k\frac{\pi}{4}d_{1t}^2\rho_{01}c_{01}, \tag{14.7}$$

with $A_1 = \frac{\pi}{4}(d_{1t}^2 - d_{1h}^2) = \frac{\pi}{4}k\,d_{1t}^2$.

From the inlet velocity triangle follows:

$$\Omega\,d_{1t} = 2u_{1t} = 2M_{1t}\frac{c_1}{c_{01}}\sin(\beta_{1t})\,c_{01}. \tag{14.8}$$

We take the diameter ratio at the rotor entry $d_{1h}/d_{1t} = 0.4$, so that the obstruction factor k is about 0.85.

For given M_{1t}, the value of $\cos(\beta_{1t})$ for maximum mass flow rate, under the condition (14.8), can be determined analytically. Some results are listed in Table 14.1. For a tip Mach number around unity, the optimum inflow angle is about $-60°$. With these values, the mass flow rate (14.7) and tip diameter (14.8) are:

$$\dot{m} \approx 0.288\rho_{01}\,c_{01}\,d_{1t}^2 \quad \text{and} \quad \Omega\,d_{1t} \approx 1.69\,c_{01}.$$

The corresponding entry flow factor is ($\rho_{00} \approx \rho_{01}$):

$$\Phi_1 = \frac{\dot{m}}{\rho_{00}\,\Omega\,d_{1t}^3} \approx 0.171.$$

With $Q_{00} = \Phi_1\Omega\,d_{1t}^3$ and $\Delta h_{0s} = \eta_{tt}\,\psi u_2^2$,

the specific speed is:

$$\Omega_s = \frac{\Omega\sqrt{Q_{00}}}{(\Delta h_{0s})^{3/4}} = \frac{\sqrt{8\,\Phi_1}}{(\eta_{tt}\,\psi)^{3/4}}\left(\frac{r_{1t}}{r_2}\right)^{3/2}. \tag{14.9}$$

Table 14.1 Optimum inflow angle as a function of tip Mach number

M_{1t}	0.25	0.9	1.0	1.2	1.5
$(\beta_{1t})_{opt}$	$-55.2°$	$-59.7°$	$-60.6°$	$-62.6°$	$-65.4°$

With radial-end blades, $\psi \approx 0.9$ and $\eta_{tt} \approx 0.82$, the coefficient in (14.9) is about 1.47. To $r_{1t}/r_2 = 0.75$, which is about the maximum possible diameter ratio, corresponds then $\Omega_s \approx 0.95$. A lower specific speed is attained with a lower diameter ratio and a lower flow factor. A higher specific speed is attained with a higher flow factor and larger blade backward sweep at the impeller exit.

The specific speed ranges from about 0.4–1.4. The best efficiency is reached in the 0.7–1.1 range, with closed rotors with backward sweep 25°–50° [1, 3]. A still higher specific speed is possible with a mixed-flow impeller, but mixed-flow machines are rare, as typically the aim of a radial compressor is a much larger pressure ratio than with an axial compressor.

The tip Mach number of the relative entry flow can be further reduced by positive pre-whirl by inlet guide vanes (Fig. 14.3). But positive pre-whirl reduces the rotor work. As the aim with centrifugal rotors is mostly a high work coefficient, inlet guide vanes are not often used, unless they are meant for flow rate variation.

A final remark is that the optimum inflow angles at the tip are similar to those found for axial compressors (Tables 13.2 and 13.4).

14.3   Pressure Ratio

First, we take a compressor with radial-end blades, as sketched in Fig. 14.3, as an example. The slip velocity at the rotor exit (difference between the actual and geometrical tangential velocity components v_{2u} and v_{2u}^b) may be expressed by the slip factor, σ, so that $v_{2u} = \sigma u_2 + w_{2u}^b$. With Wiesner's formula (Eq. 3.21 in Chap. 3), the slip factor is estimated by

$$\sigma = 1 - \frac{\sqrt{\cos \beta_2^b}}{Z^{0.7}}, \tag{14.10}$$

with β_2^b the outlet blade angle. For radial-end blades $\beta_2^b = 0$. With $Z = 20$, it follows that $\sigma \approx 0.9$.

Rotor work is $\Delta W = u_2 v_{2u} = \sigma u_2^2$.

The pressure ratio follows from $\frac{p_{03}}{p_{00}} = \left(\frac{T_{03s}}{T_{00}}\right)^{\frac{\gamma}{\gamma-1}}$.

Total-to-total efficiency is $\eta_{tt} = \frac{T_{03s}-T_{00}}{T_{03}-T_{00}}$.

Thus:

$$T_{03s} = T_{00} + \eta_{tt}(T_{03} - T_{00}) = T_{00} + \eta_{tt}\frac{\Delta W}{C_p} = T_{00}\left(1 + \eta_{tt}\frac{\Delta W}{C_p T_{00}}\right).$$

Further: $C_p T_{00} = \frac{\gamma}{\gamma-1} R T_{00} = \frac{1}{\gamma-1} c_{00}^2$.

For radial-end blades:

$$r = \frac{p_{03}}{p_{00}} = \left[1 + \eta_{tt}(\gamma - 1)\frac{\sigma u_2^2}{c_{00}^2}\right]^{\frac{\gamma}{\gamma-1}}.$$

Example: $T_{00} = 288$ K, $\gamma = 1.4$, $R = 287$ J/kgK, $c_{00} = 340$ m/s, $u_2 = 450$ m/s, $\eta_{tt} = 0.82$, $\sigma = 0.9$.
This combination gives :

$$r = \frac{p_{03}}{p_{00}} \approx 4.3.$$

The peripheral rotor speed is commonly expressed by a Mach number, called the *machine Mach number*, as

$$M_{u0} = u_2/c_{00}.$$

To the given example corresponds $M_{u0} = 1.32$. The rotor exit Mach number may, with radial-end blades, be determined with the approximation $v_2 \approx u_2$ (Fig. 14.3). So: $h_{02} = h_{00} + \Delta W$ or $h_2 + 1/2\, v_2^2 = h_{00} + \sigma u_2^2$,

$$h_2 \approx h_{00} + \left(\sigma - \frac{1}{2}\right)u_2^2, \quad \text{and} \quad \frac{1}{\gamma - 1}c_2^2 \approx \frac{1}{\gamma - 1}c_{00}^2 + \left(\sigma - \frac{1}{2}\right)u_2^2.$$

Thus:

$$M_2^2 = \frac{v_2^2}{c_2^2} \approx \frac{u_2^2}{c_{00}^2 + (\gamma - 1)(\sigma - \frac{1}{2})u_2^2} = \frac{M_{u0}^2}{1 + (\gamma - 1)(\sigma - \frac{1}{2})M_{u0}^2}.$$

With $M_{u0} = 1.32$ corresponds $M_2 \approx 1.17$. This means, with pressure ratio equal to 4.3, that the rotor outflow is slightly supersonic. Some deceleration within a vaneless space may reduce the Mach number to just below 1 ($M \approx 0.9$). This allows leading the flow into a vaned diffuser, just avoiding choking.

The previous example is not optimal. A higher pressure ratio is attained with a comparable Mach number at the rotor exit by moderate backsweep combined with a higher peripheral speed. We illustrate this strategy with: $u_{1t} = 300$ m/s, $u_2 = 500$ m/s, $\beta_{1t} = -60°$, $\beta_2 = -30°$ (flow angles).

With inflow in the meridional plane: $v_1 = 173.2$ m/s, $w_{1t} = 346.4$ m/s. With $T_{00} = T_{01} = 288$ K and properties of dry air follows $T_1 = 273.1$ K, $c_1 = 331.2$ m/s, $M_{w1t} = 1.05$. For the exit triangle, we take $v_{2r} = v_1$. Then: $w_{2u} = -100$ m/s, $w_2 = 200$ m/s, $v_{2u} = 400$ m/s. The rotor deceleration ratio is $w_{1t}/w_2 = 1.73$.

$\Delta W = u_2 v_{2u} = 200.0$ kJ/kg, $T_{02} = 487.1$ K, $T_2 = 392.5$ K, $c_2 = 397.1$ m/s, $M_{v2} = 1.10$. With a total-to-total isentropic efficiency of 0.84, the isentropic temperature increase is 167.3 K and the corresponding total pressure ratio is $p_{03}/p_{01} \approx 5$.

The Flow at the rotor exit is slightly supersonic, comparable to the first example. The Mach number at the impeller entry was not calculated in the first example, but, clearly, due to the higher peripheral speed it is now higher and brought to the level of the Mach number at the rotor exit. The machine Mach number is now about 1.5.

As a third example, we take an impeller with total pressure ratio 11, with large peripheral speed and large backsweep, documented by Higashimori et al. [7]. The peripheral speeds at rotor entry and exit are 460 m/s and 680 m/s. Blade angles are not specified, but from figures, one may estimate the flow angles $\beta_{1t} = -65°$ and $\beta_2 = -45°$. We calculate in the same way as with the previous example.

From the inlet triangle follows: $v_1 = 214.5$ m/s and $w_{1t} = 507.5$ m/s. With $T_{00} = T_{01} = 288$ K and properties of dry air follows: $T_1 = 265.1$ K, $c_1 = 326.4$ m/s, $M_{w1t} = 1.56$. This Mach number is near to the value 1.6, given by the authors.

For calculation of the exit triangle, we take $v_{2r} = v_1$. Then: $w_{2u} = -214.5$ m/s, $w_2 = 303.3$ m/s, $v_{2u} = 465.5$ m/s. The deceleration ratio in the rotor is $w_{1t}/w_2 = 1.67$. The rotor work is $\Delta W = u_2 v_{2u} = 316.5$ kJ/kg, and with properties of dry air: $T_{02} = 603.1$ K, $T_2 = 472.4$ K, $c_2 = 435.6$ m/s, $M_{v2} = 1.18$. This Mach number is near to the value 1.2, given by the authors.

From $T_{02}/T_{01} = 2.09$ follows that the isentropic total pressure ratio of the impeller is 13.25. With the specified pressure ratio of 11 corresponds a polytropic impeller efficiency of about 93%, which is thus realistic. With an estimated total-to-total isentropic efficiency of the complete compressor of 80%, the total pressure ratio of the compressor is about 9.0. The machine Mach number is about 2.0.

The conclusion is that a pressure ratio equal to 5 may be attained by a single-stage radial compressor with moderate backsweep, about 30°, and a peripheral speed of about 500 m/s. The inflow at the tip of the inducer and the outflow of the rotor are then just supersonic. A pressure ratio larger than 5 is attainable since the peripheral speed can exceed 500 m/s, but with a pressure ratio above 5, supersonic entry to the inducer and the diffuser become unavoidable. Shock waves occur then, lowering the efficiency. Moreover, the operating range becomes then narrower (see Sect. 14.7). Single-stage radial compressors with a pressure ratio 6, even 9, exist, but they can only be used for a limited number of applications.

14.4   Rotor Shape

14.4.1   Number of Blades

Determination of the minimum required number of blades with radial rotors was already discussed in Chap. 3 (Sects. 3.3.4 and 3.3.5). We repeat some elements of the reasoning, but adapted to centrifugal compressors. In Chap. 3, we analysed fans, which are machines of low rotor solidity and therefore we used an estimate of the velocity difference between the suction and pressure sides of a blade passage based on Pfleiderer's reasoning for rotors with small overlap of blades. With compressors,

Fig. 14.9 Average
streamline within a rotor and
transverse velocity variation:
Coriolis force Co,
Centrifugal force by
streamline curvature Cu

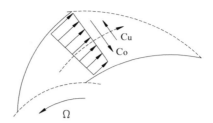

the rotor solidity is high and the reasoning according to Stodola, considering the
rotor as composed of blade channels, is more appropriate.

The equation for the velocity difference across a blade channel at a given radius
(Eq. 3.14 in Chap. 3) is

$$w_s - w_p = \Delta\theta \cos\beta \left[2\Omega r + \frac{d}{dr}(r w_u) \right], \qquad (14.11)$$

with $\Delta\theta = 2\pi/Z$ and Z the number of blades. This equation was derived in Chap. 3
from flow analysis in the rotating frame and considering the circulation of the relative
velocity on two different contours.

The first term between the square brackets originates from the Coriolis force,
while the second term comes from the streamline curvature, i.e. the lift. For radial-
end blades (when including slip) and blades with backward sweep, the second term
is negative at the rotor exit. The equation may also be derived directly, albeit with
the necessity of some approximations, from a moment of momentum balance on a
streamtube with infinitesimal radial length, spanning a blade passage (see Exercise
14.8.1).

Based on Eq. (14.11), we may write at the rotor exit:

$$(w_s - w_p)_2 = \frac{2\pi}{Z} \cos\beta_2 2 u_2 f_R,$$

and

$$(w_s - w_2)_2 = (w_2 - w_p)_2 = \frac{2\pi}{Z} \cos\beta_2 u_2 f_R, \qquad (14.12)$$

where w_2 is the average relative flow velocity at the rotor exit and f_R is a factor smaller
than unity, expressing the effect of the curvature of the streamlines in backward sense.
Figure 14.9 sketches an average streamline, the intervening forces and the resulting
flow velocity variation transverse to the streamlines.

From the flow pattern follows that two criteria for avoiding separation at the rotor
exit have to be met. Firstly, there must be a sufficient number of blades, so that the
Coriolis force cannot cause flow reversal at the pressure side. Secondly, the blade

loading must be limited to avoid boundary layer separation near the trailing edge at the suction side, as a consequence of the local flow deceleration.

Flow reversal at the pressure side is avoided for

$$w_2 > \frac{2\pi}{Z} \cos \beta_2 u_2 f_R.$$

(14.13)

Avoidance of boundary layer separation at the suction side may be expressed by a local diffusion factor criterion as

$$\frac{w_s - w_2}{w_s} = D_{loc} < 0.5 \quad \text{or} \quad w_s - w_2 < \frac{(D_{loc})_{\max}}{1 - (D_{loc})_{\max}} w_2.$$

(14.14)

For $(D_{loc})_{\max} = 0.5$ this criterion also results in (14.13). This value of D_{loc} is the critical value for separation in axial cascades. Deceleration in the boundary layer at the suction side is more disadvantageous with a centrifugal rotor, due to the jet-wake flow. This phenomenon weakens the suction side boundary layer (see Sect. 14.4.3). Therefore, the critical value of D_{loc} is rather 0.45. So, in practice, separation at the suction side occurs rather than flow reversal at the pressure side.

Therefore, we further apply criterion (14.14) as

$$\frac{2\pi}{Z} \cos \beta_2 u_2 f_R < 0.80 \, w_2.$$

(14.15)

The streamline curvature in the vicinity of the rotor exit is quite modest with centrifugal compressors. In practice, only rotors with radial-end blades ($\beta_2 = -15°$, after slip), with slight backward sweep ($\beta_2 = -20°$ to $-30°$) and with moderate backward sweep ($\beta_2 = -45°$) are used. The curvature factor f_R in (14.15) may therefore, by way of simplification, be taken as 0.80, so that approximately holds

$$\frac{2\pi}{Z} \cos \beta_2 u_2 < w_2.$$

(14.16)

The simplification spares us then deriving an expression for the curvature effect, as in Chap. 3. We note that the reasoning in Chap. 3 is not applicable to centrifugal compressors for high work due to the presence of the inducer. The effect of the curvature is rather a local rotor exit phenomenon with a centrifugal compressor, but the reasoning in Chap. 3 derives the curvature effect from the overall rotor through-flow. This is only justified for centrifugal fans since these have small or modest rotor solidity. For the same reason, Wiesner's formula for slip is better justified for centrifugal compressors than Pfleiderer's formula.

With the velocity triangles of Fig. 14.3, we may estimate the ratio w_2/u_2 to about 0.35. With $\beta_2 \approx -15°$, it follows from Eq. (14.16) that $Z_{\min} \approx 17$. For $\beta_2 \approx -45°$ ($\beta_2^b \approx -40°$), w_2/u_2 may be up to 0.45, which results in $Z_{\min} \approx 10$. Values used in practice are often 20 and 14 or 16.

14.4.2 Rotor Entry and Exit

Rotors for large work require a high peripheral speed (Figs. 14.2 and 14.7). This necessitates appropriate shapes of rotor entry and exit. Large bending stress by the centrifugal force at the impeller entry is avoided by an axial part, termed an *inducer*, as sketched in Fig. 14.10. Within the inducer, the relative velocity w_1 is turned towards the meridional plane. In the figure, the relative velocity at the end of the inducer has no longer a tangential component. Downstream of the inducer, the blade may be purely radial, eliminating bending stress by the centrifugal force. The inducer itself may be formed by radial filaments, thus avoiding bending stress.

In reality, the rotor does not have to be completely free from bending stress. The inducer shape may deviate from radial filaments and its exit velocity may still have a tangential component. Some backward sweep of the blades at the impeller exit is allowed. With large backsweep, bending stress at rotor exit may be reduced by *forward lean* (Fig. 14.2), by which we mean shifting the tip of the blades in the sense of the rotation. With such lean, the blade shape comes closer to a shape formed by radial filaments. This type of lean is commonly used for reduction of stresses. It effects somewhat the efficiency (see Sect. 14.6).

An inducer functions as an axial compressor, with the difference that there is no trailing edge. Figure 14.11 sketches the pressure distribution with an axial compressor

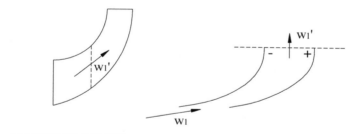

Fig. 14.10 Inducer of a high-speed centrifugal compressor; left: meridional view; right: mean circumferential streamsurface section

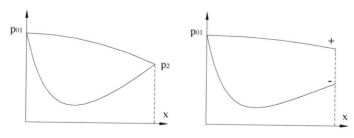

Fig. 14.11 Pressure distribution with an axial compressor (left) and an inducer of a radial compressor (right)

and with an inducer. The boundary layer loading at the suction side is much more advantageous with an inducer due to the pressure difference between the pressure and the suction sides at the outflow. Maintaining this pressure difference, when the flow passes to the radial part, is essential. This is attained by providing a radial component to the velocity w_1' (Fig. 14.10). The associated Coriolis force generates then a pressure difference in the radial part of the rotor.

The inducer part is thus matched to the radial part, so that the pressure difference by the lift force in the inducer smoothly shifts to the pressure difference by the Coriolis force in the radial part. The term radial part is used here to denominate the exit part of the impeller, because it looks approximately radial in a meridional view, but it is not fully radial with backward swept blades near the exit.

The diffusion factor within the inducer part is

$$DF = 1 - \frac{w_1'}{w_1} + \frac{\Delta w_u}{2\sigma w_1}.$$

With outflow in the meridional plane:

$$\Delta w_u = u_1; \quad \frac{u_1}{w_1} = |\sin \beta_1| \approx 0.766; \quad (\beta_1 \approx -50°).$$

For $\beta_1 = -50°$ as a mean value at the entry, the contribution of the blade loading to the diffusion factor is $0.383/\sigma$. A general deceleration is not required for inducer functioning. So, the magnitudes of w_1' and w_1 may even be equal. Further, the diffusion factor may exceed that of an axial compressor due to absence of a trailing edge. For equal velocity magnitudes and $DF = 0.6$ follows $\sigma = 0.64$, which is a low value. This implies that a general deceleration is possible ($w_1' < w_1$). Even with strong deceleration, e.g. $w_1' = 0.7 w_1$, the solidity still does not have to be large.

An inducer may thus be made with much fewer blades than needed for the radial part. Application of splitter blades is thus customary. Generally, the inducer has half the number of blades of the radial rotor part (Figs. 14.2 and 14.7). But, impellers exist with main blade channels split into three or four parts. By using splitter blades, the displacement by the blades after rotor entry is lower, which increases the design flow rate and the choking flow rate.

14.4.3 Secondary Flow in the Rotor

Some secondary flow patterns are similar to these with axial compressors, as sketched in Fig. 13.14. But, there is one additional secondary motion.

The passage vortex in axial compressors is due to the centrifugal force associated to the turning in the circumferential direction. In the inducer of a radial compressor, the cause is similar. In the radial part of the rotor, the tangential pressure difference in a blade passage is due to the Coriolis force, weakened somewhat by a centrifugal force due to meridional curvature by slip or backsweep (Fig. 14.9). But the generation mechanism of the passage vortex stays similar. Thus, a passage vortex develops from rotor entry to rotor exit due to a blade-to-blade force (F_{bb}).

Due to the flow turning in the meridional plane, there is a centrifugal force directed from casing to hub (F_{ch}). This force causes a secondary motion near the blade surfaces, called the *blade surface vortex*. This flow pattern does not occur with an axial compressor. The mechanism is smaller velocity in boundary layers, similar to that of the passage vortex.

Figure 14.12 sketches the secondary vortices in an open rotor, in a transition section between the axial and radial parts, omitting the horseshoe vortex. Hereafter, we denominate this transition part by the term axial-to-radial part. As with an axial compressor, the leakage and scraping vortices merge towards the exit of the rotor and a corner vortex forms in the corner of the hub and the suction side.

With Fig. 14.12, one sees that low-momentum fluid from the hub boundary layer moves by the passage vortex and the blade surface vortex to the corner of the blade suction surface and the casing. Low-momentum fluid from the blade suction side is added. It is then further spread over the casing by the leakage and scraping vortices (otherwise said: by the rotor motion). Low-momentum fluid from the blade pressure surface moves, in principle, by the blade surface vortex to the corner of the blade pressure surface and the casing, but this movement is counteracted by the scraping vortex. So, there is no accumulation of low-momentum fluid in the corner of the blade pressure surface and the casing. Core fluid is driven by the secondary flow patterns to the corner of the blade pressure surface and the hub.

The secondary flow thus moves low-momentum fluid to the corner of the blade suction surface and the casing and moves high-momentum fluid to the corner of the pressure surface and the hub. This mechanism is even stronger with a shrouded

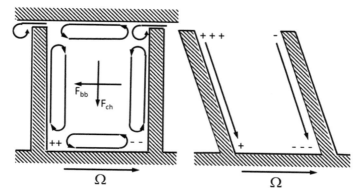

Fig. 14.12 Left: Secondary vortices in an unshrouded radial impeller; axial-to-radial part; view from downstream; scraping vortex, leakage vortex, blade surface vortices, passage vortex. Suction-side leg of the horseshoe vortex is not drawn. Right: backward lean with the objective to reduce the strength of the blade surface vortices (see Sect. 14.6)

rotor, because there is then a passage vortex at the shroud with sense opposite to the scraping vortex and there is no leakage vortex. The through-flow thus splits in a zone of high through-flow velocity, called jet, and a zone of low through-flow velocity, called wake. We call this flow type a *segregated flow*. The precise locations of the jet and the wake are, of course, somewhat dependent on geometric ratios of the blade channel, the backsweep angle and the ratio of through-flow velocity to blade speed.

Splitting into high-velocity and low-velocity flow parts leads to an almost lossless jet flow, while the wake flow systematically accumulates the losses. A segregated flow thus resembles a separated one. The jet flow only decelerates moderately in a diverging blade channel, as the space made available by widening is filled by the wake. This is once more analogous to a separated flow. A particular feature is that, once the segregation has started, the magnitudes of the Coriolis force and the centrifugal force by meridional flow turning become different in the wake and jet zones, which maintains the distinction between the zones.

14.5   Diffusers

14.5.1   Mixing Zone

Jet and wake flows mix rapidly in circumferential direction downstream of the rotor, due to strongly different velocity directions in the absolute frame, as sketched in Fig. 14.13. The collision of the two flow parts leads to a rather large mixing loss. The effect of the Coriolis force is at its strongest with radial-end blades. Backward sweep generates a lift force opposing the Coriolis force. So, backward sweep softens the segregation, causing a decrease of the mixing loss. But, backward sweep decreases rotor work. Thus, the rotor must run faster to compensate the decrease. Backward sweep also diminishes the number of blades required, resulting in less friction loss. The velocity magnitude at the diffuser entry decreases as well.

Fig. 14.13 Mixing of jet flow and wake flow downstream of a rotor

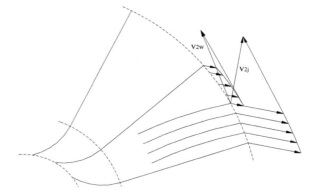

Taking all these effects together, regarding efficiency, some backward sweep of the blades is advantageous (see the examples in Sect. 14.3).

Impellers with radial-end blades are used only very exceptionally, only when maximum possible work is essential. Examples are compressors of light-weight gases, e.g. helium.

Usually, a large-work impeller has about 30° backward sweep, as in Figs. 14.2 and 14.7. But also with some backward sweep, there is non-homogeneity of the outflow of the impeller. This means that a radial space without vanes is necessary for the mixing. Downstream of this space, a vaned diffuser may be mounted.

14.5.2 Vaneless Diffusers

Vaneless diffusers with parallel walls are applied in compressors with moderate rotor work. The diffuser entry velocity is then subsonic. They function also as an entry part of a vaned diffuser in compressors with large rotor work. The absolute velocity at the impeller exit may then be supersonic. The vaneless space is then meant to mix the jet and wake flows and to decrease the Mach number level.

In a vaneless diffuser, streamlines are long due to the small flow angle with respect to the tangential direction, and the flow angle decreases when the flow rate decreases. As streamlines run rather tangentially and the pressure gradient is in the radial direction, the boundary layers do not have much momentum in the pressure gradient direction. There is thus a strong tendency towards flow reversal in the radial direction within the boundary layers. With low flow rate, radial backflow occurs within the boundary layers. The backflow forms in cells rotating in the rotor running sense, analogous to rotating stall in the stator of an axial compressor.

Choking requires that the radial velocity component attains the speed of sound, which never occurs in practice. Shock waves are not possible and there is no incidence loss. All these features together make that a vaneless diffuser is well suited when the velocity reduction downstream of the rotor is moderate. For strong deceleration, the necessary radial dimensions become large, which means large losses and strong risk for flow instability.

14.5.3 Vaned Diffusers

Vanes force the streamlines to a more radial direction than attained by a vaneless diffuser. With a large rotor work, the tangential velocity component is much larger than the radial component in the entry flow of a diffuser (see Fig. 14.3). Therefore, reduction of the tangential velocity component becomes increasingly advantageous as the impeller work is larger.

Figure 14.14 sketches the two common forms, called *cascade diffusers* and *channel diffusers*. The diffuser is normally followed by a collector scroll (volute).

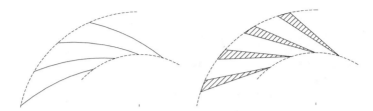

Fig. 14.14 Cascade diffuser (left) and channel diffuser (right)

Cascade diffusers are built with curved plates or profiled vanes. Cascade diffusers are not appropriate for very large pressure ratio. The divergence angle of the channels formed by the vanes must be limited to about 10° to 12° in order to avoid boundary layer separation. So, strong velocity reduction requires a large radius ratio, which makes channel diffusers then more efficient.

Channel diffusers mostly start from a square throat section, but with rounded corners, with an increasing width in an orthogonal plane and constant axial width. The blades are called vane islands or blades with blunt trailing edge. With a given divergence angle and radius ratio, these diffusers attain a higher velocity reduction, as the tangential velocity component decreases more strongly. The sudden widening at the outlet generates a dump diffusion loss. But, this dump loss is not very significant, as it only concerns the radial velocity component.

The cascade diffuser is the most efficient one with a weaker velocity reduction in the diffuser. The channel diffuser is better for a stronger velocity reduction. Some designs come in between, being cascade diffusers with blades with finite trailing edge thickness or curved channel diffusers. The pipe diffuser is a variant of the channel diffuser. Pipe diffusers are made with a cylindrical throat and a conical diverging pipe. They are seldom used, as their construction is more complicated and they do not have advantages compared to channel diffusers.

Boundary layer separation occurs under off-design conditions with reduced flow rate. The separation may be rotating or pulsating. A flow rate increase generates choking when sonic speed is attained in the entrance region of a vane channel.

14.6 Three-Dimensional Blade Shaping

Flow phenomena in radial compressors are less understood than in axial ones and not much has yet been done in practice on truly three-dimensional shaping of rotor blades and diffuser vanes. But there is a potential for efficiency improvement by lowering the strength of shock waves at the entry of impellers and vaned diffusers, lowering the strength of the secondary flow and lowering the leakage flow rate.

There are already some studies on the topic, but these come partly to conflicting conclusions. One reason is that often blade surfaces in the impeller are defined as ruled surfaces because then the impeller may be made by flank milling, which is the usual manufacturing method. This way, some flow features are connected because changes of blade lean or blade sweep cause changes of streamline curvature. Many studies have even been done with only one degree of freedom, e.g., with circumferential displacement by a constant angle of the blade profile at the shroud.

3D shaping is still far from mature. Some principles may be formulated based on a simplified numerical analysis on blade lean [6] and two combined numerical and experimental optimisation studies done without the restriction of ruled surfaces [5, 8]. But, noting is definite yet on fully optimised shapes.

A symbolic representation of the pressure distribution by global *backward lean* is given in Fig. 14.12 (right). The term backward lean is used here for lean by displacement of the tip against the sense of the rotation. Such lean is often called negative lean with centrifugal compressors, but this term is confusing because such lean is called negative at the hub and positive at the shroud with axial compressors (Sect. 13.3.6). The plus and minus signs express a relative value of the pressure, with two signs for values without lean. The section shown is in the axial-to-radial part. The pressure difference between the pressure and suction sides of the blade passage is due to the centrifugal force by the flow turning in the circumferential direction and by the Coriolis force associated to the radial component of the relative velocity. By the meridional flow turning, there is a centrifugal force in the sense of the casing to the hub. This force is smaller in the blade boundary layers, leading to the creation of the blade surface vortices, shown in Fig. 14.12 (left).

Blade lean results in the pressure changes, as sketched in Fig. 14.12 (right), because there is approximately force equilibrium in the core flow on lines in meridional planes drawn perpendicularly to the hub and shroud surfaces (no significant velocity component along these lines). The pressure changes cause forces along the blade surfaces, counteracting the blade surface vortices. The tangential pressure differences in the blade channel stay the same. There is thus no influence on the passage vortex. So, one may expect reduction by lean of the intensity of the blade surface vortices in an axial-to-radial section. This actually happens, as shown by He and Zheng [6], but they do not obtain efficiency improvement, because other flow features change too by the global lean obtained by circumferential blade displacement by a constant angle at the shroud.

For efficiency improvement, independent choices of sweep, lean and streamline curvature are necessary, as shown in the studies by Elfert et al. [5] and Mosdzien et al. [8]. From these studies emerge three main shape features that improve efficiency. A first is compound sweep at the leading edge of the inducer of a transonic centrifugal compressor (supersonic inflow at the shroud and subsonic inflow at the hub), in the same way as with an axial fan (Fig. 13.18). Compound sweep is forward sweep at the blade tip, compensated by backward sweep at a somewhat smaller radius. The sweep lowers the effective inflow Mach number and thus the strength of the bow shock. Forward sweep pushes the passage shock downstream because a 3D shock surface has to be approximately perpendicular to an end wall with as consequence that the shock position at a lower radius influences the shock position on the outer end wall (backward sweep advances the shock position). With the more downstream position, the shock strength of the passage shock is lower and the margin against stall is larger.

The second feature is compound lean in the axial-to-radial part of the impeller, backward near to the hub, but compensated by forward lean towards the shroud, such that the streamline curvature in circumferential direction is not too much reduced in the inducer part. The third one is inversed blade curvature near the shroud in the vicinity of the impeller exit. By inversing the flow curvature, the pressure difference over a blade is reduced, which reduces the tip leakage flow. The blade tip at the exit is then advanced in the sense of the rotation. We recall that such a form of lean is traditionally done for reduction of bending stress (Sect. 14.4.2).

14.7   Performance Characteristics

14.7.1   Flow Instability

Reduced flow rate compared to the design flow rate, at constant rotational speed, may generate rotating stall within the inducer or within a vaned diffuser. Rotating flow reversal may occur within a vaneless diffuser. These phenomena generate losses, causing a decrease of the pressure ratio of the compressor, which generates a maximum in the characteristic of pressure ratio as a function of mass flow rate, and thus the possibility of surge. There is also a maximum in the characteristic with radial-end blades or with small backward sweep, without these phenomena. This implies that surge may then occur without rotating stall or rotating flow reversal (partial separation).

14.7.2   Choking

The choking conditions of the rotor and stator are the same as with axial compressors. So, expressions (13.23) and (13.25) describe approximately the choking mass flow rates, with appropriate values of the blade speed at rotor entry and rotor exit. The choking flow rate depends on the rotational speed for both rotor and stator.

The mass flow rate with choking in the rotor is

$$\dot{m}_c = A_1 \rho_{01} c_{01} \left(\frac{2}{\gamma + 1} \right)^{\frac{1}{\gamma-1}+\frac{1}{2}} \left(1 + (\gamma - 1) \frac{\frac{1}{2} u_1^2}{c_{01}^2} \right)^{\frac{1}{\gamma-1}+\frac{1}{2}}, \qquad (14.17)$$

where A_1 is the through-flow area. The inducer entry is the most critical since the further the flow enters the impeller, the lower is the velocity w and the higher is the enthalpy h. With an average value of the blade speed at the entry, we obtain an approximation of the choking mass flow rate.

The choking mass flow rate for a vaned diffuser is

$$
\dot{m}_c = A_2 \rho_{01} c_{01} \left(\frac{2}{\gamma + 1} \right)^{\frac{1}{\gamma - 1} + \frac{1}{2}} \left(1 + (\gamma - 1) \frac{\psi u_2^2}{c_{01}^2} \right)^{\frac{1}{n - 1} + \frac{1}{2}}, \tag{14.18}
$$

where A_2 is the through-flow area and n is the polytropic exponent of the total-to-total process in the rotor.

The expressions are approximate because an entry Mach number above unity does not necessarily mean appearance of choking. With a high entry Mach number, shock waves reduce the Mach number below unity. This velocity reduction creates a margin with respect to choking, but the margin may be small. Therefore, for high peripheral rotor speed, it becomes imperative to match the choking mass flow rates of rotor and stator.

The second and third examples of Sect. 14.3 illustrate the tendencies. For a discussion on the topic, we refer to Casey and Rusch [2]. They demonstrate that with a machine Mach number of 1.6, an operating range of about 10% at design speed can be reached by optimal matching, with operating range defined as the ratio of the difference of choking mass flow rate and stall mass flow rate to the choking mass flow rate. The operating range of the third example in Sect. 14.3 is 9% at a machine Mach number of 2, which thus means that rotor and stator choking are very well matched [7].

14.7.3 Operating Characteristics and Operating Range

The term characteristic is normally used without any further specification for the dependence of the pressure ratio on the mass flow rate at constant rotational speed. Figure 14.15 sketches the field of characteristics of a centrifugal compressor with a design pressure ratio 5. The shape of the characteristics is similar to that of an axial compressor, but with a lower efficiency. The characteristic at a given rotational speed may be changed by adjustable inlet guide vanes or adjustable diffuser vanes (see Exercise 14.8.2). With adaptation to another operating point by adjustable vanes, the efficiency decreases.

The operating range of a centrifugal compressor decreases when the pressure ratio in the design point is higher. The reason is higher Mach numbers at rotor and diffuser entry, causing a smaller margin against choking. Hence the necessity to match the choking mass flow rates of rotor and stator, as discussed above. With a pressure ratio of 5, as in Fig. 14.15, the operating margin is about 20%. With lower pressure ratio, it becomes larger.

A narrow operating range is practicable for some applications, but not for all. A process compressor (Fig. 14.1) requires a large operating range. The blades are

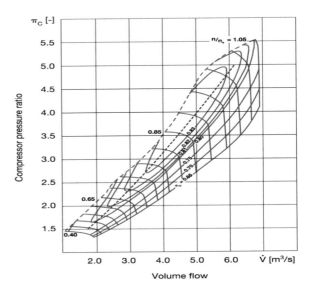

Fig. 14.15 Characteristics of a turbocharger centrifugal compressor with a large design stage pressure ratio; total-to-total isentropic efficiency levels; operating line imposed by the engine (*Courtesy* ABB Turbocharging)

swept moderately backwards (blade angle − 40°) and the rotor exit Mach number is much lower than unity. A wide operating range is generally not necessary with air compressors, small turbo-shaft engines (typical for helicopters) and turbo-expanders. Small turbo-shaft engines may feature axial and radial compressors, but two radial stages as well. The pressure ratio of a radial stage may exceed 5, with a small operating range at the design point. This is acceptable, because the engine is seldom used at its maximum power. A turbo-expander is an expansion turbine combined with a compressor driven by the turbine, similar to a turbocharger.

This type of machine is applied in process engineering. The gas cools down by expansion, allowing the separation of moisture. The dried gas gets compressed again. A large operating range is normally not required. The compressor of a diesel engine turbocharger requires a moderately large operating range.

The surge margin of a turbocharger compressor is often enlarged by a shroud port, as shown in Fig. 14.16. It consists of an annular chamber in the casing (struts not shown in the figure) with two connected circumferential slots, one in the inducer mid region and one in the intake. Turbocharger compressors have mostly a vaneless diffuser and have a large inducer tip diameter for realisation of a large flow rate. They enter into stall with a recirculation zone at the tip of the inducer.

The principle is that near stall, when the inducer is highly loaded, the pressure difference drives flow from the inducer mid region to the intake. The recirculation zone is then sucked away, postponing stall in the inducer. When the compressor operates near choke, the pressure difference drives flow from the intake to the inducer mid region, which increases the choking mass flow rate of the rotor. The operation margin is enlarged, but the efficiency decreases somewhat.

Fig. 14.16 Shroud port for enlarging the surge margin

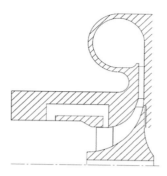

With vaned diffusers, stall may occur in the diffuser. Increase of the stall margin then requires vanes in the annular chamber creating a whirl component in opposite sense to the impeller rotation in the flow led by the shroud port to the rotor tip. The rotor work is then enlarged, causing an increase of the steepness of the characteristic near stall, which means a stabilising effect, delaying stall.

Axial grooves, distributed over the periphery of the casing are often suggested as an alternative, also for use with axial compressors. These make a similar connection between the mid region of the rotor blades and the intake and are made with lean such that the flow injected in the intake near stall, has a tangential component opposite to the rotation. But their functioning is more complex due to the varying pressure by the passing blades. Because of the associated efficiency reduction, they are not used in practice.

14.8 Exercises

14.8.1 Derive the expression (14.11) for the velocity difference circumferentially across a blade channel from a moment of momentum balance on a streamtube with infinitesimal radial length spanning a blade passage as sketched in the figure.

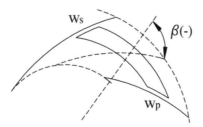

A: The mass flow rate through the blade channel may be approximated by

$$\dot{m}_b = \rho \frac{2\pi r}{Z} w \cos \beta,$$

where ρ is an average value of the density and w is an average value of the relative velocity at the position of the infinitesimal streamtube and β is the local angle of the mean streamline with respect to the radial direction. The moment of momentum transferred to the flow in the infinitesimal streamtube is

$$M = \dot{m}_b \frac{d}{dr}(r v_u)\, dr,$$

with v_u averaged over the blade passage.

This is equal to the moment of the pressure forces on the blades (neglecting a possible moment by friction forces):

$$M = (p_p - p_s)\, dr\, r,$$

where the subscripts p and s refer to pressure and suction sides. The Bernoulli equation on a streamline in a rotor is

$$d\tfrac{1}{2}u^2 = d\tfrac{1}{2}w^2 + \tfrac{1}{\rho}dp + dq_{irr}.$$

Therefore, on a constant radius, neglecting losses, and assuming that the integration constant of the Bernoulli equation is the same on all streamlines (which is satisfied for constant enthalpy and free vortex flow upstream of the rotor inlet; see Chap. 3, Sect. 3.3.2):

$$\frac{1}{2}\left(w_s^2 - w_p^2\right) = \frac{1}{\rho}\left(p_p - p_s\right),$$

with an average value of the density. From the last equation follows an approximation of the pressure difference circumferentially across the blade channel:

$$p_p - p_s = \rho\, w\, (w_s - w_p).$$

Combination of the equations above results in

$$\rho w(w_s - w_p)\, r\, dr = \rho\, \Delta\theta\, r\, w \cos \beta [d(r u) + d(r w_u)],$$

with $\Delta\theta = \frac{2\pi}{Z}$ and $v_u = u + w_u$. From this follows (14.11).

14.8.2 (Variable geometry). Derive from the velocity triangles in Fig. 14.3 that turning adjustable inlet guide vanes in the sense of the rotation is mainly a way to reduce the work transferred with unchanged mass flow rate. Observe that incidence occurs at the impeller entry. Derive also that turning adjustable diffuser vanes to a more tangential position is mainly a way to reduce the mass flow rate with unchanged work transferred (through-flow area is reduced). Again, incidence at the impeller entry reduces the efficiency.

14.8.3 (Peripheral speed for high pressure ratio). The basic shape of a centrifugal compressor rotor for high flow rate and high pressure ratio is with an axial entry part, called inducer. In the meridional view, the rotor can be considered as composed of an axial part (inducer), an axial-to-radial part and a radial part.

For lowest bending stress due to centrifugal force, rotor blades should be made by radial filaments. A basic shape realising this condition is with inducer blades that end in a meridional plane, followed by axial-to-radial and radial blade parts in the same meridional plane. But for reasons of efficiency improvement, by lower mixing loss at the impeller exit and lower velocity at the diffuser entry, some backsweep is necessary. A simple geometry is obtained by applying gradual backsweep in the axial-to-radial and radial parts, with the same value on axial lines (no lean).

The basic impeller of the compressor used for demonstration of the effect of lean by He and Zheng [6] is constructed this way. Of course, this is not an optimal impeller shape. The value of the backsweep at the rotor exit is not mentioned, but from a figure, one can deduce that is it about 25°. The same figure shows that the primary blade channels are subdivided by two splitter blades. The number of blades is very high: $13 + 13 + 13 = 39$. Slip is thus small. Around the impeller are a vaneless space and a vaned diffuser. The CFD simulations are done on a periodic part of the rotor (one complete rotor channel), extended with the corresponding part of the vaneless space and imposing a uniform backpressure. The mixing loss at the exit of the rotor is thus included, but not the diffuser flow.

The total pressure ratio of the impeller at peak efficiency is $p_{02}/p_{01} = 3.3$. The pressure coefficient (ψ) is 0.68, defined as the ratio of the isentropic total enthalpy increase through the impeller ($h_{02s} - h_{01}$) to the square of the tip speed. Take as ambient conditions 100 kPa and 288 K and as properties of air $R = 287$ J/kgK and $\gamma = 1.40$. Calculate the peripheral speed of the impeller (**A:** $u_2 = 415.9$ m/s). Observe that the tip speed is not extremely high so that, very likely, the impeller can run at a higher rotational speed. Calculate the total pressure ratio for peripheral speed equal to 500 m/s (**A:** $p_{02}/p_{01} \approx 5.0$).

14.8.4 (Impeller analysis). The one-dimensional flow representation at the entry and exit of the impeller of the previous exercise may be constructed with the published data, an estimate of the blade angle at the impeller exit (about $-25°$) and an estimate of the blade thickness obstruction at the impeller exit.

Calculate the velocity at the entry to the impeller (just before entry: no blade obstruction counted), assuming axial inflow in the absolute system, from the specified radius ratios: $r_{1t}/r_2 = 0.56$, $r_{1h}/r_2 = 0.41$ and the flow coefficient defined with the static density at rotor inlet: $\varphi = \dot{m}/(\rho_1 u_2 d_2^2) = 0.05$ (**A:** $v_1/u_2 = 0.4375$, $v_1 = 182.0$ m/s).

Calculate the blade speed at the impeller tip, the relative velocity at the impeller tip, and the corresponding Mach number (**A:** $u_{1t} = 232.9$ m/s, $w_{1t} = 295.5$ m/s, $M_{w1t} = 0.895$). So, observe that the flow at the tip of the inducer entry is supercritical, but not transonic. This is, once more, an indication that the impeller is not used at its maximum rotational speed (the authors say in the title of the paper that the compressor is transonic). Determine the flow angle of the relative flow at the tip of the entry to the impeller (**A**: $\beta_{1t} = -52°$).

Calculate the work input to the rotor with an estimated polytropic total-to-total rotor efficiency of 0.9 (**A**: $\Delta W = 133.3$ kJ/kg). Calculate the mixed-out velocity triangles at the rotor exit. Assume as ratio of the meridional components of entry and exit velocities: $v_{2r}/v_1 = 0.85$ (**A**: $v_{2u} = 320.6$ m/s, $v_{2r} = 154.7$ m/s, $M_{v2} = 0.94$, $w_{1t}/w_2 = 1.63$).

Calculate the densities at rotor entry and exit. Determine the velocity ratio v_{2r}/v_1 from the mass flow balance through the rotor, using the specified geometric ratios: $r_{1t}/r_2 = 0.56$, $r_{1h}/r_2 = 0.41$, $b_2/r_2 = 0.05$ (**A**: $v_{2r}/v_1 = 0.834$). The previous calculations may now be iterated with the updated velocity ratio (not done).

Construct the velocity triangle just before rotor exit with an estimated blade angle $\beta_2^b = -25°$ and an estimated velocity ratio $v_{2r}/v_1 = 1$. Use this last estimate because there is no information on blade thickness (**A**: $w_{2u}^b = -84.85$ m/s). Construct the velocity triangle just after rotor exit with the slip factor by the formula of Wiesner (Eq. 14.10), with $\beta_2^b = -25°$ and $Z = 39$, and the velocity ratio $v_{2r}/v_1 = 0.85$. Estimate the work transferred and derive the polytropic total-to-total rotor efficiency (**A**: $\sigma = 0.927$, $v_{2u} = 300.5$ m/s, $\eta_\infty = 0.95$). With the updated polytropic efficiency, the previous calculations may be iterated (not done). The updated estimate ($\eta_\infty = 0.95$) seems somewhat optimistic and the initial estimate ($\eta_\infty = 0.90$) seems somewhat pessimistic. A precise value cannot be determined.

14.8.5 (Degree of reaction). Calculate, for the impeller of the previous exercises, at the casing and at the hub, the static enthalpy increase due to centrifugal force, $\frac{1}{2}(u_2^2 - u_1^2)$, the static enthalpy increase due to deceleration, $\frac{1}{2}(w_1^2 - w_2^2)$, and the complete static enthalpy increase (**A**: casing: 59.36 kJ/kg and 27.17 kJ/kg with sum 86.53 kJ/kg; hub: 71.94 kJ/kg and 14.59 kJ/kg with sum 86.53 kJ/kg). Observe that the increase of static enthalpy, and thus pressure, is mainly due to the centrifugal force (68.6 and 83.1%). Calculate the increase of kinetic energy $\frac{1}{2}(v_2^2 - v_1^2)$ (**A**: 46.80 kJ/kg). Verify that the sum of the three terms is the work input (133.32 kJ/kg) and determine the degree of reaction (**A**: $R = 0.649$).

14.8.6 (Entry Mach number of a transonic impeller). The impeller of the radial compressor used by Elfert et al. [5] for demonstration of the potential of efficiency optimisation by sweep and lean has 13 main blades and 13 splitter blades. The rotor diameter is 224 mm and the rotational speed is 50,000 rpm. From figures in the paper, one can deduce the radius ratios of the entry: $r_{1t}/r_2 \approx 0.70$; $r_{1h}/r_2 \approx 0.265$. The outlet width and the backsweep angle of the rotor can also be seen, but one has to be careful with these parameters, because the authors explicitly say that they hide exact values of rotor parameters. The observed parameters are: $b/r_2 \approx 0.0875$; $\beta_2^b \approx -45°$. The radius ratios of the entry seem realistic and, therefore, we only use these for verification of some operation parameters.

Before optimisation, the compressor realises a total pressure ratio $p_{03}/p_{01} = 5.70$ and a mass flow rate of 2.87 kg/s at the maximum efficiency point, with total-to-total isentropic efficiency 0.85. After optimisation, the maximum efficiency becomes 0.863 with slightly higher values of pressure ratio and mass flow rate. Use the parameters before optimisation for calculations.

Calculate the velocity at entry to the impeller (just before entry: no blade obstruction) in the absolute frame (axial inflow). Take 100 kPa and 288 K as ambient conditions and $R = 287$ J/kgK and $\gamma = 1.40$ as properties of air. Ignore losses in the intake of the compressor. Observe that the mass flow rate determines the product of density and axial velocity and that conservation of energy determines a relation between density and axial velocity. Solution of this system of equations requires iteration. On may start with an estimation of the density equal to 90% of the density in the atmosphere (**A**: $v_1 = 160.76$ m/s, $\rho_1 = 1.079$ kg/m^3). Calculate the blade speed at the impeller tip, the relative velocity at the impeller tip, the corresponding Mach number and flow angle (**A**: $u_{1t} = 410.50$ m/s, $w_{1t} = 440.86$ m/s, $M_{w1t} = 1.326$, $\beta_{1t} = -68.6°$). Observe the supersonic flow at the tip of the inducer, with a quite high Mach number and a large flow angle.

14.8.7 (Exit Mach number and backsweep angle). Determine the work input to the rotor of the previous exercise from the specified pressure ratio and isentropic efficiency. Calculate the velocity triangle just after rotor exit. Assume as velocity ratio $v_{2r}/v_1 = 0.9$. (**A**: $\Delta W = 219.3$ kJ/kg, $v_{2u} = 373.9$ m/s). Determine the Mach number of the exit flow. (**A**: $M_{v2} = 0.97$). So, observe that the exit Mach number is just subsonic.

Estimate the backsweep angle of the impeller from the velocity triangle at rotor exit and the slip factor. Determine the slip factor by the formula of Wiesner (Eq. 14.10). Assume $v_{2r}/v_1 = 1$ for the velocity triangle at rotor exit, just before exit (relative flow follows the blade angle), (do this because there is no information on blade thickness). Iteration is necessary. One may start from a slip factor equal to 0.9. Compare with the flow angle (**A**: $\beta_2^b = -45.3°$; $\beta_2 = -55.75°$, $w_{1t}/w_2 = 1.71$).

14.8.8 (Impeller and diffuser efficiencies). Determine the static and total pressure at the rotor exit of the compressor of the previous exercises. Assume 0.95 as polytropic total-to-total rotor efficiency (**A**: $p_{02}/p_{01} = 6.53$, $p_2/p_{01} = 3.57$). Calculate the pressure loss coefficient of the diffuser (**A**: $\xi_p = 0.28$). Observe the rather high pressure loss coefficient.

The isentropic efficiency of the complete compressor is the ratio of the isentropic enthalpy increase ($h_{03ss} - h_{00}$ in Fig. 14.8) to the actual enthalpy increase through the compressor ($h_{03} - h_{00}$). Here, states "00" and "01" are identical. The isentropic efficiency of the rotor is, similarly, defined as the ratio of the isentropic enthalpy increase ($h_{02ss} - h_{00}$ in Fig. 14.8) to the work input ($h_{02} - h_{00}$).

A possible, but not common, definition of the isentropic efficiency of the diffuser is the ratio of ($h_{03ss} - h_{00}$) to ($h_{02ss} - h_{00}$). The isentropic compressor efficiency is then the product of the isentropic efficiencies of the rotor and the diffuser. Calculate these efficiencies (**A**: $\eta_{srot} = 0.9355$, $\eta_{sdif} = 0.9086$). Remark the rather high diffuser efficiency because part of the static enthalpy increase by the rotor is a lossless contribution to both the denominator and the numerator. But, observe that the commonly used definition of the isentropic rotor efficiency has a similar feature due to lossless contributions.

References

1. Balje OE (1981) Turbomachines: a guide to design, selection and theory. Wiley, ISBN 0-471-06036-4
2. Casey M, Rusch D (2014) The matching of a vaned diffuser with a radial compressor impeller and its effect on the stage performance. J Turbomachinery 136:121004
3. Cumpsty NA (1989) Compressor aerodynamics. Longman Scientific and Technical, ISBN 0-582-01364-X
4. Dixon SL, Hall CA (2014) Fluid mechanics and thermodynamics of turbomachinery, 7th edn. Elsevier, ISBN 978-0-12-415954-9
5. Elfert M, Weber A, Wittrock D, Peters A, Voss C, Nicke E (2017) Experimental and numerical verification and optimization of a fast rotating high-performance radial compressor. J Turbomachinery 139:101007
6. He X, Zheng X (2016) Mechanisms of lean on the performance of transonic centrifugal impellers. J Propulsion Power 32:1220–1229
7. Higashimori H, Hasagawa K, Sumida K, Suita T (2004) Detailed flow study of Mach number 1.6 high transonic flow with a shock wave in a pressure ratio 11 centrifugal compressor impeller. J Turbomachinery 116:473–481
8. Mosdzien M, Enneking M, Hehn A, Grates D, Jeschke P (2018) Influence of blade geometry on secondary flow development in a transonic centrifugal compressor. J Glob Power Propul Soc 2:429–441
9. Van den Braembussche R (2019) Design and analysis of centrifugal compressors. Wiley-ASME, ISBN 978-1-119-42409-3
10. Whitfield A, Baines NC (1990) Design of radial turbomachines. Longman Scientific and Technical, ISBN 0-582-49501-6

Chapter 15
Axial and Radial Turbines for Gases

Abstract Fundamentals of axial turbines were discussed in the chapter on steam turbines (Chap. 6). Turbine parts in gas turbines follow the same principles. When analysing the performance of axial turbines, we assumed, by way of a simplification, a given stator outflow angle ($\alpha_1 = 72°$ and $75°$). The analysis is generalised here and completed with a discussion of blade design and operating characteristics. The fundamental theory of radial turbines was treated in the chapter on hydraulic turbines (Chap. 9). The specific aspects for a compressible fluid are discussed in the present chapter. In particular, the rotor has an exducer part. Operating characteristics of radial turbines are derived. The chapter concludes with the non-dimensional form of the operating characteristics of turbomachines with a compressible fluid.

15.1 Axial Turbines

15.1.1 Kinematic Parameters

Figure 15.1 sketches the mean-line flow through a repeating stage of an axial turbine.

As with the analysis of axial compressors, flow angles are measured with respect to the axial direction, positive in the rotation sense (α for stator, β for rotor). Constant axial velocity through the cascade is assumed. The fundamental relations are similar to these with compressors.

Work coefficient (or stage loading coefficient), flow coefficient and degree of reaction are

$$\psi = \frac{\Delta W}{u^2} = \frac{v_{1u} - v_{2u}}{u}, \ \phi = \frac{v_a}{u}, \ R = \frac{h_1 - h_2}{h_{01} - h_{02}} = -\frac{w_{mu}}{u}.$$

Thus: $\frac{w_{1u}}{u} - \frac{w_{2u}}{u} = \psi$ and $\frac{w_{1u}}{u} + \frac{w_{2u}}{u} = -2R$.
From this follows

$$\frac{w_{1u}}{u} = -R + \frac{\psi}{2}, \ \frac{w_{2u}}{u} = -R - \frac{\psi}{2}, \ \frac{v_{1u}}{u} = 1 - R + \frac{\psi}{2}, \ \frac{v_{2u}}{u} = 1 - R - \frac{\psi}{2}.$$

$$(15.1)$$

E. Dick, *Fundamentals of Turbomachines*, Fluid Mechanics and Its Applications 130,
https://doi.org/10.1007/978-3-030-93578-8_15

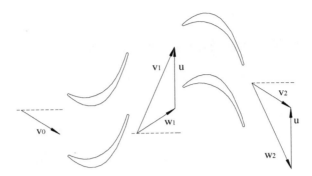

Fig. 15.1 Mean-line flow through a repeating axial turbine stage ($\psi = 1.8, \phi = 0.6$, $R = 0.5$)

Figure 15.2 sketches some examples of velocity triangles for $\phi = 2/3$. The conventions are as for axial turbines in Chap. 6. The vertical direction is the axial direction (x) with the through-flow from top to bottom. The blade speed direction is horizontal (y) with the positive sense from left to right and the radial direction (z) is perpendicular to the drawing, upward. The xyz-frame is right-handed, but the machine is left-turning.

There is symmetry in the relations: replacing R by $1 - R$ goes with replacing w_1 by $-v_2$ and w_2 by $-v_1$. For $\psi = 2(1 - R)$, the cascade inflow and outflow velocities lie in the meridional plane. For larger values of ψ, there is whirl at the stage exit in the contra-rotation sense ($v_{2u} < 0$). With a high degree of reaction, this already occurs with a moderately high work coefficient.

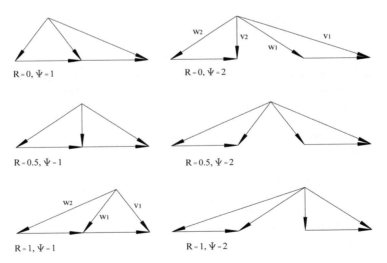

Fig. 15.2 Axial turbine velocity triangles with varying degree of reaction and work coefficient

Fig. 15.3 h–s diagram of an
axial turbine

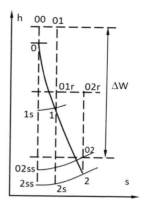

Figure 15.3 is a sketch of an h–s diagram.
Using stage repetition, the total-to-total isentropic efficiency is

$$\eta_{tt} = \frac{\Delta W}{h_{00} - h_{02ss}} = \frac{h_{00} - h_{02}}{h_{00} - h_{02ss}} \approx \frac{h_0 - h_2}{h_0 - h_{2ss}}. \tag{15.2}$$

Ignoring the divergence of the isobars, it follows that

$$h_0 - h_{2ss} \approx h_0 - h_2 + (h_1 - h_{1s}) + (h_2 - h_{2s}) = \Delta W + \xi_s \frac{v_1^2}{2} + \xi_r \frac{w_2^2}{2},$$

so that

$$\eta_{tt} \approx \frac{\Delta W}{\Delta W + \xi_s \frac{v_1^2}{2} + \xi_r \frac{w_2^2}{2}}. \tag{15.3}$$

The coefficients ξ_s and ξ_r are the enthalpy loss coefficients of stator and rotor.
With (15.1), (15.3) results in

$$\eta_{tt} \approx \frac{\psi}{\psi + \frac{1}{2}\xi_s \left[\phi^2 + \left(1 - R + \frac{\psi}{2}\right)^2 \right] + \frac{1}{2}\xi_r \left[\phi^2 + \left(R + \frac{\psi}{2}\right)^2 \right]}.$$

The optimum solidity of a turbine cascade (ratio of chord c to spacing s) may
be derived from Zweifel's tangential force coefficient (Eq. 2.26 in Chap. 2), written
here for a rotor:

$$C_{Fu} = \frac{|F_u|}{\frac{1}{2}\rho_2 w_2^2 c_a} = \frac{\rho_2 w_{2a} s |w_{1u} - w_{2u}|}{\frac{1}{2}\rho_2 w_2^2 c_a} = \frac{2|\Delta w_u| w_{2a}}{\sigma_a w_2^2}. \tag{15.4}$$

$|F_u|$ is the magnitude of the blade force in the tangential direction per unit of
span. The term $\frac{1}{2}\rho_2 w_2^2$ is the incompressible lossless value of $p_{01r} - p_2$, which is

the difference between the total pressure at entry and the static pressure at exit. c_a is the axial chord and σ_a the axial solidity. We recall that the optimum value of the tangential force coefficient of a turbine cascade is quite universally around unity.

The Zweifel coefficient is almost always used in its incompressible lossless form (15.4). The reason is simply the observation that the validity of the optimum solidity is best with the constant-density lossless formulation.

Loss coefficients may be determined by Soderberg's correlation (Eq. 6.14 in Chap. 6), representing the losses in an accelerating cascade with optimal solidity (Zweifel tangential force coefficient near to unity) as a fraction of the exit kinetic energy, with

$$\xi = \xi_1 + \xi_2, \quad \xi_1 = 0.025(1 + (\delta^o/90)^2), \quad \xi_2 = 3.2 \ (c_a/h)\xi_1. \tag{15.5}$$

Here, δ^o represents the flow turning within the cascade in degrees. The coefficient ξ_1 estimates the losses on the blades, with factor 0.025 for a Reynolds number based on the hydraulic diameter and the outlet velocity equal to 10^5. With a lower Reynolds number, the factor is somewhat higher. The coefficient ξ_2 estimates the losses in the end wall boundary layers and the losses by secondary flows.

Henceforth, we will take the aspect ratio, which is the ratio of the blade height h to the axial chord c_a, equal to 4 as an example. Soderberg's loss formulae are intended for subsonic flow. If necessary, losses for shock waves have to be added. The loss formulae do not encompass clearance losses and losses by film cooling either. On the other hand, the formulae overestimate somewhat the blade losses and secondary losses for modern blade rows. So, we write:

$$\xi = 0.045\left(1 + \left(\frac{\delta^o}{90}\right)^2\right).$$

with

$$\delta = \Delta\alpha = \alpha_1 - \alpha_2$$

or

$$\delta = \Delta\beta = \beta_1 - \beta_2.$$

Angles follow from Eq. (15.1) with

$$\tan\alpha_1 = \frac{v_{1u}}{v_a}, \quad \tan\alpha_2 = \frac{v_{2u}}{v_a}, \quad \tan\beta_1 = \frac{w_{1u}}{v_a}, \quad \tan\beta_2 = \frac{w_{2u}}{v_a}.$$

Efficiency contour lines for $R = 0, 0.25, 0.5$ and 0.75 are plotted in Fig. 15.4. The dashed straight lines with strong slope represent $\alpha_1 = 70°$ and $\alpha_1 = 60°$. The other

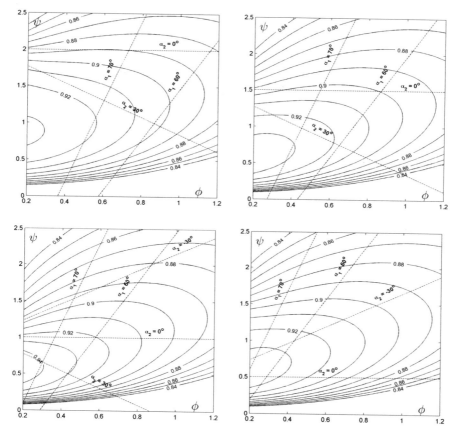

Fig. 15.4 Efficiency dependent on work coefficient and flow coefficient: top left: R = 0; top right: R = 0.25; bottom left: R = 0.50; bottom right: R = 0.75

dashed straight lines represent $\alpha_2 = 0$ and $\alpha_2 = \pm 30°$. The efficiency contours for $R = 0.25$ and $R = 0.75$ are identical. We find, in accordance with the results in Chap. 6, that the efficiency is maximum for $R = 0.50$. But, the corresponding work and flow coefficients are small: $\psi \approx 0.60, \phi \approx 0.30$.

For larger values of ψ and ϕ, the efficiency is almost independent of the degree of reaction. This insensitivity to the degree of reaction is illustrated in Table 15.1, with the maximum efficiencies derived from Fig. 15.4 for chosen values of the stage loading coefficient ψ, but where α_1 is not allowed to exceed 70°. The efficiency decreases with increasing stage loading.

With gas turbines, the work coefficient in the turbine part is always large. This certainly applies to aero-engines, where limitation of weight is crucial. But it applies to power gas turbines as well, where the number of stages in the turbine part is taken as low as possible, limiting in this way the total blade surface area and thus the necessary amount of cooling air.

Table 15.1 Maximum total-to-total efficiency weakly dependent on degree of reaction R for chosen values work coefficient ψ

ψ	R = 0	R = 0.25	R = 0.50	R = 0.75
1.0	0.921	0.923	0.922	0.923
1.5	0.909	0.903	0.900	0.903
2.0	0.890	0.883	0.882	0.883
2.5	0.869	0.865	0.864	0.865

The weak dependence of the efficiency to the degree of reaction for higher values of ψ and ϕ allows representation of efficiency as a function of ψ and ϕ only. Manufacturers use such a diagram composed from turbine data that are publicly available and from their own data. This type of diagram is mostly called a Smith chart [4].

The optimum value of ϕ as a function of ψ for degrees of reaction between 25 and 75%, according to Fig. 15.4, is approximately a straight line through the couples ($\phi = 0.3$; $\psi = 1$) and ($\phi = 1$; $\psi = 2.5$). In reality, the larger values of ϕ are taken somewhat lower due to Mach number effects.

It is not possible to derive practical values of ψ and ϕ from efficiency considerations. With gas turbines, values of ψ and ϕ are much higher than these for optimum efficiency. Extreme values are $\psi \approx 2.20$ together with $\phi \approx 0.9$ on the mean radius, typical in the LP turbine part of aero-engines. The objective is limiting the number of stages (low mean blade speed). In HP and IP turbine parts of aero-engines and in power gas turbines, $\psi \approx 1.5$ to 2.0 together with $\phi \approx 0.5$ to 0.6. The blade speed in HP and IP parts of aero-engines is very high, up to 450 m/s on the mean radius. With $\psi = 1.80$ and $u = 450$ m/s, the stage work is about 350 kJ/kg. In power gas turbines, the blade height is larger and the mean blade speed is lower. With $\psi = 1.80$ and $u = 375$ m/s on the mean radius, the stage work is about 250 kJ/kg.

The degree of reaction follows from efficiency optimisation for chosen values of ψ and ϕ. With Table 15.1, there is no conclusion on an optimal value. But Soderberg's correlation does not include clearance losses and these influence quite strongly the optimum value. The optimum degree of reaction is normally somewhat below 50%, but the precise value is case-dependent, as discussed hereafter. The cause is the difference in leakage loss in rotor and stator parts. In a stator, there is some leakage at the hub side, but it is kept small by sealing (typically a labyrinth sealing) between the vane end bands and the hub. The leakage is quite large at the tips of unshrouded rotor blades.

Leakage loss encompasses two parts. The first is by flow from the pressure to the suction side of a blade or vane. This part depends on the pressure difference between the pressure and the suction sides, which does not specifically depend on the degree of reaction. The overflow does not exist with stator vanes (end bands) and shrouded blades. The second part originates from the through-flow, so from the pressure drop in the axial direction. The axial component of the clearance flow is strongly reduced, but does not disappear completely, with shrouded blades. This component decreases considerably with unshrouded rotor blades by lowering the degree of reaction and thus lowering the pressure drop through the rotor. Also with shrouded rotor blades, leakage losses decrease somewhat with a degree of reaction lower than 50%, because the radius at which leakage occurs is smaller in a stator than in a rotor [8].

With shrouded rotor blades, the optimal degree of reaction on the mean radius is close to 50%, but somewhat smaller, about 40–45%. With short unshrouded rotor blades, thus considerable leakage loss, the optimum value is lower. HP and IP blades of aero-engines are rather short and may be shrouded or unshrouded (Rolls-Royce practice is shrouded blades; e.g. Tent 800, Fig. 12.14). In power gas turbines, blades are not shrouded in cooled stages. They are shrouded in non-cooled stages or have a *tip treatment* (a partial form of shroud: e.g. SGT6-5000F, Fig. 11.7). With long shrouded rotor blades as in the LP-part of an aero-engine, the optimum degree of reaction on the mean radius is again somewhat lower than 50%, because leakage loss represents only a small fraction of the total loss.

Combining all considerations, the conclusion is that there is no universal optimum of the kinematic parameters at the mean radius of a turbine stage in a gas turbine. The above analysis only allows demonstration of tendencies, but not derivation of precise results. It should also be noticed that stages in a real turbine are not always repeating stages. The entry to a first stage is normally in the meridional plane and the exit has large negative post-whirl, enlarging the stage work (an example is Fig. 11.6 in Chap. 11; even more post-whirl may be chosen). Further, the degree of reaction may be chosen lower than the aerodynamic optimum for lowering the rotor entry temperature.

We keep in mind that the work coefficient is always very large. Referring to Figs. 15.2 and 15.4, this means that there is typically strong negative post-whirl at the exit of a stage (Fig. 15.1). At the exit of a turbine part, sometimes outlet guide vanes are required for turning the flow to the meridional direction.

15.1.2 Radial Variation of Flow Parameters

The equation for simple radial equilibrium (Eq. 13.19) reads:

$$\frac{v_u^2}{r} + \frac{d}{dr}\left(\frac{1}{2}v_a^2 + \frac{1}{2}v_u^2\right) = 0 \text{ or } \frac{v_u}{r}\frac{d}{dr}(r\,v_u) + \frac{d}{dr}\left(\frac{1}{2}v_a^2\right) = 0. \tag{15.6}$$

The equation is satisfied for $r\,v_u = cst$, together with $v_a = cst$. If satisfied this way at rotor inflow and outflow, the work is constant over the radius. The blade form is then called a free-vortex blade. The disadvantage with a large radius variation is a rather strong variation of the stator outflow angle α_1 from a higher value at the hub to a lower value at the casing.

This rather strong variation cannot be matched with the weak variation on the maximum efficiency line in the Smith chart: Fig. 15.4. In using this figure, one has to take into account that high values of ψ and ϕ occur at the hub, so where the degree of reaction is the lowest, and low values at the casing, so where the degree of reaction is the highest. Examples are $\phi = 1$ and $\psi = 2.5$ at $R = 0.25$ (hub) and $\phi = 0.3$ and $\psi = 1$ at $R = 0.75$ (casing). For each of these combinations, α_1 is close to 65°. These sets do not go together with free-vortex blades. But the radial equilibrium equation can be satisfied in other ways, e.g. for constant stator outflow angle.

With $tg\alpha_1 = v_{1u}/v_{1a} =$ constant, Eq. (15.6) becomes for the stator outflow:

$$\frac{v_{1a}^2}{r}\tan^2\alpha_1 = -\frac{d}{dr}\left(\frac{1}{2}v_{1a}^2(1+\tan^2\alpha_1)\right) \; or \; \frac{v_{1a}}{r}\sin^2\alpha_1 = -\frac{d(v_{1a})}{dr}.$$

The equation is satisfied for $v_{1a} = C\,r^{-\sin^2\alpha_1}$.

The axial velocity v_{1a} is then higher at the hub and lower at the casing.

With a very high radius ratio, as with the final stage of the LP part of a steam turbine, the stator outflow velocity variation is in between the results for free-vortex flow and constant stator angle, but close to this last one (Exercise 6.10.3 in Chap. 6 shows that the efficiency is close to optimum on every radius for $\alpha_1 =$ ct.). With gas turbine stages with large radius ratio, a similar angle distribution is chosen.

15.1.3 Secondary Flow

The secondary flow generation is similar to that with axial compressors (Sect. 13.2 in Chap. 13; [3]). The blade passage vortex and the horseshoe vortex are the main components. The vortices just downstream of the exit of a rotor blade passage are sketched in Fig. 15.5, but the precise positions are case-dependent.

At a hub, the pressure-side leg of the horseshoe vortex is swept by the developing passage vortex towards the suction side of the adjacent blade and merges with the passage vortex, resulting in a vortex close to this suction side. This vortex pushes the

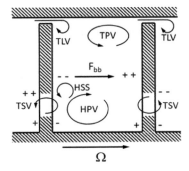

Fig. 15.5 Secondary vortices at the exit of an axial turbine rotor passage (view from downstream): hub passage vortex (HPV), horseshoe suction-side vortex leg (HSS), trailing shed vortex (TSV), tip leakage vortex (TLV), tip passage vortex (TPV). Not drawn are corner vortices in case of separation in the corner of a suction side and an end wall. Blade-to-blade force F_{bb}

suction-side horseshoe vortex leg, which has the opposite rotation sense, towards the mid-height of the blade passage.

At a hub, there is always a rather strong trailing shed vortex near the endwall, downstream of the trailing edge, due to the load variation along the blade span (lower loading in the endwall boundary layer), with rotation sense opposite to the passage vortex. The horseshoe suction-side leg mostly merges with this vortex downstream of the blade passage (merging is not represented in Fig. 15.5). Further, in case of corner separation, there may be two vortices in the separation zone in the corner of a suction side and an endwall, one with the same rotation sense as the passage vortex and one with opposite rotation sense (not represented in Fig. 15.5).

Somewhat downstream of the blade passage, two vortices are mostly visible near an endwall without blade clearance. One is the passage vortex, with the same-sense separation vortex merged with it. The other is the trailing shed vortex, with the two other vortices with the same rotation sense merged with it. This merged vortex is usually called the counter-vortex.

In presence of tip clearance, a tip leakage vortex forms. With a turbine, the scraping of the endwall generates vorticity in the rotation sense of the blade passage vortex so that a combined scraping vortex and passage vortex forms (see Fig. 15.5). Horseshoe vortices, corner vortices and trailing shed vortices normally cannot be observed in presence of a clearance.

15.1.4 Blade Profiles

The general flow in turbines is accelerating. This facilitates the blade profile design compared to that of a compressor. Compressor blade profiles are strongly determined by the limitation of the loading due to the strong adverse pressure gradient in the rear part of the suction side. Turbines allow, due to the generally accelerating flow, much higher loading (work coefficient around 1.80 compared to 0.4 for compressors). This results in much higher flow turnings with turbines.

Subsonic and supercritical cascades

Rotor and stator cascades are subsonic or supercritical in the low-pressure turbine (LP) part of a turbofan engine with direct fan drive, due to low blade speed in the LP turbine. Within the high pressure (HP) part, the blade speed is higher and rather determined by strength limitations, resulting in transonic cascades as a rule. Rotor cascades are mostly also subsonic or supercritical in the high-pressure and the intermediate-pressure parts of steam turbines. Steam turbine blades are manufactured with cheaper materials (steel alloys), which results in lower blade speeds.

Figure 15.6 (left) is a sketch of the optimum Mach number distribution (boundary layer edge) of a subsonic rotor cascade. The right-hand part of the figure represents the corresponding momentum loss thickness distribution [1].

Fig. 15.6 Mach number and
momentum loss thickness
distribution with a subsonic
turbine rotor cascade

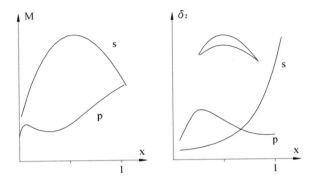

Optimisation, in principle, means minimisation of the ratio of the sum of the momentum loss thicknesses to the blade distance perpendicular to the outflow. This implies that the spacing should be as large as possible, just avoiding separation at the suction side. Moreover, the weight of the turbine is then reduced because fewer blades are required. For the basic aspects concerning separation and transition, we refer to Sect. 13.4.1 in Chap. 13.

The pressure side features a strong acceleration immediately downstream of the stagnation point. This reduces the loading capacity of the profile, but is unavoidable if the profile must have a certain thickness. After the initial acceleration, a weak deceleration is built in, in order to increase the loading capacity. This may continue to nearly laminar separation. Then, acceleration follows towards the trailing edge, sufficiently strong for keeping the boundary layer laminar. The objective is to keep the boundary layer laminar on the whole pressure side. Friction is then minimal. The boundary layer thickness at the pressure side decreases during the last phase.

At the suction side, there is first a strong acceleration. This acceleration continues over a long distance, becomes then weaker and is followed by a deceleration. The objective of the strong acceleration is keeping the boundary layer laminar over the longest possible distance. Friction is then minimal. The boundary layer becomes transitional when the acceleration decreases. In order to withstand the following deceleration, the weak acceleration phase must be such that the boundary layer becomes turbulent before the velocity maximum. If not, the deceleration could generate laminar separation. The turbulent boundary layer grows strongly in the deceleration phase.

Optimisation of the turbine blade is strongly determined by the suction side laminar-to-turbulent transition position. Control of this position is difficult in the LP turbine part of an aero-engine, due to the low Reynolds number. The gas temperature is high, generating a large dynamic viscosity. Moreover, the pressure within an LP turbine is low in cruise conditions, generating a low density. Both effects result in a large kinematic viscosity. Depending on the engine dimension, Reynolds numbers may be as low as 250,000–50,000 [1].

With the lower Reynolds numbers, laminar separation on the suction side is unavoidable. Care has to be taken then that transition occurs within the separated boundary layer with turbulent reattachment and formation of a separation bubble.

Driving the laminar separated boundary layer into turbulence is often realised by impact of wakes of an upstream blade row (relative motion between stator and rotor). Tuning the impact is difficult. So, optimising a turbine cascade is mainly a matter of boundary layer control. Optimum profiles are therefore termed *controlled boundary layer* (CBL) profiles.

Within a turbine, the turbulence level in the core flow is very high. The degree of turbulence, i.e. the magnitude of the fluctuating velocity compared to the average velocity, typically exceeds 5%. Transition in an attached boundary layer is then directly induced by fluctuations in the main flow, without necessity for instability of the boundary layer flow, as with a very quiet main flow (natural transition by Tollmien-Schlichting waves). Transition is then termed *bypass transition*, meaning that the instability is bypassed (see discussion in Sect. 13.4.1).

With wake impact, transition is very quick in an attached boundary layer due to the kinematic distortion together with a high turbulence level within the wake. The boundary layer may become laminar again after the wake has passed. With wake impact on a separated boundary layer, the separated shear layer quickly becomes turbulent due to instability as a consequence of the inflectional velocity profile (Kelvin–Helmholtz instability), strongly increasing the potential for reattachment.

With a low Reynolds number, the momentum loss thickness at the end of the suction side is mainly determined by the size of the separation bubble. It is then favourable to choose a blade profile shape with the velocity peak rather far forward and a high wake impact frequency. The separation bubble decreases then. The turbulent part of the boundary layer increases, causing larger friction loss in that part, but there is an advantage by the strong decrease of the mixing loss behind the separation bubble. With a higher Reynolds number the bubble size is already small spontaneously. It is then favourable that the velocity peak lies rather far backward and that the frequency of the wake impact is lower. So the optimum blade profile shape depends strongly on the Reynolds number [2].

Adaptation to supercritical conditions is done the same way as with an aerofoil. The Mach number distribution of Fig. 15.6 does not change essentially. The flow becomes locally supersonic on the suction side and the supersonic zone ends with a normal shock. The suction side shape must be such that the Mach number just upstream of the normal shock is as low as possible.

Optimisation of a stator cascade does not principally differ from that of a rotor cascade, but the general acceleration in a stator cascade is often stronger (degree of reaction lower than 50%; axial inflow). With a limitation on the maximum Mach number at the suction side, the decelerating zone in Fig. 15.6 becomes then weaker.

Transonic cascades and supersonic cascades

A transonic cascade means subsonic inflow and supersonic outflow. These cascades occur in both rotors and stators. Figure 15.7 (left), sketches the shock waves within a rotor cascade with high flow turning. Such cascades feature oblique shock patterns starting from the trailing edge. One of the shocks impacts on the suction side of the adjacent blade. The boundary layer should not separate under the impact. Therefore, the suction side is designed such that the boundary layer is turbulent in the impact

Fig. 15.7 Shock patterns
with transonic rotor turbine
cascades; left: high flow
turning; right: rotor tip
section of the last stage of a
steam turbine

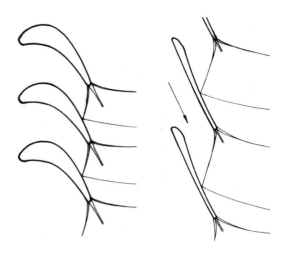

zone. Figure 15.7 (right) sketches the shock waves at the tip section of a rotor blade
of the last stage of an LP part of a steam turbine.

Supersonic cascades are only used in impulse turbines. These cascades are
designed with free-vortex flow in the blade passage. Transition segments at the
inflow and outflow edges generate the appropriate acceleration and deceleration (see
Sect. 6.4.4 in Chap. 6).

15.1.5 Three-Dimensional Blade Shaping

High flow turning with turbines has as a consequence that secondary flow loss may
be significant for low aspect ratio vanes and blades, although the boundary layers in
turbines are thinner than in compressors. Lowering the losses in end wall regions of
low aspect ratio stages is mainly achieved by bowing the stator vanes. The effect of
bowing is the same as discussed for compressors in Sect. 13.3.6 in Chap. 13. The
possible increase of isentropic efficiency by bowing is in the order of two percentage
points.

Combined sweep and lean of high aspect ratio stator vanes is typically used to
generate radial forces for limiting the radial variation of the degree of reaction in the
LP turbine part of an aero gas turbine, similarly as in the LP part of steam turbines
(Sect. 6.9.2 in Chap. 6). Observe the combined sweep and lean of some stator vanes
at the casing in the LP turbine part in Fig. 12.9 (Trent 800). Sweep is also visible in
Fig. 12.14 (Trent 800).

15.1.6 Vane and Blade Clocking

Clocking is setting the relative position of two subsequent stators or rotors with an equal number of blades. Maximum efficiency is obtained when wake segments from the upstream row impinge the leading edge of the downstream row. The efficiency is the lowest when the wake segments follow a path midway the downstream blades as then wake mixing is the strongest and so is the mixing loss. The principle is also used with axial compressors. A subtle difference is that with turbines it is mostly advantageous to shift the impacting wake somewhat to the suction side. This helps in promoting transition in the suction side boundary layer. The possible isentropic efficiency improvement with clocking is in the order of 0.5 percentage points, similarly with turbines and compressors.

15.1.7 Operating Characteristic of Axial Turbines

In order to express that the outflow angles of the vanes and the blades α_1 and β_2 are approximately constant with varying flow rate, we note

$$\psi = \frac{v_{1u} - v_{2u}}{u} = \frac{v_{1u} - u - w_{2u}}{u} \text{ with } \tan\alpha_1 = \frac{v_{1u}}{v_a} \text{ and } \tan\beta_2 = \frac{w_{2u}}{v_a}$$

Thus:

$$\psi = \phi(\tan\alpha_1 - \tan\beta_2) - 1. \tag{15.7}$$

The angle α_1 is about $60°$–$70°$ and the angle β_2 about $-60°$ to $-70°$. So, the term $(\tan\alpha_1 - \tan\beta_2)$ is always strongly positive with an order of magnitude of 5. Increasing the flow rate increases the stage work. In other words, the flow rate increases with increasing pressure ratio through the turbine (p_{00}/p_{02}).

The relation (15.7) further implies that with increased rotational speed, the pressure ratio increases stronger than the flow rate (work $\sim u^2$, flow rate $\sim u$). Thus, for a higher rotational speed, the characteristic curve of mass flow rate as a function of pressure ratio lies under that for a lower rotational speed (see Fig. 15.8).

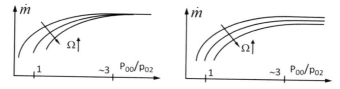

Fig. 15.8 Mass flow rate of an axial turbine as a function of pressure ratio; left: choking in stator; right: choking in rotor

Choking may occur with sufficiently large pressure ratio.

Within the stator: $h_{00} = h_{01} = h_1 + \frac{1}{2}v_1^2$.

Flow speed equal to the speed of sound at the stator exit is reached for $v_1^2 = c_1^2 = (\gamma - 1)h_1$, thus $h_{00} = \frac{1}{2}(\gamma + 1)h_1$.

The corresponding pressure ratio and mass flow rate are

$$\frac{p_1}{p_{00}} = \left(\frac{h_1}{h_{00}}\right)^{\frac{n}{n-1}} = \left(\frac{2}{\gamma + 1}\right)^{\frac{n}{n-1}}, \tag{15.8}$$

$$\dot{m}_c = A_1 \rho c = A_1 \rho_{00} c_{00} \left(\frac{h_1}{h_{00}}\right)^{\frac{1}{n-1} + \frac{1}{2}}, \tag{15.9}$$

where n is a polytropic exponent and A_1 the stator blade passage area. With $\gamma = 1.33$ and $n = 1.30$ it follows that $p_1/p_{00} \approx 0.5$. With a very low degree of reaction, the corresponding choking pressure ratio p_{00}/p_{02} of the turbine stage is slightly above 2. It increases with increasing degree of reaction and the value due to the stator becomes about $(1/0.5)^2 = 4$ with 50% degree of reaction. The choking mass flow rate determined by the stator does not depend on the rotational speed.

Within the rotor: $h_{01} - h_{02} = u_1 v_{1u} - u_2 v_{2u}$,

or $h_{00} = h_{01} = u_1 v_{1u} + h_2 + \frac{1}{2}v_2^2 - u_2 v_{2u}$.

For the exit velocity triangle (Fig. 15.1): $w_2^2 = u_2^2 + v_2^2 - 2u_2 v_{2u}$.

Thus: $h_{00} = u_1 v_{1u} + h_2 + \frac{1}{2}w_2^2 - \frac{1}{2}u_2^2$.

This expression is noted more generally ($u_1 \neq u_2$) than needed for an axial turbine, because we reuse the result in a later similar analysis for a radial turbine.

Flow speed equal to the speed of sound at the rotor exit is reached for

$$w_2^2 = (\gamma - 1)h_2.$$

$$h_{00} = \frac{\gamma + 1}{2}h_2 + u_1 v_{1u} - \frac{1}{2}u_2^2$$

and

$$\frac{h_2}{h_{00}} = \frac{2}{\gamma + 1}\left(1 - \frac{u_1 v_{1u} - \frac{1}{2}u_2^2}{h_{00}}\right)$$

The term $u_1 v_{1u} - \frac{1}{2}u_2^2$ is positive for sufficiently large ψ (Fig. 15.2). The pressure ratio and mass flow rate with choking are

$$\left(\frac{p_2}{p_{00}}\right) = \left(\frac{h_2}{h_{00}}\right)^{\frac{n}{n-1}} = \left(\frac{2}{\gamma + 1}\right)^{\frac{n}{n-1}}\left(1 - \frac{u_1 v_{1u} - \frac{1}{2}u_2^2}{h_{00}}\right)^{\frac{n}{n-1}}, \tag{15.10}$$

$$\dot{m}_c = A_2 \rho c = A_2 \rho_{00} c_{00} \left(\frac{h_2}{h_{00}}\right)^{\frac{1}{n-1} + \frac{1}{2}}. \tag{15.11}$$

With choking in the rotor, the pressure ratio and mass flow rate depend on the rotational speed.

For a low degree of reaction, the stage pressure ratio corresponding to Eq. (15.8) is lower than that corresponding to Eq. (15.10). The stator then determines the choking. This is also obvious from the higher magnitude of v_1 than that of w_2 for a low degree of reaction in Fig. 15.2. The choking mass flow rate is then not dependent on the rotational speed and the choking pressure ratio is slightly above 2.

In Fig. 15.2 one sees that with 50% degree of reaction, the magnitudes of v_1 and w_2 are equal, but the Mach number of w_2 is larger. The rotor then determines the choking. This happens already for a somewhat lower degree of reaction and the corresponding stage pressure ratio is then somewhat lower than 4, around 3.50. From Eq. 15.10 follows that the stage pressure ratio for rotor choking decreases somewhat with increasing degree of reaction. The rotational speed dependence following from the relations (15.10) and (15.11) implies that with a higher rotational speed, the choking pressure ratio increases somewhat, with a decrease of the corresponding mass flow rate.

Characteristics are thus as sketched in Fig. 15.8, for fixed p_{00} and T_{00}. Without choking, a characteristic curve shifts with increasing rotational speed as drawn and derived earlier in this section. The horizontal parts of the curves for choking in the rotor may lie very close together, since the term $u_1 v_{1u} - \frac{1}{2} u_2^2$ is typically much smaller than h_{00}. From the velocity triangles in Figs. 15.1 and 15.2, it follows that for low mass flow rate (u = constant, direction of w_2 is approximately constant), $v_{2u} > 0$, so that $v_{2u} - v_{1u} > 0$, which means that the stage functions as compressor. Thus, for very low mass flow rate, $p_{00}/p_{02} < 1$.

We note that a value around 3 of the total pressure ratio, for obtaining choking, is not very high with an axial turbine:

$$p_{00}/p_{02} = (T_{00}/T_{02})^{n/(n-1)}.$$

E.g., with $\Delta W = 250$ kJ/kg is $\Delta T = T_{00} - T_{02} \approx 220$ K. With $n = 1.30$, $T_{00} = 1500$ K, follows $p_{00}/p_{02} \approx 2$. With $T_{00} = 1000$ K, follows $p_{00}/p_{02} \approx 3$. Thus, turbines with a high stage work often operate near to choking or with choking. Normally, this is no drawback. The flow rate is then proportional to the total entry values of density and velocity of sound, and thus, at a given inlet temperature, proportional to the entry pressure. Usually, this is a perfectly practicable characteristic (see Fig. 11.5).

15.2 Radial Turbines

15.2.1 Shape and Functioning

Figure 15.9 shows the rotor of a small gas turbine (2 MW, pressure ratio 7) with a centrifugal compressor (at the right-hand side) mounted back to back with a radial

Fig. 15.9 Rotor of a small
gas turbine with radial
compressor (right) and radial
turbine (left) (Courtesy
Siemens Energy)

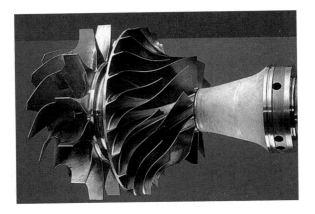

turbine (at the left-hand side). The compressor rotor has splitter blades and the blades
are about 30° swept back at the periphery and have a small lean. The turbine rotor
has a lower number of blades (12 in the example) and the blades are purely radial
at the inflow. Figure 15.10 sketches a meridional section and an orthogonal view of
a radial or centripetal turbine for gas expansion. The gas is supplied from a volute,
which may be followed by a stator vane ring. As with an axial turbine, the flow at
the exit of the stator is quite near to the tangential direction. The flow leaving the
volute is already spontaneously near to the tangential direction.

The figure shows stator vanes that direct the flow somewhat more to the radial
direction. From this also emerges that radial turbines may be built without a stator
vane ring. Nozzle guide vanes (stator vanes) are used when the desired flow angle
cannot be obtained by the volute due to dimension restrictions or when variable stator
vanes are wanted. The rotor blades are purely radial at the entry. The gas is deflected to
the axial direction in the meridional plane. At the rotor exit, the flow has a significant
tangential velocity component in the relative frame. The diffuser downstream of the

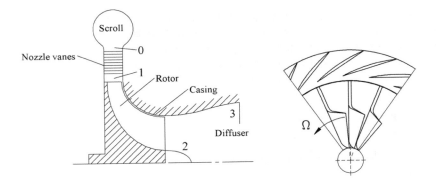

Fig. 15.10 Radial turbine components

rotor recovers pressure from the exit kinetic energy and so decreases the pressure at the rotor exit.

The principal functioning of a centripetal turbine for gas expansion does not differ from that of a hydraulic turbine of Francis type. Some features differ, however. The largest difference is the rotor shape. The rotor is open and the blades end in an orthogonal plane. There is an *exducer* part, analogous to the inducer part of a radial compressor. An expansion turbine is normally designed for maximum rotor work. This is commonly achieved by a large blade speed, typically around 400 m/s at the rotor periphery. This requires an open rotor with purely radial blade form at the rotor entry and an exducer part formed by near-radial filaments. As a consequence of the expansion, the outflow area must be larger than the inflow area. A centripetal gas expansion turbine thus features a rather high ratio of exit diameter to entry diameter, typically around 0.7 (see below).

Figure 15.11 shows the velocity triangle at the rotor entry (tip of the blades) and the exit and a sketch of the entry flow. The operating point with optimum efficiency is obtained with a stator exit velocity that is somewhat smaller than that for perfect alignment with the blades of the relative flow at the rotor entry.

This feature is a particularity of rotors with radial entry. The relative vortex in the blade channels deflects the relative velocity w_1 from the radial direction, generating a phenomenon analogous to the slip at the exit of a centrifugal rotor. There is thus no incidence loss with the velocity triangle in Fig. 15.11. The tangential component of the absolute inlet velocity for incidence-free entry may be expressed with a slip factor as $v_{1u} = \sigma\, u_1$. The slip factor can be determined by the same formulae as with compressors [4, 5, 7].

In order to prevent separation, the number of blades of a centripetal rotor has to be similar to that of a centrifugal rotor. As Fig. 15.11 demonstrates, the pressure difference across a blade channel is mainly generated by the Coriolis force. The flow curvature generates a centrifugal force or lift, opposite to the Coriolis force, but this force is weak. With an exactly radial entry, about 20 blades are required to avoid separation. This may be determined in the same way as with centrifugal rotors. Adapting Eq. (14.13) to turbine inflow, the condition for prevention of flow reversal is

Fig. 15.11 Flow at rotor entry and exit of a radial turbine

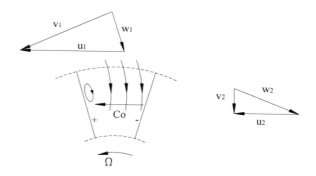

$$w_1 > \frac{2\pi}{Z} u_1.$$

Approximately: $\tan \alpha_1 = \frac{\sigma u_1}{w_1}$ so that $Z > 2\pi \frac{\tan \alpha_1}{\sigma}$.

With $\alpha_1 \approx 70°$ and $\sigma \approx 0.8$ to 0.9 it follows that $Z \approx 20$.

The difference with a centrifugal rotor is that a separation bubble at the entry of a centripetal rotor is acceptable, because the flow reattaches deeper in the rotor, as is clear from Fig. 15.11. Typically, far fewer blades are chosen than needed to prevent separation: $Z \approx 12$. This results in smaller blade surface area and, consequently, less friction. The axial rotor part, the exducer, may thus have the same number of blades as the radial part. In other words, splitter blades are not necessary. The consequence of a lower number of blades is a slip factor that is lower than typical with a centrifugal rotor: $\sigma \approx 0.8$

Figure 15.11 shows a velocity triangle at the rotor exit. Kinetic energy supplied to the diffuser can only be recovered partially, i.e. to about 50%. The outflow should thus be near to axial and the magnitude of the exit velocity v_2 as low as possible. So the exit angle β_2 has to be large, typically about $-70°$ at the tip.

15.2.2 Kinematic Parameters

Total enthalpy is constant in the volute and the nozzles: $h_{00} = h_1 + \frac{1}{2} v_1^2$.

Rothalpy is constant in the rotor: $I = h + \frac{1}{2} w^2 - \frac{1}{2} u^2 = cst$.

The entry velocity triangle (Fig. 15.11) is approximately rectangular, as $\alpha_1 \approx 70°$:

$$w_1^2 + v_1^2 \approx u_1^2.$$

To the entry thus applies $I \approx h_1 - \frac{1}{2} v_1^2$.

To the exit applies, with exact axial v_2:

$$I = h_2 + \frac{1}{2} w_2^2 - \frac{1}{2} u_2^2 = h_2 + \frac{1}{2} v_2^2 = h_{02}.$$

Rotor work is $\Delta W = \sigma u_1^2$.

With $\sigma \approx 0.8$, it follows that ΔW is about v_1^2.

The degree of reaction follows from

$$1 - R = \frac{\frac{1}{2} v_1^2 - \frac{1}{2} v_2^2}{u_1 v_{1u}}.$$

With $v_2 \approx v_{1r}$: $R \approx 1 - \frac{1}{2} \frac{v_{1u}^2}{u_1 v_{1u}} = 1 - \frac{1}{2} \frac{v_{1u}}{u_1} = 1 - \frac{\sigma}{2} \approx 0.6$.

The work coefficient is $\psi = \frac{u_1 v_{1u}}{u_1^2} = \sigma \approx 0.8$.

An h–s diagram rendering these relations is shown in Fig. 15.12.

Fig. 15.12 h–s diagram for expansion in a centripetal turbine

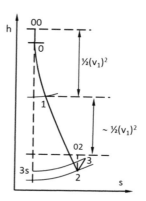

Regarding work, a radial turbine is not very advantageous, as the work coefficient is lower than with a typical axial turbine, for which it may even reach values above 2. The peripheral speeds are comparable, namely about 400 m/s. The observation that there is no strict limitation to the flow turning of an accelerating flow implies that the work coefficient is not better. Within an axial turbine, the flow turning in the rotor may amount to about 140°. There is a limitation to about 70° with a radial turbine.

It is typical with a radial turbine that backpressure is imposed and exit kinetic energy is lost. The total-to-static isentropic enthalpy drop $\Delta h_s = h_{00} - h_{3s}$ should thus be used for the definition of the spouting velocity v_s and the speed ratio, and the total-to-static efficiency has to be maximised.

The rotor work is then

$$\Delta W = \sigma\, u_1^2 = \eta_{ts}\, \Delta h_s.$$

With the total-to-static efficiency of about 0.8, it follows that

$$u_1^2 \approx \Delta h_s = v_s^2/2 \ \text{ thus } \ \lambda = \frac{u_1}{v_s} = \frac{1}{\sqrt{2}} \approx 0.7.$$

Two kinematic parameters are thus determined: R and ψ (or λ). The third parameter, the flow coefficient ϕ, follows from efficiency optimisation.

Losses are low within the volute and the vane ring. Losses within the rotor are significantly higher, due to the separation bubble and the flow segregation as a consequence of the Coriolis force and the centrifugal force associated to the flow turning. A significant loss further occurs due to the clearance between the open rotor and the casing and to the high friction velocity on the casing (absolute velocity). A closed rotor would be more advantageous regarding these aspects. The losses within the diffuser are also very high.

Optimisation thus means, in principle, minimising the sum of rotor and diffuser losses. This optimisation cannot be achieved with simple means, as the result depends strongly on the loss coefficient values, which cannot be expressed by simple correlations. The result is similar to that with Francis turbines. The optimum outflow angle approximately corresponds to [4, 7]:

$$\phi = \frac{v_{2m}}{u_1} \approx 0.25.$$

Specific speed is

$$\Omega_S = \frac{\Omega \sqrt{Q}}{\Delta h_s^{3/4}},$$

with $Q = k \pi r_{2t}^2 v_{2m}$ being the volume flow rate at the rotor exit and k being an obstruction factor. Thus:

$$\Omega_S \approx \frac{\Omega r_{2t} \sqrt{k\pi\, 0.25}\, \sqrt{u_1}}{(u_1)^{3/2}} \approx 0.8 \frac{r_{2t}}{r_1}.$$

This result demonstrates that the attainable specific speed for good efficiency is 0.5–0.7. The optimum efficiency is about 0.85. With a flow coefficient somewhat deviating from the optimal value and some reduction of the efficiency (efficiency $\approx$ 0.80), the realisable range is $\Omega_S \approx 0.3$–0.9.

15.2.3 Operating Characteristic of Radial Turbines

The analysis is entirely analogous to that of axial turbines. The operating characteristic follows from

$$\psi = \frac{u_1 v_{1u} - u_2 v_{2u}}{u_1^2} = \frac{u_1 v_{1u} - u_2 (u_2 + w_{2u})}{u_1^2} = \frac{v_{1u}}{u_1} - \frac{u_2^2}{u_1^2} - \frac{u_2}{u_1}\frac{w_{2u}}{u_1},$$

with $\tan \alpha_1 = \frac{v_{1u}}{v_{1m}}$ and $\tan \beta_2 = \frac{w_{2u}}{v_{2m}}$.
With $v_{2m} = v_{1m}$:

$$\psi = \phi(\tan \alpha_1 - \frac{u_2}{u_1} \tan \beta_2) - \frac{u_2^2}{u_1^2}.$$

This relation is similar as with axial turbines. The conditions for choking (15.8)–(15.11) are also similar. Characteristics are thus similar as sketched in Fig. 15.8.

Figure 15.11 shows that the magnitude of w_2 is smaller than that of v_1, but the velocity of sound is lower at the rotor exit. With types with the design Mach number of v_1 sufficiently larger than the design Mach number of w_2, the stator determines the choking. Because the degree of reaction is around 0.6, the stage pressure ratio p_{00}/p_2 for choking is then around 5, independent of the rotational speed.

But there are also types, meant for large mass flow rate, with comparable design values of the Mach numbers of v_1 and w_2. The rotor may then determine the choking. A special case with rotor choking is a turbine without stator vanes. The pressure ratio for choking in the rotor depends on the rotational speed and may be rather low or high, depending on the case. No universal value can be attributed to this pressure ratio. E.g., with $T_{00} = 500$ K, $u_1 = 400$ m/s, $u_2 = 200$ m/s, $\psi = 0.8$, $\gamma = 1.33$ and $n = 1.30$, Eq. (15.10) gives:

$$\frac{p_2}{p_{00}} = \left(\frac{2}{\gamma + 1}\right)^{\frac{n}{n-1}} \left(1 - \frac{0.8 \times 400^2 - 0.5 \times 200^2}{1150 \times 500}\right)^{\frac{n}{n-1}} \approx 0.20 = 1/5.$$

But for $T_{00} = 1500$ K, this is $p_{00}/p_2 \approx 2.55$.

15.2.4   Radial Turbine Applications

The work coefficient of radial turbines is lower than that of axial turbines. The efficiency is lower as well. Therefore, a radial turbine is only applied if an axial type is not possible, which happens with small dimensions. Efficiency then becomes disadvantageous with axial turbines, due to relatively large clearances and large blade thickness at the trailing edge compared to the spacing. A typical application is the turbine of a turbocharger, except for the large ones. A similar application is the turbine of a turbo-expander in a chemical plant. Radial turbines are also used with small gas turbines (e.g. Fig. 15.9).

15.3   Dimensional Analysis with Compressible Fluids

15.3.1   Independent and Dependent Π-groups

A flow within a turbomachine with a compressible fluid features 8 independent parameters: R, C_p, μ, p_{00}, T_{00}, $\dot{m}$, Ω, D. The gas is determined by the gas constant R, specific heat capacity C_p and viscosity μ. Gas properties C_p and μ are temperature-dependent and thus should be given at a reference temperature (e.g. at T_{00}). Dynamic viscosity μ may be substituted by kinematic viscosity ν. The initial state is determined by p_{00} and T_{00}. Mass flow rate $\dot{m}$ and rotational speed Ω; determine the operating point. D is a characteristic diameter allowing identification of the machine within a family of similar machines.

The operating point may also be determined by the pressure ratio and the rotational speed. It is sometimes a matter of point of view to consider mass flow rate as independent and pressure ratio as dependent (the standard choice with compressors), or vice versa (the standard choice with turbines). We opt for mass flow rate here, as this choice is the most complex one for a dimensional analysis, the pressure ratio spontaneously being a dimensionless group.

The intervening dimensions are length, mass, time and temperature. There are thus 4 independent Π-groups of general form

$$\Pi = R^a \, C_p^b \, \mu^c \, p_{00}^d \, T_{00}^e \, \dot{m}^f \, \Omega^g \, D^h.$$

The choice of independent dimensionless groups is not unambiguous. It is customary to form dimensionless groups that characterize, as much as possible, the fluid (R, C_p and μ) and the operating point ($\dot{m}$ and Ω).

R and C_p feature the same dimension. Their ratio thus is a Π-group. The usual group is the heat capacity ratio:

$$\Pi_1 = \gamma = C_p/C_v, \ R = C_p - C_v.$$

A dimensionless expression of viscosity is the Reynolds number:

$$\Pi_2 = \mathrm{Re} = \Omega \, D^2/\nu.$$

The two remaining groups may be formed by dimensionless expressions of the mass flow rate and the blade speed based on the density and the speed of sound at the machine entry.

$$\Pi_3 = \frac{\dot{m}}{\rho_{00} \sqrt{\gamma R T_{00}} D^2} = \frac{\dot{m} R T_{00}}{p_{00} \sqrt{\gamma R T_{00}} D^2} \ \text{and} \ \Pi_4 = \frac{\Omega \, D}{\sqrt{\gamma R T_{00}}}.$$

Since γ is already listed as independent parameter, $\sqrt{\gamma}$ is usually left out, resulting in

$$\Pi_3 = \frac{\dot{m} \sqrt{R T_{00}}}{p_{00} \, D^2} \ \text{and} \ \Pi_4 = \frac{\Omega \, D}{\sqrt{R T_{00}}}.$$

Sometimes, R is replaced by C_p.

The dimensionless representation of the mass flow rate, as derived here and applied typically, is different from that with constant-density fluids. The flow coefficient customary with constant-density fluids is found by dividing Π_3 by Π_4.

Dimensionless dependent groups are e.g., pressure ratio p_{03}/p_{00}, temperature ratio T_{03}/T_{00} (for a compressor) and efficiency η.

Dependent dimensionless groups are expressed as functions of independent dimensionless groups according to

$$\frac{p_{03}}{p_{00}} = f\left[\frac{\dot{m}\sqrt{R T_{00}}}{p_{00} D^2}, \ \frac{\Omega D}{\sqrt{R T_{00}}}, \ \mathrm{Re}, \ \gamma \right].$$

For a given gas, γ is approximately constant. Further, if the Reynolds number is sufficiently large, it does not constitute a strong condition for similarity. The preceding relation thus mostly may be simplified to

$$\frac{p_{03}}{p_{00}} = f\left[\frac{\dot{m}\sqrt{RT_{00}}}{p_{00}D^2}, \frac{\Omega D}{\sqrt{RT_{00}}}\right].$$

15.3.2 Dimensionless Compressor and Turbine Characteristics

Performances characteristics do not change in general appearance when drawn in dimensionless form. For compressors, charts may be as in Fig. 13.28 with pressure ratio and efficiency as functions of dimensionless mass flow rate for constant values of dimensionless rotational speed. An alternative representation is with efficiency by contour lines, as in Fig. 14.15. For turbines, the chart is dimensionless mass flow rate as a function of pressure ratio for constant values of dimensionless rotational speed (see Fig. 15.8). The chart may be completed with efficiency as a function of pressure ratio. In practice, flow rate curves for various rotational speeds are so close together that representing the efficiency by contour lines is not possible.

15.3.3 Corrected Quantities

The characteristics obtained by testing of a compressor or a gas turbine (D constant) depend on the ambient pressure and temperature. This complicates the comparison of test results under different atmospheric conditions. It is therefore advisable to adjust the measured characteristics to a standard temperature (e.g., 288.15 K) and a standard pressure (e.g., 101.325 kPa). The adjustment is called the correction.

The real flow is then substituted by a dynamically similar flow, occurring under standard atmospheric conditions. According to the above formulae, the following expressions remain constant:

$$\frac{\Omega}{\sqrt{RT_{00}}}, \quad \frac{\dot{m}\sqrt{RT_{00}}}{p_{00}}, \quad \frac{p_{03}}{p_{00}}, \quad \frac{T_{03}}{T_{00}}, \quad \frac{P}{p_{00}\sqrt{RT_{00}}}, \quad \eta.$$

The expressions for corrected rotational speed and mass flow rate are

$$\Omega^* = \Omega\sqrt{\frac{288.15\,K}{T_{00}}} \quad \text{and} \quad \dot{m}^* = \dot{m}\sqrt{\frac{T_{00}}{288.15\,K}}\left(\frac{101.325\,kPa}{p_{00}}\right).$$

Fan testing requires a similar correction. The above formulae may be applied, but a simpler method is possible. With a constant-density fluid, meaning that density changes very little within the machine, flow rate is proportional to density and rotational speed. Rotational speed may thus be kept constant and mass flow rate corrected according to

$$\dot{m}^* = \dot{m}\,\frac{\rho_{00}^*}{\rho_{00}} = \dot{m}\left(\frac{T_{00}}{288.15\ K}\right)\left(\frac{101.325\ kPa}{p_{00}}\right).$$

15.4   Exercises

15.4.1 Consider an axial turbine composed of a stator ring of nozzle vanes with a flow entering in axial direction, turned over a large angle (approximately 70°), followed by two contra-rotating rotors with large flow turning (blade shapes as on Fig. 15.1). In case of post-whirl at the outflow of the second rotor, the rotors may be followed by a stator ring of outlet guide vanes bringing the flow to the axial direction. Take as fluid a combustion gas, considered as an ideal gas, with properties $R = 288$ J/kgK and $\gamma = 1.30$ at 1600 K at the turbine entry. Take on the mean radius as blade speeds $u_1 = 300$ m/s and $u_2 = -300$ m/s (thus equal magnitude $u = 300$ m/s; but opposite running senses) and constant axial velocity through the turbine with value 400 m/s.

- Sketch the velocity triangles. Take as notation 1–2 for the first rotor and 3–4 for the second rotor. Take arbitrary kinematic parameters at this stage, but large flow turning in the nozzle ring and the rotors.
- Determine the velocity triangles for impulse blades on both rotors, meant as $|w_1| = |w_2|$ and $|w_3| = |w_4|$, thus kinematic degree of reaction equal to zero, such that the outflow of the second rotor is in the axial direction. Remark that such a turbine is equivalent with a Curtis turbine, resulting in $\psi_1 = 6$ and $\psi_2 = 2$ (work coefficient:$\psi = \Delta W / u^2$).
- Determine the relative inflow Mach numbers of both rotors (**A**: $M_{w1} = 1.644$, $M_{w3} = 0.834$).
- Determine the velocity triangles for degree of reaction of the first rotor equal to 50% and work coefficients $\psi_1 = 5$ and $\psi_2 = 3$ (so same total work as with the impulse blades). Observe that there is post-whirl at the exit of the second rotor (**A**: $\alpha_4 = 36.87°$). Determine the degree of reaction of the second rotor (**A**: 50%).
- Determine the relative inflow Mach numbers of both rotors (**A**: $M_{w1} = 1.071$, $M_{w3} = 0.805$). Remark that the inflow Mach number of the first rotor is much lower than with impulse blades. This is a big advantage for the 50% degree of reaction turbine.
- Determine the deceleration ratio in the outlet guide vanes (**A**: 0.80). The necessity for a deceleration in the outlet guide vane ring is a small disadvantage of the reaction turbine (but see next exercise).

15.4.2 Consider once more the axial turbine of the previous exercise, composed of a stator ring of nozzle vanes, followed by two contra-rotating rotors with large flow turning and a stator ring of outlet guide vanes. Take the same values for blade speed, axial through-flow velocity and gas properties. We will adapt the machine in three steps into a turbine with only rotating components.

- Determine the velocity triangles for $\psi_1 = 4$ and $\psi_2 = 4$, such that the couple of the two contra-rotating rotors forms a repeating system. Remark that these conditions do not determine the system uniquely. Take as particular choice identical blade shapes, symmetrically positioned, in both rotors (**A**: obtained for $v_{2u} = 2u$).
- Determine the degree of reaction of both rotors (**A**: 100%).
- Determine the relative inflow Mach numbers of both rotors (**A**: $M_{w1} = 0.693$, $M_{w3} = 0.778$). Remark that these are lower than the values of the 50% degree of reaction turbine of the previous exercise.
- Determine the deceleration ratio in the outlet guide vanes (**A**: 0.555). Remark that the deceleration ratio is much too strong for practical realisation of the vanes.
- Replace the nozzle vane ring with a rotor rotating together with the second rotor of the repeating couple (same speed and same sense) such that the inflow and outflow velocities remain unchanged (**A**: this requires $\psi = 2$). Determine the degree of reaction of this rotor (**A**: 200%: this rotor generates an increase of kinetic energy in the absolute frame equal to the work done).
- Replace the outlet guide vane ring with a rotor rotating together with the first rotor of the repeating couple (same speed and same sense) such that the inflow and outflow velocities remain unchanged (**A**: this requires $\psi = 2$ and $R = 0$).
- The resulting turbine is solely composed of rotors. The inner repeating couple may be repeated a number of times. A turbine of this type may be used in a prop-fan or unducted fan aero-engine for driving the contra-rotating propellers.

15.4.3 The paper by Mueller et al. [6] is about mathematical optimisation of a small radial turbocharger turbine rotor with outer diameter 50 mm. In the basic design, the rotor has 10 blades and is composed by radial fibres. The shape is similar as shown in Fig. 15.9 (left).

The total-to-static pressure ratio is 1.68 and the total-to-static isentropic efficiency is 0.75 in the design operating point. The geometric features are: $\alpha_1 = 73°$, $(d_2/d_1)_{tip} = 0.769$, $(\beta_2)_{tip} = -60°$, $(d_2/d_1)_{hub} = 0.240$, $(\beta_2)_{hub} = -20°$. The performance parameters of the basic design are: $\psi = \Delta h_0/u_1^2 = 0.794$; $\phi = (v_{2m}/u_1)_{av} = 0.39$ (av = quadratic average). Dimensionless velocity triangles (normalized by u_1) can be constructed from the geometric data and the performance parameters.

Verify first the specific speed. Observe that it is high (larger than 0.9) due to the high flow coefficient (larger than 0.25). Construct the rotor exit velocity triangles at tip and hub, assuming a constant meridional component of the exit velocity and assuming perfect alignment of the relative flow with the blades. Calculate then quadratic averages of u_2 and v_{2u}, and determine v_{1u} from the work coefficient. Calculate the degree of reaction with the quadratic averaged values.

The optimisation brings the total-to-static efficiency to 0.80. An important factor contributing to the increase is the reduction of the flow coefficient to about 0.25. **A**: $\Omega_S = 0.96$, $\beta_1 = -30.5°$, $v_{1u}/u_1 = 0.85$, $(v_{2u}/u_1)_{av} = 0.096$, $(u_2/u_1)_{av} = 0.57$, $R = 0.61$.

References

1. AGARD-LS-167 (1989) Blading design for axial turbomachines. AGARD, ISBN 92-835-0512-3
2. Coull JD, Thomas RL, Hodson HP (2010) Velocity distributions for low pressure turbines. J Turbomachinery 132:041006
3. Coull JD (2017) Endwall loss in turbine cascades. J Turbomachinery 139:081004
4. Dixon SL, Hall CA (2014) Fluid mechanics and thermodynamics of turbomachinery, 7th edn. Elsevier, ISBN 978-0-12-415954-9
5. Moustapha H, Zelesky M, Baines C, Japikse D (2003) Axial and radial turbines. Concepts NREC, ISBN 0-933283-12-0
6. Mueller L, Alsalihi Z, Verstraete T (2013) Multidisciplinary optimization of a turbocharger radial turbine. J Turbomachinery 135:021022
7. Whitfield A, Baines NC (1990) Design of radial turbomachines. Longman scientific & technical. ISBN 0-582-49501-6
8. Yoon S (2013) The effect of the degree of reaction on the leakage loss in steam turbines. J Eng Gas Turbines and Power 135:022602

Index